Applied Mathematical Sciences
Volume 181

T0212017

For further volumes:
www.springer.com/series/34

Maoan Han · Pei Yu

Normal Forms, Melnikov Functions and Bifurcations of Limit Cycles

 Springer

Maoan Han
Department of Mathematics
Shanghai Normal University
Shanghai, China, People's Republic

Pei Yu
Dept. Applied Mathematics
University of Western Ontario
London, Ontario, Canada

Additional material to this book can be downloaded from http://extras.springer.com
Password: [978-1-4471-2917-2]

ISSN 0066-5452   Applied Mathematical Sciences
ISBN 978-1-4471-5830-1   ISBN 978-1-4471-2918-9 (eBook)
DOI 10.1007/978-1-4471-2918-9
Springer London Dordrecht Heidelberg New York

British Library Cataloguing in Publication Data
A catalogue record for this book is available from the British Library

Mathematics Subject Classification: 34C07, 34C20, 34C23, 34D10, 34D20, 34E10, 34E13

Printed on acid-free paper

Springer is part of Springer Science+Business Media (www.springer.com)

Preface

In the past half century, dynamical system theory, also more grandly called "nonlinear science", has developed rapidly. In particular, the discovery of chaos has revolutionized the field. The study of chaos is closely related to research in the area of bifurcation and stability, and, in particular, the theory of limit cycles has had a significant impact since chaos may be considered to be a summation of motions with an infinite number of frequencies (such as the scenario of "period doubling", which leads to chaos). Although limit cycles involve a comparatively simple motion, they appear in almost all disciplines of science and engineering including mechanics, aeronautics, electrical circuits, control systems, population problems, economics, financial systems, stock markets, ecological systems, etc. In fact, most of the early work in the theory of limit cycles was stimulated by practical problems displaying periodic behavior. Therefore, the study of the bifurcation of limit cycles is not only significant for its theoretical development, but also plays an important role in solving practical problems.

The bifurcation of limit cycles may be generally classified into three categories: (i) Hopf bifurcation from a center or a focus; (ii) Poincaré bifurcation from closed orbits; and (iii) separatrix cycle bifurcations from a homoclinic or a heteroclinic loop. Although the study of Hopf bifurcation has been ongoing for more than half century, and one of the main tools for the study of bifurcation, normal form theory, dates back 100 years, efficient computation of the normal form is still a major challenge in this field. Computation becomes extremely difficult for higher-order normal forms or focus values, which is particularly important in studying the well-known Hilbert 16th problem. The second and third categories of limit cycle bifurcation, on the other hand, are comparatively more difficult; not only are there few techniques available, but there is also no general computational method developed for the Melnikov function, which is the main tool. It is the case that researchers new to the field often have difficulty in obtaining systematic material for study. Frontier research results can be difficult to access and even an experienced researcher in the field can be unaware of them. It appears necessary and significant that monographs on the topic of the bifurcation of limit cycles be available. The present monograph is intended to introduce the most recent new developments, and to provide major advances in the fundamental theory of limit cycles.

This monograph comes mainly from our recent research results and experience of teaching the topics of the stability and bifurcation of limit cycles to graduate students. It is divided into two parts: the first part (Chaps. 2–5) is focused on limit cycles bifurcating from a Hopf singularity and the use of normal form theory, while the second part (Chaps. 6–10) considers near-Hamiltonian systems and the main mathematical tool used is the Melnikov function.

Chapter 1 is an introduction, presenting the background for nonlinear dynamics, bifurcation and stability, the normal form method, the Melnikov function and Hilbert's 16th problem.

In Chap. 2, the computation of normal forms is discussed. First, a general approach which combines center manifold theory with computation of the normal form is presented. Then, a perturbation method which has proved computationally efficient is discussed in detail.

Chapter 3 is devoted to studying the computational efficiency of existing methods for computing focus values. Three typical methods: the Poincaré method or Takens method, a perturbation technique, and the singular point value method are particularly discussed. It is shown that these three methods have the same order of computational complexity, and no method has so far been developed for computing the "minimal singular point values".

In Chap. 4, Hopf bifurcation and the computation of normal forms are applied to consider planar vector fields and focused on the well-known Hilbert 16th problem. General cubic- and higher-order systems are considered to find the maximum number of limit cycles that such systems can have, i.e., to find a lower bound for the Hilbert number for certain vector fields. The Liénard system is investigated, and also, the critical periods of bifurcating periodic solutions from two special types of planar system are studied.

Chapter 5 is focused on the application of Hopf bifurcation theory and normal form computation to practical problems, including those from engineering and biological systems, as well as problems arising from the area of Hopf bifurcation control.

Chapter 6 introduces the fundamental theory of the Melnikov function method. Basic definitions and fundamental lemmas are presented and the main theory on the number of limit cycles is given.

In Chap. 7, particular attention is given to the bifurcation of limit cycles near a center. After normalizing the Hamiltonian function, detailed steps for computing the Melnikov function are described, and formulas are explicitly given. Maple programs for computing the coefficients of the Melnikov function are developed and illustrative examples are presented.

Chapter 8 considers the bifurcation of limit cycles near a homoclinic or heteroclinic loop. The method for computing the Melnikov function near a homoclinic or heteroclinic loop is developed and explicit formulas for the coefficients in the expansion of the Melnikov function are derived. Double homoclinic loops are also studied in this chapter.

In Chap. 9, an idea for finding more limit cycles is introduced, which combines the bifurcation of limit cycles from centers, homoclinic and heteroclinic loops.

A generalized theorem is presented. In particular, two polynomial systems are studied. By using the theorems and results obtained in Chaps. 6–9 it is shown that one system can have seven limit cycles while the other can have five.

Chapter 10 investigates the bifurcation of limit cycles in equivariant systems, including S-equivariant vector fields, Z_q-equivariant vector fields and S_q-reversible vector fields. In particular, an S-equivariant quadratic system, a Z_3-equivariant system and a cubic $(2\pi/3)$-equivariant system are studied.

The Maple programs for the semisimple case (see Sect. 2.2) and Hopf bifurcation (see Sect. 2.3) can be found on "Springer Extras" as *Electronic Supplementary Material*. These can be accessed and downloaded online anytime, anywhere. To use the content on Springer Extras, please visit extras.springer.com and search using the book's ISBN. You will then be asked to enter a password, which is given on the copyright page of this print book. More Maple programs for other singularities can be found from http://pyu1.apmaths.uwo.ca/~pyu/pub/index/software.html.

We acknowledge with gratitude the support received from Shanghai Normal University (School of Mathematical and Information Sciences) and the University of Western Ontario (Department of Applied Mathematics). The support for our research program by the Natural Science Foundation of China, and the Natural Sciences and Engineering Research Council of Canada is also greatly appreciated. We are very grateful to Ms Junmin Yang for her preparation of part of the manuscript. And, of course, we owe special thanks to our respective families for their tolerance of the obsession and the late nights that seemed necessary to bring this monograph to completion. Looking ahead, we would greatly appreciate users of this monograph contacting us to call our attention the imperfections that may have slipped into the final version.

Shanghai, China　　　　　　　　　　　　　　　　　　　　　　　　Maoan Han
London, Ontario, Canada　　　　　　　　　　　　　　　　　　　　　　Pei Yu

Contents

Chapter 1
Introduction

In this chapter, we present some background for nonlinear dynamics, bifurcation and stability, the normal form method, the Melnikov function and Hilbert's 16th problem, which are needed in the following chapters.

Nonlinear dynamics, also more grandly called "nonlinear science" or "chaos theory", is a rapidly-growing area, which plays an important role in the study of almost all disciplines of science and engineering, including mathematics, mechanics, aeronautics, electrical circuits, control systems, population problems, economics, financial systems, stock markets, ecological systems, etc. In general, any dynamical system contains certain parameters (usually called *bifurcation parameters* or *control parameters*) and thus it is vital to study the dynamic behavior of such systems as the parameters are varied. The complex dynamical phenomena include instability, bifurcation and chaos [6, 9, 19, 34, 72, 98, 105, 142, 143, 206, 225]. Studies of nonlinear dynamical systems may be roughly divided into two main categories: local analysis and global analysis. For instance, post-critical behavior such as saddle-node bifurcation and Hopf bifurcation can be studied locally in the vicinity of a critical point, while heteroclinic and homoclinic orbits, and chaos are essentially global behavior and have to be studied globally. These two categories need to be treated with different theories and methodologies.

The phenomenon of limit cycles was first discovered and studied by Poincaré [178] who presented the breakthrough qualitative theory of differential equations. In order to determine the existence of limit cycles, and their properties, for a given differential equation, Poincaré introduced the now well-known method of the Poincaré map, which is still the most basic tool for studying the stability and bifurcations of periodic orbits. The driving force behind the study of limit cycle theory was the invention of the triode vacuum tube which was able to produce stable, self-excited oscillations of constant amplitude. It was noted that such an oscillation phenomenon could not be described by linear differential equations. At the end of the 1920s, van der Pol [196] developed a differential equation to describe oscillations of constant amplitude in a triode vacuum tube:

$$\ddot{x} + \mu(x^2 - 1)\dot{x} + x = 0, \quad \mu \neq 0, \tag{1.1}$$

M. Han, P. Yu, *Normal Forms, Melnikov Functions and Bifurcations of Limit Cycles*,
Applied Mathematical Sciences 181,
DOI 10.1007/978-1-4471-2918-9_1, © Springer-Verlag London Limited 2012

where the dot denotes differentiation with respect to time, t, and μ is a parameter. Equation (1.1) is now called *van der Pol's equation*. Later a more general equation called the *Liénard equation* [144] was developed, of which van der Pol's equation is a special case.

Limit cycles are a common phenomenon, existing in almost all disciplines of science and engineering, including applied mathematics, physics, chemistry, mechanics, electrical circuits, control systems, economics, financial systems, ecological systems, etc. In fact, most early history in the theory of limit cycles was stimulated by practical problems displaying periodic behavior. Limit cycles are generated through bifurcations in many different ways, though mainly via Hopf bifurcation from a center or a focus, Poincaré bifurcation from closed orbits, or separatrix cycle bifurcation from homoclinic or heteroclinic orbits. Computing limit cycles and determining their stability is not only theoretically significant, but also practically important.

Related to limit cycle theory, the well-known 23 mathematical problems proposed by Hilbert in 1900 [99] have had significant impact on mathematics in the 20th century. Two of the 23 problems remain unsolved, one of which is the 16th problem. This problem has two parts: the first is about the relative positions of separate branches of algebraic curves, and the second is about the upper bound of the number of limit cycles and their relative locations in polynomial vector fields. A general result concerning these configurations was obtained in [150]. Today, "Hilbert's 16th problem" usually refers to the second part. Many, many studies have been made of this problem, especially for quadratic and cubic systems; see [7, 14, 48, 50–52, 66, 74–78, 81–85, 87–97, 99, 101, 106, 109, 111, 114, 128, 131–136, 138–140, 149, 151, 159, 163, 183, 187, 190, 204, 205, 209, 210, 214, 216, 217, 237–239, 243, 257, 259]. A detailed introduction and related literature can be found in Li [128], Schlomiuk [187], Ilyashenko [106] and Han [82]. Up to now, the problem has not been solved completely, even for the case $n = 2$. As was said in Smale [190], "Except for the Riemann hypothesis, it seems to be the most elusive of Hilbert's problems."

The second part of Hilbert's 16th problem is to find the maximum number and relative positions of the limit cycles of a polynomial system of degree n,

$$\dot{x} = P_n(x, y), \qquad \dot{y} = Q_n(x, y), \tag{1.2}$$

where $P_n(x, y)$ and $Q_n(x, y)$ represent nth-degree polynomials in x and y. More precisely, the second part of Hilbert's 16th problem is to find the upper bound, known as the Hilbert number, $H(n)$, of polynomial planar vector fields on the number of limit cycles that the systems can have. Although it has not been possible to obtain the precise number for $H(n)$, a great deal of effort has been made in finding the maximum number of limit cycles and raising the lower bound of the Hilbert number, $H(n)$, for some specific systems of certain degrees, hoping to get close to the real upper bound of $H(n)$.

In order to help understand and attack the problem, the weakened Hilbert 16th problem was posed by Arnold [7]. The problem is to ask for the maximum number of isolated zeros of the Abelian integral or Melnikov function,

$$M(h, \delta) = \oint_{H(x,y)=h} Q_n \, dx - P_n \, dy, \tag{1.3}$$

where $H(x, y)$, P_n and Q_n are real polynomials in x and y with $\deg(H) = m + 1$, $\deg(P_n)$, $\deg(Q_n) \leq n$. It is obvious that the problem of finding the maximum number of zeros consists of two parts: first find an upper bound and second find a lower bound for the number. The weakened Hilbert 16th problem is very interesting and important and has been studied by many researchers. The reason is, as we will see later, that the problem is closely related to the maximum number of limit cycles of the following near-Hamiltonian system:

$$\dot{x} = H_y + \varepsilon P_n(x, y), \qquad \dot{y} = -H_x + \varepsilon Q_n(x, y). \tag{1.4}$$

When the problem is restricted to a neighborhood of an isolated fixed point, then the question reduces to studying degenerate Hopf bifurcations, which gives rise to the study of weak focus points (or fine focus points). The basic idea of finding multiple limit cycles around the fixed point is to compute the focus values and solve the coupled polynomial equations to find the conditions under which multiple limit cycles can bifurcate from the fixed point; see [232, 253] for discussions on the efficient computation of normal forms and focus values. Suppose the origin of system (1.3), $(x, y) = (0, 0)$, is a fixed point of the system, at which the eigenvalues of the Jacobian of the system are a purely imaginary pair, $\pm i\omega_c$. Further, assume that the focus values are obtained as $v_i, i = 0, 1, 2, \ldots$. The basic idea for finding k small limit cycles around the origin involves two steps: first, find the conditions such that $v_0 = v_1 = v_2 = \cdots = v_{k-1} = 0$, but $v_k \neq 0$, and then second perform appropriate small perturbations to prove the existence of k limit cycles (e.g., see [230, 238, 239]).

In 1952, Bautin [14] proved that a quadratic planar polynomial vector field could have a maximum of 3 small limit cycles. Later, around the end of the 1970s, concrete examples were constructed to show the existence of 4 limit cycles [38, 189], i.e., $H(2) \geq 4$. In the past few years, great progress has been achieved in obtaining better estimates of the lower bounds of $H(n)$ for $n \geq 3$: $H(3) \geq 13$ [137, 147, 215], $H(4) \geq 20$ [86], $H(5) \geq 25$ [216], $H(6) \geq 35$ [203], $H(7) \geq 49$ [138], $H(9) \geq 80$ [205], and $H(11) \geq 121$ [204].

Hopf bifurcation [100] is perhaps the most popular and important one among the three categories leading to limit cycles. Consider the following general nonlinear system:

$$\dot{x} = f(x, \mu), \quad x \in R^n, \mu \in R, \qquad f : R^{n+1} \to R^n, \tag{1.5}$$

where x is an n-dimensional state vector, while μ is a scalar parameter called a *bifurcation parameter*. (Note that in general one may assume that μ is an m-dimensional vector for $m \geq 1$.) The function f is assumed to be analytic with respect to both x and μ. Equilibrium solutions of system (1.5) can be found by solving the nonlinear algebraic equation $f(x, \mu) = 0$ for all real μ. Let x^* be an equilibrium (or fixed point) of the system; i.e., $f(x^*, \mu) \equiv 0$ for any real value of μ. Further, suppose that the Jacobian of the system evaluated at the equilibrium x^* has a pair of complex conjugates, denoted by $\lambda_{1,2}(\mu)$ with $\lambda_1 = \bar{\lambda}_2 = \alpha(\mu) + i\omega(\mu)$ such that $\alpha(\mu^*) = 0$ and $\frac{d\alpha(\mu^*)}{d\mu} \neq 0$. Here, the second condition is usually called the *transversality condition*, implying that the crossing of the complex conjugate pair at the imaginary axis is not tangent to the imaginary axis. Then Hopf bifurcation occurs at the critical point $\mu = \mu^*$, giving rise to the bifurcation of a family of limit cycles. Other local bifurcations such as double-zero, Hopf-zero, double-Hopf, etc., may result in more complex dynamical behaviors.

In studying the Hopf bifurcation of limit cycles, center manifold theory and normal form theory are two main tools. Although normal form theory for differential equations can be traced back to the work of one hundred years ago, most credit should be given to Poincaré [178]. A lot of the early applications of normal form theory were in models from celestial mechanics. Later normal form theory was successfully applied to a broad spectrum of problems from different disciplines, including oscillation and vibration, control systems, etc. Birkhoff is commonly referred to as the first one who developed normal form theory for Hamiltonian systems [25]. There are many other researchers who have made contributions to normal form theory in the 1970s such as Arnold [5] and Takens [193]. More recently, normal form theory has been widely applied in the study of the singularities of vector fields and in bifurcation theory (e.g., see [8, 15, 16, 26, 54, 55, 69, 72, 79, 124, 125, 156, 191, 193, 208, 211]), which can be used to analyze the local dynamic behavior of a system such as a limit cycle, quasiperiodic motion, and more complex bifurcation solutions. In order to facilitate the application of normal forms, their symbolic computation using computer algebra systems has received considerable attention (e.g., see [45, 68, 223, 224, 226, 233, 234, 247–250, 252–254]). Although the Conventional (Classical) Normal Form (CNF) was developed to simplify the original system "as much as possible", it has been found that the CNF can be further simplified [5, 47, 186, 193, 195, 212, 213, 251], since only the linear part of the vector field has been used in simplifying nonlinear terms. For systems without parameters, a near-identity nonlinear transformation similar to the one used for CNF works, leading to the so-called *Simplest Normal Form* (SNF), which is also called the *unique normal form* or *minimal normal form* (e.g., see [10–13]). However, for systems with parameters (which is usually the case in solving real problems), computation of the SNF becomes much more involved. Moreover, the method used in CNF theory by adding unfolding to the normal form of the simplified system is no longer applicable for the SNF. Thus, one has to derive the parametric SNF directly from the original system with parameters, and introduce additional time rescaling and parameter rescaling. So far, only three cases have been investigated: a single-zero singularity [223], Hopf bifurcation [68, 247] and generalized Hopf bifurcation [233]. The slow progress in this direction is due to the extreme difficulty in computing the SNFs with perturbation parameters.

For the weakened Hilbert 16th problem (1.4), we assume that the Hamiltonian function at level $h \in (h_1, h_2)$ (i.e., $H(x, y) = h$) contains at least one family of closed orbits, γ^h, and when $\varepsilon = 0$ the integrable Hamiltonian vector field (1.4) does not have any limit cycles. Assume moreover that all the cycles are nonisolated, either extending to infinity or ending at centers or equilibrium curves. Then we define an integral around the closed orbit, γ^h, which equals zero for $\varepsilon = 0$:

$$I(h) = \oint_{\gamma^h} p(x, y, 0)\, dx + q(x, y, 0)\, dy = 0, \quad \gamma \subseteq H(x, y) = h, \qquad (1.6)$$

where $I(h)$ is called the *Poincaré–Melnikov–Abelian integral*, which is a multivalued function corresponding to different ovals at the same Hamiltonian level h with fixed polynomials H, p and q.

When $\varepsilon > 0$ (i.e., when a perturbation is added), energy is accumulated inside the families of closed orbits, and limit cycles will be generated from the existing family of closed orbits, γ^h. Choose $h_0 \in (h_1, h_2)$, and let ℓ be a cross section passing a point on γ^{h_0}, denoted by $A(h_0)$. Then for h near h_0 the periodic orbit, γ^h, has a unique intersection point with ℓ near $A(h_0)$, denoted by $A(h)$; i.e., $A(h) = \gamma^h \cap \ell$. Let $B(h, \varepsilon)$ be the first intersection point of the orbit with ℓ. Then we obtain

$$H(B) - H(A) = \int_{\widehat{AB}} H_x\, dx + H_y\, dy = \varepsilon\big[M(h) + O(\varepsilon)\big], \qquad (1.7)$$

where $M(h)$ is called the *Melnikov function* [164], given by

$$M(h) = \oint_{\gamma^h} (H_y q + H_x p)|_{\varepsilon=0}\, dt$$

$$= \oint_{\gamma^h} (q\, dx - p\, dy)|_{\varepsilon=0}$$

$$= \iint_{H \le h} (p_x + q_y)|_{\varepsilon=0}\, dx\, dy. \qquad (1.8)$$

The resulting map from $A(h)$ to $B(h, \varepsilon)$ is called a *Poincaré map*. Assume that the coefficients involved in the Hamiltonian, H, and polynomials, p and q, can be varied, then we can choose appropriate values for these coefficients such that the function $M(h)$ has as many zeros as possible, implying that the study of the bifurcation of limit cycles in this case is transferred to the study of the zeros of the function $M(h)$. The main difficulty is computational efficiency since it becomes much more difficult (as for high-order focus values) when $M(h)$ contains higher-order expressions of these coefficients. We shall develop, with the aid of Maple, efficient computational methods to study limit cycles bifurcating near centers, near homoclinic and heteroclinic orbits.

For the application of the normal form method to higher-dimensional systems, we particularly consider Hopf bifurcation and study the stability of bifurcating limit cycles. In practical problems, a system which exhibits limit cycles usually has high dimension; for example, a mechanical system may have more than 20 degrees of freedom. For such a system, the usual symbolic calculation for computing the focus values of lower-dimensional systems (e.g., 2 or 3 dimensions) is no longer applicable. Instead, symbolic and numerical computations must be combined to compute equilibrium solutions. For instance, consider the general nonlinear system,

$$\dot{x} = f(x, \mu), \quad x \in R^n, \ \mu \in R^m, \qquad f : R^{n+m} \to R^n, \tag{1.9}$$

where x is a vector state variable and μ is a vector parameter variable. First, solve numerically the equation $f(x, \mu) = 0$, with additional equations resulting from bifurcation conditions, to determine the critical points of the system. Then, for each critical point, apply a computer algebra system (such as Maple) and the normal form method to symbolically compute the corresponding normal form, and thus bifurcation and stability analysis can be easily carried out.

Recently, software packages (e.g., AUTO-07P [4] and MATCONT [56, 57]) have been developed to numerically determine bifurcation diagrams. Moreover, the latest versions of MATCONT support the numerical normal form computations at the generalized Hopf and other codimension 2 singular points. Although numerical approaches do not provide useful analytical formulas, they can help to provide information on the bifurcation of limit cycles for further analytical study. In determining very high-order multiple limit cycles (e.g., 9 limit cycles around a singular point), analytical and efficient computational methods are still essential.

With regard to the application of normal form theory to higher-dimensional systems, so far, except for the results obtained by using numerical methods (e.g., with the AUTO or MATCONT software packages) and our few results [113, 227, 234] on Hopf bifurcation, to the best of our knowledge, no theoretical results exist in this research area.

Chapter 2
Hopf Bifurcation and Normal Form Computation

In this chapter, we discuss the computation of normal forms. First we present a general approach which combines center manifold theory with computation of the normal form. Then, we focus on a perturbation method which has proved efficient in computation. Efficient computation is also discussed, and a sufficient and necessary condition for determining Hopf critical points of high-dimensional systems is given.

Note that for computing the normal form of systems with bifurcation (perturbation) parameters, one usually takes two steps. First, at a critical point (at which the dynamic system has a singularity) one sets the parameters to zero to obtain a so-called "reduced" (or "simplified") system and then normal form theory is applied to this system to obtain the normal form. Having found the normal form of the reduced system, one adds "unfolding" terms to get a parametric normal form for bifurcation analysis. However, this way one usually does not know the relationship between the original system parameters and the unfolding.

2.1 Hopf Bifurcation

Now suppose that system (1.5), which is rewritten below for convenience,

$$\dot{x} = f(x, \mu), \quad x \in R^n, \mu \in R, \qquad f : R^{n+1} \to R^n, \qquad (2.1)$$

has an equilibrium, given by $x = p(\mu)$. Suppose the Jacobian, $Df(\mu_0)$, of the system evaluated on the equilibrium at a critical point μ_0 has a simple pair of purely imaginary eigenvalues, $\pm i\omega$ ($\omega > 0$), and no other eigenvalues with zero real part. The implicit function theorem guarantees (since $Df(\mu_0)$ is invertible) that for each μ near μ_0 there will be an equilibrium $p(\mu)$ near $p(\mu_0)$ which varies smoothly with μ. Nonetheless, the dimensions of stable and unstable manifolds of $p(\mu)$ do change if the eigenvalues of $Df(p(\mu))$ cross the imaginary axis at μ_0. This qualitative change in the local flow near $p(\mu)$ must be marked by some other local changes in the phase portraits not involving fixed points.

M. Han, P. Yu, *Normal Forms, Melnikov Functions and Bifurcations of Limit Cycles*,
Applied Mathematical Sciences 181,
DOI 10.1007/978-1-4471-2918-9_2, © Springer-Verlag London Limited 2012

A clue to what happens in the generic bifurcation problem involving an equilibrium with purely imaginary eigenvalues can be gained from examining linear systems in which there is a change of this type. For example, consider the system

$$\dot{x} = \mu x - \omega y,$$
$$\dot{y} = \omega x + \mu y,$$

(2.2)

whose solutions have the form

$$\begin{pmatrix} x(t) \\ y(t) \end{pmatrix} = e^{\mu t} \begin{bmatrix} \cos \omega t & -\sin \omega t \\ \sin \omega t & \cos \omega t \end{bmatrix} \begin{pmatrix} x_0 \\ y_0 \end{pmatrix}.$$

(2.3)

When $\mu < 0$, solutions spiral into the origin, and when $\mu > 0$, solutions spiral away from the origin. When $\mu = 0$, all solutions are periodic. Even in a one-parameter family of equations, it is highly special to find a parameter value at which there is a whole family of periodic orbits, but there is still a surface of periodic orbits which appears in the general problem.

The normal form theorem gives us the required information about how the generic problem differs from system (2.2). By smooth changes of coordinates, the Taylor series of degree 3 for the general problem can be brought to the following form (see next section):

$$\dot{x} = [d\mu + a(x^2 + y^2)]x - [\omega + c\mu + b(x^2 + y^2)]y,$$
$$\dot{y} = [\omega + c\mu + b(x^2 + y^2)]x + [d\mu + a(x^2 + y^2)]y,$$

(2.4)

which is expressed in polar coordinates as

$$\dot{r} = (d\mu + ar^2)r,$$
$$\dot{\theta} = \omega + c\mu + br^2.$$

(2.5)

Since the $\dot{r}$ equation in (2.5) separates from θ, we see that there are periodic orbits of (2.4) which are circles $r = \text{const.}$, obtained from the nonzero solutions of $\dot{r} = 0$ in (2.5). If $a \neq 0$ and $d \neq 0$ these solutions lie along the parabola $\mu = -(\frac{a}{d})r^2$. This implies that the surface of periodic orbits has a quadratic tangency with its tangent plane $\mu = 0$ in $R^2 \times R$.

In the following, we first introduce the Hopf bifurcation theorem and then discuss in detail the computation of normal forms associated with various singularities (including Hopf bifurcation). It should be mentioned that in the original paper [100] (with English translation included in [162]) that Hopf did not know the terminology of the Poincaré normal form.

Theorem 2.1 ([100]) *Suppose that the system* $\dot{x} = f(x, \mu)$, $x \in R^n$, $\mu \in R$, *has an equilibrium* (x_0, μ_0) *at which the following properties are satisfied.*

(H1) $D_x f(x_0, \mu_0)$ *has a simple pair of purely imaginary eigenvalues and no other eigenvalues with zero real parts.*

Then (H1) *implies that there is a smooth curve of equilibria* $(x(\mu), \mu)$ *with* $x(\mu_0) = x_0$. *The eigenvalues* $\lambda(\mu)$, $\bar{\lambda}(\mu)$ *of* $D_x f(x(\mu), \mu_0)$, *which are imaginary at* $\mu = \mu_0$, *vary smoothly with* μ.

If, moreover,

(H2) $\qquad \dfrac{d}{d\mu} \big(\operatorname{Re} \lambda(\mu) \big) \big|_{\mu = \mu_0} = d \neq 0,$

then there is a unique three-dimensional center manifold passing through (x_0, μ_0) *in* $\mathbf{R}^n \times \mathbf{R}$ *and a smooth system of coordinates (preserving the planes* $\mu = const.$) *for which the Taylor expansion of degree 3 on the center manifold is given by* (2.4). *If* $a \neq 0$, *there is a surface of periodic solutions in the center manifold which has quadratic tangency with the eigenspace of* $\lambda(\mu_0)$, $\bar{\lambda}(\mu_0)$ *agreeing to second order with the paraboloid*

$$\mu = -\left(\frac{a}{d}\right)(x^2 + y^2). \tag{2.6}$$

If $a < 0$, *then these periodic solutions are stable limit cycles, while if* $a > 0$, *the periodic solutions are repelling.*

Fig. 2.1 Transversality of Hopf bifurcation

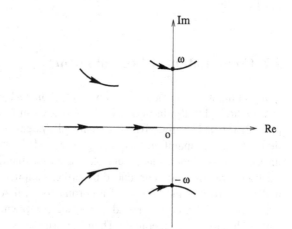

Fig. 2.2 Post-critical bifurcation path for a Hopf bifurcation: (**a**) stable; (**b**) unstable

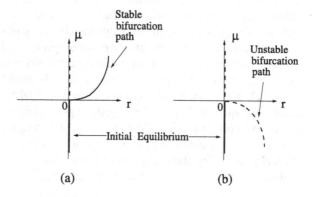

Fig. 2.3 Bifurcating periodic
solutions (limit cycles):
(**a**) stable; (**b**) unstable

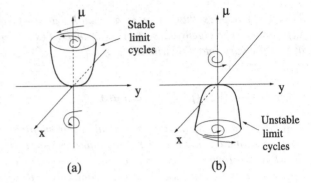

This theorem can be proved by a direct application of the center manifold and normal form theorems. The transversality conditions given in (H1) and (H2) are illustrated in Fig. 2.1. The parameter–amplitude relation (2.6) is shown in Fig. 2.2, where $r = \sqrt{x^2 + y^2}$, and the bifurcating periodic solutions depicted in three-dimensional space are given in Fig. 2.3.

Computation of center manifolds and normal forms will be discussed in the following sections.

2.2 Computation of Normal Forms

Many computational methods such as the Poincaré method (also called the Takens method), briefly discussed in the last section have been developed (e.g., see [47, 72, 166]). Usually, given a dynamical system described in differential equations, the center manifold theory (e.g., see [27]) is first applied to obtain a locally invariant small-dimensional manifold—a center manifold. Then additional nonlinear transformations are introduced to further simplify the center manifold to a normal form. To find the "form" of a normal form, first a homogeneous polynomial vector field of degree k is found in a space complementary to the range of the so-called "homological operator". Then the original vector field is decomposed into two parts: one of them, called the *nonresonant* terms, is eliminated and the other, called the *resonant* terms, is kept in the normal form. This simple form can be used conveniently for analyzing the local dynamic behavior of the original system.

For a practical system, not only the possible qualitative dynamical behavior of the system of concern, but also the quantitative relationship between the normal forms and the equations of the original system needs to be established. Normal forms are, in general, not uniquely defined and finding a normal form for a given system of differential equations is not a simple task. In particular, finding the explicit formulas for normal forms in terms of the coefficients of the original nonlinear system is not easy. Therefore, the crucial part in computing a normal form is the computational efficiency in finding the coefficients of the normal forms and the corresponding nonlinear transformations. Furthermore, the algebraic manipulation becomes very involved as the order of approximation increases. Thus, symbolic computations using symbolic computer languages such as Maple, Mathematica and Macsyma have been introduced for computing normal forms. However, it seems that even with a symbolic manipulator the computation of normal forms is still limited to lower-order approximations. This is because a program which computes a normal form usually quickly runs out of computer memory as the order of approximation increases. Therefore, computationally efficient methodology and software packages need to be developed for computing higher-order normal forms.

The idea of normal form theory is to use successive nonlinear transformations to derive a new set of differential equations by removing as many nonlinear terms from the system as possible. The terms remaining in the normal form are called the *resonant* terms. If the Jacobian matrix of the linearized system evaluated at an equilibrium can be transformed into diagonal form, then the bases of the nonlinear transformations are decoupled from each other. However, for a general singular vector field such as a system with non-semisimple double-zero or triple-zero eigenvalues, these bases are coupled. Such a coupling makes computation of the normal forms complicated.

First, we introduce a general approach for computing normal forms based on the work of [20, 21, 207, 226]. Consider a system described by the general nonlinear ordinary differential equation,

$$\dot{y} = Ay + F(y), \quad y \in R^n, \quad F(y): R^n \to R^n, \qquad (2.7)$$

where a dot indicates differentiation with respect to time t, Ay represents the linear part, and F is a nonlinear vector function and assumed to be analytic, satisfying $F(0) = 0$ and $D_y F(0) = 0$, $i = 1, 2, \ldots, n$. By introducing the linear transformation,

$$y = Tx, \qquad (2.8)$$

(2.7) can be transformed into

$$\dot{x} = Jx + f(x), \qquad (2.9)$$

where J is in Jordan canonical form. If the eigenvalues of A with zero real parts are denoted by $\lambda_1, \lambda_2, \ldots, \lambda_{n_0}$, and those with nonzero real parts are given by $\lambda_{n_0+1}, \lambda_{n_0+2}, \ldots, \lambda_n$, then $J = T^{-1}AT = \mathrm{diag}(J_0, J_1)$, where $J_0 = \mathrm{diag}(\lambda_1, \lambda_2, \ldots, \lambda_{n_0})$ and $J_1 = \mathrm{diag}(\lambda_{n_0+1}, \lambda_{n_0+2}, \ldots, \lambda_n)$. Next, let $x = (x_1, x_2)^T$, where x_1 and x_2 are variables associated with the eigenvalues with zero real parts and nonzero real parts, respectively, then (2.9) can be written as

$$\dot{x}_1 = J_0 x_1 + f_1(x_1, x_2),$$
$$\dot{x}_2 = J_1 x_2 + f_2(x_1, x_2). \qquad (2.10)$$

Now the center manifold of the system may be defined in the form

$$x_2 = N(x_1), \qquad N(0) = 0, \qquad DN_{x_1}(0) = 0, \qquad (2.11)$$

which satisfies

$$D_{x_1} N(x_1) \big[J_0 x_1 + f_1(x_1, N(x_1)) \big] = J_1 N(x_1) + f_2(x_1, N(x_1)). \qquad (2.12)$$

Once $N(x_1)$ is determined from (2.12), the differential equation describing the dynamics on the center manifold is given by

$$\dot{x}_1 = J_0 x_1 + f_1(x_1, N(x_1)). \qquad (2.13)$$

Further, (2.13) may be transformed into a set of simpler differential equations—the so-called *normal form*,

$$\dot{u} = J_0 u + C(u), \qquad (2.14)$$

under a series of nonlinear transformations given by

$$x_1 = u + H(u) \equiv u + \sum_{k \geq 2} H_k(u), \qquad (2.15)$$

where $H_k(u)$ is a kth-order vector of homogeneous polynomials in u. Substituting (2.15) into (2.13) results in

$$(I + D_u H(u))\dot{u} = J_0(u + H(u)) + f_1(u + H(u), N(u + H(u))), \qquad (2.16)$$

and then substituting (2.14) into (2.16) yields

$$(I + D_u H(u))(J_0 u + C(u))$$
$$= J_0(u + H(u)) + f_1(u + H(u), N(u + H(u))), \qquad (2.17)$$

which can be rearranged as

$$D_u H(u) J_0 u - J_0 H(u)$$
$$= f_1(u + H(u), N(u + H(u))) - D_u H(u)C(u) - C(u). \qquad (2.18)$$

To find the expression of the center manifold, one may substitute (2.15) into (2.12) to obtain

$$D_{x_1} N(x_1) \big[J_0(u + H(u)) + f_1(u + H(u), N(u + H(u))) \big]$$
$$= J_1 N(u + H(u)) + f_2(u + H(u), N(u + H(u))). \qquad (2.19)$$

Now rewrite (2.18) as

$$J_0 H(u) + f_1(u + H(u), N(u + H(u)))$$
$$= D_u H(u) J_0 u + D_u H(u)C(u) + C(u), \qquad (2.20)$$

and substitute this into (2.19) to obtain

$$D_{x_1} N(x_1)(I + D_u H(u))(J_0 u + C(u))$$
$$= J_1 N(u + H(u)) + f_2(u + H(u), N(u + H(u))) \qquad (2.21)$$

which can be written as

$$D_u N\big(u + H(u)\big) J_0 u - J_1 N\big(u + H(u)\big)$$
$$= f_2\big(u + H(u), N\big(u + H(u)\big)\big) - D_u N\big(u + H(u)\big) C(u). \qquad (2.22)$$

Combining (2.18) and (2.22) yields

$$D_u \begin{pmatrix} H(u) \\ h(u) \end{pmatrix} J_0 u - \begin{bmatrix} J_0 & 0 \\ 0 & J_1 \end{bmatrix} \begin{pmatrix} H(u) \\ h(u) \end{pmatrix}$$
$$= \begin{pmatrix} f_1(u + H(u), h(u)) \\ f_2(u + H(u), h(u)) \end{pmatrix} - D_u \begin{pmatrix} H(u) \\ h(u) \end{pmatrix} C(u) - \begin{pmatrix} C(u) \\ 0 \end{pmatrix}, \qquad (2.23)$$

where $h(u) = N(u + H(u))$. Furthermore, let

$$\overline{H}(u) = \begin{pmatrix} H(u) \\ h(u) \end{pmatrix}, \qquad f(u) = \begin{pmatrix} f_1(u + H(u), h(u)) \\ f_2(u + H(u), h(u)) \end{pmatrix}, \qquad \overline{C}(u) = \begin{pmatrix} C(u) \\ 0 \end{pmatrix},$$

then one can write (2.23) in a more compact form:

$$D_u \overline{H}(u) J_0 u - J \overline{H}(u) = f(u) - D_u \overline{H}(u) C(u) - \overline{C}(u). \qquad (2.24)$$

If (2.24) can be solved for $\overline{H}(u)$ and $\overline{C}(u)$ explicitly, one then will have the normal form as well as the nonlinear transformation. In general, however, the exact solution of (2.24) cannot be obtained. Thus, approximate solutions may be sought and assumed in the form

$$\overline{H}(u) = \sum_{m \geq 2} \overline{H}^m(u) = \sum_{m \geq 2} \left(\sum \overline{H}_{\tilde{m}} u_1^{m_1} u_2^{m_2} \cdots u_{n_0}^{m_{n_0}} \right),$$
$$C(u) = \sum_{m \geq 2} C^m(u) = \sum_{m \geq 2} \left(\sum C_{\tilde{m}} u_1^{m_1} u_2^{m_2} \cdots u_{n_0}^{m_{n_0}} \right), \qquad (2.25)$$

where $\overline{H}_{\tilde{m}} = (H_{\tilde{m}}, h_{\tilde{m}})^T$, $\tilde{m} \overset{\Delta}{=} m_1 m_2 \cdots m_{n_0}$, denoting a choice of the values of $m_1, m_2, \ldots, m_{n_0}$ which satisfies $m_1 + m_2 + \cdots + m_{n_0} = m$ and $m_i \geq 0, i = 1, 2, \ldots, n_0$. The summation given in the parentheses of (2.25) represents the sum of all possible choices for $\tilde{m}$. Similarly, suppose that $f(u) - D_u \overline{H}(u) C(u)$ can be expressed as

$$\tilde{f}(u) \overset{\Delta}{=} f(u) - D_u \overline{H}(u) C(u) = \sum_{m \geq 2} \tilde{f}^m(u)$$
$$= \sum_{m \geq 2} \left(\sum \tilde{f}_{\tilde{m}} u_1^{m_1} u_2^{m_2} \cdots u_{n_0}^{m_{n_0}} \right). \qquad (2.26)$$

Then, it can be shown that $\tilde{f}^m(u)$ only depends upon $\overline{H}^{m'}(u)$ and $C^{m'}(u)$ for $m' < m$. Therefore, once $\overline{H}^{m'}(u)$ and $C^{m'}(u)$ $(m' < m)$ are determined, the expression of $\tilde{f}^m(u)$ is definitely defined. To explicitly calculate the coefficients $\overline{H}_{\tilde{m}}$ and $C_{\tilde{m}}$, we have the following result.

Theorem 2.2 *If some of the eigenvalues of A with zero real parts are non-semisimple, then (2.24) can be written as*

$$(\lambda_0 - J)\overline{H}_{\tilde{m}} + \sum_{j=1}^{n_0-1} (J_0)_{j(j+1)}(m_j + 1)\overline{H}_{m_1 m_2 \cdots (m_j+1)(m_{j+1}-1)\cdots m_{n_0}} = \tilde{f}_{\tilde{m}} - \overline{C}_{\tilde{m}}, \tag{2.27}$$

where $\lambda_0 = m_1\lambda_1 + m_2\lambda_2 + \cdots + m_{n_0}\lambda_{n_0}$, n_0 is the number of eigenvalues with zero real parts and $(J_0)_{j(j+1)}$ denotes the element of J_0 located on the jth row and $(j+1)$th column. In addition, $m_j \leq n_0 - 1$ and $m_{j+1} \geq 1$.

Proof We can prove this result as follows. First note that if some of the eigenvalues of A with zero real parts are non-semisimple, then J_0 can be decomposed into

$$J_0 = S + N, \tag{2.28}$$

where $S = \mathrm{diag}(\lambda_1, \lambda_2, \ldots, \lambda_{n_0})$ and N is given in the form

$$N_{ij} = \begin{cases} 1 \text{ or } 0 & \text{if } i+1 = j, \\ 0 & \text{otherwise.} \end{cases} \tag{2.29}$$

It is clear that some of the nondiagonal elements of J_0 are nonzero, which causes some of the equations to be coupled. Therefore,

$$D_u \overline{H}_{m_1 \cdots m_{j-1}(m_j+1)(m_{j+1}-1)m_{j+2}\cdots m_{n_0}}$$

$$\times u_1^{m_1} \cdots u_{j-1}^{m_{j-1}} u_j^{m_j+1} u_{j+1}^{m_{j+1}-1} u_{j+2}^{m_{j+2}} \cdots u_{n_0}^{m_{n_0}} N u$$

$$= \sum_{j=1}^{n_0-1} \frac{\partial}{\partial u_j} \left(\overline{H}_{m_1 \cdots m_{j-1}(m_j+1)(m_{j+1}-1)m_{j+2}\cdots m_{n_0}} \right.$$

$$\left. \times u_1^{m_1} \cdots u_j^{m_j+1} u_{j+1}^{m_{j+1}-1} \cdots u_{n_0}^{m_{n_0}} \right) N_{j(j+1)} u_{j+1}$$

$$= \sum_{j=1}^{n_0-1} N_{j(j+1)}(m_j + 1)\overline{H}_{m_1 \cdots m_{j-1}(m_j+1)(m_{j+1}-1)m_{j+2}\cdots m_{n_0}}$$

$$\times u_1^{m_1} \cdots u_{j-1}^{m_{j-1}} u_j^{m_j} u_{j+1}^{m_{j+1}} u_{j+2}^{m_{j+2}} \cdots u_{m_{n_0}}^{m_{n_0}},$$

where $m_j \leq n_0 - 1$ and $m_{j+1} \geq 1$. In fact, the two elements m_j and m_{j+1} in the original subscripts of $\overline{H}$, $m_1 m_2 \cdots m_{j-1} m_j m_{j+1} m_{j+2} \cdots m_{n_0}$, have been replaced by $m_j + 1$ and $m_{j+1} - 1$, respectively. Thus, for a fixed m, the expression of $D_u \overline{H}^m(u) J_0 u$ becomes

$$D_u \overline{H}^m(u) J_0 u$$

$$= D_u \overline{H}^m(u) S u + D_u \overline{H}^m(u) N u$$

$$= \sum \left[\lambda_0 \overline{H}_{\tilde{m}} u_1^{m_1} u_2^{m_2} \cdots u_{n_0}^{m_{n_0}} \right.$$

$$+ \sum_{j=1}^{n_0-1} N_{j(j+1)} (m_j + 1) \overline{H}_{m_1 m_2 \cdots (m_j+1)(m_{j+1}-1) \cdots m_{n_0}}$$

$$\left. \times u_1^{m_1} \cdots u_j^{m_j} u_{j+1}^{m_{j+1}} \cdots u_{n_0}^{m_{n_0}} \right]$$

$$= \sum \left[\lambda_0 \overline{H}_{\tilde{m}} + \sum_{j=1}^{n_0-1} N_{j(j+1)} (m_j + 1) \overline{H}_{m_1 m_2 \cdots (m_j+1)(m_{j+1}-1) \cdots m_{n_0}} \right]$$

$$\times u_1^{m_1} \cdots u_{n_0}^{m_{n_0}}.$$

Now substituting the above equation into (2.24) and balancing the corresponding coefficients of $u_1^{m_1} u_2^{m_2} \cdots u_{n_0}^{m_{n_0}}$ in the resulting equation gives

$$(\lambda_0 - J) \overline{H}_{\tilde{m}} + \sum_{j=1}^{n_0-1} (J_0)_{j(j+1)} (m_j + 1) \overline{H}_{m_1 m_2 \cdots (m_j+1)(m_{j+1}-1) \cdots m_{n_0}} = \tilde{f}_{\tilde{m}} - \overline{C}_{\tilde{m}},$$

which is (2.27).   □

For the semisimple case, it can be seen that the summation term on the left-hand side of (2.27) disappears. Thus, if the eigenvalues of A with zero real parts are all semisimple, then (2.24) can be expressed in a simpler form

$$(\lambda_0 I - J) \overline{H}_{\tilde{m}} = \tilde{f}_{\tilde{m}} - \overline{C}_{\tilde{m}}, \tag{2.30}$$

or

$$(\lambda_0 I - J_0) H_{\tilde{m}} = \tilde{f}_{1\tilde{m}} - C_{\tilde{m}}, \tag{2.31}$$

$$(\lambda_0 I - J_1) h_{\tilde{m}} = \tilde{f}_{2\tilde{m}}. \tag{2.32}$$

For non-semisimple cases, since the summation term also contains the $\overline{H}$ coefficients, the computation becomes more involved.

In the following, we show how to compute the normal form for semisimple cases [21], and present the computation for two non-semisimple cases—a double-zero singularity [20] and a 1:1 resonant double-Hopf singularity [226].

2.2.1 Semisimple Case

If we can find all the coefficients $\overline{H}_{\tilde{m}}$, $C_{\tilde{m}}$ and $C_{m_1 m_2 \cdots m_{n_0}}$ satisfying (2.31) and (2.32), then we have solved the problem. To achieve this, first note that the diagonal matrix $\lambda_0 I - J_1$ is nonsingular since the real part of every component of $\lambda_0 I$ is zero while every component of J_1 is not zero. Therefore, (2.32) can be solved to uniquely determine $h_{\tilde{m}}$:

$$h_{\tilde{m}} = (\lambda_0 I - J_1)^{-1} \tilde{f}_{2\tilde{m}}. \tag{2.33}$$

For (2.31), we need to treat the nonresonant and resonant terms separately.

(1) Nonresonant terms. In this case, $\lambda_0 I - J_0$ is nonsingular, i.e., $\det(\lambda_0 I - J_0) \neq 0$. Therefore, $H_{\tilde{m}}$ can always be determined for a given $C_{\tilde{m}}$. In order to obtain a normal form for this particular order as simply as possible, one may choose $C_{\tilde{m}} = 0$, and thus,

$$C_{\tilde{m}} = 0 \quad \text{and} \quad H_{\tilde{m}} = (\lambda_0 I - J_0)^{-1} \tilde{f}_{1\tilde{m}}. \tag{2.34}$$

(2) Resonant terms. For this case, $\lambda_0 I - J_0$ is singular, i.e., $\det(\lambda_0 I - J_0) = 0$, implying that some components of $\lambda_0 I - J_0$ equal zero (remember that $\lambda_0 I - J_0$ is a diagonal matrix). Note that (2.31) has n_0 linear equations. This equation can be rewritten in component form:

$$(\lambda_0 I - J_0)_k H_{\tilde{m}k} = \tilde{f}_{1\tilde{m}k} - C_{\tilde{m}k}, \tag{2.35}$$

where the subscript k denotes the kth component of the coefficient matrix and the kth components of the vectors $H_{\tilde{m}}$, $\tilde{f}_{1\tilde{m}}$ and $C_{\tilde{m}}$. Thus, it is easy to find the following rules.
(i) When $(\lambda_0 I - J_0)_k \neq 0$,

$$C_{\tilde{m}k} = 0, \qquad H_{\tilde{m}k} = \frac{\tilde{f}_{1\tilde{m}k}}{(\lambda_0 I - J_0)_k}. \tag{2.36}$$

(ii) When $(\lambda_0 I - J_0)_k = 0$,

$$H_{\tilde{m}k} = 0, \qquad C_{\tilde{m}k} = \tilde{f}_{1\tilde{m}k}. \tag{2.37}$$

Summarizing the above results we have the following theorem.

Theorem 2.3 *For a fixed m, the coefficients, $C_{\tilde{m}}$, of the normal form and the corresponding coefficients, $H_{\tilde{m}}$ and $h_{\tilde{m}}$, of the nonlinear transformation are determined from the following formulas.*

(i) $$h_{\tilde{m}} = (\lambda_0 I - J_1)^{-1} \tilde{f}_{2\tilde{m}}.$$

(ii) *If $\lambda_0 I - J_0$ is nonsingular, then*

$$C_{\tilde{m}} = 0, \qquad H_{\tilde{m}} = (\lambda_0 I - J_0)^{-1} \tilde{f}_{1\tilde{m}}.$$

(iii) *If $\lambda_0 I - J_0$ is singular, then*

$$C_{\tilde{m}k} = 0, \qquad H_{\tilde{m}k} = \frac{\tilde{f}_{1\tilde{m}k}}{(\lambda_0 I - J_0)_k} \quad \text{for } (\lambda_0 I - J_0)_k \neq 0,$$

and

$$H_{\tilde{m}k} = 0, \qquad C_{\tilde{m}k} = \tilde{f}_{1\tilde{m}k} \quad \text{for } (\lambda_0 I - J_0)_k = 0,$$

where the subscript k indicates the kth linear equation of (2.31).
(iv) *The center manifold $N(u + H(u))$ is given by*

$$N\left(u + H(u)\right) \equiv h(u) = \sum_{m \geq 2}\left(\sum h_{\tilde{m}} u_1^{m_1} u_2^{m_2} \cdots u_{n_0}^{m_{n_0}}\right).$$

To end this section, we show several numerical examples and the corresponding normal forms, obtained using the Maple programs, which can be found on "Springer Extras" (by visiting extras.springer.com and searching for the book using its ISBN).

Example 2.4 First, let us consider a simple nonlinear oscillator, given by

$$\begin{aligned}
\dot{x} &= -y + x\left[1 - \left(x^2 + y^2\right)\right], \\
\dot{y} &= x + y\left[1 - \left(x^2 + y^2\right)\right].
\end{aligned} \tag{2.38}$$

By using the polar coordinate transformations $x = r\cos\theta$ and $y = r\sin\theta$, one can transform (2.38) into

$$\begin{aligned}
\dot{r} &= r\left(1 - r^2\right), \\
\dot{\theta} &= 1.
\end{aligned} \tag{2.39}$$

Note: multiplying the r equation by $2r$ results in

$$\frac{dr^2}{dt} = 2r\dot{r} = 2r^2\left(1 - r^2\right), \tag{2.40}$$

which becomes the logistic equation if we let $r^2 = R > 0$.

Solving the θ equation yields $\theta = t + \theta_0$. The phase portrait in R^2 is shown in Fig. 2.4.

It is easy to show that the equilibrium point $r = 0$ (i.e., $(x, y) = (0, 0)$) is unstable and the solution $r = 1, \theta = t + \theta_0$ approaches the *unit circle*, which is called a *limit cycle*.

Note that (2.39) is the normal form of system (2.38), which is an exact result. In fact, if we apply the Maple programs (which can be found on "Springer Extras" by visiting extras.springer.com and searching for the book using its ISBN) to execute system (2.38), we would obtain

$$\begin{aligned}
\dot{r} &= r + v_1 r^3 + v_2 r^5 + v_3 r^7 + \cdots, \\
\dot{\theta} &= 1 + \tau_1 r^2 + \tau_2 r^4 + \tau_3 r^6 + \cdots,
\end{aligned}$$

Fig. 2.4 The phase portrait
of system (2.38)

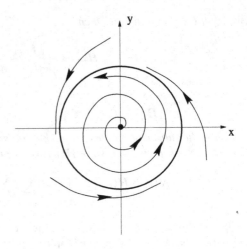

where $v_1 = -1$, $v_i = 0$, $i = 2, 3, \ldots$, and $\tau_i = 0$, $i = 1, 2, \ldots$, agreeing with the form (2.39).

Example 2.5 The first example is a simple two-dimensional system with a Hopf-type singularity, described by

$$\dot{x}_1 = x_2 + x_1^2,$$
$$\dot{x}_2 = -x_1 + x_2^2 + x_2^3, \tag{2.41}$$

whose Jacobian evaluated at the origin is in Jordan canonical form, with a purely imaginary pair of eigenvalues: $\pm i$. Executing the Maple program yields the normal form given in polar coordinates as

$$\dot{r} = \frac{3}{8}r^3 + \frac{5}{16}r^4 + \cdots,$$
$$\dot{\theta} = 1 - \frac{1}{3}r^2 + \frac{2023}{6912}r^4 + \cdots, \tag{2.42}$$

up to fifth order.

Example 2.6 The second example is an expansion of the above example, leading to a five-dimensional system described by

$$\dot{x}_1 = x_2 + x_1^2 - x_1 x_3,$$
$$\dot{x}_2 = -x_1 + x_2^2 + x_1 x_4 + x_2^3,$$
$$\dot{x}_3 = -x_3 + x_1^2, \tag{2.43}$$
$$\dot{x}_4 = -x_4 + x_5 + x_1^2,$$
$$\dot{x}_5 = -x_4 - x_5 + x_2^2.$$

The Jacobian of the system is in Jordan canonical form, with eigenvalues $\pm i, -1$, $-1 \pm i$, indicating that the system has a Hopf singularity at the equilibrium point $x_i = 0$. Executing the Maple program yields the normal form of the reduced system in polar coordinates as

$$\dot{r} = \frac{3}{40}r^3 - \frac{14867}{68000}r^5 + \cdots,$$

$$\dot{\theta} = 1 - \frac{1}{72}r^2 + \frac{5691403}{14688000}r^4 + \cdots, \tag{2.44}$$

up to fifth order. This example, as well as the one above, will be discussed again when we study the perturbation method in Sect. 2.3.

Example 2.7 Consider the following six-dimensional system:

$$\dot{x}_1 = -(x_1 - x_2 - x_4)^2,$$

$$\dot{x}_2 = x_3 - (x_1 - x_2 + x_5)^2,$$

$$\dot{x}_3 = -x_2 - (x_2 - x_3 + x_4)^2,$$

$$\dot{x}_4 = -x_4 + (x_1 - x_5)^2, \tag{2.45}$$

$$\dot{x}_5 = -x_5 + x_6 + (x_1 - x_4)^2,$$

$$\dot{x}_6 = -x_5 - x_6 + (x_2 + x_5)^2.$$

The system has an equilibrium at the origin, and the eigenvalues of the Jacobian of the system evaluated at the origin are $0, \pm i, -1, -1 \pm i$, implying that the dimension of the center manifold is 3. Executing our Maple programs gives the following normal form up to fifth order:

$$\dot{y} = -y^2 - \frac{1}{2}r^2 + 2y^3 - 2yr^2 + 5y^4 - \frac{411}{160}r^4 + \frac{33}{8}y^2r^2 - \frac{11}{2}y^5$$

$$+ \frac{1013}{40}y^3r^2 - \frac{503}{800}yr^4 + \cdots,$$

$$\dot{r} = yr - \frac{1}{40}r^3 + \frac{1}{2}y^2r - \frac{53}{20}y^3r + \frac{83}{50}yr^3 - \frac{75597}{136000}r^5$$

$$+ \frac{759}{100}y^4r + \frac{7423}{1000}y^2r^3 + \cdots, \tag{2.46}$$

$$\dot{\theta} = 1 - \frac{67}{40}r^2 + \frac{3}{2}y^2 + \frac{33}{10}y^3 + \frac{7}{100}yr^2 - \frac{219}{50}y^4 - \frac{9022679}{816000}r^4$$

$$+ \frac{192391}{6000}y^2r^2 + \cdots.$$

The nonlinear transformation is not listed, but can be obtained from the computer output.

Example 2.8 The equations are described by

$$\dot{x}_1 = x_2 + x_1^3 - x_1^2 x_5 + x_1^2 x_7,$$
$$\dot{x}_2 = -x_1 - 2x_1 x_3^2,$$
$$\dot{x}_3 = \sqrt{2}x_4 + x_1^2 x_3 - 4x_5^3,$$
$$\dot{x}_4 = -\sqrt{2}x_3, \tag{2.47}$$
$$\dot{x}_5 = -x_5 + (x_1 - x_5)^2,$$
$$\dot{x}_6 = -x_6 + x_7 + (x_1 - x_4)^2,$$
$$\dot{x}_7 = -x_6 - x_7 + (x_2 - x_6)^2.$$

The Jacobian of this system evaluated at the origin has two pairs of purely imaginary eigenvalues: $\pm i$ and $\pm\sqrt{2}i$, and three noncritical eigenvalues: $-1, -1 \pm i$. This is an *internal nonresonant* case. The normal form up to fifth order given in polar coordinates is

$$\dot{r}_1 = \frac{3}{8}r_1^3 + \frac{157}{1360}r_1^5 - \frac{9}{40}r_1^3 r_2^2 + \cdots,$$
$$\dot{\theta}_1 = 1 + \frac{1}{2}r_2^2 - \frac{5543}{21760}r_1^4 - \frac{1}{16}r_2^4 - \frac{3}{80}r_1^2 r_2^2 + \cdots,$$
$$\dot{r}_2 = \frac{1}{4}r_1^2 r_2 - \frac{1}{16}r_1^2 r_2^3 + \cdots, \tag{2.48}$$
$$\dot{\theta}_2 = \sqrt{2} - \frac{\sqrt{2}}{32}r_1^4 + \cdots.$$

It should be noted from the normal form (2.48) that the two equations describing the amplitudes r_1 and r_2 are decoupled from the two equations describing the phases θ_1 and θ_2, as expected from the characteristics of nonresonance.

More examples can be found in [21].

2.2.2 A Double-Zero Eigenvalue

For a system with non-semisimple double-zero eigenvalues, Takens [194] and Bogdanov [22] obtained the "form" of the normal forms and gave a very detailed bifurcation analysis. (That's why this case is also called the Takens–Bogdanov zero singularity.) Their results have been extended to the study of global bifurcations and chaos. Baider and Sanders [13] gave a complete formal classification of such a system, while Broer [23] presented a universal local bifurcation analysis for such a system. Yu and Huseyin [244], on the other hand, studied the static and dynamic bifurcations for this case, and obtained explicit formulas for the bifurcation solutions and stability conditions. But the results are limited to first-order approximations.

It is well known that the two standard normal forms for a double-zero singularity are given by (e.g., see [72])

$$\dot{x} = y + \sum_{j=2}^{k} a_j x^j, \qquad \dot{y} = \sum_{j=2}^{k} b_j x^j, \tag{2.49}$$

and

$$\dot{x} = y, \qquad \dot{y} = \sum_{j=2}^{k} (a_j x^j + b_j x^{j-1} y), \tag{2.50}$$

where the a_j's and b_j's are constant coefficients.

In the following, we apply Theorem 2.2 to derive the explicit formulas for the normal forms and associated nonlinear transformations of a system characterized by a non-semisimple double-zero eigenvalue.

In this case, $n_0 = 2$ and the Jacobian, J, of (2.9) is now in the form

$$J = \begin{bmatrix} J_0 & 0 \\ 0 & J_1 \end{bmatrix} = \begin{bmatrix} 0 & 1 & 0 \\ 0 & 0 & 0 \\ 0 & 0 & J_1 \end{bmatrix}, \tag{2.51}$$

where J_1 has eigenvalues with nonzero real parts, given in Jordan canonical form. By using (2.25) and (2.26), one may assume the following forms:

$$C(u) = \sum_{m \geq 2} C^m(u) = \sum_{m \geq 2} \sum_{j=0}^{m} C_{j(m-j)} u_1^j u_2^{m-j},$$

$$\overline{H}(u) = \sum_{m \geq 2} \overline{H}^m(u) = \sum_{m \geq 2} \sum_{j=0}^{m} \overline{H}_{j(m-j)} u_1^j u_2^{m-j}, \tag{2.52}$$

$$\tilde{f}(u) = \sum_{m \geq 2} \tilde{f}^m(u) = \sum_{m \geq 2} \sum_{j=0}^{m} \tilde{f}_{j(m-j)} u_1^j u_2^{m-j},$$

and thus, (2.24) can be written as

$$\sum_{m \geq 2} \left\{ \sum_{j=0}^{m-1} [(j+1)\overline{H}_{(j+1)(m-j-1)} - J\overline{H}_{j(m-j)}] u_1^j u_2^{m-j} - J\overline{H}_{m0} u_1^m \right\}$$

$$= \sum_{m \geq 2} \sum_{j=0}^{m} [\tilde{f}_{j(m-j)} - \overline{C}_{j(m-j)}] u_1^j u_2^{m-j}. \tag{2.53}$$

Comparing the coefficients of the components $u_1^j u_2^{k-j}$ in (2.53) yields

$$(j+1)\overline{H}_{(j+1)(m-j-1)} - J\overline{H}_{j(m-j)} = \tilde{f}_{j(m-j)} - \overline{C}_{j(m-j)},$$
$$j = 0, 1, \ldots, m-1, \tag{2.54}$$

and

$$-J\overline{H}_{m0} = \tilde{f}_{m0} - \overline{C}_{m0}. \tag{2.55}$$

Let

$$\overline{H}_{ij} = (\overline{H}_{1,ij}, \overline{H}_{2,ij}, \overline{H}_{3,ij}, \ldots, \overline{H}_{n,ij})^T,$$
$$\tilde{f}_{ij} = (\tilde{f}_{1,ij}, \tilde{f}_{2,ij}, \tilde{f}_{3,ij}, \ldots, \tilde{f}_{n,ij})^T, \tag{2.56}$$
$$\overline{C}_{ij} = (C_{1,ij}, C_{2,ij}, 0, \ldots, 0)^T,$$

then (2.55) can be written as

$$-\overline{H}_{2,m0} = \tilde{f}_{1,m0} - C_{1,m0},$$
$$0 = \tilde{f}_{2,m0} - C_{2,m0}, \tag{2.57}$$

and

$$-\lambda_s \overline{H}_{s,m0} = \tilde{f}_{s,m0}, \quad s = 3, 4, \ldots, n. \tag{2.58}$$

On the other hand, the following equations can be found from (2.54):

$$(j+1)\overline{H}_{1,(j+1)(m-j-1)} - \overline{H}_{2,j(m-j)} = \tilde{f}_{1,j(m-j)} - C_{1,j(m-j)}, \tag{2.59}$$

$$(j+1)\overline{H}_{2,(j+1)(m-j-1)} = \tilde{f}_{2,j(m-j)} - C_{2,j(m-j)}, \tag{2.60}$$

and

$$(j+1)\overline{H}_{s,(j+1)(m-j-1)} - \lambda_s \overline{H}_{2,j(m-j)} = \tilde{f}_{s,j(m-j)}, \quad s = 3, 4, \ldots, n, \tag{2.61}$$

where $j = 1, 2, \ldots, m-1$. Note that the coefficients associated with the Jacobian J_1, determined by (2.58) and (2.61) are decoupled (2.59) and (2.60), and thus can be solved independently. Now, let us first find the coefficients associated with the Jacobian J_0, which are determined by (2.59) and (2.60). To achieve this, solving (2.60) for $\overline{H}_{2,(j+1)(m-j-1)}$ and then replacing j by $j-1$ for $j \geq 2$ in the resulting expression yields

$$\overline{H}_{2,j(m-j)} = \frac{1}{j}[\tilde{f}_{2,(j-1)(m-j+1)} - C_{2,(j-1)(m-j+1)}]. \tag{2.62}$$

Then, substituting (2.62) into (2.59) gives

$$\overline{H}_{1,(j+1)(m-j-1)} = \frac{1}{j+1}\left\{ \tilde{f}_{1,j(m-j)} - C_{1,j(m-j)} \right.$$
$$\left. + \frac{1}{j}[\tilde{f}_{2,(j-1)(m-j+1)} - C_{2,(j-1)(m-j+1)}] \right\}, \tag{2.63}$$

where $j = 2, 3, \ldots, m$. Setting $j = m$ in (2.62) and $j = m - 1$ in (2.63) results in

$$\bar{H}_{2,m0} = \frac{1}{m}[\tilde{f}_{2,(m-1)1} - C_{2,(m-1)1}], \tag{2.64}$$

$$\bar{H}_{1,m0} = \frac{1}{m}\left\{\tilde{f}_{1,(m-1)1} - C_{1,(m-1)1} + \frac{1}{m-1}[\tilde{f}_{2,(m-2)2} - C_{2,(m-2)2}]\right\}. \tag{2.65}$$

By using (2.64) to eliminate $\bar{H}_{2,m0}$ from (2.57), one can rewrite (2.57) as

$$\frac{1}{m}[\tilde{f}_{2,(m-1)1} - C_{2,(m-1)1}] + \tilde{f}_{1,m0} - C_{1,m0} = 0, \tag{2.66}$$

$$\tilde{f}_{2,m0} - C_{2,m0} = 0.$$

Since no assumptions have been made in the above derivations, all coefficients $C_{1,ij}$ and $C_{2,ij}$, except those involved in (2.66), can be chosen arbitrarily. In order to find as simple a normal form as possible, these coefficients may be chosen to equal zero. On the other hand, the coefficients $C_{1,m0}$, $C_{2,(m-1)1}$ and $C_{2,m0}$, which must satisfy (2.66), cannot be chosen arbitrarily and may have to be kept in the resulting normal form. Thus, the best choice of the values of these coefficients is one such that the normal form is as simple as possible and (2.66) is also satisfied. Obviously, there are two possible choices:

$$\text{if } C_{2,(m-1)1} = 0 \quad \text{then } C_{1,m0} = \frac{1}{m}\tilde{f}_{2,(m-1)1} + \tilde{f}_{1,m0};$$

$$\text{if } C_{1,m0} = 0 \quad \text{then } C_{2,(m-1)1} = \tilde{f}_{2,(m-1)1} + m\tilde{f}_{1,m0}.$$

Combining these two choices for different order normal forms may result in infinitely many different normal forms. However, if these two choices are used consistently for all orders of normal form, then one may have the following two different normal forms.

Theorem 2.9 *The normal form of the double-zero singularity up to an arbitrary order k is given by*

$$\dot{u}_1 = u_2 + \sum_{m=2}^{k} C_{1,m0}u_1^m,$$

$$\dot{u}_2 = \sum_{m=2}^{k} C_{2,m0}u_1^m, \tag{2.67}$$

or

$$\dot{u}_1 = u_2,$$

$$\dot{u}_2 = \sum_{m=2}^{k}(C_{2,(m-1)1}u_1^{m-1}u_2 + C_{2,m0}u_1^m). \tag{2.68}$$

The normal form coefficients $C_{2,(m-1)1}$ and $C_{2,m0}$ for the form (2.68) are then given by

$$
\begin{aligned}
C_{2,(m-1)1} &= \tilde{f}_{2,(m-1)1} + m\,\tilde{f}_{1,m0}, \\
C_{2,m0} &= \tilde{f}_{2,m0}.
\end{aligned}
\tag{2.69}
$$

The above two forms, (2.67) and (2.68), are the same as those given by (2.49) and (2.50), respectively. Usually, the second form, (2.68), is chosen for bifurcation analysis. Moreover, the coefficients of the corresponding nonlinear transformation are determined as

$$
\overline{H}_{2,(j+1)(m-j-1)} = \frac{1}{j+1}[\tilde{f}_{2,j(m-j)} - C_{2,j(m-j)}],
$$

$$
j = 0, 1, \ldots, m-1,
$$

$$
\overline{H}_{1,(j+1)(m-j-1)} = \frac{1}{j+1}\left[\tilde{f}_{1,j(m-j)} + \frac{1}{j}\tilde{f}_{2,(j-1)(m-j+1)}\right],
\tag{2.70}
$$

$$
j = 1, 2, \ldots, m-1,
$$

$$
\overline{H}_{2,m0} = -\tilde{f}_{1,m0}.
$$

Here, one may observe from (2.70) that the three coefficients $\overline{H}_{1,1(m-1)}$, $\overline{H}_{1,0m}$ and $\overline{H}_{2,0m}$ are not defined in these equations. In addition, since $\overline{H}_{1,0m}$ does not appear in any of these equations, it can be chosen arbitrarily. The other two coefficients, $\overline{H}_{1,1(m-1)}$ and $\overline{H}_{2,0m}$, must satisfy an equation obtained from (2.54) by setting $j = 0$, i.e.,

$$
\overline{H}_{1,1(m-1)} - \overline{H}_{2,0m} = \tilde{f}_{1,0m},
\tag{2.71}
$$

which indicates that these two coefficients are not independent. Thus, one of them can be chosen arbitrarily and the other is then determined. To make the normal form definite, we adopt the following choices:

$$
\begin{aligned}
\overline{H}_{1,0m} &= 0, \\
\overline{H}_{2,0m} &= 0, \\
\overline{H}_{1,1(m-1)} &= \tilde{f}_{1,0m}.
\end{aligned}
\tag{2.72}
$$

Finally, the coefficients of the nonlinear transformation associated with the Jacobian J_1 can be uniquely determined as follows. First, from (2.58), one can find

$$
\overline{H}_{s,m0} = -\frac{1}{\lambda_s}\tilde{f}_{s,m0}, \quad s = 3, 4, \ldots, n.
\tag{2.73}
$$

Then, an iteration equation can be derived from (2.61) as

$$\overline{H}_{s,j(m-j)} = -\frac{1}{\lambda_s}\left(\tilde{f}_{s,j(m-j)} - (j+1)\overline{H}_{s,(j+1)(m-j-1)}\right) \qquad (2.74)$$

for $j = 1, 2, \ldots, (m-1)$ and $s = 3, 4, \ldots, n$.

Therefore, the normal form and the associated nonlinear transformation are uniquely determined. A symbolic computation software package has been developed for (non-semisimple) double-zero and triple-zero singularities [20]. An example of the application of a double-zero singularity is shown in Sect. 5.2.2.

2.2.3 1:1 Resonant Hopf Bifurcation

Hopf and generalized Hopf bifurcations have been extensively studied by many researchers (e.g., see [47, 54, 55, 63, 72, 102, 166, 218]), and are associated with a pair of purely imaginary eigenvalues at an equilibrium. If the Jacobian of a system evaluated at a critical point involves two pairs of purely imaginary eigenvalues, the so-called "double-Hopf bifurcation" may occur. Such bifurcations may exhibit more complicated and interesting dynamic behavior such as quasiperiodic motions on tori, and chaos (e.g., see [21, 72, 102, 166, 221, 223, 226, 231]). A bifurcation is called *nonresonant* if the ratio of the two eigenvalues is not a rational number, otherwise it is called *resonant*. The most important resonance is the 1:1 non-semisimple case, which is also usually called *1:1 double-Hopf bifurcation*. In this subsection, the method developed above is applied to consider the 1:1 non-semisimple case for general n-dimensional systems which are not necessarily described on a four-dimensional center manifold.

For this case, $J = \text{diag}[J_0\ J_1]$, and both J_0 and J_1 are assumed to be in Jordan canonical form. In particular, J_0 has eigenvalues $\pm i\omega$. Without loss of generality, we may assume $\omega = 1$, and thus J_0 is given in the form

$$J_0 = \begin{bmatrix} i & 1 & 0 & 0 \\ 0 & i & 0 & 0 \\ 0 & 0 & -i & 1 \\ 0 & 0 & 0 & -i \end{bmatrix}. \qquad (2.75)$$

It should be noted that the third and fourth equations of (2.9) are the complex conjugates of the first and second equations, respectively, and thus $x_3 = \bar{x}_1$ and $x_4 = \bar{x}_2$, where the bar denotes complex conjugate. J_1, on the other hand, is assumed to have eigenvalues with nonzero real parts. One may assume that J_1 is given in the following general block form:

$$J_1 = \begin{bmatrix} J_{15} & 0 & \cdots & 0 \\ 0 & J_{16} & \cdots & 0 \\ \vdots & \vdots & \ddots & \vdots \\ 0 & 0 & \cdots & J_{1i} \end{bmatrix}, \qquad (2.76)$$

where

$$
J_{1j} = \begin{bmatrix} \lambda_j & s & \cdots & 0 & 0 \\ 0 & \lambda_j & \cdots & 0 & 0 \\ \vdots & \vdots & \ddots & \vdots & \vdots \\ 0 & 0 & \cdots & \lambda_j & s \\ 0 & 0 & \cdots & 0 & \lambda_j \end{bmatrix},
\tag{2.77}
$$

for $j = 5, 6, \ldots, i$. Note that the dimension of J_1 is $(n-4) \times (n-4)$. Also note that the "s" given in (2.77), on the shoulder of the λ_j's, only takes the value of either 1 or 0. This indicates that both J_0 and J_1 are non-semisimple.

Now, (2.23) or (2.24) can be rewritten as

$$
\sum_m \left(\begin{array}{c} [\lambda_0 I - J_0] H_{m_1 m_2 m_3 m_4} \\ [\lambda_0 I - J_1] h_{m_1 m_2 m_3 m_4} \end{array} \right) u_1^{m_1} u_2^{m_2} u_3^{m_3} u_4^{m_4}
$$

$$
+ \sum_m m_1 \left(\begin{array}{c} H_{m_1 m_2 m_3 m_4} \\ h_{m_1 m_2 m_3 m_4} \end{array} \right) u_1^{m_1 - 1} u_2^{m_2 + 1} u_3^{m_3} u_4^{m_4}
$$

$$
+ \sum_m m_3 \left(\begin{array}{c} H_{m_1 m_2 m_3 m_4} \\ h_{m_1 m_2 m_3 m_4} \end{array} \right) u_1^{m_1} u_2^{m_2} u_3^{m_3 - 1} u_4^{m_4 + 1}
$$

$$
= \sum_m \left(\begin{array}{c} \tilde{f}_{1 m_1 m_2 m_3 m_4} - C_{m_1 m_2 m_3 m_4} \\ \tilde{f}_{2 m_1 m_2 m_3 m_4} \end{array} \right) u_1^{m_1} u_2^{m_2} u_3^{m_3} u_4^{m_4}
\tag{2.78}
$$

where

$$
\lambda_0 = (m_1 + m_2 - m_3 - m_4)\mathrm{i} \quad \text{and} \quad m_1 + m_2 + m_3 + m_4 = m.
\tag{2.79}
$$

Equation (2.78) can be used for solving for the coefficients C_m, H_m and h_m, order by order, starting from $m = 2$, for all possible choices such that $m_1 + m_2 + m_3 + m_4 = m$. Here, $\tilde{f}_{1m}$ and $\tilde{f}_{2m}$ denote the mth-order coefficients of $f_1(u + H(u), h(u)) - DH(u)C(u)$ and $f_2(u + H(u), h(u)) - Dh(u)C(u)$, respectively. In the following we shall present a procedure for solving (2.78).

Firstly, it is noted from (2.23) that the mth-order coefficients $\tilde{f}_{1m}$ and $\tilde{f}_{2m}$ contain C, H and h coefficients whose orders are less than m. Therefore, for computing the mth-order coefficients, we only need to consider the terms on the left-hand side of (2.78) and the C_m term on the right-hand side. Secondly, since λ_0 only contains the eigenvalues of J_0 (with zero real parts) and all eigenvalues of J_1 have nonzero real parts, $[\lambda_0 I - J_1]$ cannot equal zero in any of its components. This suggests that h_m can be uniquely determined from (2.78). However, note that the second and third terms on the left-hand side of (2.78) do not have the same index as other terms. Thus, the solution procedure is not as simple as the semisimple case [226]. Making the index consistent in (2.78) yields

$$[\lambda_0 I - J_0] H_{m_1 m_2 m_3 m_4} + (m_1 + 1) H_{(m_1+1)(m_2-1)m_3 m_4}$$
$$+ (m_3 + 1) H_{m_1 m_2 (m_3+1)(m_4-1)}$$
$$= \tilde{f}_{1m_1 m_2 m_3 m_4} - C_{m_1 m_2 m_3 m_4} \tag{2.80}$$

and

$$[\lambda_0 I - J_1] h_{m_1 m_2 m_3 m_4} + (m_1 + 1) h_{(m_1+1)(m_2-1)m_3 m_4}$$
$$+ (m_3 + 1) h_{m_1 m_2 (m_3+1)(m_4-1)}$$
$$= \tilde{f}_{2m_1 m_2 m_3 m_4}. \tag{2.81}$$

Note that (2.81) does not contain the normal form coefficients $C_{m_1 m_2 m_3 m_4}$. In addition, it is easy to show that the matrix $[\lambda_0 I - J_1]$ is nonsingular. Therefore, one can uniquely determine $h_{m_1 m_2 m_3 m_4}$ from (2.81) by using the following recursive formulas, where $m_1 + m_2 + m_3 + m_4 = m$, $m_i \geq 0$, $i = 1, 2, 3, 4$. For convenience, let $k_1 = m_1 + m_2$, $k_2 = m_3 + m_4$ and so $k_1 + k_2 = m$.

(1) For $m_2 = m_4 = 0$:

$$h_{k_1 0 k_2 0} = [\lambda_0 I - J_1]^{-1} \tilde{f}_{2k_1 0 k_2 0}. \tag{2.82}$$

(2) For $m_4 = 0$, $m_2 = 1, 2, \ldots, k_1$:

$$h_{(k_1-m_2)m_2 k_2 0}$$
$$= [\lambda_0 I - J_1]^{-1} \left(\tilde{f}_{2(k_1-m_2)m_2 k_2 0} - (k_1 - m_2 + 1) h_{(k_1-m_2+1)(m_2-1)k_2 0} \right). \tag{2.83}$$

(3) For $1 \leq m_4 \leq k_2$, $m_2 = 0, 1, 2, \ldots, k_1$:

$$h_{k_1 0 (k_2-m_4)m_4}$$
$$= [\lambda_0 I - J_1]^{-1} \left(\tilde{f}_{2k_1 0 (k_2-m_4)m_4} - (k_2 - m_4 + 1) h_{k_1 0 (k_2-m_4+1)(m_4-1)} \right) \tag{2.84}$$

when $m_2 = 0$, and

$$h_{(k_1-m_2)m_2(k_2-m_4)m_4} = [\lambda_0 I - J_1]^{-1} \left(\tilde{f}_{2(k_1-m_2)m_2(k_2-m_4)m_4} \right.$$
$$- (k_1 - m_2 + 1) h_{(k_1-m_2+1)(m_2-1)(k_2-m_4)m_4}$$
$$\left. - (k_2 - m_4 + 1) h_{(k_1-m_2)m_2(k_2-m_4+1)(m_4-1)} \right) \tag{2.85}$$

when $m_2 = 1, 2, \ldots, k_1$.

Now, we solve (2.80) for $H_{m_1m_2m_3m_4}$ and $C_{m_1m_2m_3m_4}$. Since we want to choose the normal form as simply as possible, we may set all the coefficients $C_{m_1m_2m_3m_4}$ to zero if the matrix $[\lambda_0 I - J_0]$ is nonsingular, and then the coefficients $H_{m_1m_2m_3m_4}$ are uniquely determined from (2.80). However, the matrix $[\lambda_0 I - J_0]$ can be singular, giving rise to the so-called *resonant* terms which have to be retained in the normal form. Because the third and fourth equations of (2.7) are the complex conjugates of the first and second equations, respectively, thus the third and fourth components of H_m and C_m are, respectively, the complex conjugates of the first and second components of H_m and C_m. Therefore, we only need to consider the first two equations of (2.80) to solve $H_m^* = (H_{1m_1m_2m_3m_4}, H_{2m_1m_2m_3m_4})^T$ and $C_m^* = (C_{1m_1m_2m_3m_4}, C_{2m_1m_2m_3m_4})^T$.

Since

$$[\lambda_0 I - J_0] = \begin{bmatrix} \lambda_0 i & 1 & 0 & 0 \\ 0 & \lambda_0 i & 0 & 0 \\ 0 & 0 & -\lambda_0 i & 1 \\ 0 & 0 & 0 & -\lambda_0 i \end{bmatrix}, \tag{2.86}$$

where $\lambda_0 = k_1 - k_2 - 1$, there exist two cases: (i) $k_1 - k_2 \neq 1$ (i.e., $\lambda_0 \neq 0$) which is called the *nonresonant* case; and (ii) $k_1 - k_2 = 1$ (i.e., $\lambda_0 = 0$) called the *resonant* case.

(A) Nonresonant Case: $m_1 + m_2 \neq m_3 + m_4 + 1$ For the nonresonant case, $k_1 - k_2 \neq 1$, i.e., $m_1 + m_2 \neq m_3 + m_4 + 1$, we may set all $C_{m_1m_2m_3m_4}^* = 0$, and then similarly determine $H_{m_1m_2m_3m_4}^*$ from (2.80). The detailed steps for solving for $H_{m_1m_2m_3m_4}^*$ are given below.

(1) For $m_2 = m_4 = 0$:

$$H_{k_10k_20}^* = [\lambda_0 I - J_0]^{-1} \tilde{f}_{1k_10k_20}^*. \tag{2.87}$$

(2) For $m_4 = 0$, $m_2 = 1, 2, \ldots, k_1$:

$$H_{(k_1-m_2)m_2k_20}^* = [\lambda_0 I - J_0]^{-1} \big(\tilde{f}_{1(k_1-m_2)m_2k_20}^* - (k_1 - m_2 + 1) $$
$$\times H_{(k_1-m_2+1)(m_2-1)k_20}^* \big). \tag{2.88}$$

(3) For $1 \leq m_4 \leq k_2$, $m_2 = 0, 1, 2, \ldots, k_1$:

$$H_{k_10(k_2-m_4)m_4}^* = [\lambda_0 I - J_0]^{-1} \big(\tilde{f}_{1k_10(k_2-m_4)m_4}^* - (k_2 - m_4 + 1) $$
$$\times H_{k_10(k_2-m_4+1)(m_4-1)}^* \big) \tag{2.89}$$

when $m_2 = 0$, and

$$H_{(k_1-m_2)m_2(k_2-m_4)m_4}^* = [\lambda_0 I - J_0]^{-1} \big(\tilde{f}_{1(k_1-m_2)m_2(k_2-m_4)m_4}^* $$
$$- (k_1 - m_2 + 1) H_{(k_1-m_2+1)(m_2-1)(k_2-m_4)m_4}^* $$
$$- (k_2 - m_4 + 1) H_{(k_1-m_2)m_2(k_2-m_4+1)(m_4-1)}^* \big) \tag{2.90}$$

when $m_2 = 1, 2, \ldots, k_1$.

(B) Resonant Case: $m_1 + m_2 = m_3 + m_4 + 1$ When $m_1 + m_2 = m_3 + m_4 + 1$ or $k_1 = k_2 + 1$, the matrix $[\lambda_0 I - J_0]$ is singular and we may not be able to set all $C^*_{m_1 m_2 m_3 m_4}$ to zero. These nonzero C^*_m terms consist of the normal form. From $k_1 + k_2 = m$ and $k_1 - k_2 = 1$, we obtain $k_1 = \frac{1}{2}(m+1)$ and $k_2 = \frac{1}{2}(m-1)$. Since both k_1 and k_2 are non-negative integers, m must be a positive odd integer. Let $m = 2k+1$, then $k_1 = k+1$ and $k_2 = k$ for $k = 1, 2, \ldots$. Thus, (2.80) becomes

$$(k + 2 - m_2) H_{1(k+2-m_2)(m_2-1)(k-m_4)m_4} - H_{2(k+1-m_2)m_2(k-m_4)m_4}$$

$$= \tilde{f}_{11(k+1-m_2)m_2(k-m_4)m_4} - C_{1(k+1-m_2)m_2(k-m_4)m_4}$$

$$- (k + 1 - m_4) H_{1(k+1-m_2)m_2(k+1-m_4)(m_4-1)}, \tag{2.91}$$

and

$$(k + 2 - m_2) H_{2(k+2-m_2)(m_2-1)(k-m_4)m_4}$$

$$= \tilde{f}_{12(k+1-m_2)m_2(k-m_4)m_4} - C_{2(k+1-m_2)m_2(k-m_4)m_4}$$

$$- (k + 1 - m_4) H_{2(k+1-m_2)m_2(k+1-m_4)(m_4-1)}. \tag{2.92}$$

It seems that we may similarly find recursive formulas from (2.91) and (2.92). However, the procedure is actually not as straightforward as the nonresonant case since one has to determine which components of C^*_m should be retained.

Nevertheless, we can still follow a procedure similar to that used for the nonresonant case to obtain the following results.

(1) $m_4 = 0, m_2 = 0, 1, \ldots, k+1$:

$$(k + 2 - m_2) H_{1(k+2-m_2)(m_2-1)k0} - H_{2(k+1-m_2)m_2 k0}$$

$$= \tilde{f}_{11(k+1-m_2)m_2 k0} - C_{1(k+1-m_2)m_2 k0}, \tag{2.93}$$

$$(k + 2 - m_2) H_{2(k+2-m_2)(m_2-1)k0}$$

$$= \tilde{f}_{12(k+1-m_2)m_2 k0} - C_{2(k+1-m_2)m_2 k0}, \tag{2.94}$$

when $m_2 = 1, 2, \ldots, k+1$; and

$$-H_{2(k+1)0k0} = \tilde{f}_{11(k+1)0k0} - C_{1(k+1)0k0}, \tag{2.95}$$

$$0 = \tilde{f}_{12(k+1)0k0} - C_{2(k+1)0k0}, \tag{2.96}$$

when $m_2 = 0$.

It follows from (2.96) that

$$C_{2(k+1)0k0} = \tilde{f}_{12(k+1)0k0}. \tag{2.97}$$

Then, a careful consideration of (2.93)–(2.96) shows that the following three equations:

$$(k + 1) H_{2(k+1)0k0} = \tilde{f}_{12k1k0} - C_{2k1k0}, \tag{2.98}$$

$$-H_{2(k+1)0k0} = \tilde{f}_{11(k+1)0k0} - C_{1(k+1)0k0}, \qquad (2.99)$$

$$H_{11kk0} - H_{20(k+1)k0} = \tilde{f}_{110(k+1)k0} - C_{10(k+1)k0}, \qquad (2.100)$$

are unsolved, while the other equations are used to obtain

$$C_{1(k+1-m_2)m_2k0} = 0 \quad \text{for } m_2 = 1, 2, \ldots, k,$$
$$\qquad\qquad\qquad\qquad\qquad\qquad\qquad (2.101)$$
$$C_{2(k+1-m_2)m_2k0} = 0 \quad \text{for } m_2 = 2, 3, \ldots, k+1,$$

and

$$H_{2(k+2-m_2)(m_2-1)k0} = \frac{1}{k+2-m_2} \tilde{f}_{12(k+1-m_2)m_2k0} \qquad (2.102)$$

for $m_2 = 2, 3, \ldots, k+1$; and then

$$H_{1(k+2-m_2)(m_2-1)k0}$$
$$= \frac{1}{k+2-m_2} [\tilde{f}_{11(k+1-m_2)m_2k0} + H_{2(k+1-m_2)m_2k0}], \qquad (2.103)$$

for $m_2 = 1, 2, \ldots, k$.

Next, we shall consider (2.98)–(2.100) and determine how to choose the C_m coefficients which must be used to solve these equations. First note that (2.100) is decoupled from (2.98) and (2.99), so we may set $C_{10(k+1)k0}$ to zero and then use either H_{11kk0} or $H_{20(k+1)k0}$ to solve the equation. Hence, we obtain

$$H_{11kk0} = \tilde{f}_{110(k+1)k0},$$
$$\qquad\qquad\qquad\qquad\qquad (2.104)$$
$$C_{10(k+1)k0} = H_{20(k+1)k0} = 0.$$

Equations (2.98) and (2.99) can eliminate only one of $C_{1(k+1)0k0}$ and C_{2k1k0}, however, since there is only one H_m coefficient involved in these two equations (i.e., $H_{2(k+1)0k0}$). Further, note that $H_{10(k+1)0k0}$ does not appear in (2.93)–(2.96) and can thus be set to zero. In summary, for $m_4 = 0$, we have two choices:

(i) $C_{1(k+1)0k0} \neq 0, C_{2k1k0} = 0$;
(ii) $C_{1(k+1)0k0} = 0, C_{2k1k0} \neq 0$.

The other components of C_m^* and H_m^* are

$$C_{1(k+1-m_2)m_2k0} = 0 \quad \text{for } m_2 = 1, 2, \ldots, k,$$
$$C_{2(k+1-m_2)m_2k0} = 0 \quad \text{for } m_2 = 2, 3, \ldots, k+1,$$
$$C_{2(k+1)0k0} \neq 0 \quad \text{(see (2.97))},$$
$$H_{1(k+2-m_2)(m_2-1)k0} \neq 0 \quad \text{for } m_2 = 1, 2, \ldots, k+1,$$
$$H_{2(k+2-m_2)(m_2-1)k0} \neq 0 \quad \text{for } m_2 = 1, 2, \ldots, k+1$$
$$\text{(see (2.98), (2.99), (2.103) and (2.104))},$$
$$H_{10(k+1)k0} = H_{20(k+1)k0} = 0.$$

(2) When $1 \leq m_4 \leq k$, we have similar formulas to $0 \leq m_2 \leq k+1$ (like those for case $m_4 = 0$). The detailed formulations are omitted here but the results are listed below. There are two choices:

   (i) $C_{1(k+1)0(k-m_4)m_4} \neq 0, C_{2k1(k-m_4)m_4} = 0$;

  (ii) $C_{1(k+1)0(k-m_4)m_4} = 0, C_{2k1(k-m_4)m_4} \neq 0$.

     The other components of C_m^* and H_m^* are

$$C_{1(k+1-m_2)m_2(k-m_4)m_4} = 0 \quad \text{for } m_2 = 1, 2, \ldots, k,$$
$$C_{2(k+1-m_2)m_2(k-m_4)m_4} = 0 \quad \text{for } m_2 = 2, 3, \ldots, k+1,$$
$$C_{2(k+1)0(k-m_4)m_4} \neq 0,$$
$$H_{1(k+2-m_2)(m_2-1)(k-m_4)m_4} \neq 0 \quad \text{for } m_2 = 1, 2, \ldots, k+1,$$
$$H_{2(k+2-m_2)(m_2-1)(k-m_4)m_4} \neq 0 \quad \text{for } m_2 = 1, 2, \ldots, k+1,$$
$$H_{10(k+1)(k-m_4)m_4} = H_{20(k+1)(k-m_4)m_4} = 0.$$

Summarizing the above results yields the following theorem.

Theorem 2.10 *The two normal forms associated with non-semisimple 1:1 Hopf bifurcation are*

$$\dot{u}_1 = iu_1 + u_2 + \sum_{k \geq 1} \sum_{j=0}^{k} C_{1(k+1)0(k-j)j} u_1^{k+1} \bar{u}_1^{k-j} \bar{u}_2^{j},$$

$$\dot{u}_2 = iu_2 + \sum_{k \geq 1} \sum_{j=0}^{k} C_{2(k+1)0(k-j)j} u_1^{k+1} \bar{u}_1^{k-j} \bar{u}_2^{j}, \tag{2.105}$$

and

$$\dot{u}_1 = iu_1 + u_2,$$

$$\dot{u}_2 = iu_2 + \sum_{k \geq 1} \sum_{j=0}^{k} C_{2(k+1)0(k-j)j} u_1^{k+1} \bar{u}_1^{k-j} \bar{u}_2^{j} \tag{2.106}$$

$$+ \sum_{k \geq 1} \sum_{j=0}^{k} C_{2k1(k-j)j} u_1^{k} u_2 \bar{u}_1^{k-j} \bar{u}_2^{j}.$$

The form (2.106) is usually used in applications since its first equation has only two terms, which greatly simplifies finding equilibrium solutions.

To end this section, we give an example to illustrate the application of Theorem 2.10. The results are obtained directly by executing a Maple program, developed on the basis of Theorem 2.10.

Example 2.11 A nonlinear electrical circuit, shown in Fig. 2.5, consists of a DC source E, two capacitors C_1, C_2, two inductors L_1, L_2, a resistor R, and a conductance. L_1 and C_1 are connected in parallel, while L_2, C_2 and R are in series. All five

Fig. 2.5 A nonlinear electrical circuit

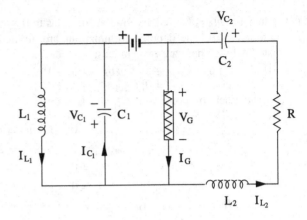

elements, L_1, L_2, C_1, C_2 and R are assumed to be linear time-invariant elements, but C_1 and R may be varied as control parameters. The conductance, however, is a nonlinear element with the characteristics

$$I_G = -\alpha V_G + \beta V_G^3, \quad \alpha > 0, \beta > 0, \tag{2.107}$$

where I_G and V_G represent the current and voltage of the conductance, respectively. The two parameters α and β may also be considered as control parameters.

The voltages across the capacitors and the currents in the inductors are chosen as the state variables (as shown in Fig. 2.5), leading to the equations of the circuit, given by

$$\dot{z}_1 = \frac{1}{C_1}\left(\alpha z_1 + z_2 - z_4 - \beta z_1^3\right),$$

$$\dot{z}_2 = -\frac{1}{L_1}z_1,$$

$$\dot{z}_3 = \frac{1}{C_2}z_4, \tag{2.108}$$

$$\dot{z}_4 = \frac{1}{L_2}(E + z_1 - z_3 - Rz_4),$$

where z_1, z_2, z_3 and z_4 denote the state variables $V_{C_1}, I_{L_1}, V_{C_2}$ and I_{L_2}, respectively. It is clear that $(z_1, z_2, z_3, z_4) = (0, 0, E, 0)$ is the unique equilibrium of the system. Let all the parameters $C_1, C_2, L_1, L_2, R, E, \alpha$ and β equal one unit, then the Jacobian of system (2.108) evaluated at the equilibrium has eigenvalues $\lambda_{1,2} = \lambda_{3,4} = \pm i$, indicating a double-Hopf bifurcation. Applying the linear transformation, given by

$$\begin{pmatrix} z_1 \\ z_2 \\ z_3 \\ z_4 \end{pmatrix} = \begin{pmatrix} 0 \\ 0 \\ 1 \\ 0 \end{pmatrix} + \begin{bmatrix} 2 & 0 & 4 & 1 \\ 0 & -2 & -1 & -4 \\ 0 & 2 & 1 & 0 \\ 2 & 0 & 0 & 1 \end{bmatrix} \begin{pmatrix} x_1 \\ x_2 \\ x_3 \\ x_4 \end{pmatrix} \tag{2.109}$$

to (2.108) results in

$$\dot{x}_1 = -x_2 + x_3,$$

$$\dot{x}_2 = x_1 + x_4 + \frac{1}{8}(2x_1 + 4x_3 + x_4)^3,$$

$$\dot{x}_3 = -x_4 - \frac{1}{4}(2x_1 + 4x_3 + x_4)^3, \tag{2.110}$$

$$\dot{x}_4 = x_3.$$

Executing the Maple program gives the following complex normal form up to third-order terms:

$$\dot{u}_1 = iu_1 + u_2,$$

$$\dot{u}_2 = iu_2 - \frac{3}{4}u_1^2\bar{u}_1 - \frac{3}{2}u_1^2\bar{u}_2 - \left(3 - \frac{3}{2}i\right)u_1u_2\bar{u}_1 \tag{2.111}$$

$$- \left(\frac{3}{2} + 3i\right)u_1u_2\bar{u}_2 + \cdots,$$

which can be transformed to real form by using the polar coordinates, $u_1 = R_1e^{i\theta_1}$, $u_2 = R_2e^{i\theta_2}$, as follows:

$$\dot{R}_1 = R_2\cos\phi, \tag{2.112}$$

$$\dot{R}_2 = -\frac{3}{4}R_1^3\cos\phi - 3R_1^2R_2$$

$$- \frac{3}{2}R_1^2R_2\cos 2\phi - \frac{3}{2}R_1R_2^2\cos\phi + 3R_1R_2^2\sin\phi, \tag{2.113}$$

$$\dot{\phi} = -\frac{R_2}{R_1}\sin\phi - \frac{3}{2}R_1^2 + 3R_2R_1\cos\phi + \frac{3}{2}R_2R_1\sin\phi$$

$$+ \frac{3}{2}R_1^2\sin 2\phi + \frac{3R_1^3}{4R_2}\sin\phi, \tag{2.114}$$

where $\phi = \theta_1 - \theta_2$ is the phase difference of motion.

Usually, there exist periodic solutions bifurcating from a 1:1 double-Hopf bifurcation, and they can be determined by the steady-state solution of (2.114).

Letting $\dot{R}_1 = \dot{R}_2 = \dot{\phi} = 0$ yields nonzero steady-state solutions (periodic solutions). It follows from (2.112) that

$$\phi = \phi_+ = \frac{1}{2}\pi \quad \text{or} \quad \phi = \phi_- = \frac{3}{2}\pi, \tag{2.115}$$

where the subscripts $\pm$ indicate that $\sin\phi_\pm = \pm 1$. Substituting $\phi = \phi_\pm$ into (2.113) and (2.115) yields two sets of polynomial equations:

$$-\frac{3}{2}R_1^2 R_2 \pm 3R_1 R_2^2 = 0, \tag{2.116}$$

$$\frac{1}{4R_1 R_2}\left(\mp 4R_2^3 - 6R_1^3 R_2 \pm 6R_1^2 R_2^2 \pm 3R_1^4\right) = 0. \tag{2.117}$$

It is easy to see from (2.116) that there is no nonzero solution for (R_1, R_2) when $\phi = \phi_+ = \frac{1}{2}\pi$. When $\phi = \phi_- = \frac{3}{2}\pi$, the only nonzero solution (phase-locked solution) is

$$(R_1, R_2, \phi) = \left(\frac{2}{\sqrt{6}}, \frac{1}{\sqrt{6}}, \frac{3}{2}\pi\right). \tag{2.118}$$

To find the stability of this periodic solution, evaluating the Jacobian of (2.112)–(2.114), given by

$$J = \begin{bmatrix} 0 & 0 & -R_2 \\ -3R_1 R_2 + 3R_2^2 & -\frac{3}{2}R_1^2 + 6R_1 R_2 & \frac{3}{2}R_1 R_2^2 + \frac{3}{4}R_1^3 \\ -3R_1 + \frac{3}{2}R_2 + \frac{R_2}{R_1^2} + \frac{9R_1^2}{4R_2} & -\frac{1}{R_1} + \frac{3}{2}R_1 - \frac{3R_1^3}{4R_2^2} & -3R_1 R_2 - 3R_1^2 \end{bmatrix} \tag{2.119}$$

at the solution (2.118) yields

$$J = \begin{bmatrix} 0 & 0 & -\frac{1}{\sqrt{6}} \\ -\frac{1}{2} & 1 & \frac{3}{2\sqrt{6}} \\ \sqrt{6} & -\sqrt{6} & -3 \end{bmatrix}.$$

The three eigenvalues of the above Jacobian matrix are

$$-2.1245702691, \qquad -0.4268172555, \qquad 0.5513875246,$$

implying that this periodic solution is unstable. So we expect that a trajectory starting from a nonzero initial point would converge to the equilibrium point, with a very slow convergence speed, particularly in the neighborhood of the equilibrium point, since at the critical value $\alpha = 1$ ($\mu = 0$), the equilibrium point is an elementary center. Numerical simulation results for this case are depicted in Fig. 2.6.

Now we consider adding unfolding to (2.111). For example, let

$$\alpha = \alpha_c + \mu = 1 + \mu, \tag{2.120}$$

where μ is a bifurcation parameter, then (2.110) becomes

$$\dot{x}_1 = -x_2 + x_3,$$

$$\dot{x}_2 = x_1 + x_4 - \frac{1}{8}(2x_1 + 4x_3 + x_4)\mu + \frac{1}{8}(2x_1 + 4x_3 + x_4)^3,$$

$$\dot{x}_3 = -x_4 + \frac{1}{4}(2x_1 + 4x_3 + x_4)\mu - \frac{1}{4}(2x_1 + 4x_3 + x_4)^3, \tag{2.121}$$

$$\dot{x}_4 = x_3.$$

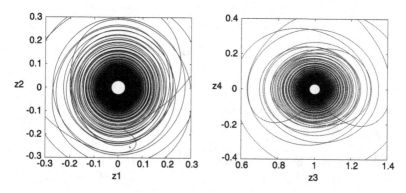

Fig. 2.6 Simulation result of system (2.108) with initial condition $(z_1, z_2, z_3, z_4) = (2.0, -1.0, 0.5, 1.0)$ for $\mu = 0$ $(\alpha = 1.0)$, converging to the equilibrium point $(z_1, z_2, z_3, z_4) = (0, 0, 1, 0)$

The corresponding real normal form is given by (2.112)–(2.114) but now the second equation $\dot{R}_2$ needs to add an unfolding term $\frac{1}{2}\mu$. Hence the Jacobian matrix given by (2.119) is not changed.

Similarly, substituting $\phi = \phi_\pm$ into (2.113) and (2.115) yields two sets of polynomial equations:

$$\frac{1}{2}\mu - \frac{3}{2}R_1^2 R_2 \pm 3R_1 R_2^2 = 0,$$

$$\frac{1}{4R_1 R_2}\left(\mp 4R_2^2 - 6R_1^3 R_2 \pm 6R_1^2 R_2^2 \pm 3R_1^4\right) = 0,$$

from which eliminating R_2 results in

$$R_2 = \frac{1}{3R_1^2(3R_1^2 + 2)}\left(\pm 9R_1^5 - 3\mu R_1^2 + 2\mu\right), \tag{2.122}$$

and a resultant,

$$F_1 = 81R_1^{10} - 54R_1^8 \mp 54\mu R_1^7 \pm 108\mu R_1^5 + 18\mu^2 R_1^4 - 24\mu^2 R_1^2 + 8\mu^2. \tag{2.123}$$

It is easy to observe from (2.123) that if for $\phi = \phi_+ = \frac{1}{2}\pi$, F_1 has a solution R_1^*, then $-R_1^*$ must be a solution for $\phi = \phi_- = \frac{3}{2}\pi$. But it is clear from (2.122) that R_2 does not change sign for these two cases. It has been shown that for $\mu < 0.063364025$, both sets ($\phi = \phi_+ = \frac{1}{2}\pi$ and $\phi = \phi_- = \frac{3}{2}\pi$) have two positive solutions, and one of them is stable; while for $\mu > 0.063364025$, only the second set (i.e., for $\phi = \phi_- = \frac{3}{2}\pi$) has one positive solution which is stable.

In the following, we take three values of $\mu = -0.1, 0.05, 0.2$ to demonstrate the above predictions. For $\mu = -0.1$, it is easy to show that the equilibrium solution $R_1 = R_2 = 0$ (i.e., $(z_1, z_2, z_3, z_4) = (0, 0, 1, 0)$) is stable. There exists one positive

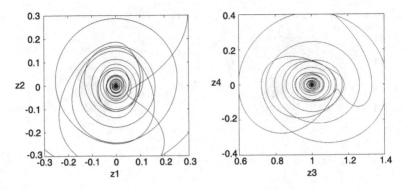

Fig. 2.7 Simulation result of system (2.108) with initial condition $(z_1, z_2, z_3, z_4) = (2.0, -1.0, 0.5, 1.0)$ for $\mu = -0.1$ ($\alpha = 0.9$), converging to the equilibrium point $(z_1, z_2, z_3, z_4) = (0, 0, 1, 0)$

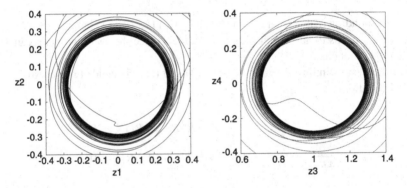

Fig. 2.8 Simulation result of system (2.108) with initial condition $(z_1, z_2, z_3, z_4) = (2.0, -1.0, 0.5, 1.0)$ for $\mu = 0.05$ ($\alpha = 1.05$), converging to a larger limit cycle from outside

solution:

$$(R_1, R_2, \phi) = \left(0.8851664606, 0.4816736981, \frac{1}{2}\pi \right),$$

for which the three eigenvalues of the Jacobian are

$$-0.3452769287, \qquad -2.8895657071, \qquad 0.9880883600,$$

indicating that the periodic solution is unstable, as expected, and so all trajectories around the equilibrium point converge to the equilibrium; this is depicted in Fig. 2.7. Note that due to the unfolding parameter $\mu < 0$, the convergence is very fast compared with that shown in Fig. 2.6.

For $\mu = 0.05$ and $\mu = 0.2$, we list the results in Table 2.1, where the solutions and their stability based on the eigenvalues of the Jacobian are given. The corresponding simulation results are shown in Figs. 2.8–2.10.

Table 2.1 Bifurcation of periodic solutions (PS) for 1:1 resonant case

μ	PS (R_1, R_2, ϕ)	Eigenvalues of the Jacobian	Stability
0.05	$(0.538140, 0.185664, \frac{\pi}{2})$	$-0.459453 \pm 0.826837i, -0.084528$	Stable
	$(0.739432, 0.336194, \frac{\pi}{2})$	$-0.808474, -1.105175, 0.199009$	Unstable
	$(0.297963, 0.108584, \frac{3\pi}{2})$	$-0.062462 \pm 0.753452i, -0.371659$	Stable
0.20	$(0.386313, 0.212635, \frac{3\pi}{2})$	$-0.071537 \pm 1.089584i - 0.774928$	Stable

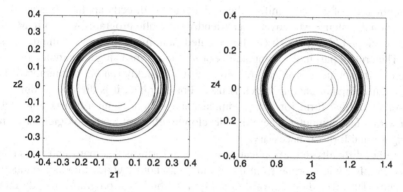

Fig. 2.9 Simulation result of system (2.108) with initial condition $(z_1, z_2, z_3, z_4) = (0.05, -0.1, 1.1, 0.05)$ for $\mu = 0.05$ ($\alpha = 1.05$), converging to a smaller limit cycle from inside

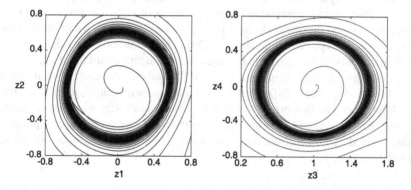

Fig. 2.10 Simulation result of system (2.108) with initial conditions $(z_1, z_2, z_3, z_4) = (2.0, 1.0, 1.5, 1.0)$ and $(0.05, -0.02, 1.02, 0.02)$ for $\mu = 0.2$ ($\alpha = 1.2$), converging to a limit cycle

Remark 2.12 It can be seen from Figs. 2.8 and 2.9 that when $\mu = 0.05$, there exist two different stable limit cycles. Therefore, depending upon the initial conditions, the trajectories may converge to different limit cycles (see the convergence in Fig. 2.8 from outside while in Fig. 2.9 it is from inside). The result for $\mu = 0.2$ is given in Fig. 2.10 showing that trajectories starting from different initial conditions converge to a unique limit cycle.

2.3 A Perturbation Method Based on Multiple Timescales

In this section, we present a perturbation technique which combines the method of multiple timescales (MTS) or simply multiple scales (MS) [166, 168] and harmonic balancing [102] to study nonlinear vibration and bifurcation problems. Huseyin and Lin [103] used this approach to obtain the explicit formulas for simplified differential equations (which are actually normal forms) up to first-order approximation. Later, this method was extended to compute the normal forms of Hopf and generalized Hopf bifurcations up to an arbitrary order [218]. This method does not need the application of center manifold theory and can be directly applied to general n-dimensional systems. Moreover, user-friendly symbolic programs written in Maple were developed [218], which can be executed "automatically" on a computer system. The crucial part in the computation of normal forms using a computer algebra system is the memory problem. A computer may quickly run out of memory if an inefficient computational approach is used. For example, it is difficult to obtain a fifth-order normal form for a three-dimensional system using a matrix approach, even with a fast computer. Therefore, developing efficient methodologies for computing normal forms is necessary.

Another difficulty in the application of normal forms is that many end users may not be familiar with normal form theory and may not be good at coding symbolic programs. However they do want to apply a method or a program to study their own specific problems which usually have large dimension. Therefore, not only the computational efficiency of a method, but also the ease of use of the method needs to be considered. A perturbation method [218] was developed to provide "automatic" symbolic programs for computing the normal forms of Hopf and generalized Hopf bifurcations. Later, this method was extended to consider double-Hopf bifurcations [221, 223] and Hopf-zero singularities [226]. The Maple programs developed in these papers only require a user to prepare a very straightforward input file.

By comparison with existing results, the MTS method is believed to yield correct normal forms (e.g., see [218]); however, no rigorous mathematical proof was given for a long time. The first proof, to the best of our knowledge, was given in [254], showing that the normal form obtained using the MTS method is indeed equivalent to that derived by Poincaré normal form theory—both are based on the concept of *resonant* terms.

2.3.1 Basic Idea of the MTS

In order to show the basic idea of using the perturbation technique based on the MTS method to find the normal form of differential equations, we first consider a simple, well-known example—van der Pol's equation (1.1)—before dealing with general n-dimensional systems. We rewrite van der Pol's equation here as

$$\ddot{x} + x + \varepsilon(x^2 - 1)\dot{x} = 0, \tag{2.124}$$

where ε is a small, non-negative, real number (i.e., $0 \le \varepsilon \ll 1$). These kinds of system are called *weakly nonlinear* systems and perturbation methods can be applied to find approximate periodic solutions.

Van der Pol's equation (2.124) has been studied by many researchers. Recently, this equation was re-investigated by using, in addition to the regular (direct) perturbation method, four frequently-used perturbation approaches: the Lindstedt–Poincaré procedure, time averaging, multiple timescales and intrinsic harmonic balancing [260]. It has been shown that the regular perturbation method yields an unbounded solution which contains secular terms. The Lindstedt–Poincaré procedure cannot be used for stability analysis though it produces the same accurate approximation as that obtained using the MTS and intrinsic harmonic balancing approaches. The first-order time-averaging method has the simplest solution procedure and can be used for stability analysis, but its solution is less accurate. The intrinsic harmonic balancing technique, unlike the other three perturbation methods, does not require the solution of differential equations. However, this approach needs the "normal form" (governing equations) to be constructed for stability analysis, which is usually not straightforward, in particular, for highly codimensional systems. Moreover, the "normal form" obtained using the intrinsic harmonic balancing is only valid up to the leading-order term. Therefore, this approach is not suitable for finding higher-order normal forms.

The MTS method can be used to find not only the approximate solutions but also the normal forms. More importantly, its procedure for finding higher-order normal forms is systematic, and its formulas can be easily implemented using computer algebra systems. It has been shown [260] that the MTS method is the best approach, among the four perturbation methods mentioned above, for the study of nonlinear oscillating systems, in particular, for computing the normal forms.

To apply the MTS method, one begins by introducing the new independent variables,

$$T_k = \varepsilon^k t, \quad k = 0, 1, 2, \ldots. \tag{2.125}$$

It follows that the derivatives with respect to t become expansions in terms of the partial derivatives with respect to T_n according to

$$\frac{d}{dt} = \frac{dT_0}{dt}\frac{\partial}{\partial T_0} + \frac{dT_1}{dt}\frac{\partial}{\partial T_1} + \frac{dT_2}{dt}\frac{\partial}{\partial T_2} + \cdots$$
$$\equiv D_0 + \varepsilon D_1 + \varepsilon^2 D_2 + \cdots, \tag{2.126}$$

$$\frac{d^2}{dt^2} = D_0^2 + 2\varepsilon D_0 D_1 + \varepsilon^2 \left(D_1^2 + 2D_0 D_2\right) + \cdots,$$

etc.,

where $D_i, i = 1, 2, \ldots$, denotes the differentiation operator $\frac{\partial}{\partial T_i}$.

Next, assume that the solution of van der Pol's equation (2.124) is represented by an expansion in the form of

$$x(t; \varepsilon) = x_0(T_0, T_1, T_2, \ldots) + \varepsilon x_1(T_0, T_1, T_2, \ldots)$$
$$+ \varepsilon^2 x_2(T_0, T_1, T_2, \ldots) + \cdots . \tag{2.127}$$

Note that the number of independent timescales used in the solution depends upon the order to which the expansion is carried out. For example, if the expansion is expanded to $O(\varepsilon^2)$, then T_0, T_1 and T_2 are needed. In general, if we want to find the approximate solution up to order $O(\varepsilon^n)$, then the scaled times $T_0, T_1, \ldots, T_n$ should be used. It should be pointed out that the same ε is used in both time and space scales. In other words, we treat the scaling of the motion of the system uniformly for time and space.

Applying formulas (2.126) and (2.127) to system (2.124) and balancing like powers of ε results in the following ordered perturbation equations:

$$\varepsilon^0 : D_0^2 x_0 + x_0 = 0, \tag{2.128}$$

$$\varepsilon^1 : D_0^2 x_1 + x_1 = -2D_1 D_0 x_0 - \left(x_0^2 - 1\right) D_0 x_0, \tag{2.129}$$

etc.

The solution of the ε^0-order equation (2.128) can be expressed as

$$x_0 = a(T_1, T_2, \ldots) \cos\left[T_0 + \phi(T_1, T_2, \ldots)\right] \equiv a \cos(\theta). \tag{2.130}$$

Then substitute this solution into the ε^1-order equation (2.129) to obtain

$$D_0^2 x_1 + x_1 = \left[2D_1 a - a\left(1 - \frac{1}{4}a^2\right)\right] \sin(T_0 + \phi)$$
$$+ 2a D_1 \phi \cos(T_0 + \phi) + \frac{1}{4}a^3 \sin 3(T_0 + \phi). \tag{2.131}$$

Eliminating the secular terms, which may appear in solution x_1, requires that

$$D_1 a = \frac{1}{2}a\left(1 - \frac{1}{4}a^2\right),$$
$$D_1 \phi = 0, \tag{2.132}$$

and thus (2.131) becomes

$$D_0^2 x_1 + x_1 = \frac{1}{4}a^3 \sin 3(T_0 + \phi), \tag{2.133}$$

which, in turn, yields the solution

$$x_1 = -\frac{1}{32}a^3 \sin 3(T_0 + \phi). \tag{2.134}$$

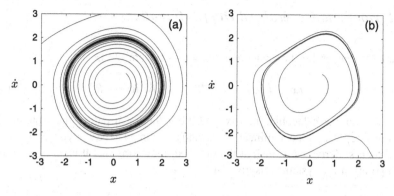

Fig. 2.11 Simulated phase portraits of van der Pol's equation (2.124) with initial conditions $(x, \dot{x}) = (0.5, 0.5)$ and $(3, -4)$ for (**a**) $\varepsilon = 0.1$; and (**b**) $\varepsilon = 0.5$

Hence, the approximate solution up to first order is obtained as

$$x(t, \varepsilon) = a \cos(t + \phi) - \frac{\varepsilon}{32} a^3 \sin 3(t + \phi). \tag{2.135}$$

It should be noted that we only find the *particular* solution from (2.133) since we can leave the *homogeneous* solution part to be included in a and ϕ. In fact, if we add the homogeneous solution, given by $a_1 \cos(t + \phi_1)$ where a_1 and ϕ_1 are determined from the initial conditions, to the particular solution (2.134), we can see that the homogeneous solution can indeed be combined with the first term of solution (2.135).

Finally, the governing equations for the amplitude a and the phase ϕ of the periodic solution above can be obtained, up to $O(\varepsilon^2)$, as follows:

$$\frac{da}{dt} = \frac{\partial a}{\partial T_1} \frac{\partial T_1}{\partial t} + O(\varepsilon^2) = \varepsilon D_1 a + O(\varepsilon^2) \approx \frac{\varepsilon}{2} a \left(1 - \frac{1}{4} a^2 \right) \tag{2.136}$$

and

$$\frac{d\theta}{dt} = 1 + \frac{d\phi}{dt} = 1 + \frac{\partial \phi}{\partial T_1} \frac{\partial T_1}{\partial t} + O(\varepsilon^2) = 1 + \varepsilon D_1 \phi + O(\varepsilon^2) \approx 1. \tag{2.137}$$

These two equations (well-known results in the existing literature) are in fact the normal form of van der Pol's equation up to order ε. The approximate amplitude of the periodic solution is $\bar{a} = 2$, confirmed by the numerical simulation result, shown in Fig. 2.11, where ε takes two values: $\varepsilon = 0.1, 0.5$. It can be seen that for the small value of ε (0.1), the limit cycle is close to a circle, while for the large value of ε (0.5), the limit cycle is no longer close to a circle, but its amplitude is still close to 2, as expected.

Now we want to extend the above procedure for a second-order differential equation to general n-dimensional systems. We use Hopf bifurcation as an example to provide a proof for the perturbation technique based on the MTS method. Other cases having similar procedures can be similarly proved.

2.3.2 Hopf and Generalized Hopf Bifurcations

Consider the general n-dimensional autonomous system (2.9),

$$\dot{x} = Jx + f(x), \quad x \in R^n, \qquad f : R^n \to R^n, \tag{2.138}$$

where J is an $n \times n$ Jacobian matrix, and Jx is the linear part of the system. The function f represents the nonlinear terms, and is assumed to be analytic. In addition, f and its first derivative vanish at the origin $\mathbf{0}$, indicating that $\mathbf{0}$ is an equilibrium (fixed point) of the system. J is given by

$$J = \begin{bmatrix} A_0 & 0 & 0 \\ 0 & A_1 & 0 \\ 0 & 0 & A_2 \end{bmatrix}, \tag{2.139}$$

where A_0 is a 2×2 matrix with a pair of purely imaginary eigenvalues, given by

$$A_0 = \begin{bmatrix} 0 & \omega_c \\ -\omega_c & 0 \end{bmatrix}, \tag{2.140}$$

and the eigenvalues of A_1 and A_2 have negative real parts, which implies that the center manifold of the system has dimension 2. The eigenvalues of J are in the Siegel domain [115] and one encounters much greater computational complexity than for a Hopf critical point (having a pair of purely imaginary eigenvalues).

Matrix A_1 is an $n_1 \times n_1$ matrix having negative real eigenvalues:

$$A_1 = \begin{bmatrix} -\alpha_3 & \cdots & 0 \\ \vdots & \ddots & \vdots \\ 0 & \cdots & -\alpha_{2+n_1} \end{bmatrix}, \tag{2.141}$$

and A_2 is an $n_2 \times n_2$ matrix whose eigenvalues are complex conjugates with negative real parts:

$$A_2 = \begin{bmatrix} -\alpha_{n_1+3} & \omega_{n_1+3} & 0 & 0 & \cdots & 0 & 0 \\ -\omega_{n_1+3} & -\alpha_{n_1+3} & 0 & 0 & \cdots & 0 & 0 \\ 0 & 0 & -\alpha_{n_1+5} & \omega_{n_1+5} & \cdots & 0 & 0 \\ 0 & 0 & -\omega_{n_1+5} & -\alpha_{n_1+5} & \cdots & 0 & 0 \\ \vdots & \vdots & \vdots & \vdots & \ddots & \vdots & \vdots \\ 0 & 0 & 0 & 0 & \cdots & -\alpha_{n-1} & \omega_{n-1} \\ 0 & 0 & 0 & 0 & \cdots & -\omega_{n-1} & -\alpha_{n-1} \end{bmatrix}, \tag{2.142}$$

where ω_c, α_p, $p = 3, 4, \ldots, n_1 + 2$, and α_q, $q = n_1 + 3$, $n_1 + 5, \ldots, n - 1$, are positive, and $2 + n_1 + 2n_2 = n$.

For the convenience of the analysis using the MTS method, one may write (2.138) in component form:

$$\dot{x}_1 = \omega_c x_2 + f_1(\boldsymbol{x}),$$
$$\dot{x}_2 = -\omega_c x_1 + f_2(\boldsymbol{x}), \tag{2.143}$$

$$\dot{x}_p = -\alpha_p x_p + f_p(\boldsymbol{x}), \quad p = 3, \ldots, n_1 + 2, \tag{2.144}$$

$$\dot{x}_q = -\alpha_q x_q + \omega_q x_{q+1} + f_q(\boldsymbol{x}),$$
$$\dot{x}_{q+1} = -\omega_q x_q - \alpha_q x_{q+1} + f_{q+1}(\boldsymbol{x}), \tag{2.145}$$
$$q = n_1 + 3, n_1 + 5, \ldots, n - 1.$$

Based on the above equations, the MTS method can be used to find the normal forms. First, assume that the solution of system (2.138) is given in the form

$$x_j(t; \varepsilon) = \varepsilon x_{j,1}(T_0, T_1, \ldots) + \varepsilon^2 x_{j,2}(T_0, T_1, \ldots) + \cdots,$$
$$j = 1, \ldots, n, \tag{2.146}$$

and then substitute (2.146) into (2.143)–(2.145) with the aid of (2.125) and (2.126) and balance like powers of ε to obtain the following ordered perturbation equations:

$$\varepsilon^1 : D_0 x_{1,1} = \omega_c x_{2,1},$$
$$D_0 x_{2,1} = -\omega_c x_{1,1}, \tag{2.147}$$

$$D_0 x_{p,1} = -\alpha_p x_{p,1}, \quad p = 3, \ldots, n_1 + 2, \tag{2.148}$$

$$D_0 x_{q,1} = -\alpha_q x_{q,1} - \omega_q x_{(q+1),1},$$
$$D_0 x_{(q+1),1} = \omega_q x_{q,1} - \alpha_q x_{(q+1),1}, \tag{2.149}$$
$$q = n_1 + 3, n_1 + 5, \ldots, n - 1;$$

$$\varepsilon^2 : D_0 x_{1,2} = \omega_c x_{2,2} - D_1 x_{1,1} + f_{1,2}(\boldsymbol{x}_1),$$
$$D_0 x_{2,2} = -\omega_c x_{1,2} - D_1 x_{2,1} + f_{2,2}(\boldsymbol{x}_1), \tag{2.150}$$

$$D_0 x_{p,2} = -\alpha_p x_{p,2} + f_{p,2}(\boldsymbol{x}_1), \quad p = 3, \ldots, n_1 + 2, \tag{2.151}$$

$$D_0 x_{q,2} = -\alpha_q x_{q,2} - \omega_q x_{(q+1),2} + f_{q,2}(\boldsymbol{x}_1),$$
$$D_0 x_{(q+1),2} = \omega_q x_{q,2} - \alpha_q x_{(q+1),2} + f_{(q+1),2}(\boldsymbol{x}_1), \tag{2.152}$$
$$q = n_1 + 3, n_1 + 5, \ldots, n - 1;$$

etc.,

where $\boldsymbol{x}_1$ represents the first-order approximation of $\boldsymbol{x}$, $\boldsymbol{x}_{i,j}$ represents the jth-order approximation of $\boldsymbol{x}_i$, and

$$f_{i,2} = \left. \frac{d^2(f_i(x_1, x_2, \ldots)/\varepsilon)}{d\varepsilon^2} \right|_{\varepsilon=0}, \tag{2.153}$$

which are functions of $x_{i,1}$ only. In general, the function $f_{i,k}$ only involves the ordered approximations $x_{i,1}, x_{i,2}, \ldots, x_{i,k-1}$, which have been found from the previous $(k-1)$ perturbation equations. It should be noted that unlike (2.127) which starts with a zero-order term, the solution form (2.146) starts with a first-order term. This is because van der Pol's equation has ε in the nonlinear term, while for the general nonlinear system (2.138), usually the first step is to use scaling $x \to \varepsilon x$ to separate different order terms, and then use the solution form starting with a zero-order term. Here the first step has been included in solution form (2.146).

To find the normal form of the system, we start with (2.146). Differentiating the first equation of (2.147) and then substituting the second equation into the resulting equation yields

$$D_0^2 x_{1,1} + \omega_c^2 x_{1,1} = 0 \tag{2.154}$$

which is a free-vibrating system with the solution

$$x_{1,1} = r(T_1, T_2, \ldots) \cos\big(\omega_c T_0 + \phi_1(T_1, T_2, \ldots)\big) \equiv r \cos\theta, \tag{2.155}$$

where r and θ are the amplitude and phase of motion, respectively. Having found $x_{1,1}$, one can easily find the solution of $x_{2,1}$ as

$$x_{2,1} = r \sin\theta \tag{2.156}$$

from the first equation of (2.147). Thus the first-order solutions for $x_{1,1}$ and $x_{2,1}$ are found.

Since we are interested in the steady-state (asymptotic) solutions of the system, the ε^1-order solutions for $x_{j,1}$, $j = 3, \ldots, n$, associated with the eigenvalues having negative real parts, are equal to 0, i.e.,

$$x_{j,1} = 0, \quad j = 3, \ldots, n. \tag{2.157}$$

It follows from solution (2.155) that

$$D_0 r = D_0 \phi = 0. \tag{2.158}$$

Next, to solve the ε^2-order perturbation equation (2.150), one may apply the above procedure and substitute the ε^1-order solution into (2.150) to obtain the following second-order nonhomogeneous ODE (Ordinary Differential Equation):

$$D_0^2 x_{1,2} + \omega_c^2 x_{1,2} = -D_1 D_0 x_{1,1} - D_1 x_{2,1} + D_0 f_{1,2} + f_{2,2}. \tag{2.159}$$

Note that the right-hand side of (2.159) is a polynomial in the first-order solutions $x_{1,2}$ and $x_{2,1}$, so the solution of $x_{1,2}$ can be expressed by a finite Fourier series

$$x_{1,2} = \sum_{j=0}^{2} C_j \cos(j\theta) + S_j \sin(j\theta). \tag{2.160}$$

In order to determine the coefficients C_j and S_j and thus solve for $x_{1,2}$, one may substitute (2.160) into (2.159) and then balance the harmonics. However, as usual, the resulting equation may involve terms which will generate secular terms in the solution of $x_{1,2}$. To eliminate the secular terms we must set their coefficient to zero. This yields two algebraic equations to determine $D_1 r$ and $D_1 \phi$ which are called *resonant* terms and will be retained in the normal form.

The solutions for $x_{j,2}, j = 3, \ldots, n$, can also be found in the same form of Fourier series as that of $x_{1,2}$, except that no secular terms would appear. Hence, they can be uniquely determined by a straightforward harmonic balancing approach.

The general solution of the n_rth-order perturbation equations of the system is given in the following theorem.

Theorem 2.13 *The solutions of the n_rth-order perturbation equations of an autonomous system can be expressed by a finite Fourier series*:

$$x_{i,n_r} = \sum_{j=0}^{n_r} C_j \cos(j\theta) + S_j \sin(j\theta). \tag{2.161}$$

Proof First, we may rewrite the first-order solution of (2.147)–(2.149) as

$$x_{i,1} = \sum_{j=0}^{1} C_j \cos(j\theta) + S_j \sin(j\theta), \tag{2.162}$$

where C_j and S_j are constants.

Next, for the second-order perturbation equations (2.150)–(2.152), it is easy to observe that the nonlinear terms are second-degree polynomials in the first-order solutions. So the highest-order terms are $x_{i,1} x_{j,1}, i, j = 1, 2, \ldots, n$. Therefore, the highest-order terms involved in the second-order solutions are $\sin^2 \theta$, $\cos^2 \theta$ and $\sin \theta \cos \theta$, which can be rewritten as $\frac{1}{2}[1 - \cos 2\theta]$, $\frac{1}{2}[1 + \cos 2\theta]$, and $\frac{1}{2} \sin 2\theta$, respectively.

Similarly, the highest-order terms in the third-order perturbation equations come from the multiplication of first- and second-order solutions, i.e., $x_{i,1} x_{j,2}, i, j = 1, 2, \ldots, n$. So the highest-order terms in third-order solutions are

$$\sin 2\theta \sin \theta, \qquad \sin 2\theta \cos \theta, \qquad \cos 2\theta \sin \theta, \qquad \cos 2\theta \cos \theta.$$

Therefore, the highest-order terms in the third-order solutions can be written as $\sin 3\theta$ or $\cos 3\theta$.

The above procedure can be easily extended to discuss higher-order perturbation equations. More rigorously, one may apply the method of mathematical induction to show that the highest-order terms in the n_rth-order solutions can be written in the form $\sin(n_r \theta)$ or $\cos(n_r \theta)$. This completes the proof. □

As discussed above, the normal form terms $D_1 r$ and $D_1 \phi$ are obtained from the second-order perturbation equations by removing the secular terms from the solutions. In general, for the n_rth-order equations, the normal form terms $D_{n_r-1}r$ and $D_{n_r-1}\phi$ are obtained by eliminating the secular terms.

Finally, the normal form of system (2.138) associated with a Hopf-type singularity, up to any order, can be written in polar coordinates as

$$\frac{dr}{dt} = \frac{\partial r}{\partial T_1}\frac{\partial T_1}{\partial t} + \frac{\partial r}{\partial T_2}\frac{\partial T_2}{\partial t} + \frac{\partial r}{\partial T_3}\frac{\partial T_3}{\partial t} + \cdots$$

$$= \varepsilon D_1 r + \varepsilon^2 D_2 r + \varepsilon^3 D_3 r + \cdots , \tag{2.163}$$

$$r\frac{d\theta}{dt} = R\left(\omega_c + \frac{\partial \phi}{\partial T_1}\frac{\partial T_1}{\partial t} + \frac{\partial \phi}{\partial T_2}\frac{\partial T_2}{\partial t} + \frac{\partial \phi}{\partial T_3}\frac{\partial T_3}{\partial t} + \cdots\right)$$

$$= r\left(\omega_c + \varepsilon D_1\phi + \varepsilon^2 D_2\phi + \varepsilon^3 D_3\phi + \cdots\right). \tag{2.164}$$

It should be noted that only odd order terms are retained in the normal form, implying that

$$D_{2k-1}r = D_{2k-1}\phi = 0, \quad \text{where } k \text{ is a positive integer.} \tag{2.165}$$

Note that the subscript $2k - 1$ is not the order of the term. More specifically, we have the following theorem.

Theorem 2.14 *The "form" (in polar coordinates) of the normal form of an autonomous system with a Hopf-type singularity is given by*

$$\dot{r} = rP(r^2), \qquad r\dot{\phi} = rQ(r^2), \tag{2.166}$$

where P and Q are polynomials in r^2.

Remark 2.15 The equations given in (2.166) are actually the Poincaré normal form for Hopf bifurcation.

Proof It is easy to prove the theorem using complex formulas. Thus, introduce the following transformations:

$$x_1 = \frac{1}{2}(z_1 + z_2), \qquad x_2 = \frac{1}{2i}(z_1 - z_2), \tag{2.167}$$

$$x_p = z_p, \quad p = 3, \ldots, n_1 + 2, \tag{2.168}$$

$$x_q = \frac{1}{2}(z_q + z_{q+1}), \qquad x_{q+1} = \frac{1}{2i}(z_q - z_{q+1}),$$

$$q = n_1 + 3, \ldots, n - 1, \tag{2.169}$$

where i is the imaginary unit satisfying $i^2 = -1$. It should be noted in the above expressions that z_2 and z_{q+1} are the complex conjugates of z_1 and z_q, respectively.

Then (2.143)–(2.145) can be transformed into complex form as follows:

$$\dot{z}_1 = i\omega_i z_1 + f_1 + i f_2,$$

$$\dot{z}_2 = -i\omega_i z_2 + f_1 - i f_2,$$ \hfill (2.170)

$$\dot{z}_p = -\alpha_p z_p + f_p, \quad p = 3, 4, \ldots, n_1 + 2,$$ \hfill (2.171)

$$\dot{z}_q = (-\alpha_q + i\omega_q) z_q + f_q + i f_{q+1},$$

$$\dot{z}_{q+1} = -(\alpha_q + i\omega_q) z_{q+1} + f_q - i f_{q+1},$$ \hfill (2.172)

$$q = n_1 + 3, n_1 + 5, \ldots, n - 1,$$

where

$$f_i = f_i\big(x_1(\mathbf{z}), x_2(\mathbf{z}), \ldots, x_n(\mathbf{z})\big),$$ \hfill (2.173)

and the $x_i(\mathbf{z})$ are given by transformations (2.167)–(2.169). Similarly, applying the MTS method to the above equations results in complex ordered perturbation equations, which are similar to the real forms (2.147)–(2.149).

The solutions to the first-order complex equations (which can be readily obtained from (2.170)–(2.172) by removing f terms) are

$$z_{1,1} = r e^{i\theta}, \qquad z_{2,1} = \overline{z}_{1,1},$$ \hfill (2.174)

$$z_{p,1} = 0, \quad p = 3, \ldots, n,$$ \hfill (2.175)

where $\overline{z}$ represents the complex conjugate of z, $r = r(T_1, T_2, \ldots)$ is real and positive, and $\theta = \omega T_0 + \phi(T_1, T_2, \ldots)$.

Similarly to the real analysis, it can been shown that the nonlinear terms on the right-hand side of the complex perturbation equations can be written as polynomials in the first-order solutions in the form of

$$F = \sum C_j z_1^{a_j} \overline{z}_1^{b_j} = \sum C_j r^{(a_j + b_j)} e^{i(a_j - b_j)\theta},$$ \hfill (2.176)

where the C_j are complex constants and a_j, b_j, are non-negative integers.

For the first equation, the secular term should be in the form $e^{i\theta}$, and thus it follows from (2.176) that the terms producing the secular terms for the first equation satisfy

$$a_j - b_j = 1.$$ \hfill (2.177)

Therefore, the powers of the normal form terms for the first equation are the solution of (2.177). Now substituting (2.177) into (2.176) results in the secular term

$$S = \sum C_j r^{2b_j + 1} e^{i\theta},$$ \hfill (2.178)

which is balanced by the term $D_{2b_j}(r e^{i\theta})$. Thus the normal form terms $D_{2b_j} r$ and $D_{2b_j} \phi$ can be found from the equation

$$D_{2b_j} r + i r D_{2b_j} \theta = S = \sum C_j r^{2b_j + 1} e^{i\theta}$$ \hfill (2.179)

which, in turn, yields

$$D_{2b_j} r = r \sum \text{Re}(C_j) r^{2b_j} = r P_{b_j}(r^2),$$

$$r D_{2b_j} \theta = R \sum \text{Im}(C_j) r^{2b_j} = r Q_{b_j}(r^2),$$

(2.180)

where Re and Im represent the real and imaginary parts, respectively, and P_{b_j} and Q_{b_j} are the b_jth-degree polynomials in r^2. Note that the above formulas indicate that only odd order terms are retained in the normal form.

The proof of Theorem 2.14 is complete.   □

Similar theorems and proofs for the equivalence of other cases such as double-Hopf and Hopf-zero singularities can be established, and are not repeated here.

Therefore, for the general system (2.138), by a back-scaling, the normal form for Hopf and generalized Hopf bifurcations can be rewritten as

$$\dot{r} = r(v_0 \mu + v_1 r^2 + \cdots + v_k r^{2k} + \cdots),$$

(2.181)

$$\dot{\theta} = \omega_c + \tau_0 \mu + \tau_1 r^2 + \tau_2 r^4 + \cdots + \tau_k r^{2k} + \cdots,$$

(2.182)

where μ is an unfolding (bifurcation parameter) when the original system involves parameters, v_k is the kth-order *focus value*. v_0 and τ_0 can be determined from linear analysis.

The Maple source program for computing the coefficients v_k, τ_k, and nonlinear transformation of the normal form, as well as sample input files, can be found on "Springer Extras" (by visiting extras.springer.com and searching for the book using its ISBN). In the following, we give two examples to illustrate the application of the method. In order to apply the Maple program to analyze a problem, the system must first be transformed such that its Jacobian evaluated at an equilibrium point should be in real Jordan canonical form, as shown in the sample file.

Example 2.16 The first example is the same as Example 2.6, which is a two-dimensional system, with a Hopf singularity. Executing the Maple program yields the following normal form given in polar coordinates:

$$\dot{r} = \frac{3}{8} r^3 + \frac{5}{16} r^5 + \cdots,$$

(2.183)

$$\dot{\theta} = 1 - \frac{1}{3} r^2 - \frac{29}{6912} r^4 + \cdots.$$

(2.184)

Note that the normal form given in the above equations is not exactly the same as that obtained by using the method developed in Sect. 2.2. The coefficient of r^4 in (2.184) is different from that given in the second equation of (2.42). This is not surprising since the normal form is not unique. The verification scheme described in [218] can be used to prove that both normal forms are correct! In fact, by the simplest normal form theory, one can further remove the r^4 term from the phase equation (e.g., see [219]).

Based on the normal form up to fifth order, it is easy to see that limit cycles do not exist since both the coefficients of r^3 and r^5 in the amplitude equation are positive.

Example 2.17 The second example is the same as Example 2.7, which is a five-dimensional system, with a Hopf singularity. Executing the Maple program yields the following normal form in polar coordinates up to fifth order:

$$\dot{r} = \frac{3}{40}r^3 - \frac{14867}{68000}r^5 + \cdots, \tag{2.185}$$

$$\dot{\theta} = 1 - \frac{7}{12}r^2 + \frac{8093503}{14688000}r^4 + \cdots, \tag{2.186}$$

which are again different from those given in (2.44) (see Sect. 2.2) by one term in the phase equation. Now, based on (2.185), we can easily find the steady-state solutions by setting $\dot{r} = 0$, yielding $\overline{r} = 0$, corresponding to the initial equilibrium point $x = 0$, and a nontrivial solution

$$\overline{r} = \sqrt{5100/14867}. \tag{2.187}$$

This nontrivial solution represents a periodic motion and its asymptotic solution is given by (obtained from the computer output):

$$x_1 = r\cos\theta + \frac{1}{6}r^2(3 + \cos 2\theta + 2\sin 2\theta) - \frac{1}{480}r^3(28\cos 3\theta - 13\sin 3\theta) + \cdots,$$

$$x_2 = -r\cos\theta - \frac{1}{6}r^2(3 - \cos 2\theta + 2\sin 2\theta)$$

$$\qquad - \frac{1}{480}r^3(260\cos\theta - 168\sin\theta + 17\cos 3\theta + 28\sin 3\theta) + \cdots,$$

$$x_3 = \frac{1}{10}r^2(5 + \cos 2\theta + 2\sin 2\theta)$$

$$\qquad - \frac{1}{12}r^3(5\cos\theta + 3\sin\theta - \cos 3\theta + \sin 3\theta) + \cdots, \tag{2.188}$$

$$x_4 = \frac{1}{10}r^2(5 + 2\cos 2\theta + \sin 2\theta)$$

$$\qquad + \frac{1}{510}r^3(119\cos\theta + 408\sin\theta - 23\cos 3\theta + 44\sin 3\theta) + \cdots,$$

$$x_5 = -\frac{1}{10}r^2(\cos 2\theta + 3\sin 2\theta)$$

$$\qquad - \frac{1}{510}r^3(68\cos\theta - 119\sin\theta - 24\cos 3\theta + 57\sin 3\theta) + \cdots.$$

The stability of the periodic solution is determined by the Jacobian of (2.185) evaluated at $r = \overline{r}$, giving

Fig. 2.12 Simulated time history for Example 2.17 with initial condition $(x_1, x_2, x_3, x_4, x_5) = (0.7, 0.3, 3.0, -2.0, 2.0)$, converging to a stable limit cycle

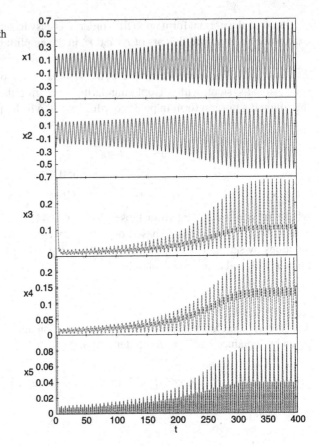

$$J(r = \overline{r}) = \left.\frac{d\dot{r}}{dr}\right|_{r=\overline{r}} = -\frac{765}{14867} < 0.$$

Therefore, the periodic solution is stable, and the frequency of the periodic motion is approximated by

$$\omega = \dot{\theta}\big|_{r=\overline{r}} = 1 - \frac{1}{12}\left(\frac{5100}{14867}\right)\left(7 - \frac{8093503}{1224000} \times \frac{5100}{14867}\right) = \frac{9174269227}{10609329072}$$

$$\approx 0.864736,$$

which indicates that the period of the motion is

$$T = \frac{2\pi}{\omega} \approx 7.266246 \text{ seconds.}$$

The numerical simulation results are shown in Figs. 2.12 and 2.13. It is observed that the convergence of the bifurcating limit cycle is very slow. In fact, the limit cycle is quite weak, in the sense that when one of the first two components of the

Fig. 2.13 Simulated phase portrait for Example 2.17 with initial condition $(x_1, x_2, x_3, x_4, x_5) = (0.7, 0.3, 3.0, -2.0, 2.0)$, projected on the x_1–x_2 plane, showing a stable limit cycle

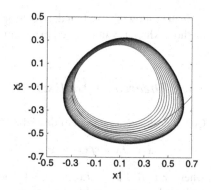

initial condition (x_1, x_2) is increased a little bit from the simulating value $(0.7, 0.3)$ to, say, $(0.7, 0.8)$, the trajectory diverges to infinity.

2.4 Efficient Computation

Since normal form computation usually requires heavy algebraic manipulation, computer algebra systems such as Maple, Mathematica, Reduce, etc., have been used extensively. A basic procedure in the symbolic computation of normal forms is to substitute an obtained lower-order ($<k$) normal form and nonlinear transformation to the original differential equation to yield an expression for the kth-order computation, which contains not only the kth-order terms, but also lower-order ($<k$) and higher-order ($>k$) terms. One must extract the kth-order terms from the expression to obtain the kth-order algebraic equation. This unnecessarily increases the computational burden and takes too much computer memory, especially when computing higher-order normal forms. Therefore, removing the unnecessary lower- and higher-order terms from the kth-order computation becomes essential in order to reduce the computation time and memory requirement. In particular, efficient algebraic methods for computing focal values have been discussed in the literature, for example, in [67] and the recent books [60, 182].

In this section, we present an efficient approach to compute the kth-order ($k \geq 2$, an arbitrary integer) algebraic equation which only contains the terms belonging to the kth-order equation. Based on the Lie bracket operator (e.g., see [108]), a recursive formula is derived which can be applied to consider any singularities. Moreover, the new method does not require solving large matrix equations; instead it solves linear algebraic equations, one by one, and is therefore computationally efficient. In addition, unlike most normal form methods which use separate nonlinear transformations at each order, the new approach uses a consistent nonlinear transformation through all order computations. This provides a convenient, one-step transformation between the original system and the normal form, which is particularly useful in real applications.

In the following, we first derive the general formula for an efficient computational method, which is summarized in a theorem, and then discuss symbolic computation.

2.4.1 Theoretical Analysis

Consider the general system described by

$$\dot{x} = Jx + f(x) \equiv v_1 + f_2(x) + f_3(x) + \cdots + f_k(x) + \cdots, \qquad (2.189)$$

where $x \in R^n$, $v_1 = Jx$ represents the linear part, and the Jacobian matrix J is, without loss of generality, in a standard Jordan canonical form. It is assumed that all eigenvalues of J have zero real parts, implying that the dynamics of system (2.189) are described on an n-dimensional center manifold. $f_k(x)$ denotes a kth-order vector of homogeneous polynomials in x. It is further assumed that system (2.189) has an equilibrium at the origin, $x = 0$.

The basic idea of normal form theory is to find a near-identity nonlinear transformation

$$x = y + h(y) \equiv y + h_2(y) + h_3(y) + \cdots + h_k(y) + \cdots, \qquad (2.190)$$

such that the resulting system

$$\dot{y} = Jy + g(y) \equiv Jy + g_2(y) + g_3(y) + \cdots + g_k(y) + \cdots, \qquad (2.191)$$

becomes as simple as possible. Here, $h_k(y)$ and $g_k(y)$ denote a kth-order vector of homogeneous polynomials in y.

Define the Lie bracket operator [72] as

$$[U_k, v_1] = Dv_1 \cdot U_k - DU_k \cdot v_1. \qquad (2.192)$$

The following theorem summarizes the results for the recursive and computationally efficient approach, which can be used to compute the kth-order normal form and the associated nonlinear transformation [225, 226, 233, 253].

Theorem 2.18 *The recursive formula for computing the coefficients of the normal form and the nonlinear transformation is given by*

$$g_k = f_k + [h_k, v_1] + \sum_{i=2}^{k-1} \{ [h_{k-i+1}, f_i] + Dh_i(f_{k-i+1} - g_{k-i+1}) \}$$

$$+ \sum_{m=2}^{[\frac{k}{2}]} \frac{1}{m!} \sum_{i=m}^{k-m} D^m f_i$$

$$\times \sum_{\substack{l_1+l_2+\cdots+l_m=k-(i-m) \\ 2 \le l_1,l_2,\ldots,l_m \le k-(i-m)-2(m-1)}} h_{l_1} h_{l_2} \cdots h_{l_m}, \qquad (2.193)$$

for $k = 2, 3, \ldots$, where f_k, h_k, and g_k are kth-order vectors of homogeneous poly-nomials in y (where y has been dropped for simplicity). f_k represents the kth-order terms of the original system, h_k is the kth-order nonlinear transformation, and g_k denotes the kth-order normal form.

Remark 2.19 The notation $D^m f_i h_{l_1} h_{l_2} \cdots h_{l_m}$ denotes the mth-order terms of the Taylor expansion of $f_i(y + h(y))$ about y. More precisely,

$$D^m f_i(y+h) = D\big(D(\cdots D((Df_i)h_{l_1})h_{l_2})\cdots h_{l_{m-1}}\big)h_{l_m}, \qquad (2.194)$$

where each differential operator D affects only function f_i, not h_{l_j} (i.e., h_{l_j} is treated as a constant vector in the process of differentiation), and thus $m \leq i$. Note that at each level of the differentiation, the D operator is actually a Frechét deriva-tive, giving rise to a matrix, which is multiplied by a vector to generate another vector, and then to another level of Frechét derivative, and so on.

Proof First differentiating (2.177) results in

$$\dot{x} = \dot{y} + Dh(y)\dot{y} = \big(I + Dh(y)\big)\dot{y}. \qquad (2.195)$$

Then substituting (2.189) and (2.191) into (2.195) yields

$$Jx + f_2(x) + f_3(x) + \cdots + f_k(x) + \cdots$$
$$= \big(I + Dh(y)\big)\big(Jy + g_2(y) + g_3(y) + \cdots + g_k(y) + \cdots\big). \qquad (2.196)$$

Next substituting (2.177) into (2.196) and rearranging the resulting equation gives

$$g_2(y) + g_3(y) + \cdots + g_k(y) + \cdots$$
$$= Jh(y) - Dh(y)Jy$$
$$\quad - Dh(y)g_2(y) - Dh(y)g_3(y) - \cdots - Dh(y)g_k(y) - \cdots$$
$$\quad + f_2\big(y + h(y)\big) + \cdots + f_k\big(y + h(y)\big) + \cdots, \qquad (2.197)$$

which can be rewritten, using the Taylor expansion about y, as

$$g_2(y) + g_3(y) + \cdots + g_k(y) + \cdots$$
$$= f_2(y) + f_3(y) + \cdots + f_k(y) + \cdots + \sum_{i=2}^{\infty}\big\{Jh_i(y) - Dh_i(y)Jy\big\}$$
$$+ \sum_{i=2}^{\infty} D\,h(y)\big\{f_i(y) - g_i(y)\big\} + \sum_{i=2}^{\infty}\big\{Df_i(y)h(y) - Dh(y)f_i(y)\big\}$$
$$+ \frac{1}{2!}\big\{D^2 f_2(y)h^2(y) + D^2 f_3(y)h^2(y) + \cdots\big\}$$

$$+ \frac{1}{3!}\left\{D^3 f_3(y)h^3(y) + D^3 f_4(y)h^3(y) + \cdots\right\} + \cdots$$

$$+ \frac{1}{k!}\left\{D^k f_k(y)h^k(y) + D^k f_{k+1}(y)h^k(y) + \cdots\right\} + \cdots. \qquad (2.198)$$

Further, one can use Lie bracket notation to rewrite the Taylor expansion in component form according to the order of the terms:

$$\sum_{i=2}^{\infty} g_i(y) = \sum_{i=2}^{\infty} f_i(y) + \sum_{i=2}^{\infty}[h_i(y), v_1(y)]$$

$$+ \sum_{\substack{p=4\, i+j=p \\ i,j\geq 2}}^{\infty}\sum^{\infty} Dh_j(y)\{f_i(y) - g_i(y)\}$$

$$+ \sum_{\substack{p=4\, i+j=p \\ i,j\geq 2}}^{\infty}\sum^{\infty} \{Df_i(y)h_j(y) - Dh_j(y)f_i(y)\}$$

$$+ \sum_{m=2}^{\infty} \frac{1}{m!} \sum_{i=m}^{\infty} D^m f_i(y)\{h_2(y) + h_3(y) + \cdots\}^m. \qquad (2.199)$$

Finally, we may round off (2.199) up to kth order, which is enough for the proof, and put it in ascending order:

$$\sum_{i=2}^{k} g_i(y) = \sum_{i=2}^{k} f_i(y) + \sum_{i=2}^{k}[h_i(y), v_1(y)]$$

$$+ \sum_{j=3}^{k}\sum_{i=2}^{j-1}[h_{j-i+1}(y), f_i(y)]$$

$$+ \sum_{j=3}^{k}\sum_{i=2}^{j-1} Dh_i(y)\{f_{k-i+1}(y) - g_{k-i+1}(y)\}$$

$$+ \sum_{j=4}^{k}\sum_{m=2}^{[\frac{j}{2}]} \frac{1}{m!} \sum_{i=m}^{j-m} D^m f_i(y)$$

$$\times \sum_{\substack{l_1+l_2+\cdots+l_m=j-(i-m) \\ 2\leq l_1,l_2,\ldots,l_m \leq j-(i-m)-2(m-1)}} h_{l_1}(y)\cdots h_{l_m}(y), \qquad (2.200)$$

where the property of the Lie bracket,

$$[X_i, Y_j] \in \mathcal{H}_{i+j-1} \quad \text{for } X_i \in \mathcal{H}_i \text{ and } Y_j \in \mathcal{H}_j, \qquad (2.201)$$

has been used, where $\mathcal{H}_k$ denotes linear vector space consisting of kth-degree homogeneous polynomials.

Now by taking the terms in (2.200) according to their order one obtains

$$g_2 = f_2 + [h_2, v_1],$$

$$g_3 = f_3 + [h_3, v_1] + [h_2, f_2] + Dh_2(f_2 - g_2),$$

$$g_4 = f_4 + [h_4, v_1] + [h_3, f_2] + [h_2, f_3] \quad (2.202)$$

$$+ Dh_2(f_3 - g_3) + Dh_3(f_2 - g_2) + \frac{1}{2}D^2 f_2 h_2^2,$$

etc.,

where the variable y has been dropped for simplicity. For general k, we have

$$g_k = f_k + [h_k, v_1] + \sum_{i=2}^{k-1}\left\{[h_i, f_{k-i+1}] + Dh_i(f_{k-i+1} - g_{k-i+1})\right\}$$

$$+ \sum_{m=2}^{[\frac{k}{2}]} \frac{1}{m!} \sum_{i=m}^{k-m} D^m f_i \sum_{\substack{l_1+l_2+\cdots+l_m=k-(i-m) \\ 2\leq l_1, l_2, \ldots, l_m \leq k-(i-m)-2(m-1)}} h_{l_1} h_{l_2} \cdots h_{l_m},$$

which is (2.193) and the proof is thus completed. □

Remark 2.20 The following observations are made from (2.193).

(1) The only operation appearing in the formula is the Frechét derivative involved in Dh_i, $D^m f_i$ and the Lie bracket $[\bullet, \bullet]$. This operation can be easily implemented on computers using a computer algebra system.
(2) The kth-order equation contains all the kth-order and only the kth-order terms. The equation is given in a recursive form.
(3) The kth-order equation depends upon the known vector homogeneous polynomials $v_1, f_2, f_3, \ldots, f_{k-1}$, and upon $h_2, h_3, \ldots, h_{k-1}$, as well as $g_2, g_3, \ldots, g_{k-1}$, which have been explicitly determined from the lower-order equations.
(4) The equation involves the coefficients of the nonlinear transformation h and the coefficients of the kth-order normal form g_k. If the jth-order ($j < k$) coefficients of h_j are completely determined from the jth-order equation, then the only unknown coefficients in the kth-order equation are h_k and g_k, which yields the normal form.
(5) If the kth-order equation contains lower-order coefficients of $h_j (j < k)$ which are undetermined in the lower-order ($<k$) equations, they may be used to eliminate some coefficients of g_k, and thus the normal form can be further simplified.
(6) For most of the approaches in computing the simplest normal forms (e.g., see [2, 3, 54, 55, 202, 219, 255]), the nonlinear vector field, $f(x)$, given in (2.189) is assumed to be a conventional normal form in order to simplify symbolic computations. All the approaches described in the above-mentioned references

generate the kth-order algebraic equation which contains lower-order ($<k$) as well as higher-order ($>k$) terms. This is extremely time consuming in symbolic computations and it also takes too much computer memory. With the above efficient recursive formulas, the kth-order equation exactly contains the kth-order terms, which greatly saves computer memory and computational time. Therefore, for our approach, the vector field $f(x)$ can be assumed to be a general analytic function, not necessarily a conventional normal form.

2.4.2 Symbolic Computation

The recursive formula given in (2.193) has been directly used to develop a symbolic computational program based on Maple. The main operation involved in the computation is the multiplication of a matrix by a vector (a Lie bracket operator consists of two such multiplications). The computations for the second term and the third term in (2.193) (i.e., the Lie bracket and the first summation) are straightforward, while the last summation in (2.193) needs careful consideration in order to achieve the minimum number of operations. First it should be noted that the variable used in functions $h_{l_1}, h_{l_2}, \ldots, h_{l_m}$, must be different (i.e., not the y variable) from that of f_i, so that the operator D can treat them as "constant" in the processing of differentiation (see (2.194)). When the differentiation is complete, the variables in these h functions should be changed back to the original variable y. Secondly, in order to achieve the minimum number of operations for the last summation, note that many terms in the summation are actually the same due to the fact that the indices $l_1, l_2, \ldots, l_m$ can be equal, and due to the fact that

$$D\big((Df_i)h_1\big)h_2 = D\big((Df_i)h_2\big)h_1, \tag{2.203}$$

which can be proved by direct calculation as follows.

$$D\big((Df_i)h_1\big)h_2$$

$$= D\left(\begin{bmatrix} f_{11} & f_{12} & \cdots & f_{1n} \\ f_{21} & f_{22} & \cdots & f_{2n} \\ \vdots & \vdots & \ddots & \vdots \\ f_{n1} & f_{n2} & \cdots & f_{nn} \end{bmatrix} \begin{pmatrix} h_{11} \\ h_{12} \\ \vdots \\ h_{1n} \end{pmatrix}\right) h_2$$

$$= D\begin{pmatrix} \sum_{j=1}^{n} f_{1j}h_{1j} \\ \sum_{j=1}^{n} f_{2j}h_{2j} \\ \vdots \\ \sum_{j=1}^{n} f_{nj}h_{nj} \end{pmatrix} \begin{pmatrix} h_{21} \\ h_{22} \\ \vdots \\ h_{2n} \end{pmatrix}$$

$$
= \begin{bmatrix} \sum_{j=1}^{n} f_{1j1}h_{1j} & \sum_{j=1}^{n} f_{1j2}h_{1j} & \cdots & \sum_{j=1}^{n} f_{1jn}h_{1j} \\ \sum_{j=1}^{n} f_{2j1}h_{1j} & \sum_{j=1}^{n} f_{2j2}h_{1j} & \cdots & \sum_{j=1}^{n} f_{2jn}h_{1j} \\ \vdots & \vdots & \ddots & \vdots \\ \sum_{j=1}^{n} f_{nj1}h_{1j} & \sum_{j=1}^{n} f_{nj2}h_{1j} & \cdots & \sum_{j=1}^{n} f_{njn}h_{1j} \end{bmatrix} \begin{pmatrix} h_{21} \\ h_{22} \\ \vdots \\ h_{2n} \end{pmatrix}
$$

$$
= \begin{pmatrix} \sum_{l=1}^{n}(\sum_{j=1}^{n} f_{1jl}h_{1j})h_{2l} \\ \sum_{l=1}^{n}(\sum_{j=1}^{n} f_{2jl}h_{1j})h_{2l} \\ \vdots \\ \sum_{l=1}^{n}(\sum_{j=1}^{n} f_{njl}h_{1j})h_{2l} \end{pmatrix}
$$

$$
= \begin{pmatrix} \sum_{j=1}^{n}(\sum_{l=1}^{n} f_{1jl}h_{2l})h_{1j} \\ \sum_{j=1}^{n}(\sum_{l=1}^{n} f_{2jl}h_{2l})h_{1j} \\ \vdots \\ \sum_{j=1}^{n}(\sum_{l=1}^{n} f_{njl}h_{2l})h_{1j} \end{pmatrix}
$$

$$
= D\big((D f_i)h_2\big)h_1, \tag{2.204}
$$

where the fact that h_{l_j} is not affected by the operator D has been used. Thus the order of differentiation in (2.194) with respect to h_{l_j} has no influence. Then the last summation in (2.193) can be rewritten as

$$
\sum_{m=2}^{\left[\frac{k}{2}\right]} \frac{1}{m!} \sum_{i=m}^{k-m} D^m f_i \sum_{\substack{l_1+l_2+\cdots+l_m=k-(i-m) \\ 2 \le l_1,l_2,\ldots,l_m \le k-(i-m)-2(m-1)}} h_{l_1}h_{l_2}\cdots h_{l_m}
$$

$$
= \sum_{m=2}^{\left[\frac{k}{2}\right]} \frac{1}{m!} \sum_{i=m}^{k-m} D^m f_i \sum_{\substack{q_1 l_1+\cdots+q_p l_p=k-(i-m) \\ 2 \le l_p<\cdots<l_1 \le (k-(i-m))/m}} \frac{m!}{q_1!\cdots q_p!} h_{l_1}^{q_1}\cdots h_{l_p}^{q_p}
$$

$$
= \sum_{m=2}^{\left[\frac{k}{2}\right]} \sum_{i=m}^{k-m} D^m f_i \sum_{\substack{q_1 l_1+q_2 l_2+\cdots+q_p l_p=k-(i-m) \\ 2 \le l_p<\cdots<l_1 \le (k-(i-m))/m}} \frac{h_{l_1}^{q_1} h_{l_2}^{q_2}\cdots h_{l_p}^{q_p}}{q_1! q_2!\cdots q_p!}, \tag{2.205}
$$

where q_j, $j = 1, 2, \ldots, p$, are nonzero positive integers.

Based on (2.193) and (2.205), Maple programs have been developed which only require simple preparation of an input file by a user. The algorithm is outlined below.

(1) Read a prepared input file. The input file lists the order of the normal form to be computed, ord, and the case of singularity to be considered.
(2) Separate the different order terms from the given input differential equations.
(3) A procedure for computing the Lie bracket operator.

(4) A procedure for computing the Jacobian matrix and the multiplication of a matrix by a vector.
(5) A procedure for solving a linear algebraic equation.
(6) Obtain the ordered terms for functions $F[k]$, $G[k]$ and $H[k]$.
(7) For order k, recursively compute the algebraic equations which will be used for finding the coefficients of the normal form and the corresponding nonlinear transformation.
  (i) Compute the Lie bracket term $[h_k, v_1]$.
  (ii) Compute the summation $\sum_{i=2}^{k-1}\{[h_{k-i+1}, f_i] + Dh_i(f_{k-i+1} - g_{k-i+1})\}$.
  (iii) Find the indices $q_1, q_2, \ldots, q_p$ for the term $h_{l_1}^{q_1} h_{l_2}^{q_2} \cdots h_{l_p}^{q_p}$, where $2 \le l_p \le \cdots \le l_1 \le (k - (i - m))/m$, satisfying $q_1 l_1 + q_2 l_2 + \cdots + q_p l_p = k - (i - m)$, which results in the coefficient $\frac{m!}{i_1! i_2! \ldots i_p!}$ for this term.
(8) Call the subroutine for a case study of the given singularity.
  (i) Obtain the kth-order algebraic equation from the main program, which is stored in the variable $\texttt{cof}[i, j, k]$.
  (ii) For complex analysis, get the real and imaginary part from the coefficient $\texttt{cof}[i, j, k]$.
  (iii) For a suborder k, determine the b coefficients and then solve for the relative c coefficients.
  (iv) Return to the main program.
(9) Write the normal form into the output file $\texttt{Nform}$.

Chapter 3
Comparison of Methods for Computing Focus Values

In the previous chapter, we studied Hopf bifurcation and normal form computation. Now, in this chapter, we turn to an important problem: how to determine the maximal number of limit cycles bifurcating from a Hopf critical point. To determine the existence of multiple limit cycles in the neighborhood of a degenerate Hopf critical point, one needs to compute the coefficients of the normal form, or more precisely, to compute the focus values of the critical point [35, 36, 146, 218, 228]. It should be noted that a *focus value* is also often called a *focal value* in the literature. Three typical methods: the Poincaré method or Takens method, a perturbation technique, and the singular point value method are particularly discussed in this chapter. It is shown that these three methods have the same order of computational complexity, though no method has been developed so far for computing the "minimal singular point values".

Many methods have been developed for computing focus values and normal forms, including the Poincaré method [178], the Takens method [193, 194], the Lyapunov–Schmidt reduction method [46], time averaging [72], a perturbation technique based on multiple timescales [166, 218], the singular point value method [35, 146], etc. The basic ideas of the Poincaré and the Takens method are similar: by using a homological operator to decompose the linear space of a vector field at each order, find the part (normal form) which cannot be removed by nonlinear transformations. Based on the Poincaré method, two pioneering books written by Marsden [162] and Hassard et al. [98] developed methods for computing the focus values associated with Hopf bifurcation. In particular, in the second book [98], an explicit formula was given for the second-order focus value when the first one is zero, even when the dimension of the system is larger than two. The Lyapunov–Schmidt reduction method [46], on the other hand, is often used by people to find an approximation for periodic solutions emerging from Hopf bifurcation. The basic idea of this method is to project the whole system into the subspace spanned by the eigenvectors associated with a pair of purely imaginary eigenvalues. This method, however, does not yield differential equations but algebraic equations. Strictly speaking, it cannot be employed for stability analysis, but can be used to compute the focus values. As a matter of fact, the idea of this method is similar

M. Han, P. Yu, *Normal Forms, Melnikov Functions and Bifurcations of Limit Cycles*, 59
Applied Mathematical Sciences 181,
DOI 10.1007/978-1-4471-2918-9_3, © Springer-Verlag London Limited 2012

to that of time averaging and multiple timescales from the viewpoint of projection into subspace. In engineering society, the time averaging and multiple timescales are widely used for computing approximate solutions of oscillators or vibrating systems. Nayfeh [166] was the first one to introduce the multiple timescales to compute the normal form of oscillating systems, described by second-order differential equations. Later, this method was combined with a perturbation technique to form a systematic and unifying procedure [218] which can be directly applied to higher-dimensional systems, without application of the center manifold reduction method.

To find the maximal number of multiple limit cycles, one needs to compute high-order normal forms or high-order focus values, related to degenerate Hopf bifurcations [64, 165]. This requires finding explicit symbolic expressions, giving rise to a crucial problem—computational efficiency—since an inefficient computer program would quickly run into a loop or terminate. Therefore, it is important to examine the existing methods for computing normal forms and focus values. In this chapter, we choose three typical methods for comparison: the Poincaré method [178] (which was also called the Takens method [193, 194]), a perturbation method using multiple timescales [166, 218], and the singular point value method [35, 36, 146]. We shall not discuss the Lyapunov–Schmidt reduction method since the main idea of this method is similar to that of the multiple timescales. We use several examples to show that all three methods have the same computational complexity. We will also discuss the so-called "minimal singular point value", which is supposed to be the best in computing focus values, and show that none of the existing methods achieves the best.

Since the perturbation method has been studied in detail in the previous chapter, here we shall focus on the Poincaré method and the singular point value method. Note that the Poincaré method and the singular point value method are only applicable to two-dimensional systems, while the perturbation method can be directly applied to n-dimensional systems. In other words, if one wants to apply the Poincaré method or the singular point value method to compute focus values, one must first apply center manifold theory, while the perturbation method combines center manifold theory and normal form theory in one unified approach [220].

3.1 The Poincaré Method

Consider the following general system described on a two-dimensional manifold:

$$\dot{x} = Lx + f(x) \equiv v_1 + f_2(x) + f_3(x) + \cdots + f_k(x) + \cdots, \qquad (3.1)$$

where $x = (x_1, x_2)^T \in R^2$, $f_k(x)$ denotes the kth-order vector of homogeneous polynomials in x, and

$$L = \begin{bmatrix} 0 & 1 \\ -1 & 0 \end{bmatrix}, \qquad (3.2)$$

with $v_1 = Lx \equiv Jx$. (Usually, J is used to denote the Jacobian matrix. Here, L is used to be consistent with the Lie bracket notation.) It is assumed that all eigenvalues of L have zero real parts, implying that the dynamics of system (3.1) are described on a two-dimensional center manifold.

The basic idea of Poincaré normal form theory is to find a near-identity nonlinear transformation,

$$x = y + h(y) \equiv y + h_2(y) + h_3(y) + \cdots + h_k(y) + \cdots, \qquad (3.3)$$

such that the resulting system,

$$\dot{y} = Ly + g(y) \equiv Ly + g_2(y) + g_3(y) + \cdots + g_k(y) + \cdots, \qquad (3.4)$$

becomes as simple as possible. Here, $h_k(y)$ and $g_k(y)$ denote some kth-order vector of homogeneous polynomials in y.

To apply normal form theory, first define an operator as follows:

$$\begin{aligned} L_k &: \mathcal{H}_k \mapsto \mathcal{H}_k, \\ U_k &\in \mathcal{H}_k \mapsto L_k(U_k) = [U_k, v_1] \in \mathcal{H}_k, \end{aligned} \qquad (3.5)$$

where $\mathcal{H}_n$ denotes a linear vector space consisting of kth-order vectors of homogeneous polynomials. The operator $[U_k, v_1]$ is called the *Lie bracket*, defined as

$$[U_k, v_1] = Dv_1 \cdot U_k - DU_k \cdot v_1. \qquad (3.6)$$

Next, define the space $\mathcal{R}_k$ as the range of L_k, and the complementary space of $\mathcal{R}_k$ as $\mathcal{K}_k = \ker(L_k)$. Thus,

$$\mathcal{H}_k = \mathcal{R}_k \oplus \mathcal{K}_k, \qquad (3.7)$$

and we can then choose bases for $\mathcal{R}_k$ and $\mathcal{K}_k$. Consequently, a vector homogeneous polynomial $f_k \in \mathcal{H}_k$ can be split into two parts: one is spanned by the basis of $\mathcal{R}_k$ and the other by that of $\mathcal{K}_k$. Normal form theory shows that the part of f_k belonging to $\mathcal{R}_k$ can be eliminated while the part belonging to $\mathcal{K}_k$ must be retained, which is called a *normal form*.

It should be pointed out here that in general, $\ker(L_k)$ does not always complement the range $\mathcal{R}(L_k)$; however, it does for the case of simple (nondegenerate) Hopf bifurcation. Certainly, it does not imply that it is always the best choice of complement. As a matter of fact, how to choose the $\mathcal{R}_k$ which best complements the range $\mathcal{R}_k$ is still an open problem, as discussed in Chap. 5.

By applying Poincaré normal form theory [178], one can find the kth-order normal form $g_k(y)$, while the part belonging to $\mathcal{R}_k$ can be removed by appropriately choosing the coefficients of the nonlinear transformation $h_k(y)$. The "form" of the normal form $g_k(y)$ is mainly determined by the linear vector v_1. However, it depends not only on the basis of the complementary space $\mathcal{K}_k$, but also on the choice of the complementary space itself. Once $\mathcal{K}_k$ is chosen, one may apply the matrix method [72] to find a basis of the space $\mathcal{R}_k$ and then determine a basis of the complementary space $\mathcal{K}_k$. Having chosen the basis of $\mathcal{K}_k$, the form of $g_k(y)$ can be determined, which actually represents the normal form.

In general, when one applies normal form theory to a system, one can find the "form" of the normal form (i.e., the basis of the complementary space $\mathcal{K}_k$), but not explicit expressions. However, in applications, solutions for the normal form and the nonlinear transformation need to be found explicitly. To achieve this, one may assume a general form of the nonlinear transformation and then substitute it back into the differential equation, with the aid of normal form theory, to obtain the kth-order algebraic equations by balancing the coefficients of the homogeneous polynomial terms. These algebraic equations are then used to determine the coefficients of the normal form and the nonlinear transformation. Thus, the key step in the computation of the kth-order normal form is to find the kth-order algebraic equations. An efficient approach for deriving the algebraic equations has been discussed in Sect. 2.4.

When the eigenvalues of a Jacobian involve one or more purely imaginary pairs, a complex analysis may simplify the solution procedure. It has actually been noted that the real analysis given in [219] yields coupled algebraic equations, while it will be seen below that complex analysis can decouple the algebraic equations.

Thus, introduce the linear transformation

$$
\begin{cases} x_1 = \dfrac{1}{2}(z+\bar{z}), \\[2mm] x_2 = \dfrac{i}{2}(z-\bar{z}), \end{cases}
\quad \text{i.e.,} \quad
\begin{cases} z = x_1 - i x_2, \\[2mm] \bar{z} = x_1 + i x_2, \end{cases}
\tag{3.8}
$$

where $i = \sqrt{-1}$, and $\bar{z}$ is the complex conjugate of z. Then the operators $\frac{\partial}{\partial z}$ and $\frac{\partial}{\partial \bar{z}}$ can be obtained by the chain rule as follows:

$$
\frac{\partial}{\partial z} = \frac{\partial}{\partial x_1}\frac{\partial x_1}{\partial z} + \frac{\partial}{\partial x_2}\frac{\partial x_2}{\partial z} = \frac{1}{2}\left(\frac{\partial}{\partial x_1} + i\frac{\partial}{\partial x_2}\right),
$$

$$
\frac{\partial}{\partial \bar{z}} = \frac{\partial}{\partial x_1}\frac{\partial x_1}{\partial \bar{z}} + \frac{\partial}{\partial x_2}\frac{\partial x_2}{\partial \bar{z}} = \frac{1}{2}\left(\frac{\partial}{\partial x_1} - i\frac{\partial}{\partial x_2}\right),
$$

and so the linear part of system (2.1), $\boldsymbol{v}_1$, becomes

$$
\begin{aligned}
\boldsymbol{v}_1 &= x_2\frac{\partial}{\partial x_1} - x_1\frac{\partial}{\partial x_2} \equiv \begin{pmatrix} x_2 \\ -x_1 \end{pmatrix} \\
&= \frac{i}{2}(z-\bar{z})\frac{\partial}{\partial x_1} - \frac{1}{2}(z+\bar{z})\frac{\partial}{\partial x_2} \\
&= \frac{1}{2}iz\left(\frac{\partial}{\partial x_1} + i\frac{\partial}{\partial x_2}\right) - \frac{1}{2}i\bar{z}\left(\frac{\partial}{\partial x_1} - i\frac{\partial}{\partial x_2}\right) \\
&= iz\frac{\partial}{\partial z} - i\bar{z}\frac{\partial}{\partial \bar{z}} \equiv \begin{pmatrix} iz \\ -i\bar{z} \end{pmatrix}.
\end{aligned}
\tag{3.9}
$$

Indeed, applying transformation (3.8) to system (3.1) yields

$$
\dot{z} = iz + f(z,\bar{z}), \qquad \frac{d\bar{z}}{dt} = -i\bar{z} + \overline{f}(z,\bar{z}),
\tag{3.10}
$$

where f is a polynomial in z and $\bar{z}$ starting from second-order terms, and $\bar{f}$ is the complex conjugate of f. Here, for convenience, we use the same notation $f = (f, \bar{f})^T$ for the complex analysis. To find the normal form of a Hopf singularity, one may use a nonlinear transformation given by

$$z = y + \sum h_k(y, \bar{y}), \qquad \bar{z} = \bar{y} + \sum \bar{h}_k(y, \bar{y}), \qquad (3.11)$$

and determine the basis, g_k, for the complementary space of $\mathcal{K}_k$, or use Poincaré normal form theory to determine the so-called "resonant terms". It is well known that the resonant terms are given in the form of $z^j \bar{z}^{j-1}$ (e.g., see [72]), and the kth-order normal form is given by

$$g_k(y, \bar{y}) = \begin{pmatrix} (b_{1k} + ib_{2k}) y^{(k+1)/2} \bar{y}^{(k-1)/2} \\ (b_{1k} - ib_{2k}) \bar{y}^{(k+1)/2} y^{(k-1)/2} \end{pmatrix}, \qquad (3.12)$$

where the b_{1k} and b_{2k} are real coefficients to be determined. Therefore, the normal form can be written as

$$\dot{y} = iy + \sum_{m=1}^{\infty} g_{2m+1}(y, \bar{y}), \qquad \dot{\bar{y}} = -i\bar{y} + \sum_{m=1}^{\infty} \bar{g}_{2m+1}(y, \bar{y}). \qquad (3.13)$$

It is easy to see from (3.13) that the normal form contains odd-order terms only, as expected. In normal form computation, the two kth-order coefficients b_{1k} and b_{2k} should, in general, be retained in the normal form.

Finally, based on (3.4), (3.10)–(3.12), one can determine the algebraic equations, order by order, starting from $k = 2$, and then apply normal form theory to solve for the coefficients b_{1k} (k is odd) explicitly in terms of the original system coefficients.

Summarizing the above results gives the following theorem.

Theorem 3.1 *For system* (3.1) *with L given by* (3.2), *the normal form is given by* (3.13), *where b_{1k} is the kth-order focus value.*

3.2 A Perturbation Technique

Since a detailed discussion of the perturbation technique based on the multiple timescales (MTS) method or simply the multiple scales (MT) method has been given in the previous chapter, we briefly outline here the procedure of the approach for the convenience of comparison. It should be noted that this technique is not only applicable to Hopf singularities [166, 218], but also to other singularities such as Hopf-zero [123, 247], double-Hopf [221, 223], etc.

Consider the general n-dimensional differential equation (2.138):

$$\dot{x} = Jx + f(x), \qquad x \in R^n, \qquad f : R^n \to R^n,$$

where Jx represents the linear terms of the system, and the nonlinear function f is assumed to be analytic with $x = 0$ being an equilibrium point of the system, i.e., $f(0) = 0$. Further, assume that the Jacobian of system (2.138), evaluated at the equilibrium point 0, contains one pair of purely imaginary eigenvalues $\pm i$, and thus the Jacobian of system (2.138) may be assumed in Jordan canonical form to be

$$J = \begin{bmatrix} 0 & 1 & 0 \\ -1 & 0 & 0 \\ 0 & 0 & A \end{bmatrix}, \quad A \in R^{(n-2)\times(n-2)}, \tag{3.14}$$

where A is *stable* (i.e., all of its eigenvalues have negative real parts).

Based on (2.166) and (2.167), we have the following theorem.

Theorem 3.2 *Suppose the general n-dimensional system* (2.138) *has a Hopf-type singular point at the origin, i.e., the linearized system of* (2.138) *has one pair of purely imaginary eigenvalues and the remaining eigenvalues have negative real parts. Then the normal form of system* (2.138) *for Hopf or generalized Hopf bifurcations up to the* $(2k + 1)$*th-order term is given by*

$$\dot{r} = r(v_1 + v_3 r^2 + v_5 r^4 + \cdots + v_{2k+1} r^{2k}), \tag{3.15}$$

$$\dot{\theta} = 1 + \frac{d\phi}{dt} = 1 + \tau_3 r^2 + \tau_5 r^4 + \cdots + \tau_{2k+1} r^{2k}, \tag{3.16}$$

where the constants v_{2k+1} *and* τ_{2k+1} *are explicitly expressed in terms of the original system parameters, and* v_{2k+1} *is the kth-order focus value.*

3.3 The Singular Point Value Method

This iterative method computes focus values by computing the singular point quantities (see [35, 36, 146] for details).

To introduce this method, consider the following planar polynomial differential system:

$$\frac{dx}{dt'} = \delta x - y + \sum_{k=2}^{\infty} X_k(x, y),$$

$$\frac{dy}{dt'} = x + \delta y + \sum_{k=2}^{\infty} Y_k(x, y), \tag{3.17}$$

where $X_k(x, y)$ and $Y_k(x, y)$ are homogeneous polynomials in x, y, of degree k. The origin $(x, y) = (0, 0)$ is a singular point of system (3.17), which is either a focus or an elementary center (when $\delta = 0$). Since we are interested in the computation of focus values, we assume $\delta = 0$ in the following analysis.

Definition 3.3 A system is said to be *concomitant* with system (3.17) if it results from (3.17) through the following transformations:

$$z = x + iy, \qquad w = x - iy, \qquad T = it', \quad i = \sqrt{-1}, \qquad (3.18)$$

given by

$$\frac{dz}{dT} = z + \sum_{k=2}^{\infty} Z_k(z, w) = Z(z, w),$$

$$\frac{dw}{dT} = - w - \sum_{k=2}^{\infty} W_k(z, w) = -W(z, w), \qquad (3.19)$$

where z, w and T are complex variables, and

$$Z_k(z, w) = \sum_{\alpha+\beta=k} a_{\alpha\beta} z^\alpha w^\beta, \qquad W_k(z, w) = \sum_{\alpha+\beta=k} b_{\alpha\beta} w^\alpha z^\beta. \qquad (3.20)$$

We usually say that systems (3.17) and (3.19) are concomitant.

If system (3.17) is a real planar differential system, then the coefficients of system (3.19) must satisfy the following conjugate conditions:

$$\overline{a_{\alpha\beta}} = b_{\alpha\beta}, \qquad \alpha \geq 0, \ \beta \geq 0, \ \alpha + \beta \geq 2. \qquad (3.21)$$

By the following transformations:

$$z = r e^{i\theta}, \qquad w = r e^{-i\theta}, \qquad T = it, \qquad (3.22)$$

system (3.19) can be transformed into

$$\frac{dr}{dt} = \frac{ir}{2} \sum_{m=1}^{\infty} \sum_{\alpha+\beta=m+2} [a_{\alpha(\beta-1)} - b_{\beta(\alpha-1)}] e^{i(\alpha-\beta)\theta} r^m,$$

$$\frac{d\theta}{dt} = 1 + \frac{1}{2} \sum_{m=1}^{\infty} \sum_{\alpha+\beta=m+2} [a_{\alpha(\beta-1)} + b_{\beta(\alpha-1)}] e^{i(\alpha-\beta)\theta} r^m. \qquad (3.23)$$

For a complex constant h, $|h| \ll 1$, we may write the solution of (3.23) satisfying the initial condition $r|_{\theta=0} = h$ as

$$r = \tilde{r}(\theta, h) = h + \sum_{k=2}^{\infty} v_k(\theta) h^k. \qquad (3.24)$$

Evidently, if system (3.17) is a real system, then $v_{2k+1}(2\pi)$, $k = 1, 2, \ldots$, is the kth-order focus (or focal) value of the origin.

For system (3.19), we can uniquely derive the following formal series:

$$\varphi(z, w) = z + \sum_{k+j=2}^{\infty} c_{kj} z^k w^j, \qquad \psi(z, w) = w + \sum_{k+j=2}^{\infty} d_{k,j} w^k z^j, \qquad (3.25)$$

such that

$$\frac{d\varphi}{dT} = \varphi + \sum_{j=1}^{\infty} p_j \varphi^{j+1} \psi^j, \qquad \frac{d\psi}{dT} = -\psi - \sum_{j=1}^{\infty} q_j \psi^{j+1} \varphi^j. \qquad (3.26)$$

Definition 3.4 Let $\mu_0 = 0$, and $\mu_k = p_k - q_k, k = 1, 2, \ldots$. Then, μ_k is called the kth-order singular point quantity of the origin of system (3.19) [35]. If $\mu_0 = \mu_1 = \cdots = \mu_{k-1} = 0$ and $\mu_k \neq 0$, then the origin of system (3.19) is called the kth-order weak critical singular point. In other words, k is the multiplicity of the origin of system (3.19). If $\mu_k = 0$ for $k = 1, 2, \ldots$, then the origin of system (3.19) is called an extended center (complex center).

If system (3.17) is a real, autonomous, differential system with the concomitant system (3.19), then for the origin, the kth-order focus quantity v_{2k+1} of system (3.17) and the kth-order quantity of the singular point of system (3.19) have the relation given in the following theorem [146].

Theorem 3.5 *Given system (3.17) ($\delta = 0$) or (3.19), for any positive integer m, the following assertion holds:*

$$v_{2k+1}(2\pi) = i\pi \left(\mu_k + \sum_{j=1}^{k-1} \xi_k^{(j)} \mu_j \right), \qquad k = 1, 2, \ldots, \qquad (3.27)$$

where v_{2k+1} and μ_k represent the kth-order focus value (see (3.15)) and singular point quantity (see Definition 3.4), respectively, and $\xi_m^{(j)}$, $j = 1, 2, \ldots, k - 1$, are polynomial functions of coefficients of system (3.19).

The following recursive formulas are used for computing the singular point quantities of system (3.19) [35]: $c_{11} = 1, c_{20} = c_{02} = c_{kk} = 0, k = 2, 3, \ldots$, and for all $(\alpha, \beta), \alpha \neq \beta$, and $m \geq 1$:

$$C_{\alpha\beta} = \frac{1}{\beta - \alpha} \sum_{k+j=3}^{\alpha+\beta+2} \big[(\alpha - k + 1) a_{k,j-1}$$

$$- (\beta - j + 1) b_{j,k-1} \big] C_{\alpha-k+1,\beta-j+1} \qquad (3.28)$$

and

$$\mu_m = \sum_{k+j=3}^{2m+4} \big[(m - k + 2) a_{k,j-1} - (m - j + 2) b_{j,k-1} \big] C_{m-k+2,m-j+2}, \qquad (3.29)$$

where $a_{kj} = b_{kj} = C_{kj} = 0$ for $k < 0$ or $j < 0$.

It is clearly seen from (3.27) that

$$\mu_0 = \mu_1 = \mu_2 = \cdots = \mu_{k-1} = 0 \quad \Longleftrightarrow \quad v_1 = v_3 = v_5 = \cdots = v_{2k-1} = 0.$$
$$(3.30)$$

(Note that here $\mu_0 = v_1 = \delta$ is the linear focus value.) Therefore, when determining the conditions such that $v_1 = v_3 = v_5 = \cdots = v_{2k-1} = 0$, one can instead use the equations $\mu_0 = \mu_1 = \mu_2 = \cdots = \mu_{k-1} = 0$. If the μ_k's are simpler than the v_{2k+1}'s, then this method is better than the method of directly computing v_{2k+1}. However, in general, these μ_k are not necessarily simpler than v_{2k+1}, as will be seen in the next section.

It should be pointed out that since the normal form is not unique, the focus values obtained by using different methods are not necessarily the same. However, the first nonzero focus value must be identical (neglecting a constant multiplier). This implies that for different focus values obtained by using different approaches, solutions to the equations $v_1 = v_3 = v_5 = \cdots = v_{2k-1} = 0$ (or $\mu_0 = \mu_1 = \mu_3 = \cdots = \mu_{k-1} = 0$) must be identical.

For the three methods described above, symbolic programs have been developed using Maple, which will be used in the following two sections.

3.4 Illustrative Examples

In this section, we present several well-known examples to demonstrate comparison of the three methods discussed in the previous section.

3.4.1 The Brusselator Model

The first example is the well-known Brusselator model [169], described by

$$\dot{w}_1 = A - (1+B)w_1 + w_1^2 w_2,$$
$$\dot{w}_2 = Bw_1 - w_1^2 w_2,$$
$$(3.31)$$

where $A, B > 0$ are parameters. The system has a unique equilibrium point,

$$w_{1e} = A, \qquad w_{2e} = \frac{B}{A}.$$
$$(3.32)$$

Evaluating the Jacobian of the system at the equilibrium point shows that a Hopf bifurcation occurs at the critical point $B = 1 + A^2$. Let

$$B = 1 + A^2 + \mu,$$
$$(3.33)$$

where μ is a perturbation parameter. Then, the Jacobian has eigenvalues $\lambda = \pm Ai$. Suppose $A = 1$, and then introduce the transformation

$$
\begin{aligned}
w_1 &= w_{1e} + x_1, \\
w_2 &= w_{2e} - x_1 + x_2,
\end{aligned}
\tag{3.34}
$$

into (3.31) to obtain the following new system:

$$
\begin{aligned}
\dot{x}_1 &= x_2 + \mu x_1 + \mu x_1^2 + 2x_1 x_2 - x_1^3 + x_1^2 x_2, \\
\dot{x}_2 &= -x_1,
\end{aligned}
\tag{3.35}
$$

such that its Jacobian evaluated at the equilibrium point $(x_1, x_2) = (0, 0)$ (i.e., $(w_1, w_2) = (1, B)$) is now in Jordan canonical form. Now, at the critical point defined by $\mu = 0$, apply the three methods described in the previous section to compute the first-order focus value. Maple programs are used to obtain the following results:

Poincaré method: $b_{13} = -\frac{3}{8}$;
perturbation method: $v_3 = -\frac{3}{8}$;
singular point value method: $\mu_1 = \frac{3}{4}i$.

It is seen that $b_{13} = v_3 = \frac{i}{2}\mu_1$. Ignoring the constant factor $\frac{i}{2}$, the three methods give an identical result for the first-order focus value: $-\frac{3}{8}$. This shows that the limit cycles bifurcating from the critical point, $\mu = 0$, in the vicinity of the equilibrium point (w_{1c}, w_{2c}) are supercritical, i.e., the bifurcating limit cycles are stable since the first-order focus value is negative.

Next, computing the second-order focus values gives:

Poincaré method: $b_{15} = -\frac{1}{96}$;
perturbation method: $v_5 = -\frac{1}{96}$;
singular point value method: $\mu_2 = -\frac{67}{48}i$.

This indicates that the Poincaré method and the perturbation method still give the same second-order focus value, but the singular point value method yields a different μ_2 (ignoring the difference factor $\frac{i}{2}$). This is not surprising since here $\mu_1 \neq 0$ and the second-order focus value is a combination of μ_1 and μ_2.

Further computation shows:

Poincaré method: $b_{17} = -\frac{2695}{36864}$;
perturbation method: $v_7 = -\frac{4543}{36864}$;
singular point value method: $\mu_3 = \frac{6239}{2304}i$.

For the third-order focus value, even the Poincaré method and the perturbation method give different results, as expected.

It should be noted that even for this very simple two-dimensional system, determining the stability of a bifurcating limit cycle is not an easy job. Now, the normal

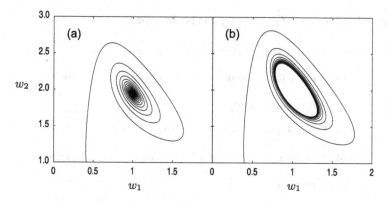

Fig. 3.1 Simulated trajectories of the Brusselator model (3.31) with $A = 1.0$ and initial point $(w_1, w_2) = (2, -1)$: (**a**) convergent to the stable equilibrium point w^+ when $B = 1.95$; and (**b**) convergent to a stable limit cycle when $B = 2.05$

form up to third order can be written in polar coordinates:

$$\dot{r} = \frac{1}{2}\mu r - \frac{3}{8}r^3 + \cdots,$$

$$\dot{\theta} = 1 - \frac{1}{24}r^2 + \cdots. \tag{3.36}$$

The normal form (3.36) clearly shows that the Hopf bifurcation is supercritical (since the third-order coefficient of the first equation is $-\frac{3}{8} < 0$), and the approximate amplitude of the limit cycle is given by

$$\bar{r} = 2\sqrt{\frac{\mu}{3}}.$$

Numerical simulation results based on the original system (3.31) are shown in Fig. 3.1, which indicates that when $A = 1.0$, $B = 1.95$, the trajectory converges to the stable equilibrium point $w_{1c} = 1$, $w_{2c} = 1.95$ (see Fig. 3.1(a)); while when $A = 1.0$, $B = 2.05$, the equilibrium point becomes unstable and a stable limit cycle bifurcates from the equilibrium point (see Fig. 3.1(b)). The simulation results based on the original system (3.31) agree well with the theoretical predictions based on the normal form (3.36).

3.4.2 A Cubic-Order System with Z_2 Symmetry

The second example is a cubic-order system with Z_2 symmetry, related to the well-known Hilbert 16th problem [99]. Since we will consider this problem again later in Chap. 4, here we focus on the comparison of computational efficiency.

The real system with Z_2 symmetry can be described by

$$
\begin{aligned}
\dot{w}_1 ={}& (a_0 + a_2)w_1 - (b_0 - b_2)w_2 + (a_1 + a_3 + a_4 + a_5)w_1^3 \\
&- (b_1 - b_3 + 3b_4 - 3b_5)w_1^2 w_2 + (a_1 + a_3 - 3a_4 - 3a_5)w_1 w_2^2 \\
&- (b_1 - b_3 - b_4 + b_5)w_2^3, \\
\dot{w}_2 ={}& (b_0 + b_2)w_1 + (a_0 - a_2)w_2 + (b_1 + b_3 + b_4 + b_5)w_1^3 \\
&+ (a_1 - a_3 + 3a_4 - 3a_5)w_1^2 w_2 + (b_1 + b_3 - 3b_4 - 3b_5)w_1 w_2^2 \\
&+ (a_1 - a_3 - a_4 + a_5)w_2^3.
\end{aligned}
\tag{3.37}
$$

The two eigenvalues of the Jacobian of (3.37) evaluated at the origin $(w_1, w_2) = (0, 0)$ are $a_0 \pm \sqrt{a_2^2 + b_2^2 - b_0^2}$, indicating that the origin $(0, 0)$ is a saddle point or a node when $a_2^2 + b_2^2 - b_0^2 \geq 0$ and a focus point or a center if $a_2^2 + b_2^2 - b_0^2 < 0$. When $a_2^2 + b_2^2 - b_0^2 = 0$, the origin is either a node or a double-zero singular point. By a parametric transformation, rename the coefficients of the resulting system to yield

$$
\begin{aligned}
\dot{x} &= ax + by + a_{30}x^3 + a_{21}x^2 y + a_{12}xy^2 + a_{03}y^3, \\
\dot{y} &= \pm bx + ay + b_{30}x^3 + b_{21}x^2 y + b_{12}xy^2 + b_{03}y^3,
\end{aligned}
\tag{3.38}
$$

where $b > 0$.

For a vector field with Z_2 symmetry, naturally the best situation is to have two symmetric focus points about the origin. Thus, if N small limit cycles are found in the neighborhood of one focus point, the whole system would have $2N$ limit cycles. Without loss of generality, the two symmetric focus points are assumed to be located on the y-axis (or the x-axis), and further assumed to be precisely located at $(0, \pm 1)$ with a proper scaling, leading to the conditions $a_{03} = -b$ and $b_{03} = -a$. Another condition, $a_{12} = a$, comes from making the two focus points of Hopf type. Furthermore, by applying proper parameter scaling and time scaling, one obtains the following new system [239]:

$$
\begin{aligned}
\frac{du}{d\tau} ={}& v + 2\bar{a}_{21}u^2 + 4auv - \frac{3}{2}v^2 + 4b\bar{a}_{30}u^3 - 2\bar{a}_{21}u^2 v \\
&- 2auv^2 + \frac{1}{2}v^3, \\
\frac{dv}{d\tau} ={}& -u - 4\bar{c}_{21}u^2 + 2(2a^2 \mp 2b^2 + 1)uv - 8\bar{c}_{30}u^3 + 4\bar{c}_{21}u^2 v \\
&- (2a^2 \mp b^2 + 1)uv^2,
\end{aligned}
\tag{3.39}
$$

where the coefficients $\bar{a}_{21}, \bar{c}_{21}, \bar{a}_{30}$ and $\bar{c}_{30}$ are expressed in terms of the original parameters $a, b, a_{21}, b_{21}, a_{30}, b_{30}$. Thus, based on (3.39), one can compute the focus values and consider the existence of small limit cycles.

Note that system (3.39) contains 6 free parameters, which suggests that we may set 6 focus values to zero, thereby obtaining 7 small limit cycles for system (3.39), so that the original system may have 14 small limit cycles. However, it has been shown in [239] that the existence of 14 limit cycles is not possible. The maximal number of small limit cycles that a cubic-order system with Z_2 symmetry can have is 12. Since here we are interested in the methods of computing focus values, we will not further discuss this interesting issue. (For details, readers are referred to [238, 239].)

We now apply the three methods reviewed in the previous section to compute the focus values of system (3.39) and obtain the following results:

```
The Poincare method:
b13:= 1/2*a-2*a^2*b21b-2*a21b*b21b-a21b*a+2*b^2*b21b
      +3/2*b*a30b-1/2*b21b:
b15 := ... (14 lines)
b17 := ... (87 lines)
b19 := ... (355 lines)
v111:= ... (1180 lines)
```

```
The perturbation method:
v3  := -2*a21b*b21b-a21b*a+1/2*a-1/2*b21b+3/2*b*a30b
       +2*b21b*b^2-2*b21b*a^2:
v5 := ... (14 lines)
v7 := ... (83 lines)
v9 := ... (344 lines)
v11:= ... (1173 lines)
```

```
The singular point value method:
mu1:= I*(-3*b*a30b-a+b21b+4*b21b*a^2-4*b21b*b^2
      +4*a21b*b21b+2*a21b*a):
mu3:= ... (14 lines)
mu3:= ... (85 lines)
mu4:= ... (355 lines)
mu5:= ... (1156 lines)
```

The numbers given in the parentheses denote the numbers of lines in the computer output files. It is easy to see that

$$b_{13} = v_3 = \frac{i}{2}\mu_1,$$

which shows that the three different methods give the same first-order focus value (differing by, at most, a constant factor), as expected. For the second-order focus values, it can be shown that $b_{15} = v_5$, but μ_2 is not equal to v_5 by a difference of a constant factor. Further, for the third-order focus values, $b_{15} \neq v_5$. However, if setting $b_{13} = v_3 = \mu_1 = 0$, which results in

$$\bar{a}_{30} = \frac{a(2\bar{a}_{21} - 1) + \bar{c}_{21}(4\bar{a}_{21} + 4a^2 - 4b^2 + 1)}{3b},$$

then one has

$$b_{15} = v_5 = \frac{i}{2}\mu_2.$$

Further, if letting $b_{15} = v_5 = \mu_2 = 0$ (which yields $\bar{c}_{30} = \bar{c}_{30}(a, b, \bar{a}_{21}, \bar{c}_{21})$), then one obtains

$$b_{17} = v_7 = \frac{i}{2}\mu_3.$$

This process can be carried out to higher-order focus values, i.e., if $b_{1(2i+1)} = v_{2i+1} = \mu_i = 0, i = 1, 2, \ldots, k - 1$, then

$$b_{1(2k+1)} = v_{2k+1} = \frac{i}{2}\mu_k.$$

Remark 3.6 The above expressions of the focus values obtained by using different methods show that their computational complexity is of the same order (see the numbers of the computer output lines, given in the parentheses).

3.4.3 A Symmetric Liénard Equation

The third example is the Liénard equation [144] described by

$$\dot{x} = y, \qquad \dot{y} = -g(x) - f(x)y, \tag{3.40}$$

where $g(x)$ and $f(x)$ are polynomial functions of x. Here, we investigate a particular class of Liénard equations with Z_2 symmetry, in which $g(x)$ is a third-degree odd polynomial, while $f(x)$ is an even function of x. To be more specific, consider the following system:

$$\dot{x} = y, \qquad \dot{y} = -\frac{1}{2}b^2x(x^2 - 1) - y\sum_{i=0}^{m}a_ix^{2i}, \tag{3.41}$$

where $b \neq 0$ and the a_i's are real coefficients. Equation (3.41) has three fixed points: $(0, 0)$ and $(\pm 1, 0)$. It is easy to use linear analysis to show that the origin $(0, 0)$ is a saddle point (with eigenvalues $\frac{1}{2}(-a_0 \pm \sqrt{a_0^2 + 2b^2})$). In order to have the two fixed points $(\pm 1, 0)$ be elementary centers, the condition

$$\sum_{i=0}^{m}a_i = 0, \quad \text{or} \quad a_0 = -\sum_{i=1}^{m}a_i, \tag{3.42}$$

must be satisfied. The eigenvalues of the Jacobian of system (3.41) evaluated at $(\pm 1, 0)$ are $\pm |b|i$. What we want to do is, for a given positive integer m, choose appropriate values of the a_i's such that system (3.41) has a maximum number of limit cycles in the neighborhood of the two fixed points $(\pm 1, 0)$. This local analysis is based on the calculation of focus values or the normal form associated with the Hopf singularity.

By introducing the scaling

$$a_i \implies ba_i, \quad i = 0, 1, \ldots, m,$$

the transformation

$$x = \pm(1 + u), \qquad y = \pm bv,$$

and the time scaling $\tau = bt$, we obtain the canonical form

$$\frac{du}{d\tau} = v,$$

$$\frac{dv}{d\tau} = -u - \frac{3}{2}u^2 - \frac{1}{2}u^3 - v \sum_{i=0}^{m} a_i(1 + u)^{2i}. \tag{3.43}$$

Now, consider the case of $m = 5$. We apply the three methods to compute the focus values (normal forms) of system (3.43) and obtain the following results.

```
The Poincare method:
b13:=  -3/4*a3-2*a4+1/4*a1-15/4*a5:
b15:=  1/8*a1+1/4*a2+7/8*a3+3/2*a4+5/8*a5-160/9*a4^3-15/4*a3^3
       -1000/9*a4*a5^2-25/36*a1^2*a5+5/9*a1^2*a2+5/12*a1^2*a3
       -20/3*a1*a4^2-5/4*a1*a3^2+5/9*a1*a2^2-625/36*a1*a5^2
       -5/3*a2^2*a3-40/9*a2^2*a4-25/3*a2^2*a5-160/9*a2*a4^2
       -5*a2*a3^2-125/3*a2*a5^2-20*a3^2*a4-125/4*a3^2*a5
       -100/3*a3*a4^2-875/12*a3*a5^2-20/9*a1*a2*a4-50/9*a1*a2*a5
       -20/3*a1*a3*a4-25/2*a1*a3*a5-200/9*a1*a4*a5-20*a2*a3*a4
       -100/3*a2*a3*a5-500/9*a2*a4*a5-100*a3*a4*a5-625/12*a5^3
       +5/36*a1^3-700/9*a4^2*a5:
b17 := ...  (40 lines)
b19 := ...  (156 lines)
b111:= ...  (507 lines)


The perturbation method:
v3  := -3/4*a3-2*a4-15/4*a5+1/4*a1:
v5  := -5*a3^2*a2+5/12*a3*a1^2-875/12*a3*a5^2-25/3*a5*a2^2
       -25/36*a5*a1^2+5/9*a2*a1^2-5/3*a3*a2^2-40/9*a4*a2^2
       -20*a4*a3^2+5/9*a2^2*a1-625/36*a5^2*a1-125/3*a5^2*a2
       -160/9*a4^2*a2-100/3*a4^2*a3-700/9*a4^2*a5-125/4*a3^2*a5
       -5/4*a3^2*a1-1000/9*a4*a5^2-20/3*a4^2*a1-160/9*a4*a3^3
       -500/9*a4*a5*a2-15/4*a3^3-625/12*a5^3-200/9*a4*a5*a1
       -50/9*a5*a2*a1-25/2*a3*a5*a1-100/3*a3*a5*a2+5/36*a1^3
       -20/9*a4*a2*a1-20*a4*a3*a2-20/3*a4*a3*a1-100*a4*a3*a5
       +1/8*a1+1/4*a2+7/8*a3+3/2*a4+5/8*a5:
v7  := ...  (40 lines)
v9  := ...  (156 lines)
v11:= ...  (501 lines)


The singular point value method:
mu1:= Complex(-1/2)*(a1-3*a3-8*a4-15*a5):
mu2:= Complex(1/96)*(81*a1-48*a2-1128*a4-483*a3-1695*a5
       -768*a2^2*a3-2048*a2^2*a4-3840*a2^2*a5-8000*a5^2*a1
       -33600*a5^2*a3-51200*a5^2*a4-15360*a4^2*a3-35840*a4^2*a5
       -8192*a4^2*a2+192*a1^2*a3-320*a1^2*a5-576*a3^2*a1
       -9216*a3^2*a4-14400*a3^2*a5+256*a2^2*a1-3072*a4^2*a1
       -9216*a3*a2*a4-46080*a3*a4*a5-25600*a4*a2*a5-2560*a1*a2*a5
```

```
        -19200*a5^2*a2+64*a1^3-1728*a3^3-8192*a4^3-24000*a5^3
        -10240*a1*a4*a5-3072*a1*a4*a3-5760*a1*a3*a5-1024*a1*a2*a4
        -15360*a3*a2*a5+256*a1^2*a2-2304*a3^2*a2):
mu3:= ...  (40 lines)
mu4:= ...  (162 lines)
mu5:= ...  (542 lines)
```

Similarly to the first example, the first-order focus values are

$$b_{13} = v_3 = \frac{i}{2}\mu_1,$$

which again indicates that the three different methods give the same first-order focus value (at most by a difference of a constant factor). For the second-order focus values, we have $b_{15} = v_5$, while μ_2 is not equal to v_5 by a difference of a constant factor. Similarly, for the third-order focus values, $b_{17} \neq v_7$. Setting $b_{13} = v_3 = \mu_1 = 0$ yields

$$a_1 = 3a_3 + 8a_4 + 15a_5,$$

and

$$b_{15} = v_5 = \frac{i}{2}\mu_2.$$

Further letting $b_{15} = v_5 = \mu_2 = 0$ results in

$$a_2 = -5a_3 - 10a_4 - 10a_5,$$

and

$$b_{17} = v_7 = \frac{i}{2}\mu_3.$$

Continuing this process yields

$$b_{1(2i+1)} = v_{2i+1} = \mu_i = 0, \quad i = 1, \ldots, k-1$$

$$\implies \quad b_{1(2k+1)} = v_{2k+1} = \frac{i}{2}\mu_k.$$

This example again shows that the three different methods have the same computational complexity.

3.4.4 A Numerical Example with 10 Small Limit Cycles

To end this section, we present a numerical example with 10 small limit cycles. We have the following focus values:

$$v_1 = -(a_0 + a_1 + a_2 + a_3 + a_4 + a_5),$$

$$v_3 = \frac{1}{4}(a_1 - 3a_3 - 8a_4 - 15a_5) \quad \text{when } v_1 = 0,$$

$$v_5 = \frac{1}{4}(a_2 + 5a_3 + 10a_4 + 10a_5) \quad \text{when } v_1 = v_3 = 0,$$

$$v_7 = -\frac{5}{16}(a_3 - 14a_5) \quad \text{when } v_1 = v_3 = v_5 = 0,$$

$$v_9 = -\frac{7}{16}(a_4 + 9a_5) \quad \text{when } v_1 = v_3 = v_5 = v_7 = 0,$$

$$v_{11} = \frac{21}{32}a_5 \quad \text{when } v_1 = v_3 = v_5 = v_7 = v_9 = 0.$$

Let $a_5 = 0.002$, and select the other parameters as

$$a_5 = 0.002,$$
$$a_4 = -9a_5 + \varepsilon_1,$$
$$a_3 = 14a_5 - \varepsilon_2,$$
$$a_2 = -5a_3 - 10a_4 - 10a_5 - \varepsilon_3,$$
$$a_1 = 3a_3 + 8a_4 + 15a_5 + \varepsilon_4,$$
$$a_0 = -a_1 - a_2 - a_3 - a_4 - a_5 + \varepsilon_5,$$

where the perturbations are chosen as

$$\varepsilon_1 = 0.1 \times 10^{-4}, \qquad \varepsilon_2 = 0.1 \times 10^{-7}, \qquad \varepsilon_3 = 0.1 \times 10^{-11},$$
$$\varepsilon_4 = 0.1 \times 10^{-16}, \qquad \varepsilon_5 = 0.1 \times 10^{-23}.$$

Then we have the perturbed focus values

$$v_{11} = 0.1309218862148051688411685519071 \times 10^{-2},$$
$$v_9 = -0.4372966689849981673958052390216 \times 10^{-5},$$
$$v_7 = 0.3124859231946314547369863240 7 \times 10^{-8},$$
$$v_5 = -0.2499987485848764534 6314 \times 10^{-12},$$
$$v_3 = 0.25 \times 10^{-17},$$
$$v_1 = -0.1 \times 10^{-23},$$

and the perturbed parameters

$$a_0 = -0.001990009999000009999999,$$
$$a_1 = -0.02992002999999999,$$
$$a_2 = 0.019900049999,$$

$$a_3 = 0.02799999,$$

$$a_4 = -0.01799,$$

$$a_5 = 0.002,$$

under which the approximate amplitudes of the 5 small limit cycles are obtained as

$$r_1 = 0.000646, \qquad r_2 = 0.003344, \qquad r_3 = 0.008842,$$

$$r_4 = 0.029766, \qquad r_5 = 0.048554.$$

The simulated phase portrait for this example is shown in Fig. 3.2, where 3 large limit cycles, which enclose all the 10 small limit cycles, are also obtained. It should be pointed out that the large limit cycles can be numerically simulated, while the small limit cycles cannot be obtained by numerical simulation. The existence of the small limit cycles must be proved theoretically.

Remark 3.7 Here we use exact perturbations, giving 5 positive roots of the amplitude equation in the normal forms, to show the existence of 5 limit cycles. More generally, the relation between the focus values and the number of limit cycles is discussed in the next chapter (see Theorems 4.1 and 4.2).

3.5 Remarks on the Comparison of Different Methods

In the previous section, we used different examples to compare the different methods for computing the focus values. It has been shown that the three typical methods have the same order of computational complexity, i.e., no method is superior to the others.

In this section, we propose the concept of the "minimal singular point value" based on the definition of a singular point value. We use an example to illustrate this concept and explain why no method among the three is superior to the others. This generalization is similar to the generalization of normal form theory to unique normal form theory, which is also called *minimal normal form theory* or *simplest normal form theory* (e.g., see [11, 13, 219, 247]).

In order to illustrate the concept, we re-investigate the third example considered in the previous section. Recall that the basic idea in computing the singular point value is to simplify the computation of the focus values. Let us look at the relation between the singular point values and the focus point values, given by (3.27):

$$v_{2k+1}(2\pi) = i\pi \left(\mu_k + \sum_{j=1}^{k-1} \xi_k^{(j)} \mu_j \right), \qquad k = 1, 2, \ldots,$$

where $\xi_k^{(j)}$, $j = 1, 2, \ldots, k-1$, are polynomial functions. The main question here is how to choose the polynomial functions $\xi_k^{(j)}$ so that μ_k becomes the simplest. In

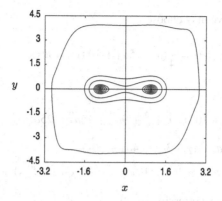

Fig. 3.2 Simulated phase portrait of Liénard equation (3.41) showing 10 small limit cycles around the weak focus points $(\pm 1, 0)$ and 3 large limit cycles under the perturbed parameter values $b = 1$, $a_0 = -0.001990009999000009999999$, $a_1 = -0.02992002999999999$, $a_2 = 0.019900049999$, $a_3 = 0.02799999$, $a_4 = -0.01799$, $a_5 = 0.002$.

other words, when we are trying to find μ_k, we should use the previous information of $\xi_k^{(j)}$ as much as possible.

Definition 3.8 The kth-order minimal singular point value, $\overline{\mu}_k$, is the one such that the expression of μ_k in (3.27) becomes the simplest under proper choices of the polynomial functions $\xi_k^{(j)}$, and then $\overline{\mu}_k = \mu_k$.

For the singular point values given for the Liénard system (see the μ_j's given in the previous section), let us check whether these singular point values are the simplest. In other words, we want to see if we can further simplify the expressions of the singular point values. For simplicity, let $\overline{\mu}_j = \frac{1}{2}\mu_j$. Then, (3.27) becomes

$$v_{2k+1} = \overline{\mu}_k + \sum_{j=1}^{k-1} \xi_k^{(j)} \overline{\mu}_j, \quad k = 1, 2, \ldots. \tag{3.44}$$

Now consider the focus values v_j. It has been shown that

$$v_3 = \overline{\mu}_1 = \frac{1}{4}(a_1 - 3a_3 - 8a_4 - 15a_5). \tag{3.45}$$

Then, we obtain

$$\begin{aligned}
v_5 = {} & \frac{1}{4}(a_2 + 5a_3 + 10a_4 + 10a_5) + \frac{1}{72}(a_1 - 3a_3 - 8a_4 - 15a_5) \\
& \times \left[9 + 10(a_1^2 + 4a_2^2 + 9a_3^2 + 16a_4^2 + 25a_5^2) \right. \\
& + 20a_1(2a_2 + 3a_3 + 4a_4 + 5a_5) \\
& \left. + 40a_2(3a_3 + 4a_4 + 5a_5) + 60a_3(4a_4 + 5a_5) + 400a_4a_5 \right],
\end{aligned} \tag{3.46}$$

which indicates that we may choose

$$\overline{\mu}_2 = \frac{1}{4}(a_2 + 5a_3 + 10a_4 + 10a_5), \tag{3.47}$$

and

$$\begin{aligned}
\xi_1^{(1)} = \frac{1}{18}\big[9 + 10\big(a_1^2 + 4a_2^2 + 9a_3^2 + 16a_4^2 + 25a_5^2\big) \\
+ 20a_1(2a_2 + 3a_3 + 4a_4 + 5a_5) \\
+ 40a_2(3a_3 + 4a_4 + 5a_5) + 60a_3(4a_4 + 5a_5) + 400a_4a_5\big]. \tag{3.48}
\end{aligned}$$

Thus, we have a simple expression for v_5:

$$v_5 = \overline{\mu}_2 + \xi_1^{(1)}\overline{\mu}_1. \tag{3.49}$$

For v_7, after a lengthy manipulation, we similarly have

$$\begin{aligned}
v_7 = {}&-\frac{5}{16}(a_3 - 14a_5) + \frac{1}{82944}(a_1 - 3a_3 - 8a_4 - 15a_5) \\
&\times \big\{(a_1 - 3a_3 - 8a_4 - 15a_5) \\
&\times \big[(a_1 - 15a_5 - 3a_3 - 8a_4) \\
&\times \big(7540(a_1 - 3a_3 - 8a_4 - 15a_5)(a_1 - 19a_3 - 40a_4 - 15a_5) \\
&\quad + 723840a_3^2 + 2895360a_3a_4 + 24489 + 2895360a_4^2\big) \\
&+ 160(a_2 + 5a_3 + 10a_4 + 10a_5) \\
&\times \big(377(a_1 - 3a_3 - 8a_4 - 15a_5)(a_1 - 15a_3 - 32a_4 - 15a_5) \\
&+ 377(a_2 + 5a_3 + 10a_4 + 10a_5) \\
&\times (3a_1 + 4a_2 - 13a_3 - 32a_4 - 5a_5) \\
&+ 477 + 18096a_3^2 + 72384a_4^2 + 72384a_3a_4\big) \\
&- \big(55296a_3 - 36864a_4 - 737280a_5 + 11581440a_4a_3^2 \\
&+ 23162880a_4^2a_3 + 15441920a_4^3 + 1930240a_3^3\big)\big] \\
&+ 16(a_2 + 5a_3 + 10a_4 + 10a_5) \\
&\times \big[(a_2 + 5a_3 + 10a_4 + 10a_5) \\
&\times \big(7540(a_2 + 5a_3 + 10a_4 + 10a_5)(a_2 - 3a_3 - 6a_4 + 10a_5) \\
&+ 5643 + 180960a_3^2 + 723840a_3a_4 + 723840a_4^2\big) \\
&- \big(7452a_3 - 3528a_4 - 92160a_5 + 1930240a_4^3 + 241280a_3^3 \\
&+ 1447680a_4a_3^2 + 2895360a_4^2a_3\big)\big]
\end{aligned}$$

$$+ \left(16605 + 61767680a_4^3 a_3 - 1080576a_4 a_3 + 46325760a_4^2 a_3^2 \right.$$

$$- 5898240a_4 a_5 + 15441920a_4 a_3^3 - 2949120a_3 a_5 - 1670400a_4^2$$

$$\left. - 122688a_3^2 + 1930240a_3^4 + 30883840a_4^4 \right) \Big\}$$

$$+ \frac{1}{576}(a_2 + 5a_3 + 10a_4 + 10a_5)$$

$$\times \left[328(a_2 + 5a_3 + 10a_4 + 10a_5)(a_2 + a_3 + 2a_4 + 10a_5) \right.$$

$$\left. + 81 + 1312a_3^2 + 5248a_3 a_4 + 5248a_4^2 \right]$$

$$\equiv \overline{\mu}_3 + \xi_2^{(1)} \overline{\mu}_1 + \xi_2^{(2)} \overline{\mu}_2, \tag{3.50}$$

where $\overline{\mu}_3 = -\frac{5}{16}(a_3 - 14a_5)$.

Similarly, we can find $\overline{\mu}_4$ and $\overline{\mu}_5$. In summary, we have

$$\overline{\mu}_1 = \frac{1}{4}(a_1 - 3a_3 - 8a_4 - 15a_5),$$

$$\overline{\mu}_2 = \frac{1}{4}(a_2 + 5a_3 + 10a_4 + 10a_5),$$

$$\overline{\mu}_3 = -\frac{5}{16}(a_3 - 14a_5), \tag{3.51}$$

$$\overline{\mu}_4 = -\frac{7}{16}(a_4 + 9a_5),$$

$$\overline{\mu}_5 = \frac{21}{32}a_5.$$

The above singular point values are much simpler than those obtained by using the singular point value method, given in the previous section. These simple values are merely very simple linear functions of the coefficients, and believed to be the minimal or the simplest singular point values. In fact, note that these simpler singular point values are exactly the ones obtained in the previous section for the third example, i.e., $\overline{\mu}_k = v_{2k+1}$ when $v_i, i = 1, 3, \ldots, 2k - 1$, is set to zero. This is certainly not surprising for this example, since each solution from $v_i = 0$ is just a simple linear expression.

However, no method has been developed so far for computing the minimal singular point values for general nonlinear dynamical systems. This is an important research topic for future study in nonlinear dynamical systems, and further research work is needed in developing efficient methods for computing focus values.

Chapter 4
Application (I)—Hilbert's 16th Problem

In this chapter, Hopf bifurcation and the computation of normal forms, developed in previous chapters, are applied to consider planar vector fields and focused on the well-known Hilbert 16th problem. General cubic- and higher-order systems are considered to find the maximal number of limit cycles that such systems can have, i.e., to find the lower bound of the Hilbert number for certain vector fields. The Liénard system is also investigated. In addition, the critical periods of bifurcating periodic solutions from two special types of planar system are studied.

In 1900 Hilbert proposed the well-known 23 mathematical problems [99], which had a significant impact on mathematics in the 20th century. One of the two unsolved problems is the 16th problem, which includes two parts. The second part of the problem considers the upper bound of the number of limit cycles and their relative locations in polynomial vector fields. This was recently chosen by Smale [191] as one of the 18 most challenging mathematical problems for the 21st century. Although the problem is still far from being completely solved, research on it has made great progress with significant contributions to the development of modern mathematics. The recent developments of Hilbert's 16th problem have been summarized in the survey articles [128, 229].

To state Hilbert's 16th problem more precisely, consider the planar vector field, described by the polynomial differential equations

$$\dot{x} = P_n(x, y), \qquad \dot{y} = Q_n(x, y), \tag{4.1}$$

where $P_n(x, y)$ and $Q_n(x, y)$ denote nth-degree polynomials in x and y. The second part of Hilbert's 16th problem is to find the upper bound on the number of limit cycles that the system can have, which is denoted by $H(n)$, known as the Hilbert number. In general, this is a very difficult problem, and it is not known whether $H(n)$ is finite. Of course, $H(1) = 0$, but the finiteness of $H(n)$ with $n \geq 2$ is still an open problem. Although it has not been possible to obtain a uniform upper bound for $H(n)$, a great deal of effort has been made in finding the maximal number of limit cycles and raising the lower bound of the Hilbert number, $H(n)$, for general planar polynomial systems or for some specific system of a certain degree, hoping to be close to the real upper bound of $H(n)$.

M. Han, P. Yu, *Normal Forms, Melnikov Functions and Bifurcations of Limit Cycles*,
Applied Mathematical Sciences 181,
DOI 10.1007/978-1-4471-2918-9_4, © Springer-Verlag London Limited 2012

One direction of research on Hilbert's 16th problem is to study the weakened Hilbert 16th problem, introduced by Arnold [6]. The weakened problem is also called the *tangential* or *infinitesimal* Hilbert 16th problem. The basic idea of the weakened problem is to consider perturbed Hamiltonian systems so that the issue of finding the number of limit cycles is transformed into finding the roots of Abelian integrals. The weakened Hilbert 16th problem with Abelian integrals and Melnikov functions will be discussed in more detail in the second part of this book (see Chap. 6).

If Hilbert's 16th problem is restricted to a neighborhood of isolated fixed points, then the problem becomes the study of degenerate Hopf bifurcations. This gives rise to the computation of focus values, which is equivalent to computing the normal form of differential equations associated with Hopf or degenerate Hopf bifurcations. The basic idea of using normal forms to consider limit cycles is as follows. Suppose the origin of system (4.1), $(x, y) = (0, 0)$, is a fixed point of the system, at which the eigenvalues of the Jacobian of the system are a purely imaginary pair, $\pm i\omega_c$. Then Hopf bifurcation occurs and a family of limit cycles bifurcates from a critical point. Assume that the associated normal form of the system is given in polar coordinates as

$$\dot{r} = r\left(v_0 + v_1 r^2 + v_2 r^4 + \cdots + v_{2k} r^{2k}\right), \tag{4.2}$$

$$\dot{\theta} = 1 + \tau_0 + \tau_1 r^2 + \tau_2 r^4 + \cdots + \tau_{2k} r^{2k}. \tag{4.3}$$

(See (3.15) and (3.16). Here, for convenience, we slightly change the notation for the focus values v_k.) To find k small limit cycles around the origin, first find the conditions such that

$$v_0 = v_1 = v_2 = \cdots = v_{k-1} = 0 \quad \text{but} \quad v_k \neq 0,$$

and then perform appropriate small perturbations to prove the existence of k limit cycles. For quadratic planar systems, in 1952, Bautin [14] proved that the maximal number of small limit cycles is 3. Globally, it was shown [38, 189] that $H(2) \geq 4$ with $(3, 1)$ distribution: 3 small limit cycles around a focus point and 1 large limit cycle around another focus point. In the past few years, great progress has been achieved in obtaining better estimations of the lower bounds of $H(n)$ for $n \geq 3$. For $n = 3$, it has been shown that cubic systems with Z_2 symmetry can have 12 small limit cycles [222, 237–239] which are distributed in the neighborhood of the two points. Recently, it has been proved that such a Z_2-equivariant system can have a 13th limit cycle at infinity [147]. Also, another cubic system with a different structure has been reported to exhibit 13 limit cycles [137, 215]. Hence, $H(3) \geq 13$. Many more results for systems with $n \geq 4$ have also been obtained (e.g., see [148] and references therein). For example, in 2004, Zhang et al. [257] proved that $H(4) \geq 15$ by perturbing a cubic-order Hamiltonian vector field with fourth-degree polynomial functions. For $n = 5$, several results have been reported, all of which are based on the study of Z_q-equivariant vector fields. In 2001, Li et al. [130] proved that fifth-order planar vector fields with Z_3 symmetry could have 23 limit cycles, by using the detection function method [133]. In 2002, the same authors [131] showed that fifth-order planar vector fields with Z_6 symmetry could have 24 limit cycles. The 29 limit cycles, found by Chen et al. [40] for fifth-order planar vector fields with Z_2 symmetry, were recently shown to be erroneous [141, 200]. It has also been shown recently that a fifth-order planar vector field with Z_5 symmetry can have 25 small limit cycles [216]. Therefore, the best result obtained so far for $n = 5$ is $H(5) \geq 25$. For $n = 6$, Wang and Yu [203] combined normal form theory with the detection function method to show that $H(6) \geq 35$ for a Z_2-equivariant vector field of degree 5 with sixth-degree polynomial perturbation. For $n = 7$, Li and Zhang [138] used the same detection function method to show that $H(7) \geq 49$ by considering a Z_8-equivariant vector field of degree 7. The result for $n = 9$ is $H(9) \geq 80$, obtained by Wang et al. [205], and that for $n = 11$ is $H(11) \geq 121$, proved by Wang and Yu [204].

Before we consider the number of limit cycles, we present two theorems which can be used to prove the existence of small limit cycles [237–239]. Without loss of generality, we may assume $v_k > 0$. Then, solving for r^2 from (4.2) such that $\frac{dr}{dt} = 0$ is equivalent to finding the roots of the equation

$$y^k + c_{k-1}y^{k-1} + \cdots + c_1 y + c_0 = 0, \tag{4.4}$$

where $y = r^2$ and $c_{i-1} = \frac{v_{i-1}}{v_k}$ for $i = 1, 2, \ldots, k$. Since we are interested in small amplitude solutions, we may assume that the roots can be put in the form $y = r^2 = O(\varepsilon)(0 < \varepsilon \ll 1)$. The following two theorems provide sufficient conditions for the existence of small limit cycles.

Theorem 4.1 *Let* $\beta_0, \beta_1, \ldots, \beta_{k-1}$, *be constants such that the equation*

$$w^k + \beta_{k-1}w^{k-1} + \cdots + \beta_1 w + \beta_0 = 0, \tag{4.5}$$

has k simple positive roots w_i, $i = 1, 2, \ldots, k$. Then for any continuous functions c_i satisfying

$$c_i(\varepsilon) = \beta_i \varepsilon^{k-i} + o(\varepsilon^{k-i}), \quad i = 0, 1, \ldots, k - 1, \tag{4.6}$$

equation (4.4) has exactly k simple positive roots of the form $y_i = \varepsilon w_i + o(\varepsilon)$ for sufficiently small $\varepsilon > 0$. Therefore, if

$$v_i = \beta_i \varepsilon^{k-i} + o(\varepsilon^{k-i}), \quad i = 0, 1, \ldots, k, \tag{4.7}$$

with $\beta_k \neq 0$ (where small o denotes higher-order terms), then (4.4) has exactly k real positive roots (i.e., system (4.1) has exactly k limit cycles) in a neighborhood of the origin for sufficiently small $\varepsilon > 0$.

Proof Let $y = \varepsilon w$ and then substitute (4.6) into (4.4) to obtain

$$\varepsilon^k (w^k + \beta_{k-1}w^{k-1} + \cdots + \beta_1 w + \beta_0) + o(\varepsilon^k) = 0,$$

or

$$w^k + \beta_{k-1}w^{k-1} + \cdots + \beta_1 w + \beta_0 + o(1) = 0, \tag{4.8}$$

where $\lim_{\varepsilon \to 0} o(1) = 0$. Then the conclusion follows by the implicit function theorem. $\square$

Note that the roots w_j and the coefficients β_j are mutually determined. However, in many cases, v_j depends on k parameters:

$$v_j = c_j(\varepsilon_1, \varepsilon_2, \ldots, \varepsilon_k), \quad j = 0, 1, \ldots, k. \tag{4.9}$$

In this case, the following theorem is more convenient in applications.

Theorem 4.2 *Suppose that condition* (4.9) *holds, and further assume that*

$$
\begin{aligned}
&c_k(0) \neq 0, \\
&c_j(0) = 0, \quad j = 0, 1, \ldots, k-1, \quad and \\
&\det\left[\frac{\partial(c_0, c_1, \ldots, c_{k-1})}{\partial(\varepsilon_1, \varepsilon_2, \ldots, \varepsilon_k)}(0)\right] \neq 0.
\end{aligned}
\tag{4.10}
$$

Then for any given $\varepsilon_0 > 0$, there exist $\varepsilon_1, \varepsilon_2, \ldots, \varepsilon_k$, and $\delta > 0$ with $|\varepsilon_j| < \varepsilon_0$, $j = 1, 2, \ldots, k$, such that (4.4) *has exactly k real positive roots (i.e., system* (4.7) *has exactly k limit cycles) in a δ-ball with center at the origin.*

Proof By the conditions given in (4.10), we can take $c_0, c_1, \ldots, c_{k-1}$, as new parameters. Hence, it is evident that (4.4) has k small real positive roots if

$$c_i(\varepsilon)c_{i+1}(\varepsilon) < 0 \quad \text{and} \quad |c_i| \ll |c_{i+1}| \ll 1, \tag{4.11}$$

for $i = 0, 1, \ldots, k-2$. $\square$

Remark 4.3 The stability of the bifurcating limit cycles can be, in general, determined from the sign of the focus values. For example, if $v_0 < 0$ ($v_0 > 0$), then the singular point is stable (unstable), and so the limit cycle closest to the singular point is unstable (stable), respectively, and so on.

In the following sections, we will pay particular attention to vector fields with Z_q symmetry for $q = 1, 2, \ldots$. For convenience, we first consider Z_q-equivariant vector fields.

4.1 Z_q-Equivariant Planar Vector Fields

Since many practical systems possess a certain degree of symmetry, studying the behavior or characteristics of such dynamical systems is not only theoretically significant, but also important in applications. On the other hand, although studying a system with symmetry is simpler than general ones, investigating such symmetric systems is also very interesting from the viewpoint of mathematics.

In this section, for convenience, we present some existing results for planar vector fields to be Z_q-equivariant, which will be needed in following sections. For their proofs, see Chap. 10. More details of the results can be found in [128].

Let G be a compact Lie group of transformations acting on $\mathbf{R}^n$.

Definition 4.4 A mapping $\Phi : \mathbf{R}^n \to \mathbf{R}^n$ is called *G-equivariant* if, for all $g \in G$ and $x \in \mathbf{R}^n$, $\Phi(gx) = g\Phi(x)$. A function $H : \mathbf{R}^n \to \mathbf{R}$ is called a *G-invariant function* if, for all $x \in \mathbf{R}^n$, $H(gx) = H(x)$. If Φ is a *G*-equivariant mapping, the vector field $\dot{x} = \Phi(x)$ is called a *G*-equivariant vector field.

Definition 4.5 Let q be an integer. A group Z_q is called a cyclic group if it is generated by a planar counterclockwise rotation through $(2\pi/q)$ about the origin. The vector field induced by the cyclic group Z_q is called a Z_q-equivariant vector field.

Introducing the transformation $z = x + iy, \bar{z} = x - iy$ into system (4.1) yields

$$\dot{z} = F(z, \bar{z}), \qquad \dot{\bar{z}} = \bar{F}(z, \bar{z}), \tag{4.12}$$

where $F(z, \bar{z}) = P(u, v) + iQ(u, v), u = \frac{1}{2}(z + \bar{z})$ and $v = \frac{1}{2i}(z - \bar{z})$.

Lemma 4.6 *A vector field defined by (4.12) is Z_q-equivariant if and only if the function $F(z, \bar{z})$ has the form*

$$F(z, \bar{z}) = \sum_{\ell=1} g_\ell(|z|^2)\bar{z}^{\ell q - 1} + \sum_{\ell=0} h_\ell(|z|^2)z^{\ell q + 1}, \tag{4.13}$$

where $g_\ell(|z|^2)$ and $h_\ell(|z|^2)$ are polynomials with complex coefficients. In addition, (4.12) is a Hamiltonian system having Z_q-equivariance if and only if (4.13) holds and

$$\frac{\partial F}{\partial z} + \frac{\partial \bar{F}}{\partial \bar{z}} \equiv 0. \tag{4.14}$$

Lemma 4.7 *A Z_q-invariant function $I(z, \bar{z})$ has the form*

$$I(z, \bar{z}) = \sum_{\ell=0} g_\ell(|z|^2)z^{\ell q} + \sum_{\ell=1} h_\ell(|z|^2)\bar{z}^{\ell q}. \tag{4.15}$$

Lemma 4.8 *For planar polynomial systems of degree* 7, *all nontrivial Z_q-equivariant vector fields defined by (4.12) have the following explicit forms:*

(1) *for $q = 2$:*

$$F(z, \bar{z}) = \left(A_0 + A_1|z|^2 + A_2|z|^4 + A_3|z|^6\right)z + \left(A_4 + A_5|z|^2 + A_6|z|^4\right)z^3$$
$$+ \left(A_7 + A_8|z|^2\right)z^5 + \left(A_9 + A_{10}|z|^2 + A_{11}|z|^4 + A_{12}|z|^6\right)\bar{z}$$
$$+ \left(A_{13} + A_{14}|z|^2 + A_{15}|z|^4\right)\bar{z}^3 + \left(A_{16} + A_{17}|z|^2\right)\bar{z}^5; \tag{4.16}$$

(2) *for* $q = 3$:

$$F(z, \bar{z}) = \left(A_0 + A_1 |z|^2 + A_2 |z|^4 + A_3 |z|^6\right)z + \left(A_4 + A_5 |z|^2\right)z^4$$
$$+ A_6 z^7 + \left(A_7 + A_8 |z|^2 + A_9 |z|^4\right)\bar{z}^2$$
$$+ \left(A_{10} + A_{11} |z|^2\right)\bar{z}^5; \tag{4.17}$$

(3) *for* $q = 4$:

$$F(z, \bar{z}) = \left(A_0 + A_1 |z|^2 + A_2 |z|^4 + A_3 |z|^6\right)z + \left(A_4 + A_5 |z|^2\right)z^5$$
$$+ \left(A_6 + A_7 |z|^2 + A_8 |z|^4\right)\bar{z}^3 + A_9 \bar{z}^7; \tag{4.18}$$

(4) *for* $q = 5$:

$$F(z, \bar{z}) = \left(A_0 + A_1 |z|^2 + A_2 |z|^4 + A_3 |z|^6\right)z$$
$$+ A_4 z^6 + \left(A_5 + A_6 |z|^2\right)\bar{z}^4; \tag{4.19}$$

(5) *for* $q = 6$:

$$F(z, \bar{z}) = \left(A_0 + A_1 |z|^2 + A_2 |z|^4 + A_3 |z|^6\right)z$$
$$+ A_4 z^7 + \left(A_5 + A_6 |z|^2\right)\bar{z}^5; \tag{4.20}$$

(6) *for* $q = 7$:

$$F(z, \bar{z}) = \left(A_0 + A_1 |z|^2 + A_2 |z|^4 + A_3 |z|^6\right)z + A_4 \bar{z}^6; \tag{4.21}$$

(7) *for* $q = 8$:

$$F(z, \bar{z}) = \left(A_0 + A_1 |z|^2 + A_2 |z|^4 + A_3 |z|^6\right)z + A_4 \bar{z}^7, \tag{4.22}$$

where the A_i's are complex.

The orbits of Hamiltonian polynomial systems given in Lemma 4.6 define different families of m algebraic curves that are Z_q-equivariant. One major concern in real algebraic geometry is: what schemes of the mutual arrangement (schemes or configurations) of ovals can be realized by the family of curves of a given degree? By using some Z_q-equivariant Hamiltonian systems, one can realize various configurations of ovals for planar algebraic curves of degree m.

4.2 Third-Order Vector Fields with Z_q Symmetry

In this section, we pay particular attention to cubic-order vector fields with Z_q symmetry for $q = 1, 2, \ldots$.

4.2.1 $q = 1$

This is a trivial case. In other words, the vector field does not hold any symmetry with regard to rotation about the origin. We consider limit cycles bifurcating from the origin of system (4.1) which is an elementary center. We show that such vector fields can have at least 9 limit cycles around the origin.

We start from a very general cubic system with a fixed point at the origin, which may be written in the form

$$
\begin{aligned}
\dot{x} &= a_{10}x + a_{01}y + a_{20}x^2 + a_{11}xy + a_{02}y^2 \\
&\quad + a_{30}x^3 + a_{21}x^2y + a_{12}xy^2 + a_{03}y^3, \\
\dot{y} &= b_{10}x + b_{01}y + b_{20}x^2 + b_{11}xy + b_{02}y^2 \\
&\quad + b_{30}x^3 + b_{21}x^2y + b_{12}xy^2 + b_{03}y^3,
\end{aligned}
\tag{4.23}
$$

where the a_{ij}'s and b_{ij}'s are real constant coefficients (parameters). It is obvious that the origin $(x, y) = (0, 0)$ is a fixed point. The system has a total of 18 parameters. However, not all of them are independent. First, note that we may use a linear transformation such that system (4.23) can be rewritten as

$$
\begin{aligned}
\dot{x} &= \alpha x + \beta y + a_{20}x^2 + a_{11}xy + a_{02}y^2 \\
&\quad + a_{30}x^3 + a_{21}x^2y + a_{12}xy^2 + a_{03}y^3, \\
\dot{y} &= \pm\beta x + \alpha y + b_{20}x^2 + b_{11}xy + b_{02}y^2 \\
&\quad + b_{30}x^3 + b_{21}x^2y + b_{12}xy^2 + b_{03}y^3,
\end{aligned}
\tag{4.24}
$$

where α and $\beta > 0$ are used to represent the eigenvalues of the linearized system of (4.23). Note that the other coefficients in (4.24) should be different from those of system (4.23), but we use the same notation for convenience. Here, when the negative sign is taken, the origin is a focus point or a center (when $\alpha = 0$); otherwise, it is a saddle point or node.

Next, suppose we are interested in small limit cycles in the neighborhood of the origin. So the negative sign is taken in (4.24), and the eigenvalues are now given by $\lambda_{1,2} = \alpha \pm \beta i$. Then we can apply a time scaling, $\tau = \beta t$, into system (4.24) to obtain

$$
\begin{aligned}
\frac{dx}{d\tau} &= \alpha x + y + a_{20}x^2 + a_{11}xy + a_{02}y^2 \\
&\quad + a_{30}x^3 + a_{21}x^2y + a_{12}xy^2 + a_{03}y^3, \\
\frac{dy}{d\tau} &= -x + \alpha y + b_{20}x^2 + b_{11}xy + b_{02}y^2 \\
&\quad + b_{30}x^3 + b_{21}x^2y + b_{12}xy^2 + b_{03}y^3,
\end{aligned}
\tag{4.25}
$$

where again the same notation for the parameters is used. Henceforth we assume that the leading β has been scaled to 1, and rename $\tau = t$. Now system (4.25) has only 15 parameters. Further, by a rotation we can remove one parameter (e.g., see [14, 152]) from system (4.25), which can be written in the general form

$$
\frac{dx}{d\tau} = \alpha x + y + Ax^2 + (B + 2D)xy + Cy^2
$$
$$
+ Fx^3 + Gx^2y + (H - 3P)xy^2 + Ky^3,
$$
$$
\frac{dy}{d\tau} = -x + \alpha y + Dx^2 + (E - 2A)xy - Dy^2 \qquad (4.26)
$$
$$
+ Lx^3 + (M - H - 3F)x^2y
$$
$$
+ (N - G)xy^2 + Py^3.
$$

This form is perhaps the simplest in the literature for cubic systems associated with a Hopf singularity at the origin [152]. The system has 14 parameters. However, since the same order terms on the right-hand side of (4.26) are homogeneous, we can further remove one more parameter. Suppose $A \neq 0$ (in the case $A = 0$ one may use another nonzero parameter in the reparametrization). Let

$$
B = bA, \qquad C = cA, \qquad D = dA, \qquad E = eA, \qquad F = fA^2,
$$
$$
G = gA^2, \qquad H = hA^2, \qquad K = kA^2, \qquad L = \ell A^2, \qquad (4.27)
$$
$$
M = mA^2, \qquad N = nA^2, \qquad P = pA^2,
$$

and then apply a spatial scaling $x \to x/A$, $y \to y/A$ to system (4.26) to obtain

$$
\frac{dx}{d\tau} = \alpha x + y + x^2 + (b + 2d)xy + cy^2 + fx^3 + gx^2y
$$
$$
+ (h - 3p)xy^2 + ky^3,
$$
$$
\frac{dy}{d\tau} = -x + \alpha y + dx^2 + (e - 2)xy - dy^2 + \ell x^3 + (m - h - 3f)x^2y \qquad (4.28)
$$
$$
+ (n - g)xy^2 + py^3,
$$

which has only 13 independent parameters. It is easy to see that the zeroth-order focus value is $v_0 = \alpha$. Other focus values are given in terms of the remaining 12 parameters. Let

$$
S = \{B, c, d, e, f, g, h, k, \ell, m, n, p\}. \qquad (4.29)
$$

Then, $v_i = v_i(S)$. In general, the maximum number of small limit cycles which exist in the vicinity of the origin is not greater than the number of independent parameters. Here it is 13. In other words, the best possibility one can have may be

$$
v_i = 0, \quad i = 0, 1, \ldots, 12 \quad \text{but} \quad v_{13} \neq 0.
$$

Then according to Theorem 4.2, the maximum number of small limit cycles which may be obtained by appropriate perturbations is 13.

When the origin is an elementary center, many results on the bifurcation of limit cycles from the origin have been obtained. Wang [199] showed 6 limit cycles bifurcating from one critical point. Another such example can be found in [152]. Later, 7 limit cycles were found (e.g., see [109, 129]). In [109], 8 limit cycles were obtained. All the results were based on the symbolic computation of focus values. Recently, it was claimed that a cubic system can have 11 limit cycles around one critical point [261], which, however, is based on perturbing a center rather than on computation of focus values.

In the remainder of the section, we will first show that the simplest cubic system given in [109] which has 7 limit cycles can actually have 8 limit cycles, and then present a cubic system which has 9 limit cycles around the origin.

(A) 8 Small Limit Cycles in a Simple Cubic System   Consider the simple cubic system given by James & Lloyd [109]:

$$\frac{dx}{d\tau} = \lambda x + y + a_3 x^2 + a_4 x^3 + a_5 x^2 y - 3b_7 xy^2 + a_7 y^3,$$

$$\frac{dy}{d\tau} = -x + \lambda y + (b_4 - a_7)x^3 + (b_5 - 3a_4)x^2 y + (b_6 - a_5)xy^2 + b_7 y^3.$$
$$(4.30)$$

In [109], an extra condition, $3a_5 = -(10a_3^2 + 11a_7)$, is imposed in order to simplify computation. In fact, lifting this restriction leads to 8 limit cycles, as we will see in the following analysis. First, choose $a_3 \neq 0$ as a scaling parameter and apply the scalings

$$a_i \rightarrow A_i a_i^2, \qquad b_i \rightarrow B_i a_i^2, \qquad x \rightarrow x/a_3, \qquad y \rightarrow y/a_3, \qquad (4.31)$$

to system (4.30) to obtain

$$\frac{dx}{d\tau} = \lambda x + y + x^2 + A_4 x^3 + A_5 x^2 y - 3B_7 xy^2 + A_7 y^3,$$

$$\frac{dy}{d\tau} = -x + \lambda y + (B_4 - A_7)x^3 + (B_5 - 3A_4)x^2 y$$
$$(4.32)$$

$$+ (B_6 - A_5)xy^2 + B_7 y^3.$$

Executing the Maple program [218] (with $\lambda = 0$ and so $v_0 = 0$) yields $v_1 = \frac{1}{8}B_5$. Setting $v_1 = 0$ gives $B_5 = 0$. Then computing v_2 results in

$$v_2 = -\frac{1}{8}B_6(A_4 - B_7).$$

There are two choices satisfying $v_2 = 0$: either $B_6 = 0$ or $B_7 = A_4$. Setting $B_6 = 0$ leads to $v_2 = v_3 = \cdots = 0$, a center. So let $B_7 = A_4$, under which v_3 becomes

$$v_3 = -\frac{1}{192}A_4 B_6(-35 + 3B_6 + 15B_4).$$

For the same reason, we must choose $B_4 = \frac{1}{15}(35 - 3B_6)$, in order to have $v_3 = 0$.

Continuing the calculation shows that

$$v_4 = \frac{1}{9600} A_4 B_6 \big[5A_7(18B_6 - 385) - 1750 - 525A_5 - 140B_6$$
$$- 6B_6^2 + 30A_5 B_6 \big],$$

from which we obtain

$$A_7 = \frac{1750 + 525A_5 + 140B_6 + 6B_6^2 - 30A_5 B_6}{5(18B_6 - 385)}.$$

Having determined the values of four parameters B_5, B_7, B_4 and A_7, executing the Maple program gives v_5 from which one solves for A_4^2 to obtain

$$A_4^2 = -\frac{1}{150} \big\{ 6912 B_6^5 - 1131660 B_6^4 + 13935075 B_6^3 - 269206875 B_6^2$$
$$+ 5571820625 B_6 + 6458046875$$
$$+ 50A_5 \big[A_5 \big(864 B_6^3 - 114660 B_6^2 + 3557400 B_6 - 32413500 \big)$$
$$- 864 B_6^4 + 115470 B_6^3 - 3149265 B_6^2 + 27878550 B_6 - 62811875 \big] \big\} \quad (4.33)$$

under which $v_5 = 0$, and then v_6 and v_7 are simplified. Now eliminating A_5 from $v_6 = 0$ and $v_7 = 0$ results in a solution for A_5,

```
A5  := 1/10*(28661513482345328640*B6^18-38320168058532606394368*B6^17
       +94353181171684211719022240*B6^16
       -570016902679901015264786700*B6^15
       + ... -3603375357373566456961845717321166992187500000*B6
       +8207895046256766703594879935424804687500000000)
       /(9051638614483322880*B6^17-730911679256885383830656*B6^16
       +1311473686075172842327920*B6^15
       -7597365875181923439673950*B6^14
       + ...  +13345336332385462657759104223828125000000000*B6
       -236716563930102676013409733886718750000000000):
```

$$(4.34)$$

and a resultant equation,

$$F = (18B_6 - 385)(8B_6 - 735)F_1(B_6), \qquad (4.35)$$

where F_1 is a 25th-degree polynomial in B_6, given by

```
F1:= 25517942739795723889201643520*B6^25
     -159615571297837113096837863170824*B6^24
     + ...
     +118646656121691880083492459766914831626518249511718750000000000*B6
     -64739758475649691824510366041125989647941589355468750000000000:
```

$$(4.36)$$

Now we can employ a numerical approach to solve for the univariate polynomial F_1. For example, we may use the built-in Maple solver fsolve to find all the real roots of the polynomial.[1] The first two real solutions, $B_6 = 385/18$ and $B_6 = 735/8$, yield two centers. The remaining condition, given by $F_1 = 0$, has seven real roots. To verify these roots, for each solution B_6, in reverse order, first use (4.34) to obtain A_5, and then A_4. Substitute each solution of (B_6, A_5, A_4) to verify that $v_5 = v_6 = v_7 = 0$ while $v_i = 0$, $i = 0, 1, 2, 3, 4$, are automatically satisfied since they are solved, one by one, using one parameter at each step. It has been shown that only two roots of $F_1 = 0$ for B_6 generate solutions which satisfy $v_5 = v_6 = v_7 = 0$. These two roots are

$$B_6 = -149.8550779008186309464167627611636111242847958047490 9,$$

$$14.57222224584475007068548678094228505795767873387819. \quad (4.37)$$

Returning to the original system (4.30), these two sets of solutions are given by (say, up to 50 decimal places)

$$a_4 = 11.72677636147392252103108673156625861130540134632217a_3^2,$$

$$a_5 = -96.28308723695332837277347016717443395057677635891191a_3^2,$$

$$a_7 = 23.87070969364661297334553081391162008758996623623660a_3^2,$$

$$b_4 = 32.30434891349705952261668588556605555819029249428315a_3^2,$$

$$b_5 = 0,$$

$$b_6 = -149.85507790081863094641676276116361112428479580474909a_3^2,$$

$$b_7 = 11.72677636147392252103108673156625861130540134632217a_3^2,$$

$$(4.38)$$

and

$$a_4 = 1.16364223535095645608539458464523114309856393525890a_3^2,$$

$$a_5 = 13.51273677593795918678747785044498166551946137068013a_3^2,$$

$$a_7 = -10.18920587951907640059427939895810711880480806331355a_3^2,$$

$$b_4 = -0.58111111583561668080376402285512367825820241344231a_3^2,$$

$$b_5 = 0,$$

$$b_6 = 14.57222224584475007068548678094228505795767873387819a_3^2,$$

$$b_7 = 1.16364223535095645608539458464523114309856393525890a_3^2.$$

$$(4.39)$$

---

[1]The command `realroot`, which uses dyadic rationals and binary splitting after a method of Collins, provides reliable intervals guaranteed to contain roots. However, `fsolve` has evolved to be equally reliable.

Using the second group of parameter values above yields the focus values $v_i \approx 0$, $i = 1, 2, \ldots, 7$, and

$$v_8 = -3860.6641354711690877558972836046612442276015500374267 8a_3^{16}.$$

One can choose an appropriate value of a_3 to make v_8 smaller. For example, choosing $a_3 = \frac{1}{2}$ gives

$$v_8 = -0.0589090596843134931603377881409402655674377677923 1.$$

Thus, system (4.30) can have at most 8 small limit cycles around the origin. Further, we may apply Theorem 4.2 to prove that the system indeed has 8 limit cycles. Since the five parameters A_4, A_7, B_4, B_7 and B_5 can be used, one by one, to perturb the focus values, v_5, v_4, v_3, v_2 and v_1, we only need to verify the Jacobian matrix obtained from the equations v_6 and v_7. Evaluating this Jacobian matrix at the second group of parameter values results in

$$\det(J_c) = \det \begin{bmatrix} 374.6133477040a_3^{10} & -151.1985716213a_3^{10} \\ 6805.7187294603a_3^{12} & -2775.8427119034a_3^{12} \end{bmatrix}$$

$$= -10852.7802551324a_3^{22} \neq 0 \quad \text{(since } a_3 \neq 0\text{)}. \tag{4.40}$$

This shows that for the second group of parameter values given, with proper perturbations, system (4.30) has exactly 8 limit cycles.

Summarizing the above results gives the following theorem.

Theorem 4.9 *For the cubic system* (4.30), *when the system parameters are properly perturbed to the critical values* $\lambda = 0$, $b_5 = 0$, $b_7 = a_4$, $b_4 = \frac{1}{15}(35a_3^2 - 3b_6)$, $a_7 = (\dfrac{1750a_3^4 + 525a_5a_3^2 + 140b_6a_3^2 + 6b_6^2 - 30a_5b_6}{5(18b_6 - 385a_3^2)})$; a_4 *and* a_5 *are given by* (4.33) *and* (4.34), *respectively, through the back scaling* (4.31); *and* $B_6 = b_6/a_3^2$ *is one of the two real roots of the polynomial equation* $F_1 = 0$ (*see* (4.37)) *satisfying the Jacobian condition* (4.40); *then system* (4.30) *has exactly 8 small limit cycles around the origin.*

Before moving on to the case of 9 limit cycles, we present a numerical example with the second group of critical values given in (4.39) at $a_3 = 1/2$. We take the following perturbations:

$$\varepsilon_1 = 0.222 \times 10^{-2}, \qquad \varepsilon_2 = 0.1 \times 10^{-2}, \qquad \varepsilon_3 = 0.1 \times 10^{-10},$$

$$\varepsilon_4 = 0.1 \times 10^{-15}, \qquad \varepsilon_5 = 0.1 \times 10^{-22}, \qquad \varepsilon_6 = 0.2 \times 10^{-30},$$

$$\varepsilon_7 = 0.1 \times 10^{-37}, \qquad \varepsilon_8 = 0.4 \times 10^{-47},$$

under which the perturbed focus values are

Fig. 4.1 The phase portrait of system (4.30) having 8 limit cycles around the origin, for $\lambda = -0.4 \times 10^{-47}$, $a_3 = 0.5$, $a_4 = 0.2935258759$, $a_5 = 3.3759641940$, $a_7 = -2.5483319254$, $b_4 = -0.1454777790$, $b_5 = 0.1 \times 10^{-37}$, $b_6 = 3.6440555615$, $b_7 = 0.2935258759$

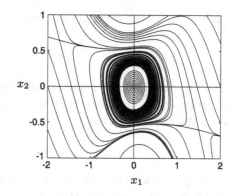

$$v_0 = -0.4 \times 10^{-47}, \qquad v_1 = 0.125 \times 10^{-38},$$

$$v_2 = -0.9110138803 \times 10^{-31}, \qquad v_3 = 0.8356442057 \times 10^{-24},$$

$$v_4 = -0.1707886684 \times 10^{-17}, \qquad v_5 = 0.5055921200 \times 10^{-12},$$

$$v_6 = -0.2408082000 \times 10^{-7}, \qquad v_7 = 0.1707722232 \times 10^{-3},$$

$$v_8 = -0.5829996790 \times 10^{-1},$$

which results in 8 positive roots of r for the amplitudes of the small limit cycles,

$$r_1 = 0.000069, \qquad r_2 = 0.000103, \qquad r_3 = 0.000364, \qquad r_4 = 0.000644,$$

$$r_5 = 0.001866, \qquad r_6 = 0.004543, \qquad r_7 = 0.011076, \qquad r_8 = 0.052743.$$

The perturbed system has three fixed points: $C_0 = (0, 0), C_1 = (1.499348, -0.602076)$ and $C_3 = (-1.726520, 0.643646)$. A linear analysis shows that C_0 is a weak stable focus point, while C_1 and C_3 are saddle points. The phase portrait of the perturbed system is shown in Fig. 4.1. It should be noted that since the origin is a high-order focus point, the dynamical behavior of the system in the vicinity of the origin is similar to that of a center. The limit cycles shown in this figure are not exact trajectories, but are used to demonstrate what they may look like.

(B) 9 Small Limit Cycles: Bifurcation From a Fixed Point So far, in the literature, the maximum number of limit cycles in the neighborhood of one singular point obtained using symbolic computation is 8 [109]. Although it was shown that 11 small limit cycles might exist around one singular point [261], the result was obtained by perturbing a center: different from the approach discussed in this paper. The difficulty in computing higher-order focus values is obvious. Also, solving coupled higher-degree multivariate polynomials is very difficult.

In the remaining part of this section, we present a cubic system which exhibits 9 small limit cycles around a single critical point. Consider the general normalized cubic system (4.28). The system has 13 independent parameters. It is easy to see that $v_0 = \alpha$, and other focus values are given in terms of the remaining 12 parameters. It has been shown that the origin is a center if $\alpha = b = e = h = n = m = 0$. We let

$$\alpha = d = e = h = 0, \tag{4.41}$$

and then compute the focus values. Here, we use the recursion formulas to compute the singular point quantities. The first-order quantity is given by $\mu_1 = \frac{1}{4}i[b(c+1) - m]$. Setting $\mu_1 = 0$ results in

$$m = b(c+1). \tag{4.42}$$

For simplicity, let $b = 0$, and so $m = 0$. Therefore, n must be nonzero, otherwise it is a center. Thus, 7 free parameters remain: c, f, g, k, ℓ, n, p.

Under the above choices of parameters, the second-order singular point quantity becomes $\mu_2 = -\frac{1}{4}in(p - f)$. In order to have $\mu_2 = 0$, the only choice is

$$p = f, \tag{4.43}$$

since $n \neq 0$. Then $\mu_3 = \frac{1}{96}ifn(45 - 30c - 35c^2 + 15\ell + 15k + 3n)$. One can choose n to set $\mu_3 = 0$, yielding

$$n = \frac{1}{3}(35c^2 + 30c - 15\ell - 15k - 45). \tag{4.44}$$

The next singular point quantity can be found as

$$\mu_4 = \frac{1}{192}ifn\big[g(7c^2 + 30c + 6k + 6\ell - 45) - 648 + 162c - 81k - 72\ell + 30k\ell$$
$$- 60c\ell - 54ck + 24k^2 + 6\ell^2 + 516c^2 - 56c^2\ell - 21c^2k + 434c^3 + 168c^4\big],$$

from which we obtain

$$g = \big(648 - 162c + 81k + 72\ell - 30k\ell + 60c\ell + 54ck - 24k^2 - 6\ell^2$$
$$- 516c^2 + 56c^2\ell + 21c^2k - 434c^3 - 168c^4\big)$$
$$/\big(7c^2 + 30c + 6k + 6\ell - 45\big). \tag{4.45}$$

Then μ_5 is given by

$$\mu_5 = -\frac{ifn}{10368(30c + 7c^2 + 6k + 6\ell - 45)^2}\bar{\mu}_5(f, c, k, \ell),$$

where $\bar{\mu}_5$ is a polynomial in f, c, k, ℓ, from which we can solve for f^2 to obtain (here £2 denotes f^2)

```
f2:=-(-54584604*c*1-39188124*c*k-27032049*c^2*1+48200751*c^2*k
    -110968380*c+19497105*k+59650425*1+1029976236*c^4
    -15231726*k^2-21311586*1^2+159042042*c^3-850276440*c^2
    -37330632*1*k-843453*1^2*k+6160698*c^2*k^2+47523483*1*c^4
    -13941396*c*1^2-9106668*c*k^2+88773246*1*c^3+173490822*k*c^3
    -3849120*c*1^3+2978208*c*k^3-2719710*c^2*1^3-6922962*c^2*k^3
    +34758558*c^4*k^2+31216050*c^4*1^2-132783063*1*c^6
    -85492743*k*c^6+477900*k*1^3+538164*k^3*1+11818224*c^3*k^2
    +44393184*c^3*1^2-46656*k^3*1^2-46656*k^2*1^3-15552*k^4*1
    -15552*1^4*k+2024946*c^6*1^2+19211742*c^2*1^2+1234197*k^2*1
    -84526605*k*c^4+42444*c^2*k^4+83592*c^2*1^4-9672012*1*c^8
    -2526636*k*c^8-284445*1^3*c^4-34587*k^3*c^4+23328*c*k^4
    +127656*c*1^4-802656*1^3*c^3-3672000*k^3*c^3+957420*k^2*1^2
    +29794716*c^5*k^2-195603336*k*c^5-105820344*1*c^5
    +11002068*c^5*1^2-58934442*1*c^7+754029*k^3+44388*k^4
    +14256*1^4-522775800*c^6-1323621*1^3+57931686*c^5
    +38574144*c^5*k*1+145800*c*1^3*k-68040*c*1^2*k^2-62856*c*1*k^3
    -7236864*1^2*c^3*k-10106208*1*c^3*k^2+7135128*c^6*k*1
    -1242675*k^2*c^4*1-1492533*k*c^4*1^2-10692*c^2*1^3*k
    -229716*c^2*k^2*1^2-92988*c^2*k^3*1+68219928*c^4*k*1
    +707616*c*k^2*1-6119712*c*k*1^2+59710608*c^3*k*1
    -13159746*c^2*k^2*1-8956494*c^2*k*1^2+21873240*c^2*1*k
    -21473424*c*1*k-105232554*c^7+154411628*c^8+12910520*c^10
    +84322630*c^9-18758754*k*c^7+5851062*c^6*k^2+239345280)
    /(24*k+24*1-495+330*c+385*c^2)/(30*c+7*c^2+6*k+6*1-45)^2/18:
```

$$(4.46)$$

Then the next three singular point quantities are

$$\mu_6 = \mu_6(c, k, \ell), \qquad \mu_7 = \mu_7(c, k, \ell), \qquad \mu_8 = \mu_8(c, k, \ell),$$

which are polynomials in c, k and ℓ, with leading (combined) degrees of 14, 22 and 26 for μ_6, μ_7 and μ_8, respectively. The Maple output files for the three polynomial equations have roughly 100, 315 and 588 lines, respectively. The three singular point quantities are coupled; we thus have to simultaneously solve the three polynomial equations, $\mu_6 = \mu_7 = \mu_8 = 0$. To find the solutions of these equations, we might first eliminate one parameter from the three equations to obtain two resultant equations, and then further eliminate one more parameter from the two equations to get a final resultant equation which is a univariate polynomial and thus we could find all possible solutions. This elimination method has been used in our other publications (e.g., see [237–239]). The method may increase the degree of the final resultant polynomial substantially.

Here, we use a numerical approach to simultaneously solve the three equations. The solution of the system of equations may be approached in several ways. A simple numerical approach, using fsolve in Maple directly, is successful in many cases. By examining the source code for fsolve, we find that this routine uses a variant of damped multivariate Newton iteration (there appears to be no published paper describing the routine), and by judicious choices of our initial guess we obtained a result (with $f^2 > 0$) after only two tries.

In general, this is not completely satisfactory, because we would like to find all real positive roots. Advanced techniques such as those described in [70] will be pursued for extensions of the work described for this case.

Finally, for system (4.28) we have used the numerical approach to obtain the critical parameter values (results from computer output using Maple give up to 1000 digits, but only 50 digits are listed here):

$$b = d = e = h = 0,$$

$$m^* = 0,$$

$$p^* = f^* = 0.57205513738312947531208972107913988960903196760780,$$

$$n^* = 1.5587366234496833705957910715181908410140471444159 8,$$

$$g^* = 0.32875002652319114041354789797311142571533487466423,$$

$$f^* = 0.57205513738312947531208972107913988960903196760780,$$

$$\ell^* = -0.28703189754662589381676063046084638581078384981160,$$

$$c^* = -0.52179929787663453432829806398960426486646485537068,$$

$$k^* = 0.027512177747988979339761212892197414525624306206 27.$$

$$(4.47)$$

Under the above critical parameter values and conditions, by taking $A = 0.3$ and the other parameter values given in (4.27), we execute the Maple program for system (4.26) to find the following focus values (again up to 1000 decimal digits, but v_9 is only given up to 50 digits here):

$$v_0 = 0, \qquad v_1 = 0.1 \times 10^{-999}, \qquad v_2 = -0.5 \times 10^{-1000},$$

$$v_3 = -0.23 \times 10^{-1000}, \qquad v_4 = 0.62 \times 10^{-999}, \qquad v_5 = -0.405 \times 10^{-999},$$

$$v_6 = 0.135 \times 10^{-998}, \qquad v_7 = 0.14 \times 10^{-998}, \qquad v_8 = 0.9 \times 10^{-999},$$

and

$$v_9 = -0.0014315972268236725754647008537844355295621179315106.$$

The above result indeed indicates that an excellent approximate solution has been obtained, and the values of $v_i, i = 0, 1, 2, \ldots, 8$, can be considered as being very close to a true zero. Thus, the maximum number of small limit cycles which can be obtained in the vicinity of the origin of system (4.28) is 9. In order to prove that the 9 small limit cycles indeed exist, we need to check the Jacobian matrix obtained from the three equations

$$v_6(\ell, c, k) = v_7(\ell, c, k) = v_8(\ell, c, k) = 0,$$

with respect to ℓ, c and k, since the other singular point quantities, $\mu_5, \mu_4, \mu_3, \mu_2$ and μ_1 can be perturbed, one after another, by the parameters, f, g, n, p and m (or b). Further, it can be shown that due to the relation between v_k and μ_k (given in (3.27)), the determinant of the Jacobian evaluated at the critical point, based on the μ_i formulas is equivalent to that based on the v_i expressions. In other words, the determinant of the Jacobian based on the former is nonzero if and only if that based on the latter is nonzero. Thus, we can use the following real equations:

$$i\mu_6(\ell, c, k) = i\mu_7(\ell, c, k) = i\mu_8(\ell, c, k) = 0,$$

to compute the determinant, where $i = \sqrt{-1}$. It should be noted that the values of these two determinants are not equal since the constant coefficients between the relations are ignored, which does not change the nonzero property of the determinant. Therefore,

$$\det(J_c) = \det \begin{bmatrix} \frac{\partial i\mu_6}{\partial \ell} & \frac{\partial i\mu_6}{\partial c} & \frac{\partial i\mu_6}{\partial k} \\ \frac{\partial i\mu_7}{\partial \ell} & \frac{\partial i\mu_7}{\partial c} & \frac{\partial i\mu_7}{\partial k} \\ \frac{\partial i\mu_8}{\partial \ell} & \frac{\partial i\mu_8}{\partial c} & \frac{\partial i\mu_8}{\partial k} \end{bmatrix}_{(\ell,c,k)=(\ell^*,c^*,k^*)}$$

$$= \det \begin{bmatrix} 0.0451319561 & 0.2439914595 & 0.0220645269 \\ 0.1223727073 & -0.0022077660 & 0.1580040454 \\ 0.0055179038 & 0.6737398583 & -0.0271869307 \end{bmatrix}$$

$$= -0.0019578515 \neq 0. \tag{4.48}$$

Then, according to Theorem 4.2 we know that 9 small amplitude limit cycles bifurcating from the origin of system (4.28) can be obtained by properly perturbing the critical values given in (4.47). The above results are summarized in the following theorem.

Theorem 4.10 *For the cubic system* (4.28), *suppose* $b = d = e = h = 0$. *When the remaining parameters are properly perturbed to the critical values:* $\alpha^* = 0$, $m^* = 0$, $p^* = f$, $n^* = \frac{1}{3}(35c^2 + 30c - 15\ell - 15k - 45)$, g^*, f^{*2} *is given by* (4.46), *while* ℓ^*, c^* *and* k^* *are given in* (4.47) *satisfying the Jacobian condition* (4.48), *then system* (4.28) *has exactly 9 small limit cycles around the origin.*

4.2.2 $q = 2$

When $q = 2$, we will show that such a vector field can have 12 small limit cycles.

To begin with, we take the complex form of a general Z_2-equivariant vector field, described by $\dot{z} = F_2(z, \bar{z})$ with $F_2(x, \bar{z})$ given by (4.16):

$$F_2(z, \bar{z}) = \left(A_0 + A_1|z|^2 + A_2|z|^4\right)z + \left(A_3 + A_4|z|^2 + A_5|z|^4\right)\bar{z}$$
$$+ \left(A_6 + A_7|z|^2\right)z^3 + \left(A_8 + A_9|z|^2\right)\bar{z}^3. \tag{4.49}$$

Here, $A_j = a_j + ib_j$ are complex, which, transformed into real form and truncated to third order, are:

$$\dot{w}_1 = (a_0 + a_3)w_1 - (b_0 - b_3)w_2 + (a_1 + a_4 + a_6 + a_8)w_1^3$$

$$- (b_1 - b_4 + 3b_6 - 3b_8)w_1^2 w_2 + (a_1 + a_4 - 3a_6 - 3a_8)w_1 w_2^2$$

$$- (b_1 - b_4 - b_6 + b_8)w_2^3,$$

$$\dot{w}_2 = (b_0 + b_3)w_1 + (a_0 - a_3)w_2 + (b_1 + b_4 + b_6 + b_8)w_1^3 \tag{4.50}$$

$$+ (a_1 - a_4 + 3a_6 - 3a_8)w_1^2 w_2 + (b_1 + b_4 - 3b_6 - 3b_8)w_1 w_2^2$$

$$+ (a_1 - a_4 - a_6 + a_8)w_2^3.$$

The two eigenvalues of the Jacobian of (4.50) evaluated at the origin, $(w_1, w_2) = (0, 0)$, are $a_0 \pm \sqrt{d_3^2 + b_3^2 - b_0^2}$, indicating that the origin is a saddle point or a node when $a_3^2 + b_3^2 - b_0^2 \geq 0$, or a focus point or a center when $a_3^2 + b_3^2 - b_0^2 < 0$. When $a_3^2 + b_3^2 - b_0^2 = 0$, the origin is either a node or a double-zero singular point.

(A) Case $a_3^2 + b_3^2 - b_0^2 \geq 0$ First, we consider $a_3^2 + b_3^2 - b_0^2 \geq 0$, under which condition the origin is a saddle point or a node. For this case, a linear transformation given by

$$\begin{pmatrix} w_1 \\ w_2 \end{pmatrix} = \begin{bmatrix} 0 & b_0 - b_3 \\ \mp\sqrt{d_3^2 + b_3^2 - b_0^2} & a_3 \end{bmatrix} \begin{pmatrix} x \\ y \end{pmatrix}, \tag{4.51}$$

can be introduced into (4.50), and the coefficients of the third-order terms in the resulting equations renamed $a_{30}, a_{21}, a_{12}, a_{03}, b_{30}, b_{21}, b_{12}$ and b_{03} to obtain

$$\dot{x} = ax + by + a_{30}x^3 + a_{21}x^2 y + a_{12}xy^2 + a_{03}y^3,$$
$$\dot{y} = bx + ay + b_{30}x^3 + b_{21}x^2 y + b_{12}xy^2 + b_{03}y^3, \tag{4.52}$$

where $a = a_0$ and $b = \pm\sqrt{d_3^2 + b_3^2 - b_0^2}$. Note that system (4.50) has 12 parameters: $a_0, a_1, a_3, a_4, a_6, a_8, b_0, b_1, b_3, b_4, b_6, b_8$, while system (4.52) has only 10 parameters: $a, b, a_{30}, a_{21}, a_{12}, a_{03}, b_{30}, b_{21}, b_{12}, b_{03}$. This is because system (4.52) has put its linear terms, via the linear transformation (4.51), in the form of two independent parameters since the vector field of a planar system in the neighborhood of a fixed point is completely determined by two independent eigenvectors associated with two eigenvalues.

It is observed from (4.52) that the system is invariant under rotation about the origin with angle π. To study the existence of small limit cycles which the system may exhibit, one needs weak focus points to generate degenerate Hopf bifurcations. For a vector field with Z_2 symmetry, naturally the best way is to have two symmetric focus points about the origin. Thus, if N small limit cycles are found in the neighborhood of one focus point, the whole system would have $2N$ limit cycles. This may put some additional constraints on the coefficients as it requires the two symmetric focus points to be Hopf-type singular points. However, it will be shown later that this still gives the maximal number of small limit cycles. If we consider one Hopf-type focus point at the origin, it may have a couple of extra free coefficients, but the total number of the limit cycles would be less than that in the case of the two symmetric focus points. This will be discussed in the next subsection for $a_3^2 + b_3^2 - b_0^2 < 0$ under which condition the origin may be a center.

It is easy to see that the Jacobian of system (4.52) evaluated at the origin $(x, y) = (0, 0)$, has eigenvalues $a \pm |b|$. When $|b| > |a|$, the origin is a saddle point, while if $|a| > |b|$, the origin is a node. Further, without loss of generality, assume that

$$a_{03} = -b, \qquad b_{03} = -a \quad \text{and} \quad a_{12} = a, \tag{4.53}$$

under which system (4.52) is symmetric about the origin and has two Hopf-type focus points at $(x, y) = (0, 1)$ and $(0, -1)$. Under the above conditions, system (4.52) can exhibit 10 small limit cycles when the origin is a saddle point [84], 5 of which are located in the neighborhood of the point $(0, 1)$ and the other 5 surround the point $(0, -1)$. Recently, it has been further shown that system (4.52) has 12 small limit cycles for a special case [237] as well as in generic cases [237–239].

Note that under the first two conditions of (4.53), the eigenvalues of the system evaluated at the two focus points $(0, \pm 1)$ are given by

$$\lambda_{\pm} = \frac{1}{2}\left[(a_{12} - a) \pm \sqrt{(a_{12} - a)^2 - 8(b^2 - a^2 + bb_{12} - aa_{12})}\right],$$

from which the real part of the eigenvalues is obtained as

$$v_0 = \frac{1}{2}(a_{12} - a) \tag{4.54}$$

if $(a_{12} - a)^2 < 8(b^2 - a^2 + bb_{12} - aa_{12})$. Furthermore, under the condition $a_{12} = a$, the two eigenvalues become a purely imaginary pair,

$$\lambda_{\pm} = \pm i\sqrt{-2(b^2 - 2a^2 + bb_{12})} \quad \text{for } b(b + b_{12}) > 2a^2.$$

Under the condition (4.53), if we further assume

$$b_{12} = \frac{1 + 4a^2}{2b} - b, \tag{4.55}$$

then the purely imaginary pair is normalized to $\lambda_{\pm} = \pm i$ with the frequency of limit cycles being 1. This seems that we have used a free parameter. Actually, this is not necessary. In other words, if we assume the frequency can be selected arbitrarily, i.e., b_{12} is free, then we can prove that this extra free coefficient does not increase the possibility of having extra limit cycles. In general, one of the coefficients b_{12}, a and b (not necessarily b_{12}), can be chosen arbitrarily, which does not affect the results.

To prove the statement above, let ω be an arbitrary frequency of the limit cycles. Then, by first introducing the parameter scaling

$$\begin{aligned}
a &\to \omega a, & b &\to \omega b, & a_{30} &\to \omega a_{30}, \\
a_{21} &\to \omega a_{21}, & b_{30} &\to \omega b_{30}, & b_{21} &\to \omega b_{21},
\end{aligned} \tag{4.56}$$

and then applying the transformation

$$\begin{pmatrix} x \\ y \end{pmatrix} = \begin{pmatrix} 0 \\ 1 \end{pmatrix} + \begin{bmatrix} 2b & 0 \\ 2a & -1 \end{bmatrix} \begin{pmatrix} u \\ v \end{pmatrix}, \tag{4.57}$$

to the resulting equation, and finally by applying a time scaling

$$\tau = \omega t, \tag{4.58}$$

one obtains the following new system:

$$\frac{du}{d\tau} = v + 2\bar{d}_{21}u^2 + 4auv - \frac{3}{2}v^2 + 4b\bar{d}_{30}u^3 - 2\bar{d}_{21}u^2v$$

$$\qquad - 2auv^2 + \frac{1}{2}v^3,$$

$$\frac{dv}{d\tau} = -u - 4\bar{b}_{21}u^2 + 2(2a^2 - 2b^2 + 1)uv - 8\bar{b}_{30}u^3 + 4\bar{b}_{21}u^2v \tag{4.59}$$

$$\qquad - (2a^2 - b^2 + 1)uv^2,$$

where

$$\bar{d}_{21} = a_{21}b - a^2,$$

$$\bar{d}_{30} = a_{30}b + a_{21}a,$$

$$\bar{b}_{21} = b_{21}b^2 - a_{21}ab - 2ab^2 + 2a^3 + a, \tag{4.60}$$

$$\bar{b}_{30} = b_{30}b^3 - a_{30}ab^2 - a_{21}a^2b + b_{21}ab^2 - a^2b^2 + a^4 + \frac{1}{2}a^2.$$

Note from (4.60) that $\bar{X}$ and X (where X represents one of the coefficients, a_{21}, a_{30}, b_{21} and b_{30}) can uniquely determine each other as long as $b \neq 0$. Therefore, once one obtains the solutions for $\bar{d}_{21}, \bar{d}_{30}, \bar{b}_{21}$ and $\bar{b}_{30}$, one can uniquely determine a_{21}, a_{30}, b_{21} and b_{30}, and vice versa. Therefore, without loss of generality, we may assume that condition (4.55) holds, since parameter and time scaling does not change the number of limit cycles.

Apply the Maple program based on the perturbation technique [218] (visit extras.springer.com and search for this book using its ISBN to find the Maple programs) to system (4.59) to obtain the focus values, v_i, explicitly expressed in terms of the system's coefficients, $v_i = v_i(a, b, \bar{d}_{21}, \bar{b}_{21}, \bar{d}_{30}, \bar{b}_{30}), i = 1, 2, \ldots, 7$. Since there are 6 free coefficients, it may be possible to set $v_i = 0, i = 1, 2, \ldots, 6$, but $v_7 \neq 0$. Hence, this seems to give some hope that the system might exhibit 14 small limit cycles by appropriately choosing the values of these coefficients.

Letting $v_1 = 0$ yields

$$\bar{d}_{30} = \frac{a(2\bar{d}_{21} - 1) + \bar{b}_{21}(4\bar{d}_{21} + 4a^2 - 4b^2 + 1)}{3b}, \tag{4.61}$$

and then substituting this $\bar{d}_{30}$ into the expression for v_2 results in

$$v_2 = \frac{1}{9}(\bar{b}_N + 6\bar{b}_D\bar{b}_{30}), \tag{4.62}$$

where

$$
\begin{aligned}
\bar{b}_N &= 2\bar{b}_{21}\{36\bar{d}_{21}^2(a^2 - b^2) + 4\bar{d}_{21}[3(a^2 - b^2)^2 - 8a^2] - 24(a^2 - b^2)^3 \\
&\quad - 2(a^2 - b^2)(11a^2 - 15b^2) + 7a^2 + 9b^2\} \\
&\quad + a[40\bar{b}_{21}^2(2\bar{d}_{21} + 2a^2 - 2b^2 - 1) + (2\bar{d}_{21} - 1)(a^2 - 9b^2)], \\
\bar{b}_D &= a(8\bar{d}_{21} + 8a^2 - 8b^2 + 1) + 10\bar{b}_{21}(2b^2 - 2a^2b - 1 - 2\bar{d}_{21}).
\end{aligned} \tag{4.63}
$$

There are two cases in solving the equation $v_2 = 0$. Since $b \neq 0$, then if $\bar{b}_D \neq 0$ one can uniquely determine $\bar{b}_{30}$, which is the generic case to be considered later. When $\bar{b}_D = 0$, which is referred as a special case, $\bar{b}_{30}$ (or b_{30}) is not used for this order.

(A-i) A Special Case   First, we outline the results for 12 limit cycles in the special case. One may solve for b_{21} from $\bar{b}_D = 0$ to find

$$b_{21} = -a\left\{\frac{2a^2}{b^2} + \frac{[2a_{21}b(10b^2 + 10ba_{21} - 1) - (40b^4 - 32b^2 + 9)]}{10b^2(2b^2 - 2ba_{21} - 1)}\right\}, \quad b \neq 0, \tag{4.64}$$

where (4.60) has been used. Note from (4.64) that it has also been assumed that $2b^2 - 2ba_{21} - 1 \neq 0$. Otherwise, if $2b^2 - 2ba_{21} - 1 = 0$ then $\bar{b}_D = 3a$. This implies that $\bar{b}_D \neq 0$ since $a = 0$ gives a center. Thus, the case $2b^2 - 2ba_{21} - 1 = 0$ is included in the generic case. Then, simplifying v_2 yields

$$
\begin{aligned}
v_2 &= \frac{1}{9}\bar{b}_N \\
&= \frac{2ab^2}{15(2b^2 - 2ba_{21} + 1)^2}[(2aa_{21} - 2ab)^2 - (2ba_{21} - 2b^2 + 1)^2] \\
&\quad \times [20a^2(2b^2 - 2ba_{21} - 1) \\
&\quad - (16b^4 + 8b^3a_{21} - 24b^2a_{21}^2 - 14b^2 - 6a_{21} + 9)].
\end{aligned} \tag{4.65}
$$

It should be noted that since one more parameter is needed to solve the equation $v_2 = 0$, thus the remaining parameters b_{30}, a_{21}, a and b must be used to solve the four equations $v_i = 0, i = 2, 3, 4, 5$. Hence, the special case gives a finite number of possible solutions for the existence of 12 limit cycles.

It is seen from (4.65) that v_2 has two factors, given in square brackets. It can be shown that the two solutions determined from the first factor lead to centers. Therefore, the second factor must be used to find, say,

$$a = \pm\left[\frac{16b^4 + 8b^3a_{21} - 24b^2a_{21}^2 - 14b^2 - 6a_{21} + 9}{20(2b^2 - 2ba_{21} - 1)}\right]^{1/2}. \tag{4.66}$$

Then, the expressions of the focus values, v_3, v_4 and v_5 can be greatly simplified. It is noted that when the order of the focus values increases, the expression becomes much more involved. Indeed, the computer outputs of v_3, v_4 and v_5 have 83, 342 and 1173 lines, respectively, while v_6 has 3376 lines. In principle, the remaining three parameters $\bar{b}_{30}$ (or b_{30}), $\bar{d}_{21}$ (or a_{21}) and b can be used to solve the three non-linear polynomial equations $v_i = 0, i = 3, 4, 5$, but $v_6 \neq 0$. Therefore, in addition, perturbing the linear coefficient a_{12} (see (4.6)) might lead to 6 small limit cycles, and thus the original system (4.52) might have a total of 12 small limit cycles for the special case.

With a given by (4.66), v_3, v_4 and v_5 can be written as

$$
v_3 = \frac{2ab^2 F F_1}{625(2b^2 - 2ba_{21} - 1)^4}, \qquad v_4 = \frac{-2ab^2 F F_2}{3515625(2b^2 - 2ba_{21} - 1)^7},
$$

$$
v_5 = \frac{db^2 F F_3}{31640625000(2b^2 - 2ba_{21} - 1)^{10}}, \tag{4.67}
$$

where F is a common factor, and $F_i = F_i(b_{30}, a_{21}, b), i = 1, 2, 3$, are ith-degree polynomials in b_{30}. Since the common factors lead to centers, we must use F_i to solve the three equations, $v_3 = v_4 = v_5 = 0$. One can first explicitly solve for b_{30} from $F_1 = 0$ to obtain $b_{30} = b_{30}(a_{21}, b)$ and then simplify F_2 and F_3 to obtain the two equations

$$
F_2^* = F_2^*(a_{21}, b) = 0, \qquad F_3^* = F_3^*(a_{21}, b) = 0, \tag{4.68}
$$

where F_2^* and F_3^* are 9th- and 14th-degree polynomials in a_{21}. Eliminating a_{21} from (4.68) results in a polynomial equation,

$$
F_4(b^2) = b\left(264b^2 + 7\right)\left(56b^2 + 3\right)\left(64b^2 + 5\right)\left(75b^2 + 11\right)\left(3b^2 - 5\right)\left(26b^2 - 45\right)F_4^*
$$

$$
= 0, \tag{4.69}
$$

where

$$
F_4^* = 10195528b^6 - 21299025b^4 + 3454965b^2 + 259405, \tag{4.70}
$$

as well as a solution for a_{21} in terms of b^2, i.e., $a_{21} = a_{21}(b^2)$.

Since $b \neq 0$, in order to have $F_4(b^2) = 0$, one may have at most five positive solutions for b^2. However, the two solutions, $b^2 = \frac{5}{3}$ and $b^2 = \frac{45}{26}$, yield centers. Therefore, the only possible expected solutions are given by the roots of the third-degree polynomial $F_4^*(b^2)$. It is easy to show that this polynomial has three different real roots for b^2, but only two of them are positive. Furthermore, it has been found that one of the two positive solutions results in $a^2 < 0$ (by (4.66)). Therefore, the only possible solutions are $b_\pm = \pm 1.3798788398$, which gives four solutions in total for the special case:

$$b = \pm 1.3798788398, \qquad a_{21} = \pm 1.0897036998,$$

$$b_{30} = \pm 0.7093364483, \qquad a = \pm 0.0877426100, \qquad (4.71)$$

$$b_{21} = \pm 0.1470131077, \qquad a_{30} = \mp 0.0284721124.$$

It is seen from (4.71) that all four solutions satisfy $|b| > |a|$, indicating that the origin is a saddle point.

In the following, we consider one of the above four solutions to show the existence of 12 small limit cycles. The critical parameter values $(b^*, a_{21}^*, b_{30}^*, a^*, b_{21}^*, a_{30}^*)$ are

$$b^* = 1.3798788398, \qquad a_{21}^* = 1.0897036998,$$

$$b_{30}^* = 0.7093364483, \qquad a^* = 0.0877426100, \qquad (4.72)$$

$$b_{21}^* = 0.1470131077, \qquad a_{30}^* = -0.0284721124,$$

and then $b_{12}^* = -1.0063695748$, obtained from (4.61). With these parameter values, executing the Maple program for calculating the focus values results in (up to 30 digits of accuracy)

$$v_1 = -0.5 \times 10^{-29},$$

$$v_2 = 0.416 \times 10^{-28},$$

$$v_3 = 0.132703026825562496489719610543 \times 10^{-27},$$

$$v_4 = 0.170977 \times 10^{-24}, \qquad (4.73)$$

$$v_5 = 0.2034487846404411042815016615 2 \times 10^{-21},$$

$$v_6 = 0.0072591372997569367227 9971391,$$

where the v_i's, $i = 1, 2, \ldots, 5$, are not exactly zero due to numerical round-off errors.

The 6 parameters, $a, b, a_{21}, b_{21}, a_{30}$ and b_{30}, can then be perturbed from the critical values given in (4.72) to find the exact 6 small amplitude limit cycles around the origin, $(u, v) = (0, 0)$ (or $(x, y) = (0, \pm 1)$).

For convenience, let us rewrite the formulas for the v_i's in reverse order:

$$v_6 = v_6(b, a_{21}, b_{30}, a, b_{21}, a_{30}),$$

$$\left. \begin{aligned} v_5 &= v_5(b, a_{21}) \\ v_4 &= v_4(b, a_{21}) \end{aligned} \right\},$$

$$v_3 = v_3(b, a_{21}, b_{30}), \qquad (4.74)$$

$$v_2 = v_2(b, a_{21}, a, b_{21}),$$

$$v_1 = v_1(b, a_{21}, a, a_{30}).$$

Note that $v_5 = v_4 = 0$ at the critical point (b^*, a_{21}^*), defined by (4.72), while $v_3 = 0$ when $b_{30} = b_{30}(b, a_{21})$, $v_2 = 0$ when $a = a(b, a_{21})$ and $b_{21} = b_{21}(b, a_{21}, a)$, and $v_1 = 0$ when $a_{30} = a_{30}(b, a_{21}, a, b_{21})$, for any values of (b, a_{21}), and $v_6(b^*, a_{21}^*, b_{30}^*, a^*, b_{21}^*, a_{30}^*) \approx 0.00726 > 0$. It should be noted that, in general, only one parameter is used for each v_i, $i = 1, 2, \ldots, 5$. But for the special case considered here, v_2 has used two parameters, a and b_{21}, since b_{21} is used to solve the additional condition $b_D = 0$ (see (4.64)). Therefore, one can either use a, or b_{21}, or both of them to perturb v_2.

The main result is summarized in the following theorem.

Theorem 4.11 *Given the cubic system (4.52) which is assumed to have a saddle point at the origin and a pair of symmetric weak focus points at $(x, y) = (0, 1)$ and $(0, -1)$, further suppose*

$$a_{12} = -b_{03} = 0.0877426100, \qquad a_{03} = -1.3798788398,$$
$$b_{12} = -1.0063695748.$$
$$(4.75)$$

Then if $b, a_{21}, b_{30}(b, a_{21}), a(b, a_{21}), b_{21}(b, a_{21}, a)$ and $a_{30}(b, a_{21}, a, b_{21})$ are perturbed as

$$b = b^* + \varepsilon_1,$$
$$a_{21} = a_{21}^* + \varepsilon_2,$$
$$b_{30} = b_{30}(b^* + \varepsilon_1, a_{21}^* + \varepsilon_2) + \varepsilon_3,$$
$$a = a(b^* + \varepsilon_1, a_{21}^* + \varepsilon_2) - \varepsilon_6, \qquad (4.76)$$
$$b_{21} = b_{21}(b^* + \varepsilon_1, a_{21}^* + \varepsilon_2, a(b^* + \varepsilon_1, a_{21}^* + \varepsilon_2) - \varepsilon_6) - \varepsilon_4,$$
$$a_{30} = a_{30}(b^* + \varepsilon_1, a_{21}^* + \varepsilon_2, a(b^* + \varepsilon_1, a_{21}^* + \varepsilon_2) - \varepsilon_6,$$
$$b_{21}(b^* + \varepsilon_1, a_{21}^* + \varepsilon_2, a(b^* + \varepsilon_1, a_{21}^* + \varepsilon_2) - \varepsilon_6) - \varepsilon_4) - \varepsilon_5,$$

where $0 < \varepsilon_6 \ll \varepsilon_5 \ll \varepsilon_4 \ll \varepsilon_3 \ll (\varepsilon_2, \varepsilon_1) \ll 1$, system (4.52) has exactly 12 small limit cycles. The notation $(\varepsilon_2, \varepsilon_1)$ means that ε_2 and ε_1 are of the same order, with $\varepsilon_2 = (\delta + \bar{\varepsilon})\varepsilon_1$ for some $\delta > 0$ and some small $\bar{\varepsilon} > 0$.

Proof First, consider v_5 and v_4 simultaneously. We want to make perturbations such that $0 < v_4 \ll -v_5 \ll v_6$. Computing the Jacobian matrix of the system consisting of the equations $v_5 = 0$ and $v_3 = 0$ with respect to b and a_{21}, evaluated at (b^*, a_{21}^*) yields

$$J(b^*, a_{21}^*) = \begin{bmatrix} J_{11} & J_{12} \\ J_{21} & J_{22} \end{bmatrix} \equiv \begin{bmatrix} \frac{\partial v_5}{\partial b} & \frac{\partial v_5}{\partial a_{21}} \\ \frac{\partial v_4}{\partial b} & \frac{\partial v_4}{\partial a_{21}} \end{bmatrix}_{(b^*, a_{21}^*)}$$

$$= \begin{bmatrix} -6.1361037671 & 0.0 \\ -2.3694030045 & 2.1866028190 \end{bmatrix}. \qquad (4.77)$$

Since $\det(J) \neq 0$, then according to Theorem 4.2, we can find perturbations to b and a_{21} such that $0 < v_4 \ll -v_5$. Letting

$$b = b^* + \varepsilon_1 \quad \text{and} \quad a_{21} = a_{21}^* + \varepsilon_2 \tag{4.78}$$

results in the following linear approximation:

$$\begin{pmatrix} v_5 \\ v_4 \end{pmatrix} \approx \begin{pmatrix} v_5(b^*, a_{21}^*) \\ v_4(b^*, a_{21}^*) \end{pmatrix} + J(b^*, a_{21}^*)\begin{pmatrix} \varepsilon_1 \\ \varepsilon_2 \end{pmatrix} = \begin{pmatrix} J_{11}\varepsilon_1 \\ J_{21}\varepsilon_1 + J_{22}\varepsilon_2 \end{pmatrix}, \tag{4.79}$$

where $\varepsilon_1 > 0, \varepsilon_2 > 0$. From the requirements $v_5 < 0$ and $v_4 > 0$, we have

$$0 < -\frac{J_{21}}{J_{22}}\varepsilon_1 < \varepsilon_2. \tag{4.80}$$

Further, in order to find sufficient conditions such that $0 < v_4 \ll -v_5 \ll v_6$, let

$$\varepsilon_2 = (\delta + \bar{\varepsilon})\varepsilon_1, \quad \text{where } \delta = -\frac{J_{21}}{J_{22}} = 1.0836000868, \tag{4.81}$$

then the linear approximation of v_4 can be rewritten as

$$v_4 \approx J_{22}\varepsilon_1\bar{\varepsilon}. \tag{4.82}$$

By noting that $v_5 \approx J_{11}\varepsilon_1$, one can take $\varepsilon_1 \ll 1$ and $\bar{\varepsilon} \ll 1$ to guarantee $0 < v_4 \ll -v_5 \ll v_6$. Also note from (4.81) that ε_2 is of the same order as ε_1.

Next, consider v_3. To estimate the value of v_3 near the critical point, calculating $\frac{\partial v_3}{\partial b_{30}}$ at $(b^* + \varepsilon_1, a_{21}^* + \varepsilon_2, b_{30}(b^* + \varepsilon_1, a_{21}^* + \varepsilon_2))$ gives

$$\frac{\partial v_3}{\partial b_{30}} \approx -0.0581541802 - 533.7000900191\varepsilon_1 + 471.0210964726\varepsilon_2, \tag{4.83}$$

which is negative for $0 < \varepsilon_1, \varepsilon_2 \ll 1$. Thus, in order to have $v_3 < 0$ after perturbation, one may choose small $\varepsilon_3 > 0$ such that

$$b_{30} = b_{30}(b^* + \varepsilon_1, a_{21}^* + \varepsilon_2) + \varepsilon_3, \tag{4.84}$$

where $b_{30}(b, a_{21})$ is given by (2.30). Then, one obtains

$$v_3 \approx (-0.0581541802 - 533.7000900191\varepsilon_1 + 471.0210964726\varepsilon_2)\varepsilon_3, \tag{4.85}$$

which is negative for sufficiently small $\varepsilon_i > 0, i = 1, 2, 3$. The additional condition $-v_3 \ll v_4$ results in $0 < \varepsilon_3 \ll (\varepsilon_2, \varepsilon_1) \ll 1$.

Similarly, we can follow the above procedure to consider v_2. Note that we may use two parameters, a and b_{21}, to perturb v_2. Here, we choose b_{21} and leave a for a later linear perturbation on v_0. Thus, computing $\frac{\partial v_2}{\partial b_{21}}$ at

$$\left(b^* + \varepsilon_1, a_{21}^* + \varepsilon_2, b_{30}\left(b^* + \varepsilon_1, a_{21}^* + \varepsilon_2\right) + \varepsilon_3, \right.$$
$$\left. b_{21}\left(b^* + \varepsilon_1, a_{21}^* + \varepsilon_2, a\left(b^* + \varepsilon_1, a_{21}^* + \varepsilon_2\right)\right)\right)$$

yields

$$\frac{\partial v_2}{\partial b_{21}} \approx -1.6661691420 + 20.9332869624\varepsilon_1$$
$$- 29.2490912085\varepsilon_2 + 6.6431544667\varepsilon_3, \tag{4.86}$$

which is negative for small $\varepsilon_i, i = 1, 2, 3$. Hence, $v_2 > 0$ after a perturbation ε_4 $(0 < \varepsilon_4 \ll 1)$ such that

$$b_{21} = b_{21}\left(b^* + \varepsilon_1, a_{21}^* + \varepsilon_2, a\right) - \varepsilon_4, \tag{4.87}$$

where b_{21} and a are given by (4.64) and (4.66), respectively, and the positive sign is taken from (4.66) for a. Then, one obtains

$$v_2 \approx (1.6661691420 - 20.9332869624\varepsilon_1$$
$$+ 29.2490912085\varepsilon_2 - 6.6431544667\varepsilon_3)\varepsilon_4, \tag{4.88}$$

which is positive for small $\varepsilon_i > 0, i = 1, 2, 3, 4$. The additional condition $|v_2| \ll |v_3|$ yields $0 < \varepsilon_4 \ll \varepsilon_3 \ll (\varepsilon_2, \varepsilon_1) \ll 1$.

For v_1, we need a perturbation on a_{30} to yield $v_1 < 0$. Since

$$\frac{\partial v_1}{\partial a_{30}} = \frac{3}{2}b^2 > 0,$$

so we perturb a_{30} to

$$a_{30}\left(b^* + \varepsilon_1, a_{21}^* + \varepsilon_2, a, b_{21}\left(b^* + \varepsilon_1, a_{21}^* + \varepsilon_2, a\right) - \varepsilon_4\right) - \varepsilon_5, \quad \varepsilon_5 > 0,$$

to obtain

$$v_1 \approx -(2.856098420 + 4.139636520\varepsilon_1)\varepsilon_5, \tag{4.89}$$

which is less than zero if $\varepsilon_1 > 0$ and $\varepsilon_5 > 0$. Applying the condition $|v_1| \ll |v_2|$ results in $0 < \varepsilon_5 \ll \varepsilon_4$.

Finally, one more perturbation on the linear part of system (4.52) is needed to have a total of 6 limit cycles in the vicinity of $(x, y) = (0, 1)$. Noting that the parameter a (which is determined by (4.66)) has not been used for perturbation. Thus, we perturb a such that the linear part $v_0 > 0$ (since $v_1 < 0$). From $v_0 = \frac{1}{2}(a_{12} - a)$ (see (4.54)), $\frac{dv_0}{da} = -\frac{1}{2}$, so we can take

$$a = a\left(b^* + \varepsilon_1, a_{21}^* + \varepsilon_2\right) - \varepsilon_6, \tag{4.90}$$

where $0 < \varepsilon_6 \ll \varepsilon_5 \ll \varepsilon_4 \ll \varepsilon_3 \ll (\varepsilon_2, \varepsilon_1) \ll 1$, to obtain

$$v_0 = \frac{1}{2}\varepsilon_6 > 0. \tag{4.91}$$

Therefore, the perturbed normal form for the Hopf bifurcation of system (4.2) arising from the points $(x, y) = (0, \pm 1)$, up to term r^{13}, is given by

$$\frac{dr}{d\tau} = r\left(v_0 + v_1 r^2 + v_2 r^4 + v_3 r^6 + v_4 r^8 + v_5 r^{10} + v_6 r^{12}\right), \tag{4.92}$$

where $0 < v_0 \ll -v_1 \ll v_2 \ll -v_3 \ll v_4 \ll -v_5 \ll v_6$.

The proof is complete. $\square$

Remark 4.12 Note that the proof given in this section is only for one of the four possible choices of the parameter values $(b, a_{21}, b_{30}, a, b_{21}, a_{30})$ such that $v_i = 0, i = 1, 2, \ldots, 5$, but $v_6 \neq 0$. The proofs for the other three choices follow the same procedure and are thus not repeated here.

To demonstrate the results obtained in Theorem 4.11, we present a numerical example, which shows that by appropriately choosing perturbations, ε_i, one can find exactly 6 positive roots for r^2, found from (4.92), i.e., $\frac{dr}{d\tau} = 0$. The results are obtained using Maple up to 30 digits. We choose $\varepsilon_1 = 0.01$ and $\bar{\varepsilon} = 0.001$. Then ε_2 is determined from (4.81). The 6 small perturbations are given by

$$
\begin{aligned}
\varepsilon_1 &= 0.01, \\
\varepsilon_2 &= 0.010846, \\
\varepsilon_3 &= 0.00000001, \\
\varepsilon_4 &= 0.00000000000006, \\
\varepsilon_5 &= 0.000000000000000000001, \\
\varepsilon_6 &= 0.0000000000000000000000002,
\end{aligned}
\tag{4.93}
$$

under which the 6 perturbed parameters are

$$
\begin{aligned}
b &= 1.389878839792798029819223084829, \\
a_{21} &= 1.100549699750322460204379462971, \\
b_{30} &= 0.723083443246592425047289189844, \\
a &= 0.088317687330421587849539710365, \\
b_{21} &= 0.148351368647488295561789258216, \\
a_{30} &= -0.028284006013282710541331939148,
\end{aligned}
\tag{4.94}
$$

and the focus values now become

$$v_0 = 0.1 \times 10^{-22},$$

$$v_1 = -0.28976447839556614981826805720 \times 10^{-17},$$

$$v_2 = 0.1008809249931814952332283330467 \times 10^{-12},$$

$$v_3 = -0.578965080933760443062740846761 \times 10^{-9}, \hspace{2cm} (4.95)$$

$$v_4 = 0.736608095896203656125558382977 \times 10^{-6},$$

$$v_5 = -0.176685951291921753599727302890 \times 10^{-3},$$

$$v_6 = 0.6881182240899440054543777118998 \times 10^{-2}.$$

Here it should be pointed out that the above small focus values are true values (not due to numerical error), while the small values of v_i, $i = 1, 2, 3, 4, 5$, given in (4.73) are due to numerical error, and should be exactly zero. In fact, using an accuracy up to 1000 digits in a Maple computation, one can show that the errors appearing in the focus values given in (4.73) are actually about the order of 10^{-999} to 10^{-993}. However, the focus values given in (4.95) are unchanged with higher accuracy (say, up to 1000 digits).

The 6 positive roots for r are then obtained by solving the polynomial equation (4.92) with the v_i's given in (4.95), as follows:

$$r_1 = 0.001998424223, \hspace{1cm} r_2 = 0.005539772256,$$

$$r_3 = 0.013978579010, \hspace{1cm} r_4 = 0.027014927622, \hspace{1cm} (4.96)$$

$$r_5 = 0.063377651980, \hspace{1cm} r_6 = 0.143875184886.$$

The above results have been verified by the Maple program developed in [218] for calculating the normal form of degenerate Hopf bifurcations. Using the perturbed parameter values given in (4.94) and executing the Maple program yields the same focus values and the same solutions given in (4.95) and (4.96), respectively. If we add v_7, which (obtained from computer output) is given by

$$v_7 = 0.0237417076196301976612909048841,$$

to the normal form (4.92), then (4.92) up to the term r^{15} can have 7 nonzero real roots for r^2, 6 of which are positive, but they are slightly different from those given by (4.96). The additional root for r^2 is, however, negative:

$$r_7^2 = -0.3138620080,$$

which does not have real solutions. This indicates that the special case of system (4.52) considered in this section can only have a maximum of 12 small limit cycles, as proved in Theorem 4.11.

By using the perturbed parameter values given in (4.94), the original system (4.52) can have exactly 6 limit cycles in the vicinity of the point $(x, y) = (0, 1)$. Due to symmetry, the system also has 6 limit cycles in the vicinity of the point

Fig. 4.2 The phase portrait of system (4.52)

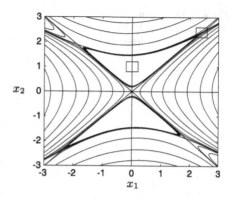

Fig. 4.3 The enlarged part of Fig. 4.2 around the saddle point C_6

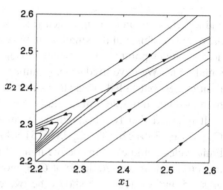

$(x, y) = (0, -1)$. Letting $\dot{x} = \dot{y} = 0$ in the perturbed system yields the following 7 real fixed points:

$$C_1 = (0, 0), \qquad C_{2,3} = (0, \pm 1),$$

$$C_{4,5} = (\pm 1.8768126014, \mp 1.8899355510), \quad \text{and}$$

$$C_{6,7} = (\pm 2.3768639365, \pm 2.4053940296),$$

and two purely imaginary fixed points: $(\pm 1.4074970614i, \pm 0.2849922368i)$. It is known that C_0 is a saddle point (with eigenvalues $\lambda_1 = 1.4781965271, \lambda_2 = -1.3015611525$) and $C_{2,3}$ are a pair of sixth-order weak focus points (with $\lambda_{1,2} = \pm i$). It is easy to apply a linear analysis to find $C_{4,5}$ (with $\lambda_1 = 2.5557639319, \lambda_2 = -3.3655252585$) and $C_{6,7}$ (with $\lambda_1 = 3.2703315475, \lambda_2 = -2.8234536559$) are two pairs of symmetric saddle points. The phase portrait is shown in Fig. 4.2, where two places (marked by two boxes: one is around the saddle point C_6 and the other around the weak focus point $(0, 1)$), are enlarged and shown in Figs. 4.3 and 4.4. Figure 4.3 clearly shows that C_6 (and C_7) is a saddle point, and Fig. 4.4 depicts the 6 small limit cycles around $(0, 1)$, three of them are stable and the other three are unstable (see Remark 4.3).

It should be pointed out that one cannot use numerical simulation to get the 6 accurate infinitesimal limit cycles in the neighborhood of a highly singular point,

Fig. 4.4 The 6 limit cycles
around the focus point (0, 1):
—— stable; · · · · · · unstable

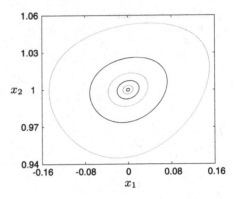

and a certain theoretical approach (like the one presented here) has to be used to prove the existence of the small limit cycles. In fact, the phase portrait of the system near the focus point (0, 1) is obtained from the normal form, rather than numerical simulations. We have tried several numerical approaches to identify the very small limit cycles, but failed. This shows that normal form theory is a powerful tool for finding small, multiple limit cycles.

It is noted that two homoclinic orbits seem to exist at the two saddle points $C_{6,7} = (\pm 2.3561270505, \pm 2.3856151461)$, and one may expect that large (global) limit cycles could exist inside the homoclinic orbits. However, a close view of the trajectories near the saddle point $C_{6,7}$ shows that it is not a homoclinic orbit. Figures 4.5 and 4.6 (Fig. 4.6 shows the two enlarged boxes given in Fig. 4.5) depict two particular trajectories passing through the saddle point C_6. (Strictly speaking, the trajectory given in part (a) approaches the saddle point as $t \to -\infty$ while the one in part (b) does so as $t \to \infty$.)

However, the possibility of a large limit cycle still exists if the largest one of the 6 small limit cycles is stable, since in this case a large unstable limit cycle must exist outside the largest small limit cycle so that a trajectory starting from outside the unstable limit cycle can diverge to nearby the saddle point C_6. Let us consider the stability of the 6 small limit cycles near the point (0, 1), and call the 6 periodic orbits from the smallest one as the first, and so on, limit cycles. Since for this case

$$0 < v_0 \ll -v_1 \ll v_2 \ll -v_3 \ll v_4 \ll -v_5 \ll v_6 \ll 1,$$

we can conclude that

1st limit cycle: stable,

2nd limit cycle: unstable,

3rd limit cycle: stable,

4th limit cycle: unstable,

5th limit cycle: stable,

6th limit cycle: unstable.

Fig. 4.5 The trajectories connecting the saddle point C_6: (**a**) starting from the saddle point; and (**b**) diverging to the saddle point from inside the region

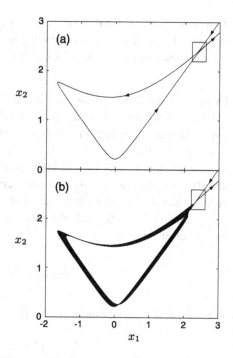

Fig. 4.6 The trajectories in the enlarged regions of Fig. 4.5: (**a**) the box in Fig. 4.5(a); and (**b**) the box in Fig. 4.5(b)

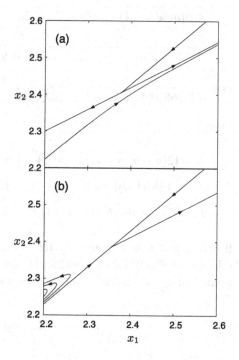

Therefore, a large (global) limit cycle in the neighborhood of the point $(0, 1)$ cannot exist. In fact, all trajectories starting inside the region (not very close to $(0, 1)$) diverge to nearby the saddle point C_6, as shown in Fig. 4.5(b).

(A-ii) Generic Case   Now we turn to the generic case in which $b_D \neq 0$ (see (4.63)). For this case, there are 6 free coefficients: $a, b, \bar{d}_{21}, \bar{b}_{21}, \bar{d}_{30}$ and $\bar{b}_{30}$, involved in the expressions of the focus values. Therefore, unlike the special case, it is now possible to have $v_i = 0, i = 1, 2, \ldots, 6$, but $v_7 \neq 0$. This indicates that it may be possible to have 7 small limit cycles around each of the focus points $(0, 1)$ and $(0, -1)$. We need to solve the four very large, coupled polynomial equations, $v_i = 0, i = 3, 4, 5, 6$. This causes very high complexity in the computation.

First, letting $v_1 = 0$ yields the solution for $\bar{d}_{30}$, given by (4.61). Then, setting $v_2 = 0$ results in

$$\bar{b}_{30} = \bar{b}_{30}(a, b, \bar{d}_{21}, \bar{b}_{21}) = -\frac{\bar{b}_N}{6\bar{b}_D}, \tag{4.97}$$

where $\bar{b}_N$ and $\bar{b}_D$ are given in (4.63). The remaining focus values are simplified as

$$v_3 = \frac{4FF_1}{9[a(8\bar{d}_{21} - 1) - 10\bar{b}_{21}(a\bar{d}_{21} + 2a^2 - 2b^2 + 1) + 8a(a^2 - b^2)]^2},$$

$$v_4 = \frac{FF_2}{405[a(8\bar{d}_{21} - 1) - 10\bar{b}_{21}(a\bar{d}_{21} + 2a^2 - 2b^2 + 1) + 8a(a^2 - b^2)]^3},$$

$$v_5 = \frac{FF_3}{233280[a(8\bar{d}_{21} - 1) - 10\bar{b}_{21}(a\bar{d}_{21} + 2a^2 - 2b^2 + 1) + 8a(a^2 - b^2)]^4},$$

$$v_6 = \frac{FF_4}{146966400[a(8\bar{d}_{21} - 1) - 10\bar{b}_{21}(a\bar{d}_{21} + 2a^2 - 2b^2 + 1) + 8a(a^2 - b^2)]^5}, \tag{4.98}$$

where

$$F = \left[2(a + b)(\bar{d}_{21} + a^2 - b^2) + b\right]\left[2(a - b)(\bar{d}_{21} + a^2 - b^2) - b\right]$$
$$\times \left[\bar{b}_{21}(2\bar{d}_{21} + a^2 - b^2 - 1) - a(2\bar{d}_{21} - 1)\right], \tag{4.99}$$
$$F_i = F_i(a, b, \bar{d}_{21}, \bar{b}_{21}), \quad i = 1, 2, 3, 4.$$

It is seen that F is a common factor of v_3, v_4, v_5 and v_6, so one cannot use any solutions of $F = 0$ since it would lead to centers. Further, it has been noted that F_1 is linear in $\bar{d}_{21}$, so one can explicitly solve for $\bar{d}_{21}$ from $F_1 = 0$ to obtain

$$\bar{d}_{21} = \bar{d}_{21}(a, b, \bar{b}_{21}) = \frac{-\bar{d}_{21N}}{20\bar{d}_{21D}}, \tag{4.100}$$

where

$$\bar{d}_{21N} = 2800\bar{b}_{21}^4(2a^2 - 2b^2 + 1) - 280a\bar{b}_{21}^3(8a^2 - 8b^2 - 9)$$
$$- 20\bar{b}_{21}^2(74a^4 - 68b^2a^2 - 6b^4 + 114a^2 + 3b^2)$$
$$+ 10a\bar{b}_{21}[12(a^2 - b^2)(3a^2 + b^2) + 7(5a^2 - 3b^2)]$$
$$- a^2(a^2 - 9b^2)(8a^2 - 8b^2 + 15),$$
$$\bar{d}_{21D} = 280\bar{b}_{21}^4 - 112a\bar{b}_{21}^3 + 6\bar{b}_{21}^2(11a^2 - b^2)$$
$$- 2a\bar{b}_{21}(5a^2 - 3b^2) + a^2(a^2 - 9b^2).$$

(4.101)

Then, substituting $\bar{d}_{21}$ above into (4.99) yields three polynomial equations:

$$F_2 = -\frac{672(10\bar{b}_{21} - a + 3b)^2 F_n^2}{25F_d^2}F_2^* = 0,$$
$$F_3 = -\frac{21504(3125\bar{b}_{21} - a + 3b)^3 F_n^3}{25F_d^5}F_3^* = 0,$$
$$F_4 = -\frac{2016(78125\bar{b}_{21} - a + 3b)^4 F_n^4}{25F_d^8}F_4^* = 0,$$

(4.102)

where

$$F_n = (10\bar{b}_{21} - a - 3b)[5\bar{b}_{21}(4\bar{b}_{21} - 4a^3 + 4ab^2 - 5a) + a^2(8a^2 - 8b^2 + 5)],$$
$$F_d = 280\bar{b}_{21}^4 - 112a\bar{b}_{21}^3 + 6(11a^2 - b^2)\bar{b}_{21}^2 - 2a(5a^2 - 3b^2)\bar{b}_{21} + a^4 - 9a^2b^2,$$

and $F_i^* = F_i^*(a, b, \bar{b}_{21}), i = 2, 3, 4$. A similar argument can be applied to show that the common factors of $F_i, i = 2, 3, 4$, lead to centers. Thus, we need to find solutions of the equations, $F_i^* = 0, i = 2, 3, 4$.

However, these three polynomial equations are coupled and have to be solved simultaneously. Similarly, we need to eliminate one of the coefficients a, b and $\bar{b}_{21}$ from the above three equations. Carefully examining these equations shows that the coefficient b gives relatively lower-order polynomials in b^2: the highest-order term for b^2 in F_2, F_3 and F_4 is 2, 5 and 10, respectively. Hence, eliminating b from the first two equations, $F_2 = 0$ and $F_3 = 0$, yields the resultant equation

$$F_5 = -\bar{b}_{21}(a + 2\bar{b}_{21})(a^2 - 16\bar{b}_{21}^2)(a^2 - 64\bar{b}_{21}^2)(a + 10\bar{b}_{21})$$
$$\times (160\bar{b}_{21}^6 - 360a\bar{b}_{21}^5 + 58a^2\bar{b}_{21}^4 - 144a^3\bar{b}_{21}^3 + 60a^4\bar{b}_{21}^2 + 2a^5\bar{b}_{21} - a^6)F_5^*$$
$$= 0,$$

(4.103)

where

$$F_5^* = 160\bar{b}_{21}^3 + 140a\bar{b}_{21}^2 - 40a^2\bar{b}_{21} + a^3,$$

(4.104)

as well as an equation for solving for b, once $\bar{b}_{21}$ is determined from (4.103):

$$F_b = (135\bar{b}_{21} + 108a)b^4 - (12600\bar{b}_{21}^3 + 5040a\bar{b}_{21}^2 - 390a^2\bar{b}_{21} - 60a^3)b^2$$
$$+ 43120\bar{b}_{21}^5 + 1400a^2\bar{b}_{21}^3$$
$$= 0. \tag{4.105}$$

Similarly, eliminating b from another two equations, $F_2 = 0$ and $F_4 = 0$, yields the resultant

$$F_6 = -\bar{b}_{21}(a + 2\bar{b}_{21})(a^2 - 16\bar{b}_{21}^2)(a^2 - 64\bar{b}_{21}^2)(a + 10\bar{b}_{21})$$
$$\times (160\bar{b}_{21}^6 - 360a\bar{b}_{21}^5 + 58a^2\bar{b}_{21}^4 - 144a^3\bar{b}_{21}^3 + 60a^4\bar{b}_{21}^2 + 2a^5\bar{b}_{21} - a^6)F_6^*$$
$$= 0, \tag{4.106}$$

and the same equation, (4.104), for solving for b. Here, F_6^* is a 23rd-degree polynomial in $\bar{b}_{21}$ (the lengthy Maple output is omitted here). Note that F_5 and F_6 have 7 common factors which give solutions leading to centers. Therefore, we must solve $F_5^* = F_6^* = 0$ for a and $\bar{b}_{21}$, which may yield possible solutions such that $v_i = 0, i = 1, 2, \ldots, 6$. It should be pointed out that the above nonlinear, variable elimination process does not miss any possible solutions. (But it may be possible to have extra solutions, and thus one has to verify all solutions using the original expressions of the v_i.)

Finally, eliminating $\bar{b}_{21}$ from the two equations, $F_5^* = F_6^* = 0$, results in

$$\bar{b}_{21} = a\frac{\bar{b}_{21N}(a)}{2\bar{b}_{21D}(a)}, \tag{4.107}$$

where

$$\bar{b}_{21N} = 4874540590576157351160568084352a^8$$
$$+ 6082006821548063186315668637808a^6$$
$$+ 2027609911961661234135846867800a^4$$
$$+ 707219153484515541381107296125a^2$$
$$+ 419461339793579454120331286700,$$
$$\bar{b}_{21D} = 11971617259908731985977164895104a^8 \tag{4.108}$$
$$+ 2512242956486215090366509936240a^6$$
$$+ 3790131072252096977549365658680a^4$$
$$+ 300269138209371121894242704325a^2$$
$$+ 445544154354000451175586895500,$$

and a resultant

$$
\begin{aligned}
F_7 = a\big(&112783323904064139036783273943 04a^{12} \\
&+ 9439268357932268111521578554849280a^{10} \\
&+ 15955549149699417710133152016699200a^{8} \\
&+ 78067338319585667946062243018556 00a^{6} \\
&+ 268967012423314994209631230770150 0a^{4} \\
&+ 131790719652429681318291214667062 5a^{2} \\
&+ 2482250875425524079421399644975 00\big) = 0.
\end{aligned}
\tag{4.109}
$$

It is obvious that (4.107) has only one real solution, $a = 0$. But if $a = 0$, then $\bar{b}_{21} = 0$, resulting in $\bar{d}_{21} = \bar{b}_{30} = \bar{d}_{30} = 0$, which can be shown to give a center. Thus, a must be nonzero. This implies that one cannot find possible nonzero values of a and $\bar{b}_{21}$ such that $F_5 = F_6 = 0$, indicating that there is no solution for $F_2 = F_3 = F_4 = 0$. Therefore, there do not exist possible nontrivial solutions for $a, b, \bar{d}_{21}, \bar{b}_{21}, \bar{d}_{30}$ and $\bar{b}_{30}$ such that $v_i = 0, i = 1, 2, \ldots, 6$. Hence, 14 *small limit cycles are not possible* even for the generic case when the origin is a saddle point, and the maximal number of small limit cycles for the generic case is also 12.

Next, we shall find all the solutions of 12 limit cycles for the generic case. It was shown in Sect. 3 that there exist four solutions for the special case; see (4.71). However, it should be noted that for each of the four solutions, there is a family of small limit cycles depending upon perturbations. The perturbation given for the special case is just one example chosen from the particular solution (4.72). For the generic case, on the other hand, since there is an extra free coefficient to be chosen for the focus values, there are infinite solutions. Finding these solutions only requires $F_2 = F_3 = 0$ ($F_4 \neq 0$), which can be obtained from (4.103) in parametric form. For example, numerically solving (4.103) for $\bar{b}_{21}$ in terms of a yields three solutions:

$$
\bar{b}_{21} = 0.2033343806a,
$$

$$
-1.1061229255a \quad \text{and}
$$

$$
0.2778854492a.
\tag{4.110}
$$

Then, for each of the above solutions, solving (4.104) gives two solutions for b^2. But for the last value of $\bar{b}_{21}$, one solution is negative. Hence, there are ten solutions in total for $(\bar{b}_{21}, b)$. However, by checking the equations $F_2 = F_3 = 0$, only four solutions are satisfied: two for the case when the origin is a saddle point:

$$b = \pm 15.7264394069a,$$

$$\bar{b}_{21} = -1.1061229255a,$$

$$\bar{d}_{21} = 0.7000000000 + 103.3880431509a^2,$$

$$\bar{b}_{30} = \frac{0.2564102564a^2(0.0089196607 + 0.0982810312a^2 + 0.2527227926a^4)}{0.0661528794 + 0.2704669566a^2},$$

$$\bar{d}_{30} = \mp(0.0806130156 - 17.7870588470a^2),$$

$$(4.111)$$

and two for the case when the origin is a node:

$$b = \pm 0.4765747114a,$$

$$\bar{b}_{21} = 0.2033343806a,$$

$$\bar{d}_{21} = 0.7000000000 + 1.0149654014a^2,$$

$$(4.112)$$

$$\bar{b}_{30} = \frac{d^2(0.0481488581 + 65.9546167690a^2 - 9379.2591506305a^4)}{0.0008286738 - 0.1076372236a^2},$$

$$\bar{d}_{30} = \pm(0.8202076319 + 2.4368685248a^2).$$

Here, a is an arbitrary real number. Note that for the special case discussed in the previous section, solutions only exist when the origin is a saddle point.

In the following, we consider one of the two cases when the origin is a node, since the case when the origin is a saddle point has been considered in the special case. Let the critical values be denoted by

$$b^* = 0.4765747114a,$$

$$\bar{b}_{21}^* = 0.2033343806a,$$

$$\bar{d}_{21}^* = 0.7000000000 + 1.0149654014a^2,$$

$$(4.113)$$

$$\bar{b}_{30}^* = \frac{d^2(0.0481488581 + 65.9546167690a^2 - 9379.2591506305a^4)}{0.0008286738 - 0.1076372236a^2},$$

$$\bar{d}_{30}^* = -(0.8202076319 + 2.4368685248a^2).$$

Then, we have the following theorem for the generic case.

Theorem 4.13 *Given the cubic system (4.52) which is assumed to have a saddle point or a node at the origin and a pair of symmetric weak focus points at $(x, y) = (0, 1)$ and $(0, -1)$, further suppose that $a_{12} = -b_{03} = a, a_{03} = -b, b_{12} = \frac{1+4a^2}{2b} - b$. Then, for an arbitrarily given $a \neq 0$, if $b, \bar{b}_{21}, \bar{d}_{21}(b, \bar{b}_{21}), \bar{b}_{30}(b, \bar{d}_{21}, \bar{b}_{21})$ and*

$\bar{d}_{30}(b, \bar{d}_{21}, \bar{b}_{21})$ *are perturbed as*

$$b = b^* + \varepsilon_1,$$

$$\bar{b}_{21} = \bar{b}_{21}^* + \varepsilon_2,$$

$$\bar{d}_{21} = \bar{d}_{21}(b^* + \varepsilon_1, \bar{b}_{21}^* + \varepsilon_2) + \varepsilon_3,$$

$$\bar{b}_{30} = \bar{b}_{30}(b^* + \varepsilon_1, \bar{b}_{21}^* + \varepsilon_2, \bar{d}_{21}(b^* + \varepsilon_1, \bar{b}_{21}^* + \varepsilon_2) + \varepsilon_3) + \varepsilon_4, \qquad (4.114)$$

$$\bar{d}_{30} = \bar{d}_{30}(b^* + \varepsilon_1, \bar{b}_{21}^* + \varepsilon_2, \bar{d}_{21}(b^* + \varepsilon_1, \bar{b}_{21}^* + \varepsilon_2) + \varepsilon_3) + \varepsilon_5,$$

$$a_{12} = a + \varepsilon_6,$$

where $0 < |\varepsilon_6| \ll |\varepsilon_5| \ll |\varepsilon_4| \ll |\varepsilon_3| \ll (|\varepsilon_2|, |\varepsilon_1|) \ll 1$, *then system* (4.52) *has exactly* 12 *small limit cycles.*

Remark 4.14 The notation $(|\varepsilon_2|, |\varepsilon_1|)$ means that ε_2 and ε_1 are of the same order, with $\varepsilon_2 = (\delta + \bar{\varepsilon})\varepsilon_1$ for some $\delta > 0$ and some small $\bar{\varepsilon} > 0$. Note that here the perturbations, ε_i, can take positive or negative values since, unlike the special case, here a is not specified.

Proof For convenience, rewrite the formulas for the v_i's in reverse order:

$$v_6 = v_6(a, b, \bar{b}_{21}, \bar{d}_{21}, \bar{b}_{30}, \bar{d}_{30}),$$

$$\left.\begin{array}{l} v_5 = v_5(a, b, \bar{b}_{21}) \\ v_4 = v_4(a, b, \bar{b}_{21}) \end{array}\right\},$$

$$v_3 = v_3(a, b, \bar{b}_{21}, \bar{d}_{21}), \qquad (4.115)$$

$$v_2 = v_2(a, b, \bar{b}_{21}, \bar{d}_{21}, \bar{b}_{30}),$$

$$v_1 = v_1(a, b, \bar{b}_{21}, \bar{d}_{21}, \bar{d}_{30}).$$

For determination, suppose $v_6 > 0$ (one can similarly prove the case when $v_6 < 0$). Then, according to the sufficient conditions established in Theorems 4.1 and 4.2, we need to find perturbations such that $0 < v_0 \ll -v_1 \ll v_2 \ll -v_3 \ll v_4 \ll -v_5 \ll v_6$. First, consider v_5 and v_4 simultaneously. Computing the Jacobian matrix of the system consisting of the equations $v_5 = 0$ and $v_4 = 0$ with respect to b and $\bar{b}_{21}$, evaluated at $(b^*, \bar{b}_{21}^*)$ yields

$$J(a) = \begin{bmatrix} \dfrac{\partial v_5}{\partial b} & \dfrac{\partial v_5}{\partial a_{21}} \\ \dfrac{\partial v_4}{\partial b} & \dfrac{\partial v_4}{\partial a_{21}} \end{bmatrix}_{(b^*, \bar{b}_{21}^*)} \equiv \begin{bmatrix} J_{11}(a) & J_{12}(a) \\ J_{21}(a) & J_{22}(a) \end{bmatrix}. \qquad (4.116)$$

Choose a such that $\det(J(a)) \neq 0$, then according to Theorem 4.2, we can find perturbations to b^* and $\bar{b}_{21}^*$ such that $0 < v_4 \ll -v_5$. Letting

$$b = b^* + \varepsilon_1 \quad \text{and} \quad \bar{b}_{21} = \bar{b}_{21}^* + \varepsilon_2 \qquad (4.117)$$

results in the following linear approximations:

$$\begin{pmatrix} v_5 \\ v_4 \end{pmatrix} \approx \begin{pmatrix} v_5(b^*, a_{21}^*) \\ v_4(b^*, a_{21}^*) \end{pmatrix} + J(a) \begin{pmatrix} \varepsilon_1 \\ \varepsilon_2 \end{pmatrix} = \begin{pmatrix} J_{11}(a)\varepsilon_1 + J_{12}(a)\varepsilon_2 \\ J_{21}(a)\varepsilon_1 + J_{22}(a)\varepsilon_2 \end{pmatrix}. \qquad (4.118)$$

It is no difficulty to find an appropriate ε_1 and ε_2 (remember that ε_i can take both positive and negative values) such that $v_5 < 0$ and $v_4 > 0$. Further, to make it more clear, let

$$\varepsilon_2 = (\delta + \bar{\varepsilon})\varepsilon_1, \quad \text{where } \delta = -\frac{J_{21}(a)}{J_{22}(a)}, \qquad (4.119)$$

then the linear approximations of v_5 and v_4 become

$$v_5 \approx \left(\frac{J_{11}(a)J_{22}(a) - J_{12}(a)J_{21}(a)}{J_{22}(a)} + J_{12}(a)\bar{\varepsilon} \right)\varepsilon_1 \quad \text{and} \quad v_4 \approx J_{22}\varepsilon_1\bar{\varepsilon}. \quad (4.120)$$

It is easy to see from (4.120) that one can always obtain $0 < v_4 \ll -v_5 \ll v_6$ by choosing $|\varepsilon_1| \ll 1$ and $|\bar{\varepsilon}| \ll 1$ with appropriate signs. Note from (4.119) that ε_2 is of the same order as ε_1.

Next, consider v_3. To estimate v_3 near the critical point, calculating $\frac{\partial v_3}{\partial \bar{d}_{21}}$ at $(b^* + \varepsilon_1, \bar{b}_{21}^* + \varepsilon_2, \bar{d}_{21}(b^* + \varepsilon_1, \bar{b}_{21}^* + \varepsilon_2))$ gives

$$\frac{\partial v_3}{\partial \bar{d}_{21}} \approx v_{30} + v_{31}\varepsilon_1 + v_{32}\varepsilon_2, \qquad (4.121)$$

which is dominated by v_{30} for $0 < |\varepsilon_1|, |\varepsilon_2| \ll 1$. Thus, in order to have $v_3 < 0$ after perturbation, one may choose ε_3 small, such that

$$\bar{d}_{21} = \bar{d}_{21}(a, b^* + \varepsilon_1, \bar{b}_{21}^* + \varepsilon_2) + \varepsilon_3, \qquad (4.122)$$

where $\bar{d}_{21}(a, b, \bar{b}_{21})$ is given by (4.101). Then, one obtains $v_3 \approx (v_{30} + v_{31}\varepsilon_1 + v_{32}\varepsilon_2)\varepsilon_3$, where $\varepsilon_3 > 0(< 0)$ if $v_{30} < 0(> 0)$, respectively. The additional condition $-v_3 \ll v_4$ results in $0 < |\varepsilon_3| \ll (|\varepsilon_2|, |\varepsilon_1|) \ll 1$.

Similarly, we can follow the procedure above to estimate v_2. Computing $\frac{\partial v_2}{\partial \bar{b}_{30}}$ at $(a, b^* + \varepsilon_1, \bar{b}_{21}^* + \varepsilon_2, \bar{d}_{21}(a, b^* + \varepsilon_1, \bar{b}_{21}^* + \varepsilon_2) + \varepsilon_3, \bar{b}_{30}(a, b^* + \varepsilon_1, \bar{b}_{21}^* + \varepsilon_2, \bar{d}_{21}(a, b^* + \varepsilon_1, \bar{b}_{21}^* + \varepsilon_2) + \varepsilon_3))$ yields

$$\frac{\partial v_2}{\partial \bar{b}_{30}} \approx v_{20} + v_{21}\varepsilon_1 + v_{22}\varepsilon_2 + v_{23}\varepsilon_3 \qquad (4.123)$$

which is dominated by v_{20} for small $\varepsilon_i, i = 1, 2, 3$. Hence, one can obtain $v_2 > 0$ after a perturbation ε_4 ($0 < |\varepsilon_4| \ll 1$) on $\bar{b}_{30}$ such that

$$\bar{b}_{30} = \bar{b}_{30}(a, b^* + \varepsilon_1, \bar{b}_{21}^* + \varepsilon_2, \bar{d}_{21}(a, b^* + \varepsilon_1, \bar{b}_{21}^* + \varepsilon_2) + \varepsilon_3) + \varepsilon_4. \qquad (4.124)$$

Then, one finds $v_2 \approx (v_{20} + v_{21}\varepsilon_1 + v_{22}\varepsilon_2 + v_{23}\varepsilon_3)\varepsilon_4$ in which $\varepsilon_4 > 0(< 0)$ if $v_{20} > 0(< 0)$, respectively. The additional condition $v_2 \ll -v_3$ produces $0 < |\varepsilon_4| \ll |\varepsilon_3| \ll (|\varepsilon_2|, |\varepsilon_1|) \ll 1$.

For v_1, we need a perturbation on $\bar{d}_{30}$ such that $v_1 < 0$. Since

$$\frac{\partial v_1}{\partial \bar{d}_{30}} = \frac{3}{2}b,$$

so we perturb $\bar{d}_{30}$ to $\bar{d}_{30}(a, b^* + \varepsilon_1, \bar{b}_{21}^* + \varepsilon_2, \bar{d}_{21}(a, b^* + \varepsilon_1, \bar{b}_{21}^* + \varepsilon_2) + \varepsilon_3) + \varepsilon_5$ to obtain $v_1 \approx (\frac{3}{2}b + v_{11}\varepsilon_1 + v_{12}\varepsilon_2 + v_{13}\varepsilon_3 + v_{14}\varepsilon_4)\varepsilon_5$, which is less than zero if $b\varepsilon_5 < 0$ (i.e., $a\varepsilon_5 < 0$). Applying the condition $-v_1 \ll v_2$ results in $0 < |\varepsilon_5| \ll |\varepsilon_4|$.

Finally, one more perturbation on the linear part of system (4.52) is needed to have a total of 6 limit cycles in the vicinity of $(x, y) = (0, 1)$. Since $v_0 = \frac{1}{2}(a_{12} - a)$ (see (4.54)), one may perturb either a or a_{12}. Since a is an arbitrary coefficient, it is better to perturb a_{12}. By letting

$$a_{12} = a + \varepsilon_6, \quad \varepsilon_6 > 0, \tag{4.125}$$

one has $v_0 = \frac{1}{2}\varepsilon_6 > 0$. Therefore, the perturbed normal form for the Hopf bifurcation of system (4.52) arising from the points $(x, y) = (0, \pm1)$, up to the term ρ^{13}, is given by (4.77) satisfying $0 < v_0 \ll -v_1 \ll v_2 \ll -v_3 \ll v_4 \ll -v_5 \ll v_6$.

This finishes the proof of Theorem 4.13. $\square$

To end this case, we present a numerical example when the origin is a node. We choose $a = -0.7$, and then

$$b^* = -0.3336022980, \quad \bar{b}_{21}^* = -0.1423340664, \quad \bar{d}_{21}^* = 1.1973330467,$$

$$\bar{b}_{30}^* = 0.0744658372, \quad \bar{d}_{30}^* = 2.0142732091,$$

for which $b_{12} = -4.1028179815$. Further, we take the perturbations

$$\varepsilon_1 = 0.3 \times 10^{-6}, \quad \varepsilon_2 = -0.4 \times 10^{-3}, \quad \varepsilon_3 = -0.7 \times 10^{-7},$$

$$\varepsilon_4 = 0.2 \times 10^{-11}, \quad \varepsilon_5 = 0.1 \times 10^{-14}, \quad \varepsilon_6 = 0.3 \times 10^{-19},$$

under which system (4.52) becomes

$$\begin{aligned}
\dot{x} = {}& -0.7000000000x - 0.3336019980y + 4.5658610164x^3 \\
& - 5.0539492766x^2y - 0.7000000000xy^2 + 0.3336019980y^3, \\
\dot{y} = {}& -0.3336019980x - 0.7000000000y - 1.3447323014x^3 \\
& - 0.8333681006x^2y - 4.1028222711xy^2 + 0.7000000000y^3.
\end{aligned} \tag{4.126}$$

Equation (4.126) has 5 real fixed points: $C_0 = (0, 0)$, $C_{1,2} = (0, \pm1)$ and $C_{3,4} = (\pm0.2398840466, \mp0.1797759006)$. A linear analysis shows that C_0 is a stable node, $C_{1,2}$ are two weakly unstable focus points, and $C_{3,4}$ are saddle points. The

Fig. 4.7 The phase portrait of system (4.52) having 12 limit cycles, when the origin is a node

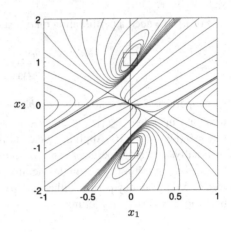

perturbed focus values for the two weak focus points are

$$v_0 = 0.1500000000 \times 10^{-19}, \qquad v_1 = -0.5004029970 \times 10^{-15},$$

$$v_2 = 0.4793370351 \times 10^{-11}, \qquad v_3 = -0.1322749898 \times 10^{-7},$$

$$v_4 = 0.2229798103 \times 10^{-5}, \qquad v_5 = -0.5916202601 \times 10^{-4},$$

and $v_6 = 0.3308305789 \times 10^{-3}$. Then the normal form (4.92) with these focus values yields the following 6 positive roots of r, which are the approximations of the 6 small limit cycles:

$$r_1 = 0.0072744336, \qquad r_2 = 0.0100774933, \qquad r_3 = 0.0150111942,$$

$$r_4 = 0.0824725838, \qquad r_5 = 0.2066015944, \qquad r_6 = 0.3591142108.$$

Note that the original coefficients b_{21}, a_{21}, b_{30} and a_{30} can be found from (4.69) as follows:

$$b_{21} = -0.8333681006, \qquad a_{21} = -5.0539492766,$$

$$b_{30} = -1.3447323014, \qquad a_{30} = 4.5658610164.$$

The phase portrait of this generic case when the origin is a node, under the above perturbations is shown in Fig. 4.7, where the two boxes contain the 12 small limit cycles near the focus points $(0, \pm 1)$. The stability of these limit cycles is the same as that for the special case, as shown in Fig. 4.2. Similarly, it can be shown that no large limit cycles can exist near the two saddle points $C_{3,4}$.

(B) Case $a_3^2 + b_3^2 - b_0^2 < 0$ Now we consider the case $a_3^2 + b_3^2 - b_0^2 < 0$, under which the origin is either a focus point (if $a_0 \neq 0$) or a center (if $a_0 = 0$). Similarly, let $a = a_0$ and $b = \sqrt{-(a_3^2 + b_3^2 - b_0^2)} > 0$, then the eigenvalues of the Jacobian of the system evaluated at the origin are $a \pm b\mathrm{i}$. Applying the linear transformation

$$
\begin{pmatrix} w_1 \\ w_2 \end{pmatrix} = \begin{bmatrix} 0 & b_0 - b_3 \\ \sqrt{-(a_3^2 + b_3^2 - b_0^2)} & a_3 \end{bmatrix} \begin{pmatrix} x \\ y \end{pmatrix} \tag{4.127}
$$

to (4.50), and renaming the coefficients of the third-order terms in the resulting equations as $a_{30}, a_{21}, a_{12}, b_{30}, b_{21}, b_{12}$ and b_{30} produces

$$
\begin{aligned}
\frac{dx}{dt} &= ax + by + a_{30}x^3 + a_{21}x^2 y + a_{12}xy^2 + a_{03}y^3, \\
\frac{dy}{dt} &= -bx + ay + b_{30}x^3 + b_{21}x^2 y + b_{12}xy^2 + b_{03}y^3.
\end{aligned} \tag{4.128}
$$

Then, applying the scalings given in (4.56) and (4.58), and the linear transformation (4.57) to (4.128) yields

$$
\begin{aligned}
\frac{du}{d\tau} &= v + 2\bar{d}_{21}u^2 + 4auv - \frac{3}{2}v^2 + 4b\bar{d}_{30}u^3 - 2\bar{d}_{21}u^2 v - 2auv^2 + \frac{1}{2}v^3, \\
\frac{dv}{d\tau} &= -u - 4\bar{b}_{21}u^2 + 2(2a^2 + 2b^2 + 1)uv - 8\bar{b}_{30}u^3 \\
&\quad + 4\bar{b}_{21}u^2 v - (2a^2 + b^2 + 1)uv^2,
\end{aligned} \tag{4.129}
$$

where $\bar{d}_{21}$ and $\bar{d}_{30}$ are given in (4.60), and $\bar{b}_{21}$ and $\bar{b}_{30}$ are

$$
\begin{aligned}
\bar{b}_{21} &= b_{21}b^2 - a_{21}ab + 2ab^2 + 2a^3 + a, \\
\bar{b}_{30} &= b_{30}b^3 - a_{30}ab^2 - a_{21}a^2 b + b_{21}ab^2 + a^2 b^2 + a^4 + \frac{1}{2}a^2.
\end{aligned} \tag{4.130}
$$

Thus, based on (4.129), we can follow the same procedures described in Part (A) of Sect. 4.2.2 to consider the existence of small limit cycles. In fact, the symbolic computation results have shown that all the formulas for this case are similar to the case when the origin is a saddle point or a node. Indeed, the maximal number of small limit cycles for this case is also 12. All the solutions of the limit cycles for the special and generic cases are similar to those given in Part (A) of Sect. 4.2.2, and thus the detailed analysis is omitted here.

One might think that in addition to the two symmetric weak focus points, it might be possible to have more small limit cycles around the origin if one could make the origin a center. However, this requires $a = 0$, for which one can easily show that all the three fixed points (the origin and the two symmetric points) become centers. Thus, this would not increase the number of limit cycles.

Finally, let us investigate the maximal number of small limit cycles around the origin, which is assumed to be an elementary center when $a = 0$ (i.e., $a_0 = 0$). Under the condition $a = 0$, (4.128) has 8 free coefficients (a_{ij}'s and b_{ij}'s), while $b = \omega > 0$ can be chosen arbitrarily, which does not affect the number of limit cycles. In general, 8 free coefficients may yield 9 small limit cycles. But a careful consideration (by following a procedure similar to that described in Part (A)) shows that the maximal number of small limit cycles around the origin is 5 (much less than 12) arising from the case of two symmetric focus points. For the more degenerate case, $a = b = 0$, giving rise to a double-zero singular point, it is obvious that 12 small limit cycles are not possible.

Summarizing the results obtained in this section, we have the following theorem.

Theorem 4.15 *Suppose the cubic system (4.52) has the property of Z_2 symmetry, then the maximal number of small limit cycles that the system can exhibit is 12, i.e., $\hat{H}_2(3) = 12$.*

Remark 4.16 As usual, $H(n)$ denotes the Hilbert number of a planar system with nth-degree polynomials. $H_2(n)$ represents the Hilbert number for systems with Z_2 symmetry, while $\hat{H}_2(n)$ denotes the number of small amplitude limit cycles bifurcating in such systems. It is obvious that in general $\hat{H}_i(n) \le H_i(n) \le H(n)$.

Remark 4.17 Recently, Liu and Li [147] have used the free parameter a to show that system (4.52) can have a limit cycle at infinity. Therefore, $H_2(3) \ge 13$.

For the last part of this section, we present a numerical example which exhibits 12 small limit cycles around the two symmetrical focus points when the origin is also a focus point. Again taking $a = -0.7$, the same as the case when the origin is a node, the critical parameter values are

$$b^* = -0.5888918635, \qquad \bar{b}_{21}^* = -0.0194519814, \qquad \bar{d}_{21}^* = 1.0122619205,$$

$$\bar{b}_{30}^* = 0.0379318682, \qquad \bar{d}_{30}^* = 0.4983879619,$$

and then $b_{12} = -3.1020867158$. Under the perturbations

$$\varepsilon_1 = -0.3 \times 10^{-4}, \qquad \varepsilon_2 = 0.3 \times 10^{-2}, \qquad \varepsilon_3 = 0.6 \times 10^{-6},$$

$$\varepsilon_4 = 0.3 \times 10^{-10}, \qquad \varepsilon_5 = 0.7 \times 10^{-14}, \qquad \varepsilon_6 = 0.6 \times 10^{-19},$$

system (4.52) can be written as

$$\dot{x} = -0.7000000000x - 0.5889218635y + 2.2161956860x^3$$
$$- 2.5682071892x^2y - 0.7000000000xy^2 + 0.5889218635y^3,$$
$$\dot{y} = 0.5889218635x - 0.7000000000y - 0.7072960219x^3$$
$$+ 2.2961669830x^2y - 3.1019886923xy^2 + 0.7000000000y^3,$$

$$(4.131)$$

which has five real fixed points: $C_0 = (0,0)$ is a stable focus point; $C_{1,2} = (0, \pm1)$ are two weakly unstable focus points; and $C_{3,4} = (\pm1.0362101038,$

Fig. 4.8 The phase portrait of system (4.52) having 12 limit cycles, when the origin is a focus point

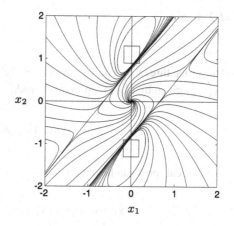

±0.4888355303) are two saddle points. The perturbed focus values are

$$v_0 = 0.3000000000 \times 10^{-19}, \qquad v_1 = -0.6183679567 \times 10^{-14},$$

$$v_2 = 0.2067107022 \times 10^{-9}, \qquad v_3 = -0.8432160503 \times 10^{-6},$$

$$v_4 = 0.5282843906 \times 10^{-3}, \qquad v_5 = -0.8736269745 \times 10^{-2},$$

and $v_6 = 0.0172191524$. By using the normal form (4.92) with the focus values above, we obtain the following 6 positive roots of r, which are the approximations of the 6 small limit cycles:

$$r_1 = 0.0024579973, \qquad r_2 = 0.0052712603,$$

$$r_3 = 0.0161047983, \qquad r_4 = 0.0366105054,$$

$$r_5 = 0.2612258576, \qquad r_6 = 0.6614266701.$$

The original coefficients b_{21}, a_{21}, b_{30} and a_{30} can be found from (4.60) and (4.130) as follows:

$$b_{21} = 2.2961669830, \qquad a_{21} = -2.5682071892,$$

$$b_{30} = -0.7072960219, \qquad a_{30} = 2.2161956860.$$

The phase portrait of this example when the origin is a focus point, is shown in Fig. 4.8, where the areas near the focus points $(0, \pm1)$ marked by the two boxes, contain the 12 small limit cycles. The existence of a large limit cycle at infinity is only for a particular value of the parameter a; see [147] for more details.

4.2.3 $q = 3$

The order three, Z_3-equivariant vector fields are described by $\dot{z} = F_3(z, \bar{z})$ with $F_3(z, \bar{z})$ given by (4.17), up to third order:

$$F_3(z, \bar{z}) = (A_0 + A_1|z|^2)z + A_2\bar{z}^2, \tag{4.132}$$

and the real form can be obtained as

$$
\begin{aligned}
\dot{x}_1 &= a_0 x_1 - b_0 x_2 + a_2 x_1^2 + 2b_2 x_1 x_2 - a_2 x_2^2 \\
&\quad + a_1 x_1^3 - b_1 x_1^2 x_2 + a_1 x_1 x_2^2 - b_1 x_2^3, \\
\dot{x}_2 &= b_0 x_1 + a_0 x_2 + b_2 x_1^2 - 2a_2 x_1 x_2 - b_2 x_2^2 \\
&\quad + b_1 x_1^3 + a_1 x_1^2 x_2 + b_1 x_1 x_2^2 + a_1 x_2^3.
\end{aligned}
\tag{4.133}
$$

Suppose system (4.133) has a fixed point at $(0, 1)$, then the other two fixed points are located at $(\pm\sqrt{3}/2, -1/2)$, which requires that

$$a_0 = -(a_1 - b_2) \quad \text{and} \quad b_0 = -(b_1 + a_2). \tag{4.134}$$

In order to let the three fixed points be elementary centers, it requires

$$b_2 = -a_1. \tag{4.135}$$

Then the frequency ω is given by $\omega = \sqrt{-3(a_2^2 + 2b_1 a_2 + 3a_1^2)}$. Hence, one may set

$$b_1 = -\frac{\omega^2 + 9a_1^2 + 3a_2^2}{6a_2}, \tag{4.136}$$

where ω can take any positive real value. Therefore, in the final normalized equations, there are only two free parameters, a_1 and a_2, which can be scaled to $a_1 \to \omega a_1$ and $a_2 \to \omega a_2$.

Now, applying the linear transformation

$$
\begin{pmatrix} x_1 \\ x_2 \end{pmatrix} = \begin{pmatrix} 0 \\ 1 \end{pmatrix} + \begin{bmatrix} 3a_1 & -1 \\ 3a_2 & 0 \end{bmatrix} \begin{pmatrix} u \\ v \end{pmatrix}, \tag{4.137}
$$

with the time scaling $\tau = \omega t$ to system (4.133) results in

$$
\frac{du}{d\tau} = v - \frac{a_1(1 + 9a_1^2 - 3a_2^2)}{a_2}u^2 + \frac{9a_2^2 + 1 + 9a_1^2}{3a_2}uv
$$

$$
- \frac{3a_1(a_1^2 + a_2^2)(1 + 9a_1^2 - 3a_2^2)}{2a_2^2}u^3
$$

$$
+ \frac{3a_1^2 + 3a_2^4 + a_2^2 + 27a_1^4 + 6a_1^2a_2^2}{2a_2^2}u^2v - \frac{a_1(1 + 9a_1^2 + a_2^2)}{2a_2^2}uv^2
$$

$$
+ \frac{1 + 9a_1^2 + 3a_2^2}{18a_2^2}v^3,
$$

$$
\frac{dv}{d\tau} = -u - \frac{9(a_1^2 + a_2^2)(1 + 9a_1^2 - 3a_2^2)}{2a_2}u^2 + \frac{2a_1(1 + 9a_1^2 + 9a_2^2)}{a_2}uv \tag{4.138}
$$

$$
- \frac{1 + 9a_1^2 + 9a_2^2}{6a_2}v^2 - \frac{9(a_1^2 + a_2^2)^2(1 + 9a_1^2 + 3a_2^2)}{2a_2}u^3
$$

$$
+ \frac{9a_1(a_1^2 + a_2^2)(1 + 9a_1^2 + 5a_2^2)}{2a_2^2}u^2v
$$

$$
- \frac{30a_1^2a_2^2 + 3a_2^4 + 27a_1^4 + a_2^2 + 3a_1^2}{2a_2^2}uv^2 + \frac{1 + 9a_1^2 + 9a_2^2}{6a_2^2}v^3.
$$

Finally, apply the Maple program to system (4.138) to obtain the focus values

$$
v_1 = -\frac{27a_1}{2}\left(a_1^2 + 9a_1^4 + 10a_1^2a_2^2 + a_2^2 + a_2^4\right),
$$

$$
v_2 = -\frac{27a_1}{8a_2^2}f_2(a_1, a_2),
$$

$$
v_3 = -\frac{3a_1}{8192a_2^2}f_3(a_1, a_2),
$$

$$
\vdots
$$

where f_2 and f_3 are polynomials in a_1 and a_2. These expressions clearly indicate that $a_2 \neq 0$, and thus the only possibility for $v_1 = 0$ is $a_1 = 0$, which yields $v_1 = v_2 = v_3 = \cdots = 0$, leading to a center. This clearly shows that the maximal number of small limit cycles around the three symmetric focus points is 3. If, instead of the 3 weak focus points, we consider the limit cycles existing in the neighborhood of the origin for (4.133), then $a_0 = 0$ ($v_0 = 0$), and the normal form computation shows that

$$
v_1 = a_1, \qquad v_2 = \frac{2}{9}a_1\left(a_2^2 + b_2^2\right), \qquad v_3 = \frac{2}{81}a_1\left(a_2^2 + b_2^2\right)\left[17(a_2^2 + b_2^2) - 42b_1\right], \ldots
$$

Fig. 4.9 The phase portrait of system (4.133) having 3 small limit cycles and 1 large limit cycle, for
$a_0 = 0.6566666667$,
$b_0 = 0.5$,
$a_1 = -a_2 = -0.3333333333$,
$b_1 = -0.8333333333$,
$b_2 = 0.3233333333$

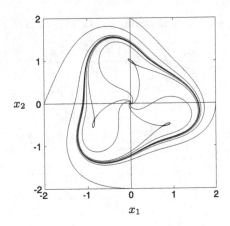

which implies that $a_1 \neq 0$, otherwise it is a center. Therefore, for this case, there is only 1 small limit cycle.

Summarizing the above results give the following theorem.

Theorem 4.18 *The maximal number of small limit cycles that a third-degree planar polynomial system with Z_3 symmetry can have is 3, i.e., $\hat{H}_3(3) = 3$.*

Further, numerical simulation shows that there exists a large limit cycle enclosing all 3 small limit cycles, as depicted in Fig. 4.9.

4.2.4 $q = 4$

Now we turn to the case $q = 4$. A similar procedure to that given in the previous subsection can be applied here. The Z_4-equivariant vector fields are given by $\dot{z} = F_4(z, \bar{z})$ with (see (4.18))

$$F_4(z, \bar{z}) = (A_0 + A_1|z|^2)z + A_2\bar{z}^3 \qquad (4.139)$$

up to third order, and the real form is

$$
\begin{aligned}
\dot{x}_1 = {} & a_0x_1 - b_0x_2 + (a_1 + a_2)x_1^3 - (b_1 - 3b_2)x_1^2x_2 \\
& + (a_1 - 3a_2)x_1x_2^2 - (b_1 + b_2)x_2^3, \\
\dot{x}_2 = {} & b_0x_1 + a_0x_2 + (b_1 + b_2)x_1^3 + (a_1 - 3a_2)x_1^2x_2 \\
& + (b_1 - 3b_2)x_1x_2^2 + (a_1 + a_2)x_2^3.
\end{aligned}
\qquad (4.140)
$$

Again suppose system (4.140) has a fixed point at $(0, 1)$. Then the other three fixed points are located at $(0, -1)$, $(\pm 1, 0)$. This yields that $a_0 = -(a_1 + a_2)$ and $b_0 = -(b_1 + b_2)$. Further requiring the four fixed points to be elementary centers needs

$a_2 = a_1$, and thus the frequency ω becomes $\omega = 2\sqrt{-2(b_1 + b_2)b_2 - 4a_1^2}$, from which one may set

$$b_1 = -\frac{16a_1^2 + 8b_2^2 + \omega^2}{8b_2}, \tag{4.141}$$

and thus there are only two free parameters, a_1 and b_2, which can be scaled as $a_1 \to \omega a_1$ and $b_2 \to \omega b_2$.

Next, with the time scaling $\tau = \omega t$, applying the linear transformation

$$\begin{pmatrix} x_1 \\ x_2 \end{pmatrix} = \begin{pmatrix} 0 \\ 1 \end{pmatrix} + \begin{bmatrix} 4a_1 & -1 \\ 4b_2 & 0 \end{bmatrix} \begin{pmatrix} u \\ v \end{pmatrix} \tag{4.142}$$

to system (4.140) yields

$$\frac{du}{d\tau} = v - \frac{a_1(1 + 24a_1^2 + 8b_2^2)}{a_2}u^2 + \frac{1 + 32a_1^2 + 1 + 32b_2^2}{4b_2}uv - \frac{a_1}{2b_2}v^2$$

$$- \frac{2a_1(a_1^2 + b_2^2)(1 + 16a_1^2 + 16b_2^2)}{2b_2^2}u^3$$

$$+ \frac{3a_1^2 + 48a_1^4 + b_2^2 + 32b_2^4 + 48a_1^2b_2^2}{2b_2^2}u^2v - \frac{a_1(3 + 48a_1^2 + 16b_2^2)}{8b_2^2}uv^2$$

$$+ \frac{1 + 16a_1^2}{32b_2^2}v^3,$$

$$\frac{dv}{d\tau} = -u - \frac{2(a_1^2 + b_2^2)(3 + 64a_1^2 - 3a_2^2)}{2b_2}u^2 + \frac{2a_1(1 + 24a_1^2 + 24b_2^2)}{b_2}uv$$

$$- \frac{1 + 32a_1^2 + 32b_2^2}{8b_2}v^2 - \frac{8(a_1^2 + b_2^2)^2(a_1^2 + 16a_1^4 + 48a_1^2b_2^2 + b_2^2)}{b_2^2}u^3$$

$$+ \frac{2a_1(a_1^2 + b_2^2)(3 + 48a_1^2 + 80b_2^2)}{b_2^2}u^2v$$

$$- \frac{80a_1^2b_2^2 + 32b_2^4 + 487a_1^4 + b_2^2 + 3a_1^2}{2b_2^2}uv^2 + \frac{1 + 16a_1^2 + 16b_2^2}{8b_2^2}v^3. \tag{4.143}$$

Then applying the Maple program to system (4.143) results in the focus values

$$v_1 = -32a_1(a_1^2 + 16a_1^4 + 16a_1^2b_2^2 + b_2^2),$$

$$v_2 = -\frac{128a_1^3}{9b_2^2}f_2(a_1, b_2),$$

$$v_3 = -\frac{a_1}{162b_2^4}f_3(a_1, b_2),$$

$$\vdots$$

Fig. 4.10 The phase portrait of system (4.140) having 4 small limit cycles and 1 large limit cycle, for $a_0 = -0.19$, $b_0 = 2.6$, $a_1 = 0.1$, $b_1 = -2.8$, $a_2 = 0.09$, $b_2 = 0.2$

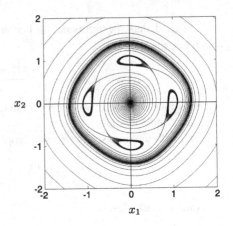

where f_2 and f_3 are polynomials in a_1 and b_2. Hence, $a_1 \neq 0$ and so $v_1 \neq 0$. Otherwise, the point $(u, v) = (0, 0)$ is a center. This shows that the Z_4-equivariant vector field (4.140) has 4 small limit cycles. A similar analysis shows that considering the origin as an elementary center leads to only 1 small limit cycle. Hence, we have the following theorem.

Theorem 4.19 *The maximal number of small limit cycles that a cubic-degree planar polynomial system with Z_4 symmetry can have is 4, i.e., $\hat{H}_4(3) = 4$.*

Further, a numerical simulation reveals that there exists a large limit cycle enclosing all 4 small limit cycles. An example is shown in Fig. 4.10.

4.2.5 $q \geq 5$

When $q \geq 5$, we have $\dot{z} = F_q(z, \bar{z})$ with $F_q(z, \bar{z})$ given in (4.19)–(4.22) up to third order:

$$F_q(z, \bar{z}) = \left(A_0 + A_1 |z|^2 \right) z. \tag{4.144}$$

Thus, the real system is given by

$$\begin{aligned}
\dot{x}_1 &= a_0 x_1 - b_0 x_2 + a_1 x_1^3 + a_1 x_1^2 x_2 - b_1 x_1 x_2^2 - b_1 x_2^3, \\
\dot{x}_2 &= b_0 x_1 + a_0 x_2 + b_1 x_1^3 + b_1 x_1^2 x_2 + a_1 x_1 x_2^2 + a_1 x_2^3.
\end{aligned} \tag{4.145}$$

In order to have q ($q \geq 5$) weak focus points with 1 at $(0, 1)$, it requires that $a_0 = -a_1$ and $b_0 = -b_1$. Then the eigenvalues of the Jacobian of system (4.145) evaluated at $(0, 1)$ are 0 and $2a_1$, indicating that it is not an elementary center, and so small limit cycles do not exist around these q fixed points. If we consider the origin of system (4.145) as an elementary center (with $a_0 = 0$), then it is easy to use

Fig. 4.11 The phase portrait of system (4.145) having 1 small limit cycle for $a_0 = 0.01, b_0 = 1.0,$ $a_1 = -0.1, b_1 = -1.5$

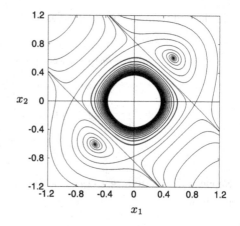

normal form to show that

$$v_1 = \frac{3}{4}a_1, \qquad v_2 = -\frac{1}{8}a_1b_1, \qquad v_3 = \frac{1}{1024}a_1\left(35a_1^2 + 40a_1b_1 + 59b_1^2\right), \dots.$$

Hence, there exists only 1 small limit cycle in the vicinity of the origin. Therefore, in general, we have the following theorem.

Theorem 4.20 *The maximal number of small limit cycles that a cubic-degree planar polynomial system with Z_q symmetry $(q \geq 5)$ can have is 1, i.e., $\hat{H}_q(3) = 1$ $(q \geq 5)$.*

A numerical example showing the existence of 1 limit cycle for system (4.145) is depicted in Fig. 4.11.

4.3 Fourth-Order Vector Fields with Z_q Symmetry

From the forms of the nontrivial Z_q-equivariant vector fields given in [128], it is easy to see that when q is even, the fourth-order planar polynomial systems are the same as that of the third-order systems. Therefore, we only need to consider the cases when q is odd. In this section, we only present the results for cases when $q \geq 5$. The case $q = 3$ is still under investigation.

4.3.1 $q = 5$

The fourth-order Z_5-equivariant vector fields are described by $\dot{z} = F_5(z, \bar{z})$ with

$$F_5(z, \bar{z}) = \left(A_0 + A_1|z|^2\right)z + A_2\bar{z}^4, \tag{4.146}$$

and the real form can be obtained as

$$\dot{x}_1 = a_0 x_1 - b_0 x_2 + a_1 x_1^3 - b_1 x_1^2 x_2 + a_1 x_1 x_2^2 - b_1 x_2^3$$
$$+ a_2 x_1^4 + 4b_2 x_1^3 x_2 - 6a_2 x_1^2 x_2^2 - 4b_2 x_1 x_2^3 + a_2 x_2^4,$$
$$\dot{x}_2 = b_0 x_1 + a_0 x_2 + b_1 x_1^3 + a_1 x_1^2 x_2 + b_1 x_1 x_2^2 + a_1 x_2^3$$
$$+ b_2 x_1^4 - 4a_2 x_1^3 x_2 - 6b_2 x_1^2 x_2^2 + 4a_2 x_1 x_2^3 + b_2 x_2^4.$$

(4.147)

Suppose that system (4.147) has a fixed point at $(0, 1)$. Then the other four fixed points are

$$\left(\pm \frac{\sqrt{10 + 2\sqrt{5}}}{4}, \frac{\sqrt{5} - 1}{4} \right) \quad \text{and} \quad \left(\pm \frac{\sqrt{10 - 2\sqrt{5}}}{4}, -\frac{\sqrt{5} + 1}{4} \right).$$

This requires that

$$a_0 = -(a_1 + b_2) \quad \text{and} \quad b_0 = -(b_1 - a_2).$$

Further requiring the five fixed points to be elementary centers needs $b_2 = a_1$, and thus the frequency ω is given by

$$\omega = \sqrt{10 a_2 b_1 - 25 a_1^2 - 15 a_2^2},$$

from which one may set

$$b_1 = \frac{25 a_1^2 + 15 a_2^2 + \omega^2}{10 a_2}.$$

Thus, there are only two free parameters, a_1 and a_2, which can be scaled as $a_1 \rightarrow \omega a_1$ and $a_2 \rightarrow \omega a_2$.

Then, with the time scaling $\tau = \omega t$, introduce the linear transformation

$$\begin{pmatrix} x_1 \\ x_2 \end{pmatrix} = \begin{pmatrix} 0 \\ 1 \end{pmatrix} + \begin{bmatrix} -5a_1 & 1 \\ 5a_2 & 0 \end{bmatrix} \begin{pmatrix} u \\ v \end{pmatrix}$$

(4.148)

into system (4.147) to obtain

$$\frac{du}{d\tau} = v - a_1\left(1 + 50a_1^2 + 30a_2^2\right)u^2 + \left(\frac{1}{5} + 15a_1^2 + 15a_2^2\right)uv - a_1 v^2$$

$$- \frac{5}{2}a_1\left(a_1^2 + a_2^2\right)\left(1 + 25a_1^2 + 85b_2^2\right)u^3$$

$$+ \left(\frac{3}{2}a_1^2 + \frac{1}{2}a_2^2 + \frac{75}{2}a_1^4 + 85a_1^2 a_2^2 + \frac{135}{2}a_2^4\right)u^2 v$$

$$- \frac{1}{10}a_1\left(3 + 75a_1^2 + 35a_2^2\right)uv^2 + \left(\frac{1}{50} + \frac{1}{2}a_1^2 + \frac{1}{2}a_2^2\right)v^3$$

$$+ 125a_1 a_2^2\left(a_1^2 + a_2^2\right)\left(a_1^2 - 3a_2^2\right)u^4 - 100a_2^2\left(a_1^4 - a_2^4\right)u^3 v$$

$$+ 30a_1 a_2^2\left(a_1^2 + a_2^2\right)u^2 v^2 - 4a_2^2\left(a_1^2 + a_2^2\right)uv^3 + \frac{1}{5}a_1 a_2^2 v^4,$$

$$\frac{dv}{d\tau} = -u - \frac{5}{2}\left(a_1^2 + a_2^2\right)\left(3 + 125a_1^2 - 15a_2^2\right)u^2 \tag{4.149}$$

$$+ 2a_1\left(1 + 50a_1^2 + 50a_2^2\right)uv - \left(\frac{1}{10} + \frac{15}{2}a_1^2 + \frac{15}{2}a_2^2\right)v^2$$

$$+ \frac{5}{2}a_1\left(a_1^2 + a_2^2\right)\left(3 + 75a_1^2 + 295a_2^2\right)u^2 v$$

$$- \frac{25}{2}\left(a_1^2 + a_2^2\right)\left(a_1^2 + a_2^2 + 25a_1^4 + 160a_1^2 a_2^2 - 25a_2^4\right)u^3$$

$$- \left(\frac{3}{2}a_1^2 + \frac{1}{2}a_2^2 + \frac{75}{2}a_1^4 + 105a_1^2 a_2^2 + 135a_2^4\right)uv^2$$

$$+ \frac{1}{10}a_1\left(1 + 25a_1^2 + a_2^2\right)v^3 + 625a_2^3\left(a_1^2 + a_2^2\right)\left[\left(a_1^2 - a_2^2\right)^2 - 4a_1^2 a_2^2\right]u^4$$

$$- 500a_1 a_2^2\left(a_1^2 + a_2^2\right)\left(a_1^2 - 3a_2^2\right)u^3 v + 150a_2^2\left(a_1^4 - a_2^4\right)u^2 v^2$$

$$- 20a_1 a_2^2\left(a_1^2 + a_2^2\right)uv^3 + a_2^2\left(a_1^2 + a_2^2\right)v^4.$$

Now applying the Maple program to system (4.149) yields the following focus values:

$$v_1 = -\frac{125}{2}a_1 a_2^2\left(a_1^2 + a_2^2\right)V_1\left(a_1^2, a_2^2\right),$$

$$v_2 = -\frac{3125}{72}a_1 a_2^2\left(a_1^2 + a_2^2\right)V_2\left(a_1^2, a_2^2\right),$$

$$v_3 = -\frac{125}{663552}a_1 a_2^2\left(a_1^2 + a_2^2\right)V_3\left(a_1^2, a_2^2\right),$$

$$\cdots$$

$$v_i = c_i a_1 a_2^2\left(a_1^2 + a_2^2\right)V_i\left(a_1^2, a_2^2\right),$$

where c_i is a constant, and $V_i(a_1^2, a_2^2)$ is a polynomial in a_1^2 and a_2^2. In particular, we have (from computer output)

```
V1 := 1+25*a1^2-3*a2^2:
V2 := 1953125*a1^10+312500*a1^8+3203125*a1^8*a2^2+18750*a1^6
      +428125*a1^6*a2^2+631250*a1^6*a2^4+101875*a1^4*a2^4
      +500*a1^4+18875*a1^4*a2^2-537750*a1^4*a2^6+275*a1^2*a2^2
      -22305*a1^2*a2^6+2960*a1^2*a2^4+5*a1^2+77625*a1^2*a2^8
      -3375*a2^10-9*a2^4+99*a2^6+405*a2^8:
V3 := 1368786125*a1^4*a2^2-1627371*a2^4+3361350*a1^2*a2^2
      -99885717*a2^6+63496101*a1^2*a2^4+8273125*a1^4+14400*a1^2
      +289537511718750*a1^10*a2^2
      + ... 50159155078125*a1^2*a2^16+15127410888671875*a1^16
      +1804302978515625*a1^14+53516387939453125*a1^18:
```

Therefore, $a_1 \neq 0$, $a_2 \neq 0$. In order to have $v_1 = 0$, i.e., $V_1 = 0$, one needs $1 + 25a_1^2 - 3a_2^2 = 0$. If this condition is satisfied, namely, $a_2 = \pm\sqrt{(1 + 25a_1^2)/3}$, then V_2 and V_3 can be simplified as

$$V_2 = -\frac{8}{9}\left(1 + 25a_1^2\right)^2\left(160a_1^2 + 7\right),$$

$$V_3 = -\frac{25600}{81}\left(1 + 25a_1^2\right)^3$$
$$\times \left(1451600000a_1^6 + 162853500a_1^4 + 6082515a_1^2 + 75443\right).$$

Hence, the best possibility is that $v_1 = 0$, but $v_2 \neq 0$, indicating that the Z_5-equivariant vector field (4.147) has 10 small limit cycles, 2 around each of the 5 weak focus points. Similarly, one can show that considering the origin as an elementary center requires $a_0 = 0$, and then normal form computation yields the focus values

$$v_1 = a_1, \qquad v_2 = v_3 = 0,$$

$$v_4 = \frac{2}{5}a_1\left(a_2^2 + b_2^2\right), \qquad v_5 = -\frac{164}{125}a_1 b_1\left(a_2^2 + b_2^2\right), \quad \ldots,$$

which implies that for this case $v_1 \neq 0$, and thus there is only 1 small limit cycle. Consequently, we have the following theorem.

Theorem 4.21 *The maximal number of small limit cycles that a fourth-degree planar polynomial system with Z_5 symmetry can have is 10, i.e., $\hat{H}_5(4) = 10$.*

To end this subsection, we present a numerical example. The simulated phase portrait is shown in Fig. 4.12. The parameter values used in this example are

$$a_0 = -0.0019999, \qquad b_0 = -0.459938866, \qquad a_1 = 0.001,$$
$$b_1 = 1.027296352, \qquad a_2 = 0.567357486, \qquad b_2 = 0.0009999.$$

Figure 4.12 shows that the system for this example has 1 stable focus point at the origin, and 5 weak focus points. In addition, there are 10 saddle points. Under the

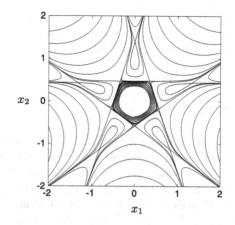

Fig. 4.12 The phase portrait of system (4.147) having 10 small limit cycles for
$a_0 = -0.0019999$,
$b_0 = -0.459938866$,
$a_1 = 0.001$,
$b_1 = 1.027296352$,
$a_2 = 0.567357486$,
$b_2 = 0.0009999$

above parameter values, the focus values for the weak focus points can be found as

$$v_0 = 0.0000001, \qquad v_1 = -0.0002223961, \qquad v_2 = 0.0222420369.$$

Then letting $\dot{r} = 0$ in the normal form,

$$\dot{r} = r\left(v_0 + v_1 r^2 + v_2 r^4\right),$$

yields the amplitudes of the limit cycles:

$$r_1 = 0.0217237573 \quad \text{and} \quad r_2 = 0.0976062839,$$

as expected.

4.3.2 $q \geq 7$ (Odd)

When $q(\geq 7)$ is an odd integer, we have $\dot{z} = F_q(z, \bar{z})$ with

$$F_q(z, \bar{z}) = \left(A_0 + A_1|z|^2\right)z, \tag{4.150}$$

which is exactly the same as (4.144). Thus, a similar discussion to that given in Sect. 4.2.5 can be applied here to obtain the following theorem.

Theorem 4.22 *The maximal number of small limit cycles that a fourth-degree planar polynomial system with Z_q symmetry ($q \geq 7$) can have is 1, i.e., $\hat{H}_q(4) = 1$ ($q \geq 7$).*

4.4 Fifth-Order Vector Fields with Z_q Symmetry

In this section, we apply normal form theory to consider the bifurcation of small limit cycles in fifth-order vector fields with Z_q symmetry. We show that when $q = 5$, there exist 25 small limit cycles; when $q = 6$, such vector fields can have 24 small limit cycles; while when $q \geq 7$, there are only 2 small limit cycles. The cases $q = 1, 2, 3, 4$, remain unsolved.

4.4.1 $q = 5$

We first consider Z_5-equivariant vector fields of order 5, and show that such vector fields can have 25 small limit cycles bifurcating from 5 symmetric degenerate Hopf singular points. Further, we show that no large limit cycles exist in such a vector field with 25 small limit cycles, i.e., $H(5) \geq 25$.

Consider the original system (4.1) and assume that the vector field has Z_5 symmetry. Then according to [128], we have

$$F_5(z, \bar{z}) = (A_0 + A_1|z|^2 + A_2|z|^4)z + A_3\bar{z}^4, \tag{4.151}$$

where $A_j = a_j + ib_j$ are complex values (with real a_j and b_j). The real vector field corresponding to function (4.151) can be written as

$$
\begin{aligned}
\dot{x} = {} & a_0x - b_0y + a_1x^3 - b_1x^2y + a_1xy^2 - b_1y^3 \\
& + a_3x^4 + 4b_3x^3y - 6b_3x^2y^2 + 4b_3xy^3 + a_3y^4 + a_2x^5 \\
& - b_2x^4y + 2a_2x^3y^2 - 2b_2x^2y^3 + a_2xy^4 - b_2y^5, \\
\dot{y} = {} & b_0x + a_0y + b_1x^3 + a_1x^2y + b_1xy^2 + a_1y^3 \\
& + b_3x^4 - 4a_3x^3y - 6b_3x^2y^2 + 4a_3xy^3 + b_3y^4 \\
& + b_2x^5 + a_2x^4y + 2b_2x^3y^2 + 2a_2x^2y^3 + b_2xy^4 + a_2y^5.
\end{aligned}
\tag{4.152}
$$

The two eigenvalues of the Jacobian of (4.152) evaluated at the origin $(x, y) = (0, 0)$ are $a_0 \pm b_0 i$, indicating that the origin is either a focus point (when $a_0 \neq 0$) or a center ($a_0 = 0$).

For this case, we have the following result.

Theorem 4.23 *The Z_5-equivariant planar vector field of order 5, described by system (4.152) can have a maximum of exactly 25 small limit cycles, among which there are 5 around each of the 5 Hopf critical points of the system, and thus, $\hat{H}_5(5) = 25$.*

Proof Since this vector field is Z_5-equivariant, if there exists 1 fixed point, there are in total 5 fixed points in the system. Without loss of generality, we may assume that one of the fixed points is located on the y-axis. Further, by a simple parameter scaling, we may suppose that this fixed point is $(0, 1)$. Therefore, the 5 fixed points are

$$(0, 1), \quad \left(\pm\cos\frac{\pi}{10}, \sin\frac{\pi}{10}\right), \quad \left(\pm\sin\frac{\pi}{5}, -\cos\frac{\pi}{5}\right), \tag{4.153}$$

which lead to the following conditions:

$$a_0 = -(a_1 + a_2 + b_3), \qquad b_0 = -(b_1 + b_2 - a_3). \qquad (4.154)$$

Next, we want the 5 fixed points to be Hopf critical points. Thus, we set

$$b_3 = a_1 + 2a_2. \qquad (4.155)$$

Then the eigenvalues of the Jacobian of system (4.152) evaluated at the 5 Hopf critical points are $\lambda_\pm = \pm i\omega$ where

$$\omega = \sqrt{5(4b_2 a_3 + 2b_1 a_3 - 5a_1^2 - 2a_1 a_2 - 20a_2^2)} > 0. \qquad (4.156)$$

Equation (4.156) shows that there are 5 free parameters which may be chosen later in perturbations. However, as shown in [239], we may use a parameter scaling and a time scaling to reduce by one more parameter. In other words, one of the 5 parameters can be chosen arbitrarily, or the frequency ω can be chosen arbitrarily. To achieve this, applying the parameter scaling

$$a_1 \longrightarrow a_1 \omega, \qquad a_2 \longrightarrow a_2 \omega, \qquad a_3 \longrightarrow a_3 \omega, \qquad b_1 \longrightarrow b_1 \omega, \qquad (4.157)$$

and time scaling $\tau = \omega t$, we obtain

$$b_2 = \frac{1}{20a_3}\left(1 - 10b_1 a_3 + 15a_3^2 + 25a_1^2 + 100a_1 a_2 + 100a_2^2\right)\omega. \qquad (4.158)$$

Since the vector field has Z_5 symmetry, we only need to consider one of the Hopf critical points, say, $(0, 1)$. Therefore, applying the transformation

$$x = -5(a_1 + 2a_2)a_3 u + v, \qquad y = 1 + 5a_3 u, \qquad (4.159)$$

to system (4.151) yields the following canonical form of equations for the Hopf critical point:

$$\dot{u} = v - \left(A_1 + 30a_3^2 A_1 - 20a_2 a_3^2 + 50A_1^3\right)x^2 + \frac{1}{5}\left(1 + 75a_3^2 + 75A_1^2\right)xy$$

$$- A_1 y^2 + \cdots + \frac{1}{100}a_3^2\left(1 + 15a_3^2 + 25A_1^2 - 10a_3 b_1\right)y^5,$$

$$\dot{v} = -u - \frac{1}{2}\left(25a_3^4 + 15A_1^2 + 75a_3^4 + 625A_1^4 - 100b_1 a_3^3\right. \qquad (4.160)$$

$$+ 800a_3^2 A_1^2 - 200a_2 a_3^2 A_1\big)x^2$$

$$+ \cdots + \frac{1}{20}a_3^2\left(A_1 + 25A_1^3 + 20a_2 a_3^2 + 15a_3^2 A_1 - 10A_1 a_3 b_1\right)y^5,$$

where $A_1 = a_1 + 2a_2$.

Now applying the Maple program [218] to system (4.160) yields the focus values, v_i, explicitly expressed in terms of the system's coefficients:

$$v_0 = A_1 - b_3, \tag{4.161}$$

$$v_1 = \frac{25}{2}a_3^2[125A_1^5 + 5A_1^3 + 15a_2a_3^4 + 160a_3^2A_1^3 + 7a_3^2A_1 + a_2a_3^2$$

$$+ 15a_3^4A_1 + 120a_3^2a_2A_1^2 + 5a_2A_1^2 + 125a_2A_1^4 - 20b_1a_3^3A_1$$

$$- 40a_2^2a_3^2A_1 - 20b_1a_3^3a_2], \tag{4.162}$$

and

$$v_2 = -\frac{625}{72}a_3^2\tilde{v}_2(A_1, a_2, a_3, b_1),$$

$$v_3 = -\frac{25}{663552}a_3^2\tilde{v}_3(A_1, a_2, a_3, b_1), \tag{4.163}$$

$$v_4 = -\frac{25}{95551488}a_3^2\tilde{v}_4(A_1, a_2, a_3, b_1),$$

where $\tilde{v}_i$, $i = 2, 3, 4$ are lengthy polynomials in A_1, a_2, a_3 and b_1.

Note that letting $v_0 = 0$ yields $b_3 = A_1$, which is the condition given by (4.155) for the fixed points to be Hopf singularities. Then there are 4 free parameters in the expressions of v_i, $i = 1, 2, 3, 4$. Therefore, the best possibility is to choose A_1, a_2, a_3 and b_1 such that $v_i = 0, i = 1, 2, 3, 4$, but $v_5 \neq 0$, leading to the possible existence of 5 small limit cycles in the vicinity of each of the 5 Hopf critical points. This suggests that the fifth-order Z_5-equivariant vector fields, described by (4.152), may have $5 \times 5 = 25$ small limit cycles.

Solving $v_1 = 0$ for b_1 yields

$$b_1 = \frac{1}{20a_3^3(A_1 + a_2)}(125A_1^5 + 5A_1^3 + 15a_2a_3^4 + 160a_3^2A_1^3 + 7a_3^2A_1 + a_2a_3^2$$

$$+ 15a_3^4A_1 + 120a_3^2a_2A_1^2 + 5a_2A_1^2 + 125a_2A_1^4 - 40a_2^2a_3^2A_1). \tag{4.164}$$

Then substituting b_1 above into v_i, $i = 2, 3, 4$, results in

$$v_2 = \frac{15(a_3^2 + A_1^2)}{2(A_1 + a_2)}\bar{v}_2(A_1, a_2, a_3),$$

$$v_3 = \frac{3000(a_3^2 + A_1^2)}{(A_1 + a_2)^3}\bar{v}_3(a_1, a_3, b_3), \tag{4.165}$$

$$v_4 = \frac{1500(a_3^2 + A_1^2)}{(A_1 + a_2)^5}\bar{v}_4(a_1, a_3, b_3),$$

where $\bar{v}_i$, $i = 2, 3, 4$, are coupled polynomials in A_1, a_2, and a_3. However, we cannot solve for the $\bar{v}_i$, one by one, by choosing A_1, a_2 and a_3. We have to solve them simultaneously.

Noting that the $\bar{v}_i$ are lower-order polynomials in a_3^2, we may eliminate a_3 from these three equations. First, eliminating a_3 from the two equations $\bar{v}_2 = 0$ and $\bar{v}_3 = 0$ yields a solution for a_3:

$$\left(a_3^{(1)}\right)^2 = \frac{a_{3n}^{(1)}(A_1, a_2)}{a_{3d}^{(1)}(A_1, a_2)}, \tag{4.166}$$

and a resultant equation:

$$F_1 = F_1(A_1, a_2) = 0. \tag{4.167}$$

Here, both $a_{3n}^{(1)}(A_1, a_2)$ and $a_{3d}^{(1)}(A_1, a_2)$ are polynomials, and $F_1(A_1, a_2)$ is a 16th-degree polynomial in a_2.

Similarly, eliminating a_3 from the other two equations, $\bar{v}_2 = 0$ and $\bar{v}_4 = 0$, yields another solution for a_3:

$$\left(a_3^{(2)}\right)^2 = \frac{a_{3n}^{(2)}(A_1, a_2)}{a_{3d}^{(2)}(A_1, a_2)}, \tag{4.168}$$

and another resultant equation:

$$F_2 = F_2(A_1, a_2) = 0, \tag{4.169}$$

where $F_2(A_1, a_2)$ is a 29th-degree polynomial in a_2. It should be noted that the final solution must satisfy $a_3^{(1)} = a_3^{(2)}$.

The remaining task is to solve the two coupled polynomial equations, $F_1 = 0$ and $F_2 = 0$. After a lengthy computation to eliminate a_2 from the two equations, we obtain

$$F_3\left(A_1^2\right) = -\frac{95175}{G_0(A_1^2)} A_1^4 \left(1 + 12A_1^2\right)^2 \left(1 + 16A_1^2\right)^3 \left(1 + 25A_1^2\right)^6$$
$$\times G_1\left(A_1^2\right) G_2\left(A_1^2\right) G_3^2\left(A_1^2\right), \tag{4.170}$$

where $G_i, i = 0, 1, 2, 3$, are polynomials in A_1^2. Now, the possible solutions for $v_1 = v_2 = v_3 = v_4 = 0$ only come from the positive real solutions of A_1^2 obtained from the following higher-degree polynomial equations: $G_1(A_1^2) = 0$, $G_2(A_1^2) = 0$ and $G_3(A_1^2) = 0$, where G_1, G_2 and G_3 are 37th-, 227th- and 257th-degree polynomials, respectively, in A_1^2. Applying the built-in Maple solver fsolve yields the following results: $G_1(a_3^2) = 0$ has only one positive solution for A_1^2, which satisfies $a_3^{(1)} = a_3^{(2)}$, while $G_2(a_3^2) = 0$ and $G_3(a_3^2) = 0$ have no real positive solutions. This unique solution results in

$$a_{0c} = -0.00094678410409569352, \qquad b_{0c} = -1.803189335128100443778,$$
$$a_{1c} = 0.00049612711650105944, \qquad b_{1c} = 2.89389952760627236198,$$
$$a_{2c} = -0.00001515670963547512, \qquad b_{2c} = -0.96198328806045136942,$$
$$a_{3c} == 0.12872690441772055479, \qquad b_{3c} = 0.00046581369723010920,$$
$$\tag{4.171}$$

where the subscript c denotes "critical value". With these critical parameter values, executing the Maple program for calculating the focus values results in (up to 50 decimal digits)

$$
\begin{aligned}
v_{1c} &= -0.5 \times 10^{-54}, \\
v_{2c} &= 0.5 \times 10^{-54}, \\
v_{3c} &= 0.275 \times 10^{-53}, \\
v_{4c} &= 0.57 \times 10^{-54}, \\
v_{5c} &= -0.117005875085799606719320810802412865701184444 \times 10^{-13},
\end{aligned}
\tag{4.172}
$$

where the v_i's, $i = 2, 3, 4$, are not exactly zero due to numerical round-off errors. In fact, we executed the program up to 1000 decimal digits to obtain

$$
\begin{aligned}
v_{1c} &= -0.5 \times 10^{-1003}, \\
v_{2c} &= -0.162705 \times 10^{-998}, \\
v_{3c} &= -0.1357285 \times 10^{-999}, \\
v_{4c} &= -0.1173415 \times 10^{-1000}, \\
v_{5c} &= -0.11700587508579960671932081080\ldots639507814500000 \times 10^{-13},
\end{aligned}
$$

from which it is noted that the first 42 decimal places of v_{5c} are identical to those given in (4.172). This clearly indicates that $v_{1c} = v_{2c} = v_{3c} = v_{4c} = 0$ but $v_{5c} \neq 0$. Hence, the maximal number of small limit cycles that system (4.152) can have is 25. To prove that the system can indeed have exactly 25 small limit cycles, we use v_2, v_3 and v_4 given in (4.165) to verify that determinant (4.10) evaluated at the critical values is nonzero:

$$
\det \begin{bmatrix}
\dfrac{\partial v_2}{\partial A_1} & \dfrac{\partial v_2}{\partial a_2} & \dfrac{\partial v_2}{\partial a_3} \\[2mm]
\dfrac{\partial v_3}{\partial A_1} & \dfrac{\partial v_3}{\partial a_2} & \dfrac{\partial v_3}{\partial a_3} \\[2mm]
\dfrac{\partial v_4}{\partial A_1} & \dfrac{\partial v_4}{\partial a_2} & \dfrac{\partial v_4}{\partial a_3}
\end{bmatrix}_{(A_1, a_2, a_3) = (A_{1c}, a_{2c}, a_{3c})}
$$

$$
= \det \begin{bmatrix}
-0.0000355524 & -0.0010914682 & 0.0000020499 \\
-0.0000025290 & -0.0000775587 & 0.0000001710 \\
-0.0000002179 & -0.0000066773 & 0.0000000148
\end{bmatrix}
$$

$$
= 0.5651106555 \times 10^{-20} \neq 0,
$$

which shows that the Z_5-equivariant vector fields of order 5, described by system (4.152), can have exactly 25 small limit cycles. Hence, $H(5) \geq 25$.

This completes the proof of Theorem 4.23.   □

The phase portrait of the nonperturbed system (4.152) is shown in Fig. 4.13. There are 21 fixed points: 1 stable focus point at $(0, 0)$; 5 weakly stable focus points

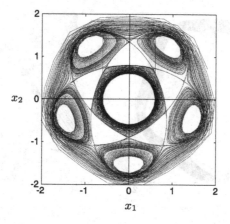

Fig. 4.13 The phase portrait of the unperturbed system (4.152) when the origin is a stable focus point for $a_{0c} = -0.0009467841$, $b_{0c} = -1.8031893351$, $a_{1c} = 0.0004961271$, $b_{1c} = 2.8938995276$, $a_{2c} = -0.0000151567$, $b_{2c} = -0.9619832881$, $a_{3c} = 0.1287269044$, $b_{3c} = 0.0004658137$

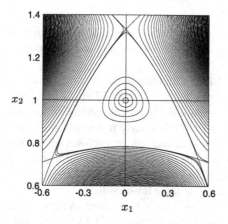

Fig. 4.14 The zoomed-in area around the point $(0, 1)$, the box marked in Fig. 4.13, showing the existence of 5 small limit cycles in the vicinity of $(0, 1)$ after an appropriate perturbation to the critical point $a_{0c} = -0.0009467841$, $b_{0c} = -1.8031893351$, $a_{1c} = 0.0004961271$, $b_{1c} = 2.8938995276$, $a_{2c} = -0.0000151567$, $b_{2c} = -0.9619832881$, $a_{3c} = 0.1287269044$, $b_{3c} = 0.0004658137$

at the locations given in (4.153); 5 unstable focus points and 10 saddle points symmetrically located, as shown in Fig. 4.13. After proper perturbations, 5 small limit cycles exist in the vicinity of each of the 5 weak focus points. The zoomed-in neighborhood of the point $(0, 1)$ is depicted in Fig. 4.14, which shows the existence of 5 limit cycles. Since $v_5 < 0$ then $v_0 > 0$, indicating that the stable weak focus points become unstable under perturbations. Thus, the smallest limit cycle is unstable, the next one is stable, and so on. The largest one is unstable.

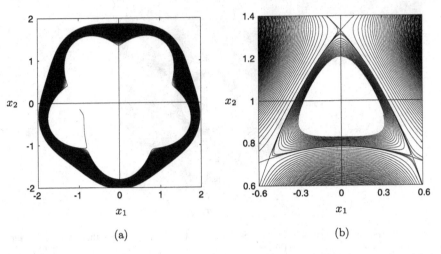

Fig. 4.15 The phase portraits of system (4.152) showing no large limit cycles near the critical point $a_{0c} = -0.0009467841$, $b_{0c} = -1.8031893351$, $a_{1c} = 0.0004961271$, $b_{1c} = 2.8938995276$, $a_{2c} = -0.0000151567$, $b_{2c} = -0.9619832881$, $a_{3c} = 0.1287269044$, $b_{3c} = 0.0004658137$: (**a**) trajectory diverging to infinity; (**b**) breaking of heteroclinic orbits

Remark 4.24 We have proved that the fifth-order Z_5-equivariant system (4.152) can have 25 small limit cycles. A natural question is: is it possible for system (4.152) to have large limit cycles?

Consider system (4.152) with the parameter values given in equation (4.171), for which we have shown that the system has 25 small limit cycles under appropriate perturbations. Now we employ numerical simulation to show that no large limit cycles exist in system (4.152). It is easy to see that one possibility for the system to have a large limit cycle is one which encloses all 21 fixed points and all 25 small limit cycles. However, as shown in Fig. 4.15(a), a trajectory starting from a region enclosing one of the weak focus points diverges to infinity, which excludes the existence of this possible large limit cycle.

Another possible way to have a large limit cycle exist is via the breaking of heteroclinic orbits. It can be seen from Fig. 4.13 that the five regions in the neighborhoods of the weak focus points are actually enclosed by heteroclinic trajectories. However, when the heteroclinic orbits break under perturbations, they do not give rise to large limit cycles, as shown in Fig. 4.15(b).

In summary, the fifth-order Z_5-equivariant system (4.152) cannot have large limit cycles. Therefore, the maximal number of limit cycles obtained in such a vector field is 25, i.e., $H(5) \geq 25$.

4.4.2 $q = 6$

In this subsection, we consider the vector field with Z_6 symmetry, and show that such vector fields can have 24 small limit cycles. Using (4.20), we have the following function:

$$F_6(z, \bar{z}) = \left(A_0 + A_1|z|^2 + A_2|z|^4\right)z + A_3\bar{z}^5, \tag{4.173}$$

where the $A_j = a_j + ib_j$ are complex values (with real a_j and b_j). The real vector field corresponding to function (4.173) can be written as

$$
\begin{aligned}
\dot{x} = &\, a_0x - b_0y + a_1x^3 - b_1x^2y + a_1xy^2 - b_1y^3 + (a_2 + a_3)x^5 \\
&- (b_2 - 5b_3)x^4y + 2(a_2 - 5a_3)x^3y^2 - 2(b_2 + 5b_3)x^2y^3 \\
&+ (a_2 + 5a_3)xy^4 - (b_2 - b_3)y^5, \\
\dot{y} = &\, b_0x + a_0y + b_1x^3 + a_1x^2y + b_1xy^2 + a_1y^3 + (b_2 + b_3)x^5 \\
&+ (a_2 - 5a_3)x^4y + 2(b_2 - 5b_3)x^3y^2 + 2(a_2 + 5a_3)x^2y^3 \\
&+ (b_2 + 5b_3)xy^4 + (a_2 - a_3)y^5.
\end{aligned}
\tag{4.174}
$$

The two eigenvalues of the Jacobian of (4.174) evaluated at the origin $(x, y) = (0, 0)$, are $a_0 \pm b_0i$, indicating that the origin is either a focus point (when $a_0 \neq 0$) or a center ($a_0 = 0$).

Since this vector field is Z_6-equivariant, if there exists 1 fixed point, there are in total 6 fixed points in the system. Without loss of generality, we may assume that one of the fixed points is located on the y-axis. Further, by a simple parameter scaling, we may suppose this fixed point is $(0, 1)$. Therefore, the 6 fixed points are

$$(0, \pm 1), \qquad \left(\pm\sqrt{3}, \frac{1}{2}\right), \qquad \left(\pm\sqrt{3}, -\frac{1}{2}\right), \tag{4.175}$$

which lead to the following conditions:

$$a_0 = a_3 - a_1 - a_2, \qquad b_0 = b_3 - b_1 - b_2. \tag{4.176}$$

Next, we want the 6 fixed points to be Hopf critical points. Thus, we set

$$a_2 = -\frac{1}{2}(a_1 + a_3). \tag{4.177}$$

Then the eigenvalues of the Jacobian of system (4.174) evaluated at the 6 Hopf critical points are

$$\lambda_\pm = \pm\omega \quad \text{where } \omega = \sqrt{12\left[b_2b_3 + b_1b_3 - 2b_3^2 - 3a_3^2\right]} > 0. \tag{4.178}$$

Equation (4.178) shows that there are 5 free parameters which may be chosen later in perturbations. However, as shown in [239], we may use a parameter scaling and a time scaling to reduce by one more parameter. In other words, one of the 5 parameters can be chosen arbitrarily, or the frequency ω can be chosen arbitrarily. To achieve this, we use the parameter scaling

$$a_1 \longrightarrow a_1\omega, \qquad b_1 \longrightarrow b_1\omega, \qquad a_3 \longrightarrow a_3\omega, \qquad b_3 \longrightarrow b_3\omega, \tag{4.179}$$

and the time scaling $\tau = \omega t$, to obtain

$$b_2 = \frac{1}{24b_3}\left(1 - 12b_1b_3 + 24b_3^2 + 36a_3^2\right)\omega. \tag{4.180}$$

Since the vector field has Z_6 symmetry, we only need to consider one of the Hopf critical points, say, $(0, 1)$. Therefore, applying the transformation

$$x = 6a_3u + v, \qquad y = 1 + 6b_3u, \tag{4.181}$$

to system (4.174) yields the following canonical form of equations for the Hopf critical point:

$$\dot{u} = v + \frac{1}{b_3}\left(a_3 + 90a_3^3 - 12a_1b_3^2 + 54a_3b_3^2\right)x^2$$

$$+ \frac{1}{6b_3}\left(1 + 144b_3^2 + 144a_3^2\right)xy + \frac{3}{2b_3}a_3y^2$$

$$+ \frac{3}{b_3^2}\left(a_3^3 + 3a_3b_3^2 + 36a_3^5 + 252a_3b_3^4 - 48a_1b_3^4 + 336a_3^3b_3^2\right.$$

$$- 24a_1a_3^2b_3^2 - 24a_3b_1b_3^3\big)x^3$$

$$+ \frac{3}{2b_3^2}\left(a_3^2 + b_3^2 + 36a_3^4 - 8b_1b_3^3 + 144b_3^4 + 156a_3^2b_3^2 - 16a_1a_3b_3^2\right)x^2y$$

$$+ \frac{1}{4b_3^2}\left(a_3 + 36a_3^3 + 12a_3b_3^2 - 8a_1b_3^2\right)xy^2 + \frac{1}{72b_3^2}\left(1 + 36a_3^2 - 96b_3^2\right)y^3$$

$$+ \frac{36}{b_3}\left(a_3^2 + b_3^2\right)\left(a_3 + 3a_3^3 - 3a_1a_3^2 + 99a_3b_3^2 - 12a_3b_1b_3 - 15a_1b_3^2\right)x^4$$

$$+ \frac{6}{b_3}\left(3a_3^2 + b_3^2 - 24a_3^4 + 144b_3^4 - 36a_1a_3b_3^2 - 12b_1b_3^3 + 72a_3^2b_3^2\right.$$

$$- 36b_1b_3a_3^2 - 12a_1a_3^3\big)x^3y$$

$$+ \frac{3}{b_3}\left(a_3 - 30a_3^3 - 42a_3b_3^2 - 6a_1b_3^2 - 12b_1b_3a_3 - 6a_1a_3^2\right)x^2y^2$$

$$+ \frac{1}{6b_3}\left(1 - 96b_3^2 - 96a_3^2 - 12a_1a_3 - 12b_1b_3\right)xy^3 - \frac{1}{12b_3}\left(a_1 + 11a_3\right)y^4$$

$$+ \frac{54}{b_3^2}\left(a_3^2 + b_3^2\right)\left(a_3^3 + a_3b_3^2 + 36a_3^5 - 12b_3b_1a_3^3 - 48b_3^2a_3^3 - 12a_1b_3^2a_3^2\right.$$

$$+ 108b_3^4a_3 - 12b_3^3b_1a_3 - 12b_3^4a_1\big)x^5$$

$$+ \frac{9}{b_3^2}\left(a_3^2 + b_3^2\right)\left(5a_3^2 + b_3^2 + 180a_3^4 - 252a_3^2b_3^2 - 60b_1b_3a_3^2\right.$$

$$- 48a_1b_3^2a_3 + 144b_3^4 - 12b_1b_3^3)x^4y$$

$$+ \frac{3}{b_3^2}\left(5a_3^3 + 3a_3b_3^2 + 180a_3^5 - 12a_1b_3^4 - 180a_3b_3^4 - 48b_3^2a_3^3\right.$$

$$- 36a_1b_3^2a_3^2 - 36b_1b_3^3a_3 - 60b_1b_3a_3^3)x^3y^2$$

$$+ \frac{1}{2b_3^2}\left(5a_3^2 + b_3^2 + 180a_3^4 - 96b_3^4 + 12a_3^2b_3^2 - 12b_1b_3^3\right.$$

$$- 24a_1b_3^2a_3 - 60b_1b_3a_3^2)x^2y^3$$

$$+ \frac{1}{24b_3^2}\left(5a_3 + 180a_3^3 + 108a_3b_3^2 - 12a_1b_3^2 - 60b_1b_3a_3\right)xy^4$$

$$+ \frac{1}{144b_3^2}\left(1 + 36a_3^2 + 48b_3^2 - 12b_1b_3\right)y^5,$$

$$\dot{v} = -u - \frac{3}{b_3}\left(3a_3^2 + 5b_3^2 + 216a_3^4 - 24a_1a_3b_3^2 - 24b_1b_3^3 + 216a_3^2b_3^2\right)x^2$$

$$- \frac{2}{b_3}a_3\left(1 + 90a_3^2 + 90b_3^2\right)xy - \frac{1}{12b_3}\left(1 + 144a_3^2 + 144b_3^2\right)y^2 \qquad (4.182)$$

$$- \frac{18}{b_3^2}\left(a_3^2 + b_3^2\right)\left(a_3^2 + 5b_3^2 + 36a_3^4 + 540a_3^2b_3^2 - 24a_1a_3b_3^2 - 48b_1b_3^3\right)x^3$$

$$- \frac{9}{b_3^2}\left(a_3^3 + 3a_3b_3^2 + 36a_3^5 + 8a_1b_3^4 + 324a_3b_3^4 + 360a_3^2b_3^3\right.$$

$$- 16a_1a_3^2b_3^2 - 24a_3b_1b_3^3)x^2y$$

$$- \frac{3}{2b_3^2}\left(a_3^2 + b_3^2 + 36a_3^4 + 144b_3^4 - 8b_1b_3^3 + 180a_3^2b_3^2 - 8a_1a_3b_3^2\right)xy^2$$

$$- \frac{1}{12b_3^2}a_3\left(1 + 36a_3^2 + 36b_3^2\right)y^3$$

$$- \frac{54}{b_3}\left(a_3^2 + b_3^2\right)\left(5a_3^2 + 5b_3^2 + 48a_3^4 - 12a_1a_3^3\right.$$

$$- 60b_1b_3^3 + 1008a_3^2b_3^2 - 60b_1b_3a_3^2 - 12a_1a_3b_3^2)x^4$$

$$- \frac{144}{b_3}\left(a_3^2 + b_3^2\right)\left(a_3 + 3a_3^3 - 3a_1a_3^2 + 117a_3b_3^2 - 12b_1a_3b_3 + 3a_1b_3^2\right)x^3y$$

$$- \frac{9}{b_3}\left(3a_3^2 + b_3^2 - 24a_3^4 + 144b_3^4 - 12a_1a_3^3 - 12b_1b_3^3 + 12a_1a_3b_3^2\right.$$

$$- 36b_1b_3a_3^2 + 120a_3^2b_3^2)x^2y^2$$

$$- \frac{2}{b_3}\left(a_3 - 30a_3^3 - 30a_3b_3^2 + 6a_1b_3^2 - 12b_1a_3b_3 - 6a_1a_3^2\right)xy^3$$

$$-\frac{1}{24b_3}\left(1 - 96a_3^2 - 96b_3^2 - 12a_1a_3 - 12b_1b_3\right)y^4$$

$$-\frac{324}{b_3^2}\left(a_3^2 + b_3^2\right)\left(a_3^4 + 2a_3^2b_3^2 + b_3^4 + 36a_3^6 - 12b_1b_3a_3^4 - 24b_3^2a_3^4\right.$$

$$\left. - 24b_1a_3^3b_3^2 + 324a_3^2b_3^4 - 12b_1b_3^5\right)x^5$$

$$-\frac{54}{b_3^2}\left(a_3^2 + b_3^2\right)\left(5a_3^3 + 5a_3b_3^2 + 180a_3^5 - 60b_1b_3a_3^3 - 168b_3^2a_3^3\right.$$

$$\left. + 12a_1a_3^2b_3^2 - 60b_1a_3b_3^3 + 612a_3b_3^4 + 12a_1b_3^4\right)x^4y$$

$$-\frac{18}{b_3^2}\left(a_3^2 + b_3^2\right)\left(5a_3^2 + b_3^2 + 180a_3^4 + 144b_3^4 - 60b_1b_3a_3^2\right.$$

$$\left. - 180b_3^2a_3^2 + 24a_1a_3b_3^2 - 12b_1b_3^3\right)x^3y^2$$

$$-\frac{3}{b_3^2}\left(5a_3^3 + 3a_3b_3^2 + 180a_3^5 + 12a_1b_3^4 - 156a_3b_3^4 + 24a_3^3b_3^2\right.$$

$$\left. + 36a_1a_3^2b_3^2 - 36b_1a_3b_3^3 - 60b_1b_3a_3^3\right)x^2y^3$$

$$-\frac{1}{4b_3^2}\left(5a_3^2 + b_3^2 + 180a_3^4 - 96b_3^4 + 84a_3^2b_3^2 - 12b_1b_3^3 + 48a_1a_3b_3^2\right.$$

$$\left. - 60b_1b_3a_3^2\right)xy^4 - \frac{1}{24b_3^2}\left(a_3 + 36a_3^3 + 36a_3b_3^2 + 12a_1b_3^2 - 12b_1a_3b_3\right)y^5.$$

Now applying the Maple program [218] to system (4.182) gives the focus values, v_i, explicitly expressed in terms of the system's coefficients,

$$v_0 = a_1 + a_3 + 2a_2, \tag{4.183}$$

$$v_1 = 18\left[-12b_3^3(a_1 + 3a_3)b_1 + \left(9a_3^3 + 3a_1a_3^2 + a_1b_3^2\right.\right.$$
$$+ 9a_3b_3^2 + 324a_3^5 + 108a_1a_3^4 + 324b_3^2a_3^3$$
$$\left.\left. + 72a_1a_3^2b_3^2 - 12a_3a_1^2b_3^2\right)\right], \tag{4.184}$$

and

$$v_2 = \frac{18}{b_3^2}\tilde{v}_2(a_1, a_3, b_1, b_3),$$

$$v_3 = \frac{1}{128b_3^4}\tilde{v}_3(a_1, a_3, b_1, b_3), \tag{4.185}$$

$$v_4 = \frac{1}{7680b_3^6}\tilde{v}_4(a_1, a_3, b_1, b_3),$$

where $\tilde{v}_i$, $i = 2, 3, 4$, are lengthy (polynomials) of a_1, a_3, b_1 and b_3.

Note that letting $v_0 = 0$ yields $a_2 = -\frac{1}{2}(a_1 + a_3)$ which is the condition given by (4.177) for the fixed points to be Hopf singularities. Then there are 4 free parameters in the expressions of v_i, $i = 1, 2, 3, 4$. Therefore, the best possibility is to choose a_1, a_3, b_1 and b_3 such that $v_i = 0$, $i = 1, 2, 3, 4$, but $v_5 \neq 0$, leading to the possible existence of 5 small limit cycles in the vicinity of each of the 6 Hopf critical points. This suggests that the fifth-order Z_6-equivariant vector fields, described by (4.174), may have $5 \times 6 = 30$ small limit cycles. If this is not possible, then the next best possibility is $v_i = 0$, $i = 1, 2, 3$, but $v_4 \neq 0$, indicating that system (4.174) may have $4 \times 6 = 24$ small limit cycles.

Solving $v_1 = 0$ for b_1 yields

$$b_1 = \frac{9a_3^3 + 3a_1a_3^2 + a_1b_3^2 + 9a_3b_3^2 + 324a_3^5 + 108a_1a_3^4 + 324b_3^2a_3^3 + 72a_1a_3^2b_3^2 - 12a_3a_1^2b_3^2}{12b_3^3(a_1 + 3a_3)}.$$

(4.186)

Then substituting the above b_1 into v_i, $i = 2, 3, 4$, results in

$$v_2 = \frac{81(a_3^2 + b_3^2)}{b_3^2(a_1 + 3a_3)}\bar{v}_2(a_1, a_3, b_3),$$

$$v_3 = \frac{81(a_3^2 + b_3^2)}{16b_3^4(a_1 + 3a_3)^3}\bar{v}_3(a_1, a_3, b_3),$$

(4.187)

$$v_4 = \frac{81(a_3^2 + b_3^2)}{640b_3^6(a_1 + 3a_3)^5}\bar{v}_4(a_1, a_3, b_3),$$

where the $\bar{v}_i$, $i = 2, 3, 4$, are coupled polynomials in a_1, a_3, and b_3. However, we cannot solve the $\bar{v}_i$, one by one, by choosing a_1, a_3 and b_3. We have to solve them simultaneously.

Noting that the $\bar{v}_i$ are lower-order polynomials in b_3^2, we may eliminate b_3 from these three equations. First, eliminating b_3 from the two equations $\bar{v}_2 = 0$ and $\bar{v}_3 = 0$ yields a solution for b_3:

$$\left(b_3^{(1)}\right)^2 = \frac{B_{3n}^{(1)}(a_1, a_3)}{B_{3d}^{(1)}(a_1, a_3)},$$

(4.188)

and a resultant equation:

$$F_1 = F_1(a_1, a_3) = 0.$$

(4.189)

Here, both $B_{3n}^{(1)}(a_1, a_3)$ and $B_{3d}^{(1)}(a_1, a_3)$ are polynomials, and $F_1(a_1, a_3)$ is a 12th-degree polynomial in a_1.

Similarly, eliminating b_3 from the other two equations, $\bar{v}_2 = 0$ and $\bar{v}_4 = 0$, yields another solution for b_3:

$$\left(b_3^{(2)}\right)^2 = \frac{B_{3n}^{(2)}(a_1, a_3)}{B_{3d}^{(2)}(a_1, a_3)},$$

(4.190)

and another resultant equation:

$$F_2 = F_2(a_1, a_3) = 0,\tag{4.191}$$

where $F_2(a_1, a_3)$ is a 20th-degree polynomial in a_1. It should be noted that the final solution must satisfy $b_3^{(1)} = b_3^{(2)}$.

The remaining task is to solve the two coupled polynomial equations, $F_1 = 0$ and $F_2 = 0$. After a lengthy computation to eliminate a_1 from the two equations, we obtain

$$F_3(a_3^2) = \frac{8100}{G_0(a_3^2)} a_3^4 (1 + 36a_3^2)(1 + 180a_3^2)(1 + 12a_3^2)^2 (1 + 24a_3^2)^2$$

$$\times (1 + 32a_3^2)^2 (3 + 160a_3^2 + 1920a_3^4) G_1(a_3^2) G_2(a_3^2) G_3^2(a_3^2),$$

$$\tag{4.192}$$

where the G_i, $i = 0, 1, 2, 3$, are polynomials in a_3^2. Now, the possible solutions for $v_1 = v_2 = v_3 = v_4 = 0$ come only from the positive real solutions of a_3^2 obtained from the following higher-degree polynomial equations:

$$G_1(a_3^2) = 0, \qquad G_2(a_3^2) = 0, \quad \text{and} \quad G_3(a_3^2) = 0,$$

where G_1, G_2 and G_3 are 39th-, 111th- and 144th-degree polynomials, respectively, in a_3^2. Applying the built-in Maple solver fsolve yields the following results: $G_1(a_3^2) = 0$ does not have positive solutions, while $G_2(a_3^2) = 0$ has 9 positive solutions and $G_3(a_3^2) = 0$ gives 8 positive solutions. However, none of these 17 solutions satisfies $b_3^{(1)} = b_3^{(2)}$. Therefore, it is not possible to have solutions such that $v_1 = v_2 = v_3 = v_4 = 0$, and thus the fifth-order Z_6-equivariant vector fields *cannot* have 30 small limit cycles.

The next best possibility is that the fifth-order Z_6-equivariant vector fields can have 24 small limit cycles. In this case, there is a free parameter and thus there exist infinitely many solutions. For example, one may choose a_3 as the free parameter. For determination, fix $a_3 = 0.01$, and obtain the values of a_1, b_3 and b_1 as follows:

$$a_{3c} = 0.01,$$

$$a_{1c} = -0.0086130251189165831322,$$

$$b_{3c} = 0.126037608541720784809,\tag{4.193}$$

$$b_{1c} = 2.53643378065134733564,$$

where the subscript c denotes "critical value". With these critical parameter values, executing the Maple program [218] for calculating the focus values results in (up to

30 decimal digits)

$$v_{1c} = 0.0,$$

$$v_{2c} = -0.15 \times 10^{-28},$$

$$v_{3c} = -0.1315 \times 10^{-26},$$ (4.194)

$$v_{4c} = 0.047881458613529651187271289056,$$

where the v_i's, $i = 2, 3$, are not exactly zero due to numerical round-off errors. This indicates that the maximal number of small limit cycles that system (4.174) can have is 24. To prove that the system can indeed have exactly 24 small limit cycles, we use v_2 and v_3 given in (4.187) to verify that the determinant (4.10) evaluated at the critical values is nonzero:

$$\det \begin{bmatrix} \frac{\partial v_2}{\partial a_1} & \frac{\partial v_2}{\partial b_3} \\ \frac{\partial v_3}{\partial a_1} & \frac{\partial v_3}{\partial b_3} \end{bmatrix}_{(a_3,a_1,b_3)=(a_{3c},a_{1c},b_{3c})}$$

$$= \det \begin{bmatrix} 4.2146332164085153345 & -0.470589921690317025966 \\ 32.24868034604072122587 & -4.10206276388842579384 \end{bmatrix}$$

$$= -2.11278602181731143854 \neq 0.$$

Summarizing the results obtained so far in this section gives the following theorem.

Theorem 4.25 *The fifth-order Z_6-equivariant vector field, described by system (4.174), can have a maximum of exactly 24 small limit cycles, among which there are 4 around each of the 6 Hopf critical points of the system.*

To end this section, we present a numerical example. Since $v_4 \approx v_{4c} > 0$, we need to choose perturbations such that $v_3 < 0$, $v_2 > 0$, $v_1 < 0$, $v_0 > 0$, and $v_0 \ll -v_1 \ll v_2 \ll -v_3 \ll v_4$. We choose the following perturbations:

$$a_3 = a_{3c}, \qquad a_1 = a_{1c} + \varepsilon_1, \qquad b_3 = b_{3c} + \varepsilon_2,$$

$$b_1 = b_1(a_1, b_3) + \varepsilon_3, \qquad a_2 = -\frac{1}{2}(a_1 + a_3) + \varepsilon_4,$$

where $b_1(a_1, b_3)$ is given by (4.186), and the ε_i's are given by

$$\varepsilon_1 = 0.001, \qquad \varepsilon_2 = 0.010846, \qquad \varepsilon_3 = 0.00886, \qquad \varepsilon_4 = 0.000000000005,$$

under which the 5 parameter values are

$$a_3 = 0.01, \qquad a_1 = -0.00761302511891658313,$$

$$b_3 = 0.13489760854172784809, \qquad b_1 = 2.29087633582649736348, \quad (4.195)$$

$$a_2 = -0.11934874355417084339,$$

and the remaining parameter values are given in (4.176) and (4.180), in which ω may be set to $\omega = 1$.

Under the parameter values given in (4.195), system (4.174) becomes

$$\dot{x} = 0.0188065126x + 2.0614758892y - 0.0076130251x^3$$
$$- 2.2908763358x^2y - 0.0076130251xy^2 - 2.2908763358y^3$$
$$+ 0.0088065126x^5 + 0.7689908806x^4y - 0.1023869749x^3y^2$$
$$- 1.1599704090x^2y^3 + 0.0488065126xy^4 + 0.2294004466y^5,$$

$$\dot{y} = -2.0614758892x + 0.0188065126y + 2.2908763358x^3 \qquad (4.196)$$
$$- 0.0076130251x^2y + 2.2908763358xy^2 - 0.0076130251y^3$$
$$+ 0.0403947704x^5 - 0.0511934874x^4y - 1.5379817610x^3y^2$$
$$+ 0.0976130251x^2y^3 + 0.5799852044xy^4 - 0.0111934874y^5.$$

System (4.196) has 25 fixed points: 1 unstable focus point at $(0, 0)$; 6 weakly unstable focus points at the locations given in (4.175); and 6 stable focus points and 12 saddle points symmetrically located, as shown in Fig. 4.16. The perturbed focus values for the 6 weak focus points are

$$v_0 = 0.1 \times 10^{-10}, \qquad v_1 = -0.118706 \times 10^{-6},$$
$$v_2 = 0.176994008259 \times 10^{-3}, \qquad v_3 = -0.38176125 \times 10^{-2},$$
$$v_4 = 0.165711379994 \times 10^{-1},$$

which can be substituted into the normal form (4.2) to obtain 4 positive solutions for r of the amplitudes of the small limit cycles

$$r_1 = 0.009937276392, \quad r_2 = 0.024096314116,$$
$$r_3 = 0.251437576450, \quad r_4 = 0.408011110035.$$

The phase portrait of the perturbed system (4.196) is shown in Fig. 4.16, where 4 limit cycles are located in the neighborhood of each of the 6 weak focus points $(0, \pm 1)$, $(\pm\sqrt{3}, \frac{1}{2})$ and $(\pm\sqrt{3}, -\frac{1}{2})$. The zoomed-in neighborhood of the point $(0, 1)$ is depicted in Fig. 4.17, which clearly shows the existence of 4 limit cycles. Since $v_4 > 0$ and $v_0 > 0$, the 6 weak focus points are unstable. Thus, the smallest limit cycle is stable, the next one is unstable, the one next to the largest one is stable, and the largest one is unstable.

Fig. 4.16 The phase portrait of system (4.196) with 24 small limit cycles, where the origin is an unstable focus point, when
$a_0 = 0.0188065126$,
$b_0 = -2.0614758892$,
$a_1 = -0.0076130251$,
$b_1 = 2.2908763358$,
$a_2 = -0.001193487436$,
$b_2 = -0.0945028381$,
$a_3 = 0.01$,
$b_3 = 0.1348976085$

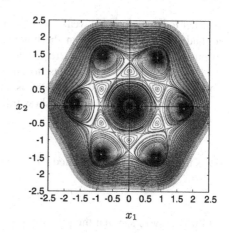

Fig. 4.17 The zoomed-in area around the point $(0, 1)$, the box marked in Fig. 4.16, showing the existence of 4 small limit cycles in the vicinity of $(0, 1)$, when
$a_0 = 0.0188065126$,
$b_0 = -2.0614758892$,
$a_1 = -0.0076130251$,
$b_1 = 2.2908763358$,
$a_2 = -0.0011934874$,
$b_2 = -0.0945028381$,
$a_3 = 0.01$,
$b_3 = 0.1348976085$

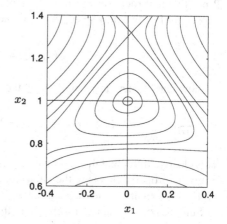

4.4.3 $q \geq 7$

Finally, we consider fifth-order symmetric vector fields for $q \geq 7$. In this case, from (4.21) we have the complex Z_q-equivariant vector field up to fifth order as

$$F(z, \bar{z}) = (A_0 + A_1|z|^2 + A_2|z|^4)z, \qquad (4.197)$$

where the A_k's are complex. Letting $A_k = a_k + ib_k, k = 0, 1, 2$, we obtain the following real form:

$$\begin{aligned}
\dot{x}_1 &= a_0 x_1 - b_0 x_2 + a_1 x_1^3 - b_1 x_1^2 x_2 + a_1 x_1 x_2^2 - b_1 x_2^3 \\
&\quad + a_2 x_1^5 - b_2 x_1^4 x_2 + 2a_2 x_1^3 x_2^2 - 2b_2 x_1^2 x_2^3 + a_2 x_1 x_2^4 - b_2 x_2^5, \\
\dot{x}_2 &= b_0 x_1 + a_0 x_2 + b_1 x_1^3 + a_1 x_1^2 x_2 + b_1 x_1 x_2^2 + a_1 x_2^3 \\
&\quad + b_2 x_1^5 + a_2 x_1^4 x_2 + 2b_2 x_1^3 x_2^2 + 2a_2 x_1^2 x_2^3 + b_2 x_1 x_2^4 + a_2 x_2^5.
\end{aligned} \qquad (4.198)$$

In order to have q ($q \geq 7$) weak focus points with 1 at $(0, 1)$, it requires $a_0 = -a_1 - a_2$ and $b_0 = -b_1 - b_2$. Then the eigenvalues of the Jacobian of system (4.145) evaluated at $(0, 1)$ are 0 and $2(a_1 + 2a_2)$, indicating that it is not an elementary center, and so small limit cycles do not exist around these q fixed points. If we consider the origin of system (4.198) as an elementary center (with $a_0 = 0$), then it is easy to use the normal form to show that

$$v_0 = a_0, \qquad v_1 = a_1, \qquad v_2 = a_2, \qquad v_i = 0, \quad i = 3, 4, \ldots. \qquad (4.199)$$

Thus the normal form associated with the origin is given by

$$\dot{r} = r\left(a_0 + a_1 r^2 + a_2 r^4\right), \qquad (4.200)$$

and it is easy to see that the polynomial equation, $a_0 + a_1 r^2 + a_2 r^4 = 0$, can have 2 positive real roots for r^2 since the three parameters a_0, a_1 and a_2 are free to be chosen. In fact, we may select the Lyapunov function

$$V(x_1, x_2) = \frac{1}{2}\left(x_1^2 + x_2^2\right)$$

to show that the derivative of V along the trajectory of the system is

$$\left.\frac{dV}{dt}\right|_{(4.198)} = x_1 \dot{x}_1 + x_2 \dot{x}_2 = \left(x_1^2 + x_2^2\right)\left[a_0 + a_1\left(x_1^2 + x_2^2\right) + a_2\left(x_1^2 + x_2^2\right)^2\right].$$

This clearly shows that the system can have a maximum of 2 limit cycles around the origin by properly choosing the values of coefficients a_0, a_1 and a_2, and the 2 limit cycles are actually circles around the origin.

Summarizing the above results gives the following theorem.

Theorem 4.26 *The maximal number of small limit cycles that a fifth-degree planar polynomial system with Z_q symmetry ($q \geq 7$) can have is 1, i.e., $\hat{H}_q(5) = 1$ ($q \geq 7$).*

In the above sections, we have considered a number of vector fields with Z_q symmetry. For convenience of reference, the results obtained in this chapter as well as those in the literature, are summarized in Table 4.1, where $\hat{H}_q(n)$ denotes the lower bound of limit cycles which exist in the corresponding system. The subscript q indicates that the vector has Z_q symmetry.

4.5 The Liénard System

The Liénard system or Liénard equation is a simplified version of Hilbert's 16th problem. However, even this simple version of Hilbert's 16th problem is still far from being completely solved, though many results have been obtained for the lower bound of the Hilbert number $H(n)$.

Table 4.1 Small limit cycles bifurcating from Hopf-type critical points

Degree (n)	Symmetry (q)	Number of limit cycles ($\hat{H}_q(n)$)
$n = 2$	$q = 1$	$\hat{H}_1(2) = 3$ ref. [14] ($H(2) \geq 4$)
$n = 3$	$q = 1$	$\hat{H}_1(3) = 9$ ref. [236] ($\hat{H}_1(3) = 13$?)
	$q = 2$	$\hat{H}_2(3) = 12$ refs. [237–239]
		($H_2(3) = 13$ ref. [147]; $H(3) \geq 13$ refs. [137, 215])
	$q = 3$	$\hat{H}_3(3) = 3$ (plus 1 large limit cycle) ref. [243]
	$q = 4$	$\hat{H}_4(3) = 4$ (plus 1 large limit cycle) ref. [243]
	$q \geq 5$	$\hat{H}_q(3) = 1$ ref. [243]
$n = 4$	q is even	same as the case $n = 3$
	$q = 3$	$\hat{H}_3(4) = 21$?
	$q = 5$	$\hat{H}_5(4) = 10$ ref. [243]
	$q \geq 7$ (odd)	$\hat{H}_q(4) = 1$ ref. [243]
$n = 5$	$q = 2$	$\hat{H}_2(5) = 38$?
	$q = 3$	$\hat{H}_3(5) = 33$?
	$q = 4$	$\hat{H}_4(5) = 36$?
	$q = 5$	$\hat{H}_5(5) = 25$ ref. [216]
	$q = 6$	$\hat{H}_6(5) = 24$ ref. [235]
	$q \geq 7$	$\hat{H}_q(5) = 2$

Note: Numbers with a question mark indicate a conjectured number of limit cycles

In this section, we will discuss the Liénard system and pay particular attention to two different types of equation.

Most of the early history in the theory of limit cycles was stimulated by practical problems displaying periodic behavior. For example, the differential equation derived by Rayleigh [180] in 1877, relating to the oscillation of a violin string, is given by

$$\ddot{x} + \varepsilon \left(\frac{1}{3} \dot{x}^2 - 1 \right) \dot{x} + x = 0, \tag{4.201}$$

where ε is a small perturbation parameter. Following the invention of the triode vacuum tube, which was able to produce stable self-excited oscillations of constant amplitude, van der Pol [196] obtained the following differential equation to describe this phenomenon:

$$\ddot{x} + \varepsilon (x^2 - 1) \dot{x} + x = 0. \tag{4.202}$$

Systems (4.201) and (4.202) can both display periodic behavior.

Perhaps the most famous class of differential equations, which generalize equation (4.202), are those first investigated by Liénard [144] in 1928:

$$\ddot{x} + f(x)\dot{x} + g(x) = 0, \tag{4.203}$$

where $f(x)$ and $g(x)$ are polynomial functions of x. Letting $\dot{x} = \overline{y}$, one obtains the above system described in the phase plane:

$$\dot{x} = \overline{y}, \qquad \dot{\overline{y}} = -g(x) - f(x)\overline{y}. \tag{4.204}$$

Further, let $\overline{y} = y - F(x)$, where $F(x) = \int_0^x f(s)\,ds$. Then we have the following equivalent system:

$$\dot{x} = \overline{y} = y - F(x),$$
$$\dot{y} = \dot{\overline{y}} + \frac{dF}{dx}\dot{x} = -g(x) - f(x)\overline{y} + f(x)\overline{y} = -g(x). \tag{4.205}$$

Taking f of degree i and g of degree j, the related Hilbert number is denoted $H(i, j)$. Liénard equations with $i = 1$ are called classical Liénard equations, and Smale's 13th problem [190] essentially concerns these. Known results are (e.g., see [59])

$$H(1, 1) = 0, \qquad H(1, 2) = H(2, 1) = H(2, 2) = H(3, 1) = 1. \tag{4.206}$$

For all other $H(i, j)$, even finiteness is not known. By compactification of both the phase plane and the Liénard family it was proven in [184] for n even and in [29] for n odd, that the finiteness question can be reduced to a study of singular perturbation problems. More information on compactification and desingularization of spaces of Liénard equations can be found in [58]. The singular perturbation problems corresponding to Liénard equations with very large coefficients are now under systematic investigation. This study has already led [61] to a counterexample of a conjecture in [145] on the number of limit cycles of classical Liénard equations. Besides the above-mentioned references, the results regarding "global" limit cycles and large amplitude limit cycles for Liénard equations can also be found in [30, 31].

4.5.1 Generalized Liénard Systems

Now, we pay particular attention to the maximum number of small amplitude limit cycles bifurcating from the origin of system (4.205) (or (4.203)), denoted by $\hat{H}(i, j)$. Then the existing results as well as the new results which we obtained recently for the Liénard system (4.205) are summarized in Table 4.2, where the numbers marked by * are obtained in [158], while those boldface numbers are our new results obtained in [241]. Note that the numbers given in this table are symmetric with respect to f and g [153], i.e., $\hat{H}(i, j) = \hat{H}(j, i)$. Thus, one only needs

Table 4.2 The values of $\hat{H}(i, j)$ for generalized Liénard systems for f and g of varying degree

Deg of f (i)	1	2	3	4	5	6	7	8	9	10	11	12	13	⋯	48	49	50
50	↑	↑	38														
49	24	33	38														
48	24	32	36														
⋮	⋮	⋮	⋮														
13	6	9	10	(13)													
12	6	8	10	12													
11	5	7	8	11													
10	5	7	8	10													
9	4	6	8	9*	(11)												
8	4	5	6	9*	10												
7	3	5	6	8*	9												
6	3	4	6	7*	(8)	(8)											
5	2	3	4	6	6*	8	9	10	11								
4	2	3	4	4	6	7*	8*	9*	9*	10	11	12	13				
3	1	2	2	4	4	6	6	6	8	8	8	10	10	⋯	36	38	38
2	1	1	2	3	3	4	5	5	6	7	7	8	9	⋯	32	33	→
1	0	1	1	2	2	3	3	4	4	5	5	6	6	⋯	24	24	→
	1	2	3	4	5	6	7	8	9	10	11	12	13	⋯	48	49	50
							Degree of g (j)										

Circled results are given in this section; an *asterisk* indicates results found in [158]; and *boldface* indicates new results

to prove results for cases $i \geq j$. It should also be noted that the notation $\hat{H}(i, j)$ as used in [153] denotes the possible maximum number of small limit cycles which may exist in the vicinity of the origin. It does not include global (large) limit cycles, nor contain other local (small) limit cycles which may appear in the neighborhood of other nonzero focus points. It will be shown in our proofs given in the next section that the maximal numbers listed in Table 4.2 can actually be reached since the sufficient conditions given in Theorems 4.1 and 4.2 can be satisfied by choosing appropriate parameter values.

It should be mentioned that a general optimal result for the maximum number of small amplitude limit cycles for classical Liénard equations can be found in [145], which is described in the so-called "Lins–De Melo–Pugh conjecture": if $g(x) = x$ and $\deg(F) = 2n+1$ or $2n+2$, $n \geq 1$, then the maximal number of limit cycles is n. In [145], this conjecture was proved only for the case $\deg(F) = 3$. A counterexample to this conjecture was recently found for the case $\deg(F) = 7$ with 4 limit cycles [61].

There has been a great deal of progress recently, when considering the bifurcation of small amplitude limit cycles from the origin for generalized Liénard systems of the form

$$\dot{x} = \phi(y) - F(x), \qquad \dot{y} = -g(x), \qquad (4.207)$$

where $\phi(y)$ is in polynomial form:

$$\phi(y) = y + c_2 y^2 + c_3 y^3 + \cdots . \qquad (4.208)$$

For definiteness, further let

$$\begin{aligned} f(x) &= a_1 + 2a_2 x + 3a_3 x^2 + 4a_4 x^3 + \cdots , \\ g(x) &= b_1 x + b_2 x^2 + b_3 x^3 + b_4 x^4 + \cdots , \end{aligned} \qquad (4.209)$$

then

$$F(x) = a_1 x + a_2 x^2 + + a_3 x^3 + \cdots . \qquad (4.210)$$

Without loss of generality, one may assume that $b_1 = 1$. Next, introduce the following transformation [157]:

$$u^2 = 2G(x) = 2 \int_0^x g(s)\,ds, \qquad u(0) = 0, \quad u'(0) > 0, \qquad (4.211)$$

so that $u^2 = x^2 + O(x^3)$ as $x \to 0$. Therefore, this transformation defines an analytical change of coordinates in a neighborhood of the origin. Let the inverse transformation be given by

$$x = \xi(u) = u + O(u^2) \quad \text{as } u \to 0, \qquad (4.212)$$

and thus,

$$\frac{g(\xi(u))}{u} \longrightarrow \frac{(g'(0))^{1/2}}{1}. \qquad (4.213)$$

Finally, in the (y, u) coordinate plane, we obtain the new system

$$\begin{aligned} \dot{y} &= -g(\xi(u)), \\ \dot{u} &= u^{-1} g(\xi(u)) [\phi(y) - F^*(u)]. \end{aligned} \qquad (4.214)$$

By condition (4.213), the local qualitative behavior of system (4.214) is identical to the following system:

$$\dot{y} = -g\big(\xi(u)\big) = -u,$$
$$\dot{u} = \phi(y) - F^*(u) = \phi(y) - \big(A_1 u + A_2 u^2 + A_3 u^3 + A_4 u^4 + \cdots\big), \tag{4.215}$$

where $A_1 = a_1$, $A_2 = a_2$, and the other A_i's are given explicitly in terms of the a_i's and b_i's. The explicit expressions for the A_i's are obtained from a Maple program; these expressions can be found in [240].

It is easy to see from (4.208) and (4.215) that the origin, $(0, 0)$, is an elementary center. Further, one can apply the Maple program [218] to system (4.215) to obtain the normal form (4.2) with the following focus values:

$$v_0 = A_1,$$

$$v_1 = -\frac{3}{8} A_3,$$

$$v_2 = -\frac{5}{16} A_5 - \frac{5}{24} A_2^2 A_3,$$

$$v_3 = -\frac{35}{128} A_7 - \frac{205}{1152} A_4^2 A_5 - \left(\frac{1885}{13824} A_4^2 + \frac{2}{3} A_2 A_4 + \frac{999}{8192} A_3^2\right) A_3,$$

$$v_4 = -\frac{63}{256} A_9 - \frac{413}{2304} A_2^2 A_7 - \left(\frac{47}{96} A_2 A_4 + \frac{2115}{4096} A_3^2 + \frac{4297}{41472} A_2^4\right) A_5$$
$$- \left(\frac{141}{160} A_2 A_6 + \frac{149}{240} A_4^2 + \frac{1093}{1152} A_2^3 A_4 + \frac{20599}{49152} A_2^2 A_3^2 + \frac{109483}{1244160} A_2^6\right) A_3, \tag{4.216}$$

$$\vdots$$

It follows from (4.216) that

$$v_0 = v_1 = v_2 = v_3 = 0, \quad v_4 \neq 0 \quad \Longleftrightarrow \quad A_1 = A_3 = A_5 = A_7 = 0, \quad A_9 \neq 0.$$

In general, one can show that

$$v_0 = v_1 = v_2 = \cdots = v_{k-1} = 0, \quad v_k \neq 0$$
$$\Longleftrightarrow \quad A_1 = A_3 = A_5 = \cdots = A_{2k-1} = 0, \quad A_{2k+1} \neq 0.$$

Therefore, in order for system (4.215) or (4.207) to have k small limit cycles, one must have that $v_0 = v_1 = v_2 = \cdots = v_{k-1} = 0$ but $v_k \neq 0$, or $A_1 = A_3 = A_5 = \cdots = A_{2k-1} = 0$ but $A_{2k+1} \neq 0$. For example, when $A_1 = 0$ but $A_3 \neq 0$, system (4.207) has a maximum of 1 limit cycle around the origin; when $A_1 = A_3 = 0$, $A_5 \neq 0$, the system (4.207) has a maximum of 2 limit cycles; etc. Since the coefficients A_i are given in terms of the a_i's and b_i's, one needs to determine the values of the a_i's and b_i's such that the above necessary conditions are satisfied. Then, with the sufficient conditions given in Theorems 4.1 and 4.2, we can apply appropriate perturbations to obtain exactly k limit cycles.

The new results we have obtained are shown in Table 4.2 with boldface numbers. There are in total 17 cases: $\hat{H}(i, j)$ for $j = 4, i = 10, 11, 12, 13$; $j = 5, i = 6, 7, 8, 9$; and $j = 6, i = 5, 6$. Due to symmetry, only 8 cases need proof. In the following, we give proofs for four cases: $\hat{H}(13, 4)$, $\hat{H}(6, 5)$, $\hat{H}(9, 5)$ and $\hat{H}(6, 6)$, which are circled in Table 4.2. Other cases can be proved in a similar manner.

We start from the case $H(6, 5)$, and then consider in sequence, $H(9, 5)$, $\hat{H}(13, 4)$ and $\hat{H}(6, 6)$.

(A) Proof of $\hat{H}(6,5) = 8$

Proof To show $\hat{H}(6, 5) = 8$, we need to solve the equations

$$A_1 = A_3 = A_5 = A_7 = A_9 = A_{11} = A_{13} = A_{15} = 0 \quad \text{but} \quad A_{17} \neq 0.$$

Note that $A_1 = a_1 = 0$ leads to $a_1 = 0$ (so $v_0 = 0$), and thus in the rest of this section we assume that $a_1 = 0$. Also note that for $\hat{H}(6, 5) = 8$, we must show that it is not possible to have $A_{17} = 0$. Therefore, once we find all solutions from the seven (since $A_1 = a_1 = 0$) nonlinear polynomial equations, we need to verify that $A_{17} \neq 0$ for all of these solutions. (Alternatively, we may add the equation $A_{17} = 0$ to the seven equations, and then show that there are no solutions for the eight polynomial equations.) For this case, $a_i = 0$ for $i \geq 8$, and $b_j = 0$ for $j \geq 6$.

It can be shown that one may use the coefficients a_3, a_5 and a_7 to solve the equations $A_3 = 0$, $A_5 = 0$ and $A_7 = 0$, respectively, to obtain

$$a_3 = \frac{2}{3}a_2 b_2,$$

$$a_5 = \frac{2}{5}a_2 b_4 + \frac{4}{3}a_4 b_2 - \frac{2}{3}a_2 b_2 b_3,$$

$$a_7 = 2b_2 a_6 - \frac{2}{3}a_2 b_2 b_5 + \left(\frac{4}{5}a_4 - \frac{2}{5}a_2 b_3\right)b_4 - \left(\frac{8}{9}b_2^3 + \frac{4}{3}b_2 b_3\right)a_4 \qquad (4.217)$$

$$+ \left(\frac{2}{3}b_3 + \frac{4}{9}b_2^2\right)a_2 b_2 b_3.$$

Then substituting the above results into the equation $A_9 = 0$ and solving for a_6 yields

$$a_6 = \frac{1}{9(81b_4 - 135b_2 b_3 - 220b_2^3)}$$

$$\times \Big[\big(810a_4 b_2 + 243a_2 b_4 - 810a_2 b_2 b_3 - 660a_2 b_2^3\big)b_5$$

$$+ \big(486a_4 b_3 + 1188a_4 b_2^2 - 594a_2 b_2^2 b_3 - 243a_2 b_3^2\big)b_4$$

$$- \big(810b_3^2 + 2520b_3 b_2^2 + 880b_2^4\big)b_2 a_4$$

$$+ \big(405b_3^2 + 1260b_2^2 b_3 + 440b_2^4\big)a_2 b_2 b_3 \Big]. \qquad (4.218)$$

Having determined a_3, a_5, a_7 and a_6, the remaining three equations $A_{11} = A_{13} = A_{15} = 0$ have four coefficients b_2, b_3, b_4 and b_5. This seems to suggest that we may use the four coefficients to find the conditions such that $A_{11} = A_{13} = A_{15} = A_{17} = 0$ but $A_{19} \neq 0$, which implies that $H(6,5) = 9$. However, a careful examination reveals that the three equations, $A_{11} = A_{13} = A_{15} = 0$, are actually homogeneous under the following substitutions:

$$b_3 = B_3 b_2^2, \qquad b_4 = B_4 b_2^3, \qquad b_5 = B_5 b_2^4. \tag{4.219}$$

Under the above transformation, A_{11}, A_{13} and A_{15} can be written as

$$A_{11} = b_2^{10} \hat{A}_{11}(B_3, B_4, B_5),$$

$$A_{13} = b_2^{12} \hat{A}_{13}(B_3, B_4, B_5),$$

$$A_{15} = b_2^{14} \hat{A}_{15}(B_3, B_4, B_5),$$

where $\hat{A}_{11}$, $\hat{A}_{13}$ and $\hat{A}_{15}$ are given as follows (from computer output):

```
hatA11 := 87750*B3^3-210600*B4*B3^2+162000*B5*B3-14040*B4^2
          -32805*B5*B4^2-56862*B4^3+54675*B5*B3*B4+189540*B4^2*B3
          +46800*B3*B4-113400*B5*B4-104000*B5+41600*B4
          +19500*B3^2-91125*B5^2:
hatA13 :=-775200*B3*B4+11372400*B4^2*B3-13219200*B4*B3^2
          +14076000*B5*B3+12393000*B3^3*B4-11153700*B3^2*B4^2
          -7290000*B3^2*B5-7030800*B5*B4-157464*B4^4+4100625*B3*B5^2
          -984150*B5*B4^2-2733750*B5^2+969000*B3^2-12240*B4^2
          +2067200*B4-3261060*B4^3+2907000*B3^3-5168000*B5
          +6743250*B5*B3*B4-5163750*B3^4+3608550*B3*B4^3
          +1476225*B3*B5*B4^2-2460375*B3^2*B4*B5:
hatA15 :=-262200000*B3*B4+919152360*B3*B4^3+2133054000*B3^3*B4
          -1580040000*B3^2*B5-2191228200*B3^2*B4^2-766764000*B5*B4
          +857530800*B4^2*B3-238698900*B5*B4^2+323190000*B3^3*B5
          -233735625*B3^2*B5^2+1415880000*B5*B3+17950896*B3*B4^4
          +23619600*B4^2*B5^2+304357500*B3*B5^2-802332000*B4*B3^2
          +1278139500*B5*B3*B4+98325000*B3^3+103831200*B4^2
          -256151160*B4^3+65550000*B3^2-20092500*B5^2+139840000*B4
          -349600000*B5+62329500*B3*B5*B4^2-339349500*B3^2*B4*B5
          -84144825*B4^2*B3^2*B5-39366000*B4*B3*B5^2
          +140241375*B4*B3^3*B5-123661728*B4^4-619447500*B3^4
          +65610000*B5^3+363588750*B3^5-265523670*B3^2*B4^3
          -872613000*B3^4*B4-49572000*B5^2*B4+785351700*B3^3*B4^2:
```

Since $b_2 \neq 0$ (otherwise it leads to a center), there exist only 3 independent coefficients, B_3, B_4 and B_5, and thus the best possibility is that $A_{11} = A_{13} = A_{15} = 0$ but $A_{17} \neq 0$, implying $\hat{H}(6,5) \leq 8$. Note that if there is a solution such that $A_{11} = A_{13} = A_{15} = 0$ and in addition, $A_{17} = 0$, then it is a center.

Now eliminating B_5 from the two equations, $\hat{A}_{11} = \hat{A}_{13} = 0$, results in the solution

```
B5 := 1/200/(2835*B3-2268*B4-1280)*(599400*B4*B3^2+328050*B3^2*B4^2
      +38400*B3*B4-51120*B4^2-102400*B4-48000*B3^2-144000*B3^3
      +19683*B4^4+194400*B4^3-364500*B3^3*B4-631800*B4^2*B3
      -131220*B3*B4^3+151875*B3^4):
```

and a resultant equation

```
f1 := (-3*B4+5*B3)*(-81*B4+220+135*B3)
    *(-35872267500*B4^3*B3^3-45448776000*B4*B3^2+29350656000*B3*B4
    -5570560000*B4+14492520000*B3^3-19137945600*B4^2
    -14921288640*B4^3-2611200000*B3^2+46920772800*B4^2*B3
    +6022998000*B4^4-5183190000*B3^4-29278462500*B3^5
    -12675852000*B3^3*B4+37539417600*B3^2*B4^2-26879104800*B3*B4^3
    -39060913500*B3*B4^4+110114575200*B3^2*B4^3+106435822500*B3^4*B4
    -153763596000*B3^3*B4^2+44840334375*B4^2*B3^4
    -29893556250*B4*B3^5+16142520375*B4^4*B3^2+387420489*B4^6
    +5490848412*B4^5+8303765625*B3^6-3874204890*B3*B4^5) :
```

Then substituting the solution of B_5 into $\hat{A}_{15}$ yields another equation:

```
f2 := 3/1600*(-81*B4+220+135*B3)*(-3*B4+5*B3)/(2835*B3-2268*B4-1280)^3
    *(-2232630864000000000*B4*B3^2+509689921536000000*B3*B4
    -47655682048000000*B4
    + ...   +367746792292968750*B3^8*B4^2-588394867668750000*B3^7*B4^3
    -136202515664062500*B4*B3^9+22700419277343750*B3^10) :
```

Note that f_1 and f_2 have common factors $(5B_3 - 3B_4)(135B_3 - 81B_4 + 220)$ which generate centers. Thus, the solutions must come from the remaining factors. Finally, eliminating B_4 from the two remaining factors yields the solution

```
B4 := 15*B3^2*(3902946741784477699875126732951050625*B3^15
    -95954718640440354934980163730112246125*B3^14
    + ...  -206225985834184357983551313500117328199968*B3
    +13178429645664855681535508030431009177760)
    /(35126520676060299298876140596559455625*B3^16
    -760183242740755501817880557294476207500*B3^15
    + ...  +64806256798829157543398224597298754197913 6*B3
    -421709748661275381809136256973792293668320) :
```

and a univariate resultant polynomial equation for B_3,

```
f3 := B3*(63*B3+496)*(189*B3+256)*(178605*B3^2-1357776*B3+241664)
    *(5779045668619503*B3^5+56315474719135758*B3^4
    +215033754040484148*B3^3+258790514446979496*B3^2
    -81098473487101056*B3-11257720644513280) :
```

Then all possible solutions are given by the roots of $f_3 = 0$. However, it is easy to check that for the solution $B_3 = 0$, A_{17} and higher A's are zero, indicating that this solution generates a center. For the next four roots obtained from the second, third and fourth factors: $B_3 = -\frac{496}{63}, -\frac{256}{189}, \frac{3592 \pm 56\sqrt{3729}}{945}$, one can show that they are not solutions of the original equations, $A_{11} = A_{13} = A_{15} = 0$. Therefore, possible solutions are only given by the last factor of f_3, which gives three real roots:

$$B_3 = -2.7751137173386343174652900810 21,$$

$$- 0.10602006876159062416178891635 8,$$

$$0.33822530891806927386942237347 4.$$

However, since b_2, a_2 and a_4 are free, there exists an infinite number of solutions such that $A_{11} = A_{13} = A_{15} = 0$ but $A_{17} \neq 0$. In other words, $\hat{H}(6, 5) \leq 8$. For example, for the critical parameter values

$$a_{2c} = a_{4c} = b_{2c} = 1, a_{3c} = 0.4,$$

$$a_{5c} = 0.7301602298161440828218867782 6,$$

$$a_{6c} = 0.3057199987940233090145352580 9,$$

$$a_{7c} = 0.0734055396720066365688165940 9,$$

$$b_{3c} = 0.1217611112105049385929920544 5,$$

$$b_{4c} = -0.5283831424913485435229099988 7,$$

$$b_{5c} = -0.0450706169660235968231029315 3,$$

(4.220)

we have $A_3 = A_5 = \cdots = A_{15} = 0$ but $A_{17} = 0.0000444424 \neq 0$, which leads to $v_1 = v_2 = \cdots = v_7 = 0$ but $v_8 \neq 0$. It should be noted that all the computations performed here with Maple are symbolic, except for the last step in which a polynomial equation in a single variable is solved numerically.

For the parameter values given in (4.220) which are up to 30 digits of accuracy, we applied our Maple program [218] to obtain the focus values as follows (using accuracy up to 30 digits):

$$v_1 = 0.0,$$

$$v_2 = -0.5 \times 10^{-30},$$

$$v_3 = -0.35 \times 10^{-29},$$

$$v_4 = 0.435 \times 10^{-28},$$

$$v_5 = -0.169045 \times 10^{-26},$$

$$v_6 = 0.8986185 \times 10^{-25},$$

$$v_7 = -0.29858535 \times 10^{-23},$$

$$v_8 = -0.0000082427511443671765959973 3.$$

Note that due to numerically accumulating errors, the focus values $v_i, i = 1, 2, \ldots, 7$, do not all have an accuracy up to 10^{-30}. When the accuracy is increased, say, to 1000 digits, the errors in $v_i, i = 1, 2, \ldots, 7$, are close to 10^{-1000}, while v_8 is not changed up to 30 digits.

The above procedure only shows $\hat{H}(6, 5) \leq 8$. In order to prove $\hat{H}(6, 5) = 8$, we need to choose proper perturbations. To achieve this, we perturb the critical parameter values given in (4.220) to obtain exactly 8 small amplitude limit cycles around the origin. The main idea is as follows: first it is observed from (4.216)–(4.218) that one may perturb v_4, v_3, v_2, and v_1 in reverse order by using the parameters a_6, a_7, a_5 and a_3, one after another. However, one cannot perturb v_7, v_6 and v_5, one by one, since the three equations $A_{11} = A_{13} = A_{15} = 0$ are solved simultaneously using b_3, b_4 and b_5. We may use Theorem 4.2 to show that such perturbations exist. To achieve this, we calculate the Jacobian of the three equations, $A_{11} = A_{13} = A_{15} = 0$, with respect to b_3, b_4 and b_5 at the critical point to obtain

$$J_c = \begin{bmatrix} 0.0133202248 & 0.0219452142 & 0.0497244039 \\ 0.0012718051 & 0.0209919708 & 0.0377906097 \\ -0.0057653018 & 0.0159251487 & 0.0075334956 \end{bmatrix}$$

and so $\det(J_c) = -0.0000038765 \neq 0$. Combining the above two perturbation steps shows that there exist perturbations, ε_i, $i = 1, 2, \ldots, 8$, such that $b_3 = b_{3c} + \varepsilon_1$, $b_4 = b_{4c} + \varepsilon_2$, $b_5 = b_{5c} + \varepsilon_3$, $a_6 = a_{6c} + \varepsilon_4$, $a_7 = a_{7c} + \varepsilon_5$, $a_5 = a_{5c} + \varepsilon_6$, $a_3 = a_{3c} + \varepsilon_7$, $a_1 = \varepsilon_8$, under which

$$v_i v_{i+1} < 0 \quad \text{and} \quad |v_i| \ll |v_{i+1}| \ll 1 \quad \text{for } i = 0, 1, 2, \ldots, 7.$$

This indicates that

$$\hat{H}(6, 5) = 8. \qquad \qquad \Box$$

Similarly, we can prove that $\hat{H}(7, 5) = 9$ and $\hat{H}(8, 5) = 10$.

(B) Proof of $\hat{H}(9, 5) = 11$

Proof For this case, $a_i = 0$ for $i \geq 10$, and $b_j = 0$ for $j \geq 6$. As for the proof of the case $\hat{H}(6, 5)$, we need to show that for properly chosen parameter values, $A_3 = A_5 = \cdots = A_{21} = 0$ but $A_{23} \neq 0$ and further, to show that it is not possible to have $A_3 = A_5 = \cdots = A_{21} = A_{23} = 0$. Thus, $\hat{H}(9, 5) = 11$. We need to solve 10 nonlinear polynomial equations.

Similarly, we can solve the equations $A_3 = A_5 = \cdots = A_{15} = 0$, one by one, by using the coefficients a_3, a_5, a_7, a_9, a_{10}, a_8 and a_6, respectively. The solutions for a_3, a_5 and a_7 are given in (4.217), while the lengthy expressions for a_9, a_{10}, a_8 and a_6 are omitted here for simplicity. Then the remaining three equations are $A_{17} = A_{19} = A_{21} = 0$. Similarly, one can verify that these equations are homogeneous under the transformation

$$b_3 = B_3 b_2^2, \qquad b_4 = B_4 b_2^3, \qquad b_5 = B_5 b_2^4,$$

which leads to the expressions

$$A_{17} = b_2^{24} \hat{A}_{17}(B_3, B_4, B_5),$$

$$A_{19} = b_2^{26} \hat{A}_{19}(B_3, B_4, B_5),$$

$$A_{21} = b_2^{28} \hat{A}_{21}(B_3, B_4, B_5).$$

Since $b_2 \neq 0$ (otherwise it leads to a center), there are only 3 free coefficients, B_3, B_4 and B_5. Thus, it is not possible to have solutions for $A_{17} = A_{19} = A_{21} = A_{23} = 0$. The best result will be $A_{17} = A_{19} = A_{21} = 0$ but $A_{23} \neq 0$, implying that $\hat{H}(9, 5) \leq 11$.

Next, eliminating B_5 from the three equations, $\hat{A}_{17} = \hat{A}_{19} = \hat{A}_{21} = 0$, results in two polynomial equations: $\tilde{f}_i = F f_i, i = 1, 2$, where F is a common factor, and $f_i, i = 1, 2$, are functions of B_3 and B_4. Then eliminating B_4 from the two equations, $f_1 = f_2 = 0$, yields an equation $f_3 = B_3 f_{31}(B_3) f_{32}(B_3) = 0$, where $f_{31}(B_3)$ is a 9th-degree polynomial in B_3 while $f_{32}(B_3)$ is a 108th-degree polynomial in B_3. With the built-in program in Maple, `fsolve`, one can show that $f_{31}(B_3)$ has 3 real solutions, and $f_{32}(B_3)$ has 24 real solutions. However, verifying the original equations shows that $A_{21} \neq 0$ for all the 24 solutions. Hence, the second polynomial $f_{32}(B_3)$ can only have solutions for $\hat{H}(9, 5) \leq 10$. The 3 real solutions obtained from $f_{31}(B_3) = 0$ do satisfy $A_{17} = A_{19} = A_{21} = 0$ but $A_{23} \neq 0$, and hence $\hat{H}(9, 5) \leq 11$. Actually, we have an infinite number of solutions since a_2, a_4 and b_2 can be chosen arbitrarily.

As an example, we choose the critical parameter values

$$a_{2c} = a_{4c} = b_{2c} = 1, \qquad a_{3c} = \frac{2}{3},$$

$$a_{5c} = 1.34796265977228916492295656132,$$

$$a_{6c} = -4.47654700356991805278499402760,$$

$$a_{7c} = -9.79768652019339892191312380939,$$

$$a_{8c} = -6.46099608912640900311153589268, \qquad (4.221)$$

$$a_{9c} = -1.41527303119400585959220295145,$$

$$a_{10c} = -0.00000029737293573499992497518,$$

$$b_{3c} = -2.97633366517505075983960102126,$$

$$b_{4c} = -4.92398279252769502075861029881,$$

$$b_{5c} = -1.94146639271750592836680707408,$$

under which $A_3 = A_5 = \cdots = A_{21} = 0$ but $A_{23} \neq 0$, and so $v_1 = v_2 = \cdots = v_{10} = 0$ but $v_{11} \neq 0$. For the parameter values given in (4.221), one can execute the Maple program [218] to obtain the focus values given below (up to 30 digits):

$$v_1 = 0.0,$$

$$v_2 = 0.125 \times 10^{-28},$$

$$v_3 = 0.4893351384740867421057560542777 \times 10^{-28},$$

$$v_4 = 0.249526710531812489005475943356 \times 10^{-27},$$

$$v_5 = 0.14935 \times 10^{-25},$$

$$v_6 = -0.1571393984810211334351668905633 \times 10^{-23},$$

$$v_7 = 0.190592875 \times 10^{-21},$$

$$v_8 = -0.2743763732146172125635687266550 \times 10^{-19},$$

$$v_9 = 0.5402669376055 \times 10^{-17},$$

$$v_{10} = -0.11456053157408102242138652 8274 \times 10^{-15},$$

$$v_{11} = 0.00000107167788911320080229 0758.$$

As the final step, as for the proof of the case $\hat{H}(6,5)$, one can choose proper perturbations to show that there exist exactly 11 small limit cycles for this case, i.e.,

$$\hat{H}(9,5) = 11. \qquad \qquad \square$$

(C) Proof of $\hat{H}(13,4) = 13$

Proof Now we turn to the case $\hat{H}(13,4)$. For this case, we need to show that there exist parameter values satisfying $A_3 = A_5 = \cdots = A_{25} = 0$ but $A_{27} \neq 0$, and it is not possible to have $A_3 = A_5 = \cdots = A_{25} = A_{27} = 0$. Thus, $\hat{H}(13,4) = 13$. Here, $a_i = 0$ for $i \geq 15$, and $b_j = 0$ for $j \geq 5$. It is necessary to solve 12 nonlinear polynomial equations.

Similarly, we can solve the equations $A_3 = A_5 = \cdots = A_{21} = 0$, one by one, using the coefficients $a_3, a_5, a_7, a_9, a_{11}, a_{13}, a_{14}, a_{12}, a_{10}$ and a_8, respectively. The solutions for a_3, a_5 and a_7 are the same as those given in (4.217) and other lengthy expressions are omitted here. The remaining two equations are $A_{23} = A_{25} = 0$. Again, an examination shows that these two equations are homogeneous under the substitutions

$$b_3 = B_3 b_2^2, \qquad b_4 = B_4 b_2^3,$$

which results in the expressions

$$A_{23} = -\frac{8 b_2^{17}(18 a_6 + 9 a_2 B_3^2 b_2^4 + 4 a_2 b_2^4 B_3 - 18 a_4 B_3 b_2^2 - 8 a_4 b_2^2) B_4^3}{11390625 F(B3, B4)} f_1(B_3, B_4),$$

$$A_{25} = \frac{1798 b_2^{19}(18 a_6 + 9 a_2 B_3^2 b_2^4 + 4 a_2 b_2^4 B_3 - 18 a_4 B_3 b_2^2 - 8 a_4 b_2^2) B_4^3}{854296875 F(B3, B4)} f_2(B_3, B_4),$$

where $F(B_3, B_4)$ is a polynomial in B_3 and B_4. Thus, it is not possible to have solutions satisfying $A_{23} = A_{25} = A_{27} = 0$ since there are only two free parameters. Hence, $\hat{H}(13,4) \leq 13$. Note from the above expressions of A_{23} and A_{25} that the common factor involves the coefficients a_2, a_4, a_6 and b_2, implying that these 4 coefficients are not used and can be chosen arbitrarily. Thus, we only need to find solutions such that $A_{23} = A_{25} = 0$ but $A_{27} \neq 0$, i.e., solve $f_1 = f_2 = 0$ for B_3 and B_4.

Eliminating B_3 from the two equations, $f_1 = f_2 = 0$, yields an equation $f_3 = B_4 f_{31}(B_4) f_{32}(B_4) = 0$ where $f_{31}(B_4)$ is a 22nd-degree polynomial in B_4 and $f_{32}(B_4)$ is a 38th-degree polynomial in B_4. With the Maple program, fsolve, one can show that both $f_{31}(B_4)$ and $f_{32}(B_4)$ have 4 real solutions. However, the 4 solutions obtained from $f_{31}(B_4) = 0$ give $A_{23} \neq 0$. Hence, the first polynomial $f_{31}(B_4)$ has solutions only for $\hat{H}(13,4) \leq 12$. The 4 real solutions obtained from

$f_{32}(B_4) = 0$ do satisfy $A_{23} = A_{25} = 0$ but $A_{27} \neq 0$, and hence $\hat{H}(13, 4) \leq 13$. Actually, we have an infinite number of solutions since a_2, a_4, a_6 and b_2 can be chosen arbitrarily.

As an example, we choose the critical parameter values

$$a_{2c} = a_{4c} = a_{6c} = 1, \qquad b_{2c} = 0.6, \qquad a_{3c} = 0.4,$$

$$a_{5c} = 0.925774938759465589222204765642,$$

$$a_{7c} = 1.316833165727700412309335663844,$$

$$a_{8c} = -0.691046315857442196616170733754,$$

$$a_{9c} = -1.420886305121482389290364186311,$$

$$a_{10c} = -4.445851614289377114006530598436,$$

$$a_{11c} = -7.952701322429532551678283165456,$$ (4.222)

$$a_{12c} = -4.386931648498706788747822342176,$$

$$a_{13c} = -0.438274375986514658811649292516,$$

$$a_{14c} = 0.0562761812649133017370873732808,$$

$$b_{3c} = -1.923875937127072195767473302856,$$

$$b_{4c} = -1.609438590228407465212354161806,$$

under which $A_3 = A_5 = \cdots = A_{25} = 0$ but $A_{27} \neq 0$, and thus $v_1 = v_2 = \cdots = v_{12} = 0$ but $v_{13} \neq 0$. For the parameter values given in (4.222), the computed focus values are given below (noting that the accuracy for this case is taken up to 100 digits since this case has more focus values):

$$v_1 = 0.0,$$

$$v_2 = 0.2 \times 10^{-99},$$

$$v_3 = 0.7 \times 10^{-99},$$

$$v_4 = 0.37 \times 10^{-98},$$

$$v_5 = 0.2165 \times 10^{-97},$$

$$v_6 = -0.385 \times 10^{-97},$$

$$v_7 = 0.130275 \times 10^{-94},$$

$$v_8 = -0.28609925 \times 10^{-92},$$

$$v_9 = 0.78087682 \times 10^{-90},$$

$$v_{10} = -0.1786023085 \times 10^{-87},$$

$$v_{11} = 0.61468531009315 \times 10^{-85},$$

$$v_{12} = -0.2114286075965054 \times 10^{-82},$$

$$v_{13} = -0.00073252213821505282085563581471476925109268994079$$

$$231794163577001601047895350909233731067569529815.$$

It should be noted that the focus values given above, $v_i, i = 1, 2, \ldots, 12$, are supposed to be zero, but they are not, due to accumulating errors. If we increase the accuracy to, say, 1000 digits, then all these focus values are close to 10^{-1000}, except for v_{13} which is not changed up to 100 digits. Finally, one can use appropriate perturbations to show that there exist exactly 13 small amplitude limit cycles. This proves that

$$\hat{H}(13, 4) = 13. \qquad\qquad \square$$

(D) Proof of $\hat{H}(6, 6) = 8$

Proof Finally, we consider the case $\hat{H}(6, 6)$ and prove that $\hat{H}(6, 6) = 8$. For this case, $a_i = 0$ for $i \geq 8$ and $b_i = 0$ for $i \geq 7$. Here we need to find parameter values which satisfy $A_3 = A_5 = \cdots = A_{15} = 0$ but $A_{17} \neq 0$, and also prove that it is not possible to have $A_3 = A_5 = \cdots = A_{15} = A_{17} = 0$. For this case, we will show that solutions do not exist that satisfy $A_3 = A_5 = \cdots = A_{15} = A_{17} = 0$. Therefore, we need to solve 8 nonlinear polynomial equations.

Similarly, we can solve the equations $A_3 = A_5 = A_7 = A_9 = 0$, one by one, using the coefficients a_3, a_5, a_7 and b_6, respectively. The solutions for a_3, a_5 and a_7 are given in (4.217), and b_6 is given below:

$$
\begin{aligned}
b_6 ={}& \frac{7}{1215(2a_4 - a_2 b_3)}\big[\big(1215 b_2 b_3 + 1980 b_2^3 - 729 b_4\big)a_6 \\
&+ \big(810 a_4 b_2 + 243 a_2 b_4 - 810 a_2 b_2 b_3 - 660 a_2 b_2^3\big)b_5 \\
&+ \big(486 b_3 b_4 + 1188 b_4 b_2^2 - 810 b_3^2 b_2 - 2520 b_3 b_2^3 - 880 b_2^5\big)a_4 - 243 b_3 b_4 \\
&+ \big(440 b_2^4 + 1260 b_2^2 b_3 + 405 b_3^2 - 594 b_2 b_4\big)a_2 b_2 b_3\big].
\end{aligned}
$$

Then the remaining four equations are $A_{11} = A_{13} = A_{15} = A_{17} = 0$. To solve these equations, we may first eliminate the coefficient a_6 from them to obtain three resultant equations:

$$\tilde{f}_1 = (a_2 b_3 - 2a_4)f_1(b_2, b_3, b_4, b_5) = 0,$$

$$\tilde{f}_2 = (a_2 b_3 - 2a_4)f_2(b_2, b_3, b_4, b_5) = 0,$$

$$\tilde{f}_3 = (a_2 b_3 - 2a_4)\big(81 b_4 - 135 b_2 b_3 - 220 b_2^3\big)f_3(b_2, b_3, b_4, b_5) = 0.$$

Ignoring the common factor, we need to solve the equations $f_1(b_2, b_3, b_4, b_5) = f_2(b_2, b_3, b_4, b_5) = f_3(b_2, b_3, b_4, b_5) = 0$. Note that now there are no a_i coefficients involved in these equations. It seems that there are four coefficients in the three equations. However, an examination shows that these equations are actually homogeneous under the substitutions

$$b_3 = B_3 b_2^2, \qquad b_4 = B_4 b_2^3, \qquad b_5 = B_5 b_2^4.$$

Then, further ignoring the free parameter b_2, we need to solve the three equations, $\hat{f}_1 = \hat{f}_2 = \hat{f}_3 = 0$, where $\hat{f}_i$ are functions of B_3, B_4 and B_5.

Eliminating B_5 from the three equations, $\hat{f}_1 = \hat{f}_2 = \hat{f}_3 = 0$, yields two equations:

$$f_4 = (5B_3 - 3B_4) f_{41}(B_3, B_4) f_{42}(B_3, B_4),$$

$$f_5 = (5B_3 - 3B_4)(81B_4 - 135B_3 - 220) f_{51}(B_3, B_4) f_{52}(B_3, B_4),$$

where f_{41}, f_{42}, f_{51} and f_{52} are 5th-, 11th-, 8th- and 15th-degree polynomials, respectively, in B_3 and B_4. To solve $f_4 = f_5 = 0$, first note that the factor $81B_4 - 135B_3 - 220$ does not give a solution. So we need to solve four groups of equations:

$$\begin{aligned}
&\text{1st group:} \quad (f_{41}, f_{51}) = (0, 0),\\
&\text{2nd group:} \quad (f_{41}, f_{52}) = (0, 0),\\
&\text{3rd group:} \quad (f_{42}, f_{51}) = (0, 0),\\
&\text{4th group:} \quad (f_{42}, f_{52}) = (0, 0).
\end{aligned}$$

For each of these groups, first eliminate B_4 and then solve the resulting equation for B_3. The first group yields the polynomial equation for B_3,

$$\begin{aligned}
F = \ &137180783939631181536668007963671256 B_3^5\\
&- 12745608963016320639029647390357350 B_3^4\\
&+ 46489461997527105443258799808067400 B_3^3\\
&- 83034261563486218004432710846072368 B_3^2\\
&+ 72584263326198471403294587088309392 B_3\\
&- 24848924199546145630434699613010656,
\end{aligned}$$

which has three real solutions, satisfying $\hat{f}_1 = \hat{f}_2 = 0$ but $\hat{f}_3 \neq 0$. Substituting these solutions back into the original equations indeed yields $A_{11} = A_{13} = A_{15} = 0$ but $A_{17} \neq 0$. Similarly, it can be shown that the second group has 16 real solutions, and the result is similar to the first group, i.e., all 16 solutions satisfy $A_{11} = A_{13} = A_{15} = 0$ but $A_{17} \neq 0$. The third and fourth groups do not have real solutions satisfying $A_{11} = A_{13} = A_{15} = 0$. Instead, they have solutions only for $A_{11} = A_{13} = 0$ but $A_{15} \neq 0$. Therefore, combining the other parameter values, the best result for this case is to choose parameter values such that $v_1 = v_2 = \cdots = v_7 = 0$ but $v_8 \neq 0$, and thus $\hat{H}(6, 6) \leq 8$.

An example for selecting the critical parameter values is given below:

$$a_{2c} = a_{4c} = b_{2c} = 1, \qquad a_{3c} = \frac{2}{3},$$

$$a_{5c} = 1.0100010639636300130642904012,$$

$$a_{6c} = 0.67235076750720532893297490162,$$

$$a_{7c} = 0.30067436107555057220956101983, \qquad (4.223)$$

$$b_{3c} = 1.19684216379096679633485259370,$$

$$b_{4c} = 1.18640626622735302655214758979,$$

$$b_{5c} = 0.87410359260179985583519841682,$$

$$b_{6c} = 0.54374053517347504259667364709,$$

under which $v_1 = v_2 = \cdots = v_7 = 0$ but $v_8 \neq 0$, with the computer output (up to 30 digits) given below:

$$v_1 = 0.0,$$

$$v_2 = 0.5 \times 10^{-30},$$

$$v_3 = 0.85 \times 10^{-29},$$

$$v_4 = -0.285 \times 10^{-29},$$

$$v_5 = 0.10419 \times 10^{-26},$$

$$v_6 = -0.4592917157219543424886974623 14 \times 10^{-25},$$

$$v_7 = 0.40509949811875989891 3218704449 \times 10^{-23},$$

$$v_8 = -0.00269870868073344640493943 1712.$$

Similarly, one can use Theorems 4.1 and 4.2 with proper perturbations to show that there exist exactly 8 small amplitude limit cycles for this case, i.e.,

$$\hat{H}(6, 6) = 8. \qquad \qquad \Box$$

4.5.2 Liénard Equation with Z_2 Symmetry

In this subsection, we reconsider the Liénard equation with Z_2 symmetry, which was studied in Sect. 3.4.3, for a comparison of different methods of computing focus values. Here, we shall discuss both local and global bifurcation of limit cycles.

For convenience, we repeat system (3.40) here:

$$\dot{x} = y,$$

$$\dot{y} = -\frac{1}{2}b^2 x (x^2 - 1) - y \sum_{i=0}^{m} a_i x^{2i}, \qquad (4.224)$$

which has three fixed points: $(0, 0)$ and $(\pm 1, 0)$. The origin $(0, 0)$ is a saddle point. The two symmetric points $(\pm 1, 0)$ are elementary centers when $\sum_{i=0}^{m} a_i = 0$. By shifting system (4.224) to the two symmetric elementary centers, we obtain the following system:

$$\frac{du}{d\tau} = v,$$

$$\frac{dv}{d\tau} = -u - \frac{3}{2}u^2 - \frac{1}{2}u^3$$

$$- v(2u + u^2) \left\{ \sum_{i=1}^{m} a_i + \sum_{i=2}^{m} a_i (1+u)^2 + \sum_{i=3}^{m} a_i (1+u)^3 + \cdots \right. \qquad (4.225)$$

$$\left. + (a_{m-1} + a_m)(1+u)^{2m-4} + a_m (1+u)^{2m-2} \right\}.$$

The Jacobian of system (4.225) evaluated at $(u, v) = (0, 0)$ (i.e., at $(x, y) = (\pm 1, 0)$) is now in Jordan canonical form. The above procedure shows that the coefficient b can be chosen as any nonzero real value, which does not affect the qualitative behavior of the system. In particular, it does not change the number of limit cycles. Thus, without loss of generality, we assume $b = 1$, and so $\tau = t$, in the following analysis.

(A) 13 Limit Cycles in System (4.224) for $m = 5$ We will prove that the Liénard equation (4.224) for $m = 5$ (i.e., $\deg(f) = 10$) can have 13 limit cycles, amongst which 10 are small limit cycles and 3 are large limit cycles.

First, we show that when $m = 5$, system (4.224) has 10 small limit cycles, 5 in the neighborhood of $(1, 0)$ and 5 in the neighborhood of $(-1, 0)$.

When $m = 5$, system (4.224) becomes

$$\dot{x} = y,$$

$$\dot{y} = -\frac{1}{2}x(x^2 - 1) - (a_0 + a_1 x^2 + a_2 x^4 + a_3 x^6 + a_4 x^8 + a_5 x^{10})y, \qquad (4.226)$$

and the condition $\sum_{i=0}^{m} a_i = 0$ becomes

$$a_0^* = -(a_1 + a_2 + a_3 + a_4 + a_5). \qquad (4.227)$$

The transformed system (4.225) for $m = 5$ can be rewritten as

$$\dot{u} = v,$$

$$\dot{v} = -u - \frac{3}{2}u^2 - 2(a_1 + 2a_2 + 3a_3 + 4a_4 + 5a_5)uv$$

$$- \frac{1}{2}u^3 - (a_1 + 6a_2 + 15a_3 + 28a_4 + 45a_5)u^2v$$

$$- 4(a_2 + 5a_3 + 14a_4 + 30a_5)u^3v \qquad (4.228)$$

$$- (a_2 + 15a_3 + 70a_4 + 210a_5)u^4v$$

$$- 2(3a_3 + 28a_4 + 128a_5)u^5v - (a_3 + 28a_4 + 210a_5)u^6v$$

$$- 8(a_4 + 15a_5)u^7v - (a_4 + 45a_5)u^8v - 10a_5u^9v - a_5u^{10}v.$$

Note that the zeroth-order focus value of system (4.226) associated with the fixed points $(\pm 1, 0)$ is given by

$$v_0 = -(a_0 + a_1 + a_2 + a_3 + a_4 + a_5). \qquad (4.229)$$

To obtain the focus values v_i $(i \geq 1)$, applying the Maple program for computing the normal forms of Hopf and generalized Hopf bifurcations, to system (4.228) yields

$$v_1 = \frac{1}{4}(a_1 - 3a_3 - 8a_4 - 15a_5). \qquad (4.230)$$

Setting $v_1 = 0$ results in

$$a_1^* = 3a_3 + 8a_4 + 15a_5. \qquad (4.231)$$

Then, v_2 can be found as

$$v_2 = \frac{1}{4}(a_2 + 5a_3 + 10a_4 + 10a_5). \qquad (4.232)$$

Hence, letting $v_2 = 0$ leads to

$$a_2^* = -(5a_3 + 10a_4 + 10a_5). \qquad (4.233)$$

Then, under the conditions (4.229), (4.231) and (4.233), one similarly obtains

$$v_3 = -\frac{5}{16}(a_3 - 14a_5), \qquad (4.234)$$

which, in turn, yields

$$a_3^* = 14a_5 \qquad (4.235)$$

in order to have $v_3 = 0$. Similarly, one may find

$$v_4 = -\frac{7}{16}(a_4 + 9a_5), \qquad (4.236)$$

and thus

$$a_4^* = -9a_5,$$ (4.237)

under which $v_4 = 0$. For convenience, we rewrite the critical values of the coefficients in reverse order as follows:

$$a_4^* = -9a_5,$$
$$a_3^* = 14a_5,$$
$$a_2^* = -(5a_3 + 10a_4 + 10a_5),$$ (4.238)
$$a_1^* = 3a_3 + 8a_4 + 15a_5,$$
$$a_0^* = -(a_1 + a_2 + a_3 + a_4 + a_5),$$

under which $v_i = 0, i = 0, 1, 2, 3, 4$.

Finally, for the critical parameter values given in (4.238), higher-order focus values are obtained as

$$v_5 = \frac{21}{32}a_5,$$

$$v_6 = a_5\left(\frac{9}{16} + \frac{464}{3}a_5^2\right),$$

$$v_7 = a_5\left(\frac{4449}{4096} - 439a_5^2 + \frac{679616}{27}a_5^4\right),$$ (4.239)

$$v_8 = a_5\left(\frac{17753}{8192} + \frac{74641}{96}a_5^2 - \frac{2849296}{15}a_5^4 + \frac{6911524864}{2025}a_5^6\right),$$

$$\vdots$$

It can be shown that $v_i = a_5 h_i(a_5^2)$ for $i \geq 5$, where $h_i(a_5^2)$ represents a polynomial in a_5^2. Thus, setting $a_5 = 0$ yields $v_5 = v_6 = v_7 = \cdots = 0$, leading to a center. So assume that $a_5 \neq 0$, then $v_6 \neq 0$, and $v_7 \neq 0$ since $v_7 = 0$ has no real solution for a_5. When a_5 is chosen such that $|a_5| \ll 1$, v_5 dominates the dynamical behavior of the system in the vicinity of the origin.

Next, we want to perform appropriate perturbations to the critical parameter values to obtain exactly 5 limit cycles around each of the 2 weak focus points $(\pm 1, 0)$. Without loss of generality, assume $0 < a_5 \ll 1$, then we need to find perturbations to a_4, a_3, a_2, a_1 and a_0 such that

$$0 < -v_0 \ll v_1 \ll -v_2 \ll v_3 \ll -v_4 \ll v_5 \ll 1.$$

Note that all the focus values $v_i, i = 0, 1, 2, 3, 4, 5$, are given in linear forms of the coefficients a_i. Further, consider the reverse order perturbations, one by one: first on a_4 for v_4, then on a_3 for v_3, on a_2 for v_2, on a_1 for v_1, and finally on

a_0 for v_0. Therefore, the perturbation procedure is straightforward. Since $\frac{\partial v_4}{\partial a_4} = -\frac{7}{16} < 0$, one may perturb $a_4 = -9a_5$ to $a_4 = -9a_5 + \varepsilon_1$ $(0 < \varepsilon_1 \ll 1)$, and thus $v_4 = -\frac{7}{16}\varepsilon_1 < 0$. Similarly, we may let $a_3 = 14a_5 - \varepsilon_2$ $(0 < \varepsilon_2 \ll \varepsilon_1)$, and thus $v_3 = -\frac{5}{16}\varepsilon_2$. This procedure can be followed until a_0 for v_0. The main results of this section are summarized in the following theorem.

Theorem 4.27 *Given the Liénard equation* (4.226) *which has a saddle point at the origin and a pair of symmetric weak focus points at* $(x, y) = (\pm1, 0)$, *under the condition* $b \neq 0$ *and* $a_0 = -(a_1 + a_2 + a_3 + a_4 + a_5)$, *further take* $b = 1$, *and perturb* a_4, a_3, a_2, a_1 *and* a_0 *as*

$$a_4 = a_4^* + \varepsilon_1 = -9a_5 + \varepsilon_1,$$

$$a_3 = a_3^* - \varepsilon_2 = 14a_5 - \varepsilon_2,$$

$$a_2 = a_2^* - \varepsilon_3 = 10a_5 - 10\varepsilon_1 + 5\varepsilon_2 - \varepsilon_3, \qquad (4.240)$$

$$a_1 = a_1^* + \varepsilon_4 = -15a_5 + 8\varepsilon_1 - 3\varepsilon_2 + \varepsilon_4,$$

$$a_0 = a_0^* + \varepsilon_5 = -a_5 + \varepsilon_1 - \varepsilon_2 + \varepsilon_3 - \varepsilon_4 + \varepsilon_5,$$

where $0 < \varepsilon_5 \ll \varepsilon_4 \ll \varepsilon_3 \ll \varepsilon_2 \ll \varepsilon_1 \ll 1$, *then system* (4.226) *has exactly* 10 *small limit cycles.*

Proof First note that

$$0 < v_5 = \frac{21}{32}a_5 \ll 1$$

since $0 < a_5 \ll 1$. Then for the given perturbation $a_4 = -9a_5 + \varepsilon_1$, we have

$$v_4 = -\frac{7}{16}\varepsilon_1,$$

where $0 < \varepsilon_1 \ll 1$. Thus, one may choose $0 < \varepsilon \ll a_5 \ll 1$ so that $0 < -v_4 = \frac{7}{16}\varepsilon_1 \ll \frac{21}{32}a_5 = v_5$.

As for $a_3 = 14a_5 - \varepsilon_2$, we obtain

$$v_3 = \frac{5}{16}\varepsilon_2,$$

where $0 < \varepsilon_2 \ll 1$. Therefore, in order to have $0 < v_3 \ll -v_4$, we should choose ε_2 such that $0 < \varepsilon_2 \ll \varepsilon_1$.

Next, for the perturbed parameter values given in (4.240), we have

$$v_2 = \frac{1}{4}(a_2 + 5a_3 + 10a_4 + 10a_5) = -\frac{1}{4}\varepsilon_3.$$

Hence, by choosing $0 < \varepsilon_3 \ll \varepsilon_2$, one obtains $0 < -v_2 \ll v_3$.

For v_1, we find

$$v_1 = \frac{1}{4}(a_1 - 3a_3 - 8a_4 - 15a_5) = \frac{1}{4}\varepsilon_4.$$

Then one may select $0 < \varepsilon_4 \ll \varepsilon_3$ so that $0 < v_1 \ll -v_2$.

Finally, substituting the parameter values given in (4.240) into v_0 yields

$$v_0 = -(a_0 + a_1 + a_2 + a_3 + a_4 + a_5) = -\varepsilon_5.$$

Thus, $0 < -v_0 \ll v_1$ as long as $\varepsilon_5 \ll \varepsilon_4$.

Summarizing the above perturbation results gives

$$0 < -v_0 = \varepsilon_5 \ll v_1 = \frac{1}{4}\varepsilon_4 \ll -v_2 = \frac{1}{4}\varepsilon_3 \ll v_3 = \frac{5}{16}\varepsilon_2$$

$$\ll -v_4 = \frac{7}{16}\varepsilon_1 \ll v_5 = \frac{21}{32}a_{10} \ll 1,$$

where $0 < \varepsilon_5 \ll \varepsilon_4 \ll \varepsilon_3 \ll \varepsilon_2 \ll \varepsilon_1 \ll a_{10} \ll 1$. Therefore, the sufficient conditions given in Theorem 4.1 are satisfied and so system (4.226) can have 5 small amplitude limit cycles near each of the two weak focus points $(\pm 1, 0)$. $\qquad\square$

Now we present a numerical example of choosing proper perturbations so as to have 10 small limit cycles. Let $b = 1$ and

$$a_5 = 0.002 \quad \Longrightarrow \quad v_5 = 0.13125 \times 10^{-2}, \tag{4.241}$$

and further choose the following perturbations:

$$\varepsilon_1 = 0.1 \times 10^{-4} \quad \Longrightarrow \quad v_4 = -0.4375 \times 10^{-5},$$
$$\varepsilon_2 = 0.1 \times 10^{-7} \quad \Longrightarrow \quad v_3 = 0.3125 \times 10^{-8},$$
$$\varepsilon_3 = 0.1 \times 10^{-11} \quad \Longrightarrow \quad v_2 = -0.25 \times 10^{-12}, \tag{4.242}$$
$$\varepsilon_4 = 0.1 \times 10^{-16} \quad \Longrightarrow \quad v_1 = 0.25 \times 10^{-17},$$
$$\varepsilon_5 = 0.1 \times 10^{-23} \quad \Longrightarrow \quad v_0 = -0.1 \times 10^{-23}.$$

Then, the normal form (4.2) associated with the weak focus points $(\pm 1, 0)$ up to the term r^{11} becomes

$$\dot{r} = r\left(-(0.1 \times 10^{-23}) + (0.25 \times 10^{-17})r^2\right.$$
$$- (0.25 \times 10^{-12})r^4 + (0.3125 \times 10^{-8})r^6$$
$$\left. - (0.4375 \times 10^{-5})r^8 + (0.13125 \times 10^{-2})r^{10}\right), \tag{4.243}$$

which, in turn, yields the following 5 positive roots for r^2:

$$r_1^2 = 0.4173252364867980937574360701 61 \times 10^{-6},$$

$$r_2^2 = 0.1118036042468552098210830436 49 \times 10^{-4},$$

$$r_3^2 = 0.7817366159749989194283448326 15 \times 10^{-4}, \qquad (4.244)$$

$$r_4^2 = 0.8860422913011401813991023335 68 \times 10^{-3},$$

$$r_5^2 = 0.2357519694773520940915530776 07 \times 10^{-2}.$$

So the amplitudes of the 5 limit cycles are approximately equal to

$$r_1 = 0.000646, \qquad r_2 = 0.003344, \qquad r_3 = 0.008842,$$

$$r_4 = 0.029766, \qquad r_5 = 0.048554.$$

Under the perturbations given in (4.242), the perturbed parameter values are

$$a_0 = -0.00199000999900000009999999,$$

$$a_2 = -0.02992002999999999,$$

$$a_4 = 0.019900049999,$$

$$a_6 = 0.02799999, \qquad (4.245)$$

$$a_8 = -0.01799,$$

$$a_{10} = 0.002,$$

under which we apply the Maple program to recompute the focus values, given below:

$$v_0 = -0.1 \times 10^{-23},$$

$$v_1 = 0.25 \times 10^{-17},$$

$$v_2 = -0.2499987485848764535084437799 9333338888875 \times 10^{-12},$$

$$v_3 = 0.3124859231946314547369856461 67768225320477149 96127 \times 10^{-8},$$

$$v_4 = -0.4372966898499816739580523967 136782648014503836 3181 \times 10^{-5},$$

$$v_5 = 0.1309218862148051688411685190 77199473105072202150 75 \times 10^{-2},$$

$$v_6 = 0.1120178869274532206517299558 44721687781633938380 20 \times 10^{-2},$$

$$v_7 = 0.2157244682967191596103078192 40012810010210590051 83 \times 10^{-2},$$

$$v_8 = 0.4317813649695795451212635906 77539724609492948077 02 \times 10^{-2},$$

which, in turn, results in the following roots of r^2:

$$r_1^2 = 0.4173251458718733431982177158 36 \times 10^{-6},$$

$$r_2^2 = 0.1118035573705153257523541991 81 \times 10^{-4},$$

Fig. 4.18 The phase portrait
of system (4.226) showing 10
small limit cycles around the
weak focus points $(\pm 1, 0)$
under the perturbed parameter
values $b = 1$,
$a_0 = -0.0019900099990000009999999$,
$a_1 = -0.02992002999999999$,
$a_2 = 0.019900049999$,
$a_3 = 0.02799999$,
$a_4 = -0.01799$, $a_5 = 0.002$

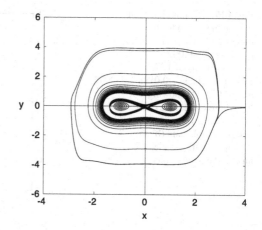

$$r_3^2 = 0.78173105386799679839415410005 \times 10^{-4},$$

$$r_4^2 = 0.88596562336463552175201695865 \times 10^{-3},$$

$$r_5^2 = 0.23568963288566441883905224986 \times 10^{-2},$$

$$r_6^2 = -0.72324961359484809706617028472,$$

$$r_{7,8}^2 = 0.110150965372161277030407639683$$

$$\pm\, 0.638988085546118611924228836755i.$$

The first 5 positive roots are identical to those given in (4.244), at least up to 6 digits
of accuracy, indicating that higher-order focus values due to the perturbations do not
affect the solutions of the limit cycles. This indeed shows that 10 small limit cycles
exist in the neighborhood of the 2 weak focus points $(\pm 1, 0)$.

For the parameter values given in (4.245), the phase portrait for system (4.226)
obtained from computer simulation is shown in Fig. 4.18, where 10 small limit cycles are depicted near the points $(\pm 1, 0)$. It should be pointed out that the trajectories
which are not near the 2 focus points $(\pm 1, 0)$ can be obtained quite accurately using
numerical simulation. However, one cannot obtain the computer simulation results
for the small limit cycles since the accuracy of some parameters is higher than the
machine precision. That is why one must use certain theoretical approaches (such
as the one presented here) to prove the existence of small limit cycles. In fact, in
the neighborhood of a highly-degenerate focus point, trajectories behave like centers, as shown in Fig. 4.18. The stability of these small limit cycles can easily be
determined by the signs of the focus values. For convenience, let these small limit
cycles be named, from smallest to largest, as l_1, l_2, l_3, l_4 and l_5. Since $v_0 < 0$, the
focus points $(\pm 1, 0)$ are stable. Then the smallest limit cycle l_1 is unstable, and thus
l_2 is stable, and so on. It can be observed from Fig. 4.18 that besides the small limit
cycles, there also exist large limit cycles. It seems that there are at least 2 large limit
cycles. In the next subsection, we shall show that 3 large limit cycles actually exist.

Fig. 4.19 The phase portrait of system (4.226) for the unperturbed parameter values $b = 1$, $a_0 = -0.002$, $a_1 = -0.03$, $a_2 = 0.02$, $a_3 = 0.028$, $a_4 = -0.018$, $a_5 = 0.002$: (**a**) showing 2 large limit cycles, the outer one is stable while the inner one is unstable; and (**b**) showing the trajectories around the fixed points

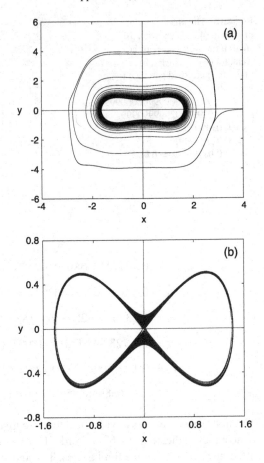

Next, we explore numerical simulations to investigate the existence of large limit cycles in system (4.226). Again we use the same parameter values for obtaining the small limit cycles, given by (4.245). Now we want to show that there are actually 3 large limit cycles exhibited by system (4.226) when the parameter values are given by (4.245). For convenience, we name the large limit cycles from the largest to the smallest, L_1, L_2 and L_3.

In order to give a comparison, we first present the results obtained from system (4.226) without perturbations. The unperturbed parameter values are

$$a_0 = -0.002, \qquad a_1 = -0.03, \qquad a_2 = 0.02,$$

$$a_3 = 0.028, \qquad a_4 = -0.018, \qquad a_5 = 0.002,$$

under which $v_i = 0$, $i = 0, 1, 2, 3, 4$, indicating that the 2 points $(\pm 1, 0)$ are unstable since the fifth-order focus value $v_5 = 0.0013125 > 0$. The simulation results are shown in Fig. 4.19. It seems that 3 large limit cycles exist: the outer one is stable while the middle one is unstable (see Fig. 4.19(a)). There is a third limit cycle which is stable, as shown in Fig. 4.19(b). To simulate unstable limit cycles (or in general

Fig. 4.20 Trajectories of
system (4.226) converging to
the largest limit cycle L_1 as
$t \to +\infty$, one from outside
with the initial point $(4, 0)$,
and the other from inside with
the initial point $(1.5, 1.5)$

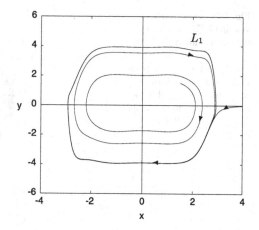

for any unstable solutions or trajectories), one needs to use reverse time evolution. In other words, negative time steps should be used in simulations. The trajectories starting nearby the two focus points diverge very slowly toward the third large limit cycle (the trajectories nearby the focus points are not shown in Fig. 4.19(b)). This implies that the unperturbed system does not have small limit cycles around the focus points, as expected.

Now, we investigate the large limit cycles existing in system (4.226) with the perturbed parameter values given in (4.245). First, consider the largest limit cycle L_1. The numerical simulation result is given in Fig. 4.20, where two typical trajectories are depicted, one from outside L_1 and the other from inside L_1. Both trajectories converge to L_1 as $t \to +\infty$. This indicates that limit cycle L_1 is stable. Then, we know that limit cycle L_2 should be unstable while limit cycle L_3 is stable.

Next consider limit cycle L_2, which is unstable. The simulation result is shown in Fig. 4.21. Figure 4.21(a) demonstrates two trajectories, one from outside L_2 and one from inside L_2, while Fig. 4.21(b) shows the final state of the trajectories. It is seen that both trajectories converge to L_2 as $t \to -\infty$, implying that limit cycle L_2 is unstable, as expected.

Finally, consider limit cycle L_3, which is stable. The numerical simulation result is shown in Fig. 4.22. Figure 4.22(a) depicts two trajectories, one from outside L_3 and one from inside L_3. Both of them converge to the stable limit cycle L_3 as $t \to +\infty$. Figure 4.22(b) shows the final state of the trajectories. It has been observed that convergence for this case is much slower than that of the larger limit cycles L_1 and L_2, because this one is close to the degenerate focus points.

Summarizing the above results for large limit cycles gives the following theorem.

Theorem 4.28 *Given the Liénard equation* (4.226) *with the parameter values* $b = 1$, $a_0 = -0.001990009999000009999999$, $a_1 = -0.02992002999999999$, $a_2 = 0.019900049999$, $a_3 = 0.02799999$, $a_4 = -0.01799$ *and* $a_5 = 0.002$, *system* (4.226) *has 3 large limit cycles. The outer one and inner one are stable while the middle one is unstable. The 3 large limit cycles enclose all 10 small limit cycles.*

Fig. 4.21 Trajectories of
system (4.226) converging to
the larger limit cycle L_2 as
$t \to -\infty$: (**a**) one from
outside with the initial point
$(1, 2)$, and the other from
inside with the initial point
$(0.0, 0.5)$; and (**b**) the final
state of the unstable limit
cycle

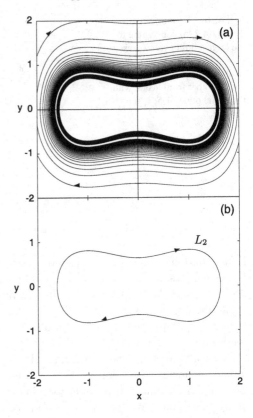

The simulation results shown in Fig. 4.23 demonstrate the trajectories around the three fixed points. Figure 4.23(a) shows the small limit cycles in the vicinity of the two weak focus points $(\pm 1, 0)$, while Fig. 4.23(b) depicts the behavior of the system near the saddle point $(0, 0)$. Note that the small limit cycles are exaggerated (larger than the true ones) for a clear view, and that the trajectories passing through the saddle point are no longer homoclinic orbits due to the perturbations.

Based on the results given in this subsection, we conclude that system (4.226) can have 13 limit cycles: 5 small limit cycles surround each of the 2 focus points and 3 large limit cycles enclose all 10 small limit cycles, as shown in Fig. 4.24(a), while Fig. 4.24(b) is a zoomed-in area around the focus point $(1, 0)$ showing 5 small limit cycles. The stability of these limit cycles is given in Table 4.3.

(B) Limit Cycles in the Liénard System (4.224) for $1 \le m \le 10$ The procedure given in the previous section can be used to consider other integer values of m. Since the proofs are similar to that of the case $m = 5$, we will omit the details but present a summary of the results. We have used the method of normal forms to prove that the exact number of small limit cycles which exist in the neighborhood of the 2 weak focus points, $(\pm 1, 0)$, is $2m$, i.e., $\overline{H}(2m, 3) = 2m$, where $\overline{H}$ is the Hilbert number of the small limit cycles in the vicinity of the two weak focus points. For the local limit cycles around the origin, as shown in Table 4.2, it has been shown

Fig. 4.22 Trajectories of
system (4.226) converging to
the large limit cycle L_3 as
$t \to +\infty$: (**a**) one from
outside with the initial point
$(0, 0.3)$, and the other from
inside with the initial point
$(0.0, 0.05)$; and (**b**) the final
state of the unstable limit
cycle

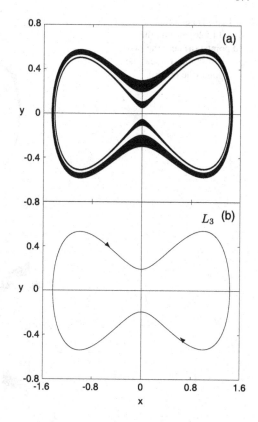

Table 4.3 Stability of the
fixed points and limit cycles
of (4.226)

$S = $ Stable, $U = $ Unstable

Saddle point $(0,0)$	Focus point $(\pm 1, 0)$	Limit cycle							
		l_1	l_2	l_3	l_4	l_5	L_3	L_2	L_1
U	S	U	S	U	S	U	S	U	S

that $\hat{H}(k, 3) = 2[3(k + 2)/8]$, where $[\cdot]$ denotes the integer part and $\deg(f) = k$,
$k = 2, 3, \ldots, 50$ [158]. The comparison of the above two different cases of local
limit cycles is given in Table 4.4. Obviously, considering the symmetry, with two
focus points we can have more limit cycles. For large limit cycles, on the other hand,
we apply a fourth-order Runge–Kutta integration scheme to show their existence, as
depicted in Figs. 4.25–4.29 for $m = 1, 2, 3, 4, 5, 6, 7, 8, 9, 10$. The numbers of large
limit cycles are also shown in Table 4.4. A careful examination of these numbers
seems to lead to the following conjecture.

Conjecture 4.29 The number of large limit cycles is determined by the following
formula:

Fig. 4.23 Trajectories of
system (4.226) around the
fixed points: (**a**) near the two
focus points; and (**b**) near the
saddle point (the origin)

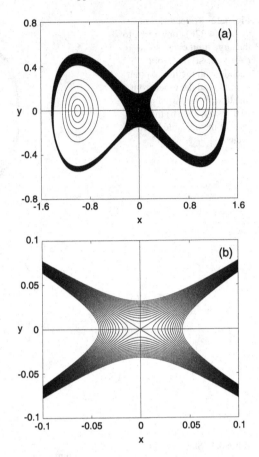

Table 4.4 Limit cycles in the Liénard system near one focus point and two focus points

$\deg(f) = k$		1	2	3	4	5	6	7	8	9	10	11	12	13	14	15	16	17	18	19	20
Focus point $(0,0)$	Small	1	2	2	4	4	6	6	6	8	8	8		10	10	12	12	12	14	14	14	16

$\frac{1}{2}\deg(f) = m$		1	2	3	4	5	6	7	8	9	10
Focus points $(\pm 1, 0)$	Small	2	4	6	8	10	12	14	16	18	20
	Large	1	1	2	2	3	3	3	3	4	4

$$H_L(2m, 3) = \begin{cases} 1 & \text{for } m = 1, \\ p & \text{for } 2^{p-1} \leq m \leq 2^p, \ p = 1, 2, 3, 4, \ldots, \end{cases} \tag{4.246}$$

where the subscript L denotes "large limit cycle".

Remark 4.30 It has been recently shown in [112] that the formula for the numbers
of small limit cycles, $\overline{H}(2m, 3) = 2m$, is true for any integer $m \geq 1$.

Fig. 4.24 The 13 limit cycles of system (4.226) obtained for parameter values $b = 1$, $a_0 = -0.001990009999000009999999$, $a_1 = -0.02992002999999999$, $a_2 = 0.019900049999$, $a_3 = 0.02799999$, $a_4 = -0.01799$, $a_5 = 0.002$: (**a**) all 13 limit cycles; and (**b**) the zoomed-in area around the focus point $(0, 0)$ having 5 small limit cycles

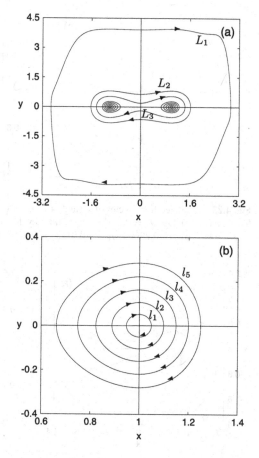

In Figs. 4.25–4.29, the large limit cycles are obtained from computer simulation. However, the small limit cycles are shown just for illustration since, in particular for larger values of m, they cannot be obtained via computer simulation. They have to be proved using a theoretical approach. In order to give more accurate information for these small limit cycles, we list below the amplitudes of the periodic solutions (obtained from the normal form) for reference.

$m = 1$: $r_1 = 0.01333333$ (see Fig. 4.25(a));

$m = 2$: $r_1 = 0.01333333$, $r_2 = 0.02$ (see Fig. 4.25(b));

$m = 3$: $r_1 = 0.00225219$, $r_2 = 0.01874441$, $r_3 = 0.37900340$
(see Fig. 4.26(a));

$m = 4$: $r_1 = 0.00022449$, $r_2 = 0.00279562$, $r_3 = 0.00581885$,
$r_4 = 0.06258961$ (see Fig. 4.26(b));

$m = 5$: $r_1 = 0.00064601$, $r_2 = 0.00334370$, $r_3 = 0.00884159$,
$r_4 = 0.02976646$, $r_5 = 0.04855430$ (see Fig. 4.27(a));

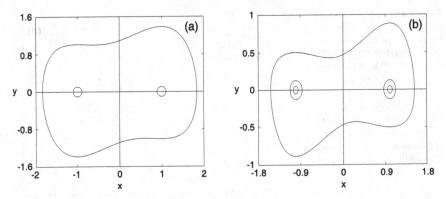

Fig. 4.25 Simulated limit cycles of system (4.224) for (**a**) $m = 1$ with perturbed parameter values $b = 1$, $a_0 = -0.299$, $a_1 = 0.2$, showing 3 limit cycles; and (**b**) $m = 2$ with perturbed parameter values $b = 1$, $a_0 = -0.29002$, $a_1 = -0.01$, $a_2 = 0.3$, showing 5 limit cycles

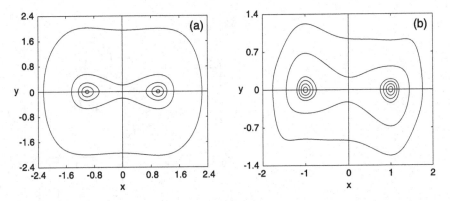

Fig. 4.26 Simulated limit cycles of system (4.224) for (**a**) $m = 3$ with perturbed parameter values $b = 1$, $a_0 = 0.0101999$, $a_1 = 0.0598$, $a_2 = -0.09$, $a_3 = 0.02$, showing 8 limit cycles; and (**b**) $m = 4$ with perturbed parameter values $b = 1$, $a_0 = 0.09009980001$, $a_1 = 0.7700002$, $a_2 = -0.9501$, $a_3 = -0.01$, $a_4 = 0.1$, showing 10 limit cycles

$$m = 6: \quad r_1 = 0.00000394, \qquad r_2 = 0.00002551, \qquad r_3 = 0.00005344,$$
$$r_4 = 0.00273615, \qquad r_5 = 0.00658136, \qquad r_6 = 0.01605414$$
(see Fig. 4.27(b));

$$m = 7: \quad r_1 = 0.00000153, \qquad r_2 = 0.00000381, \qquad r_3 = 0.00001508,$$
$$r_4 = 0.00006981, \qquad r_5 = 0.00037915, \qquad r_6 = 0.00112468,$$
$$r_7 = 0.00455979 \quad \text{(see Fig. 4.28(a));}$$

$$m = 8: \quad r_1 = 0.00020429, \qquad r_2 = 0.00105575, \qquad r_3 = 0.00308633,$$
$$r_4 = 0.00632467, \qquad r_5 = 0.01074805, \qquad r_6 = 0.02308772,$$
$$r_7 = 0.08681535, \qquad r_8 = 0.14752576 \quad \text{(see Fig. 4.28(b));}$$

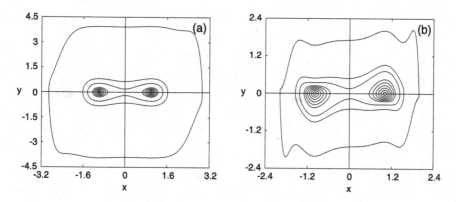

Fig. 4.27 Simulated limit cycles of system (4.224) for (**a**) $m = 5$ with perturbed parameter values $b = 1$, $a_0 = -0.00199000999900000009999999$, $a_1 = -0.02992002999999999$, $a_2 = 0.019900049999$, $a_3 = 0.02799999$, $a_4 = -0.01799$, $a_5 = 0.002$, showing 13 limit cycles; and (**b**) $m = 6$ with perturbed parameter values $b = 1$, $a_0 = -0.04801995000499990000008$, $a_1 = -1.1701598500000001$, $a_2 = -0.269800249995$, $a_3 = 2.77200005$, $a_4 = -1.33202$, $a_5 = -0.002$, $a_6 = 0.05$, showing 15 limit cycles

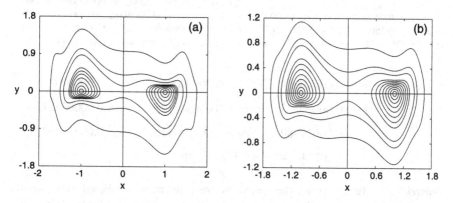

Fig. 4.28 Simulated limit cycles of system (4.224) for (**a**) $m = 7$ with perturbed parameter values $b = 1$, $a_0 = 0.04950099950004999900000003999999$, $a_1 = 1.7380149960001499999999996$, $a_2 = 2.447490004999750001$, $a_3 = -6.62201399999995$, $a_4 = 1.3365089995$, $a_5 = 1.649999$, $a_6 = -0.6495$, $a_7 = 0.05$, showing 17 limit cycles; and (**b**) $m = 8$ with perturbed parameter values $b = 1$, $a_0 = 0.01900099900019990000009999900000001$, $a_1 = 0.9252398500159997000000000001$, $a_2 = 2.75105009998000049999$, $a_3 = -4.3475598600000001$, $a_4 = -1.826730089998$, $a_5 = 3.48700001$, $a_6 = -1.02701$, $a_7 = -0.001$, $a_8 = 0.02$, showing 19 limit cycles

$$m = 9: \quad r_1 = 0.00000017, \quad r_2 = 0.00000084, \quad r_3 = 0.00000312,$$
$$r_4 = 0.00001282, \quad r_5 = 0.00010222, \quad r_6 = 0.00030878,$$
$$r_7 = 0.00124938, \quad r_8 = 0.00395682, \quad r_9 = 0.02377763$$

(see Fig. 4.29(a));

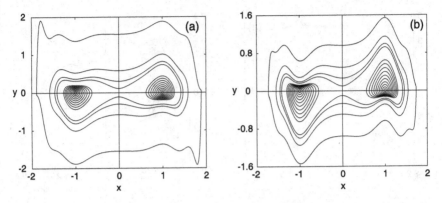

Fig. 4.29 Simulated limit cycles of system (4.224) for (**a**) $m = 9$ with perturbed parameter values $b = 1$, $a_0 = -0.0095039920039995000099996000029999999$, $a_1 = -0.6061398080599960002999999999997$, $a_2 = -2.9301959599600049999500004$, $a_3 = 2.40853155205599999999$, $a_4 = 5.1748922159640005$, $a_5 = -4.972131999996$, $a_6 = 0.494051992$, $a_7 = 0.599996$, $a_8 = -0.1695$, $a_9 = 0.01$, showing 22 limit cycles; and (**b**) $m = 10$ with perturbed parameter values $b = 1$, $a_0 = -0.00950399200399950000999996000002999999900000000002$, $a_1 = -0.76869172009599925000079999880000000000000001$, $a_2 = -5.40055960802000049999000001999997$, $a_3 = 0.35489493622399300000000004$, $a_4 = 15.15936021589200449999$, $a_5 = -8.5477037360000005$, $a_6 = -3.925792103996$, $a_7 = 3.970000008$, $a_8 = -0.841504$, $a_9 = -0.0005$, $a_{10} = 0.01$, showing 24 limit cycles

$$m = 10: \quad r_1 = 0.00000001, \quad r_2 = 0.00000002, \quad r_3 = 0.00000077,$$
$$r_4 = 0.00000285, \quad r_5 = 0.00001212, \quad r_6 = 0.00009727,$$
$$r_7 = 0.00029989, \quad r_8 = 0.00121941, \quad r_9 = 0.00388862,$$
$$r_{10} = 0.02342640 \quad \text{(see Fig. 4.29(b))}.$$

Remark 4.31 The results and the conjecture presented above are based on the small amplitude limit cycles in the neighborhood of the two symmetric focus points (obtained using an analytical approach) and other possible large limit cycles (obtained using a numerical approach). On the other hand, for classical Liénard equations, with respect to large amplitude limit cycles and limit cycles of classical Liénard equations, there are the following general results. Writing the classical Liénard system as

$$\dot{x} = y - \left(x^N + a_{N-1}x^{N-1} + \cdots + a_1 x\right),$$
$$\dot{y} = -x, \tag{4.247}$$

it is known from [29] that for $N = 2m$, under a uniform bound on the $|a_j|$, the number of limit cycles is uniformly bounded and more specifically there can be at most m large amplitude limit cycles. With a completely different approach [107] (in the case N is odd) and [118] (in the case N is even) provide for $|a_j| \le C$ an explicit but extremely large upper bound which is in the order of exponentials.

4.6 Critical Periods

Another interesting problem in the study of Hilbert's 16th problem is the bifurcation of limit cycles from equilibria of a center type, since the monotonicity of periods of closed orbit surrounding a center is a nondegeneracy condition of subharmonic bifurcation for periodically forced Hamiltonian systems [46]. Suppose the planar polynomial vector field is described by the following differential equations:

$$\dot{x} = P_n(x, y, \mu), \qquad \dot{y} = Q_n(x, y, \mu), \qquad (4.248)$$

where $P_n(x, y)$ and $Q_n(x, y)$ represent nth-degree polynomials in x and y, and $\mu \in R^k$ is a k-dimensional parameter vector. Suppose the origin of system (4.248) is a fixed point and further it is a nondegenerate center. (If the Jacobian of the system does not have a double-zero eigenvalue at the origin, then the origin is called a *nondegenerate center*.)

Now let $T(h, \mu)$ denote the minimum period of closed orbit of system (4.248) surrounding the origin for $0 < h \ll 1$, where h is the value of the Hamiltonian on the closed orbit. Then the origin is said to be a weak center of finite order k of the system for the parameter value $\mu = \mu_c$ if

$$T'(0, \mu_c) = T''(0, \mu_c) = \cdots = T^k(0, \mu_c) = 0 \quad \text{but} \quad T^{k+1}(0, \mu_c) \neq 0. \quad (4.249)$$

The origin is called an *isochronous center* if $T^k(0, \mu_c) = 0$ for all $k \geq 1$, or equivalently, $T(h, \mu_c) = \text{constant}$ for $0 < h \ll 1$. A local critical period is defined as a period corresponding to a critical point of the period function $T(h, \mu)$ which bifurcates from a weak center.

For the quadratic system, given by

$$\dot{x} = -y + \sum_{i+j=2} a_{ij} x^i y^j, \qquad \dot{y} = x + \sum_{i+j=2} b_{ij} x^i y^j, \qquad (4.250)$$

Chicone and Jacobs [44] discussed weak centers and critical periods which may bifurcate from weak centers. In the same paper, they also studied the following special Hamiltonian system:

$$\ddot{u} + V(u) = 0, \qquad (4.251)$$

where V is a $(2n)$th-degree polynomial in u. Let $u = x$ and $\dot{u} = y$. Then, the Hamiltonian of system (4.251) can be written as

$$H(x, y) = \frac{1}{2} y^2 + \int_0^x V(s) \, ds. \qquad (4.252)$$

It has been shown [44] that system (4.251) can have at most $(n - 2)$ critical periods bifurcating from the origin.

In 1993, Rousseau and Toni [185] studied a special cubic system with third-degree homogeneous polynomials only, as described below:

$$\dot{x} = -y + \sum_{i+j=3} a_{ij} x^i y^j, \qquad \dot{y} = x + \sum_{i+j=3} b_{ij} x^i y^j. \qquad (4.253)$$

They similarly discussed weak centers and the bifurcation of critical periods from weak centers.

Recently, Zhang et al. [258] gave a detailed study of cubic reversible polynomial planar systems: a system is said to be reversible if it is symmetric with respect to a line. Up to translation and rotation of coordinates, any reversible cubic differential system can be described by (e.g., see [258])

$$
\begin{aligned}
\dot{x} &= -y + a_{20}x^2 + a_{02}y^2 + a_{21}x^2 y + a_{03}y^3, \\
\dot{y} &= x + b_{11}xy + b_{30}x^3 + b_{12}xy^2,
\end{aligned}
\tag{4.254}
$$

where a_{ij} and b_{ij} are constant parameters. It has been shown [258] that system (4.254) can have at most 4 local critical periods.

The work of Mañosas and Villadelprat [161] should be also mentioned. The system considered in [161] is a Hamiltonian system with the following Hamiltonian:

$$
H(x, y) = \frac{1}{2}(x^2 + y^2) + \frac{a}{4}x^4 + \frac{b}{6}x^6,
\tag{4.255}
$$

where a and b are constants, and $b \neq 0$. It is shown that system (4.255) can at most have 1 critical period. Note that system (4.255) is not a special case of system (4.252), since the term $g(x) = \int_0^x V(s)\,ds$ in $H(x, y)$ of (4.252) is a $(2n + 1)$-th-degree polynomial.

4.6.1 Computation of Critical Periods

We apply the normal form of system (4.248) to study the bifurcation of critical periods. Suppose the normal form of (4.248) without parameters is given by

$$
\dot{r} = r\left(v_0 + v_1 r^2 + v_2 r^4 + \cdots + v_k r^{2k} + v_{k+1} r^{2k+2}\right),
\tag{4.256}
$$

$$
\dot{\theta} = 1 + b_0 + b_1 r^2 + b_2 r^4 + \cdots + b_k r^{2k} + b_{k+1} r^{2k+2}.
\tag{4.257}
$$

Note here that r and θ represent the amplitude and phase of the motion, respectively: v_0 and b_0 correspond to the linear part of the system when it contains parameters. Here, $v_0 = b_0 = 0$.

Equation (4.256) (or the focus values) can be used to determine the existence and number of limit cycles that system (4.248) can have, as shown in previous sections. Equation (4.257), on the other hand, can be applied to find the period of the periodic solutions and to determine the local critical periods of the solutions.

In the following, we describe how to use (4.257) to express the period of periodic motion and how to determine the local critical periods. For convenience, let

$$
h = r^2 > 0 \quad \text{and} \quad p(h) = b_1 h + b_2 h^2 + \cdots + b_{k+1}h^{k+1}.
\tag{4.258}
$$

Then (4.257) can be written as

$$d\theta = \left(1 + p(h)\right) dt \quad \left(b_0 = 0 \text{ for system (4.248)}\right).$$

Let the period of motion be $T(h)$. Then integrating the above equation on both sides from 0 to 2π yields

$$2\pi = \left(1 + p(h)\right) T(h),$$

which gives

$$T(h) = \frac{2\pi}{1 + p(h)} \quad \text{for } 0 < h \ll 1 \quad \left(\text{and so } 1 + p(h) \approx 1\right). \tag{4.259}$$

Now, the local critical periods are determined by $T'(h) = 0$, or

$$T'(h) = \frac{-2\pi p'(h)}{(1 + p(h))^2} = 0. \tag{4.260}$$

Thus, for $0 < h \ll 1$ (meaning that we consider small limit cycles), the local critical periods are determined by

$$p'(h) = b_1 + 2b_2 h + \cdots + k b_k h^{k-1} + (k+1) b_{k+1} h^k = 0. \tag{4.261}$$

As in the discussion about determining the number of limit cycles using focus values, we can find sufficient conditions for the polynomial $p'(h)$ to have the maximal number of zeros. If $b_1 = b_2 = \cdots = b_k = 0$ but $b_{k+1} \neq 0$, then equation $p'(h) = 0$ can have at most k real roots. Then $b_1, b_2, \ldots, b_k$ (remember that they are expressed in terms of the coefficients of the original system (4.248)) can be perturbed appropriately to have k real roots. We give a theorem below without proof (the proof is similar to Theorem 4.2), which can be used to determine the maximal number of real roots of $p'(h) = 0$.

Assume that b_i depends on k independent system parameters:

$$b_i = b_i(a_1, a_2, \ldots, a_k), \quad i = 1, 2, \ldots, k, \tag{4.262}$$

where $a_1, a_2, \ldots, a_k$, are the parameters of the original system (4.248).

Theorem 4.32 *Suppose that*

$$
\begin{aligned}
&b_i(a_{1c}, a_{2c}, \ldots, a_{kc}) = 0, \quad i = 1, 2, \ldots, k, \\
&b_{k+1}(a_{1c}, a_{2c}, \ldots, a_{kc}) \neq 0, \quad \text{and} \\
&\det\left[\frac{\partial(b_1, b_2, \ldots, b_k)}{\partial(a_1, a_2, \ldots, a_k)} (a_{1c}, a_{2c}, \ldots, a_{kc}) \right] \neq 0,
\end{aligned}
\tag{4.263}
$$

where $a_{1c}, a_{2c}, \ldots, a_{kc}$, represent critical values. Then small appropriate perturbations applied to the critical values lead to the conclusion that equation $p'(h) = 0$ has k real roots.

4.6.2 Cubic-Order Planar Reversible Systems

In this subsection, we consider the bifurcation of local critical periods from a weak center in cubic reversible polynomial planar systems (4.254). It has been recently shown that such systems can have a maximal 6 local critical periods [242], rather than 4 as claimed in [258].

In [258], the authors assumed that the 7 parameters $(a_{20}, a_{02}, a_{21}, a_{03}, b_{11}, b_{30}, b_{12})$ are independent. As a matter of fact, we can further reduce the number of parameters by one. In other words, there are in total only 6 independent parameters. To achieve this, assume that $a_{20} \neq 0$. Then, we use the scalings

$$x \to \frac{x}{a_{20}}, \qquad y \to \frac{-y}{a_{20}}, \qquad a_{02} \to m_1 a_{20}, \qquad a_{21} \to m_2 a_{20}^2,$$

$$a_{03} \to m_3 a_{20}^2, \qquad b_{11} \to n_1 a_{20}, \qquad b_{30} \to n_2 a_{20}^2, \qquad b_{12} \to n_3 a_{20}^2, \tag{4.264}$$

to obtain a new system (for $a_{20} \neq 0$):

$$\dot{x} = y + x^2 + m_1 y^2 - m_2 x^2 y - m_3 y^3,$$

$$\dot{y} = -x + n_1 xy - n_2 x^3 - n_3 xy^2. \tag{4.265}$$

System (4.265) has only 6 independent parameters, i.e., $a_{20} \neq 0$ can be chosen arbitrarily if we use the original system (4.254). This implies that for the cubic reversible polynomial planar system (4.265) (or the original system (4.254)), in general the maximal number of local critical periods that the system can have is 6.

Remark 4.33 The above scaling reduces by one more the number of system parameters. The advantage of the reduction is in making the computation (particularly, numerical computation) simpler. However, more cases need to be considered (see below), unlike the analysis based on the original system (4.254) with 7 parameters which only needs one set of parameters to be investigated.

When $a_{20} = 0$, there are only 6 parameters. We may assume $a_{02} \neq 0$, and obtain a similar system to (4.265), but now $m_1 = 1$ and there is no x^2 term, resulting in a system with only 5 independent parameters. This clearly shows that such a "degenerate" system has fewer independent parameters and so in general has fewer critical periods. By doing this, to completely analyze the system there are in total four different cases.

Case (1) $a_{20} = a_{02} = b_{11} = 0$: the corresponding system (no scaling) is given by

$$\dot{x} = y - m_2 x^2 y - m_3 y^3,$$

$$\dot{y} = -x - n_2 x^3 - n_3 xy^2, \tag{4.266}$$

where $m_2 = a_{21}, m_3 = a_{03}, n_2 = b_{30}, n_3 = b_{12}$. Note here that the advantage of not applying scaling is that it is not necessary to assume that one of the 4 parameters is nonzero, and 4 parameters can easily be handled in computation.

Case (2) $a_{20} = a_{02} = 0, b_{11} \neq 0$: the system is described by

$$\dot{x} = y - m_2 x^2 y - m_3 y^3,$$
$$\dot{y} = -x + xy - n_2 x^3 - n_3 xy^2. \tag{4.267}$$

Case (3) $a_{20} = 0, a_{02} \neq 0$: the system is given by

$$\dot{x} = y + y^2 - m_2 x^2 y - m_3 y^3,$$
$$\dot{y} = -x + n_1 xy - n_2 x^3 - n_3 xy^2. \tag{4.268}$$

Case (4) $a_{20} \neq 0$: the system is given by (4.265).

(A) Case (1): $a_{20} = a_{02} = b_{11} = 0$ (No Scaling)   The system for this case is described by (4.266) which has 4 parameters. Applying the Maple program [218], we easily obtain the coefficients b_i. In particular,

$$b_1 = \frac{1}{8}(n_3 - m_2 - 3m_3 + 3n_2). \tag{4.269}$$

Letting

$$n_3 = m_2 + 3m_3 - 3n_2, \tag{4.270}$$

we have $b_1 = 0$, and further obtain

$$b_2 = -\frac{1}{16}\left[2(m_3 + n_2)m_2 + 3\left(3m_3^2 + n_2^2\right)\right]. \tag{4.271}$$

Thus, by choosing

$$m_2 = -\frac{3(3m_3^2 + n_2^2)}{2(m_3 + n_2)}, \tag{4.272}$$

we have $b_2 = 0$, and

$$b_3 = \frac{3}{32(m_3 + n_2)^2} m_3 n_2 (m_3 - n_2)\left(m_3^2 - 10m_3 n_2 + n_2^2\right),$$

$$b_4 = -\frac{3}{128(m_3 + n_2)^3} m_3 n_2 (m_3 - n_2)^2 \tag{4.273}$$
$$\times \left(3m_3^3 - 16m_3^2 n_2 + 83m_3 n_2^2 - 6n_2^3\right),$$

where $m_3 \neq -n_2$.

It is easy to observe from (4.273) that when $m_3 = 0$, or $n_2 = 0$, or $m_3 = n_2$, in addition to $b_1 = b_2 = 0$, we have $b_3 = b_4 = \cdots = 0$, leading to the conclusion that the origin is an isochronous center.

Setting $m_3^2 - 10m_3 n_2 + n_2^2 = 0$ yields

$$m_3 = (5 \pm 2\sqrt{6})n_2, \tag{4.274}$$

which, in turn, results in $b_3 = 0$, and $b_4 = -\frac{5}{16}(49 \pm 20\sqrt{6})n_2^4$. It is obvious that $n_2 = 0$ leads to a trivial case: a linear system. If $n_2 \neq 0$, then at the critical point,

$$(m_{3c}, m_{2c}, n_{3c}) = \left((5 \pm 2\sqrt{6})n_2, -(21 \pm 8\sqrt{6})n_2, -(9 \pm 2\sqrt{6})n_2\right), \quad n_2 \neq 0,$$

system (4.266) for Case (1) can have at most 3 local critical periods. Since here we can have perturbation, one by one, on m_3 for b_3, on m_2 for b_2 and on n_3 for b_1, we know that the system can have 3 local critical periods after proper small perturbations. Alternatively, it is not difficult to show that when $n_2 \neq 0$,

$$\det\left[\frac{\partial(b_1, b_2, b_3)}{\partial(m_3, m_2, n_3)}\right]_{(m_3, m_2, n_3) = (m_{3c}, m_{2c}, n_{3c})} = \frac{3(2\sqrt{6} \pm 5)}{256}n_2^3 \neq 0.$$

So, according to Theorem 4.2, we know that system (4.266) for Case (1) can have 3 local critical periods.

Summarizing the above results, we have the following theorem.

Theorem 4.34 *For the reversible system (4.266), there exist 3 local critical periods bifurcating from the weak center (the origin) at the critical point, $m_3 = (5 \pm 2\sqrt{6})n_2$, $m_2 = -(21 \pm 8\sqrt{6})n_2$, $n_3 = -(9 \pm 2\sqrt{6})n_2$, $n_2 \neq 0$. Moreover, the origin is an isochronous center if one of the following conditions is satisfied:*

(i) $m_3 = 0, n_3 = 3m_2 = -\frac{9}{2}n_2$;
(ii) $n_2 = 0, m_2 = 3n_3 = -\frac{9}{2}m_3$;
(iii) $m_3 = n_2, m_2 = n_3 = -3n_2$.

Remark 4.35

(i) The linear isochronous center is included in the above as a special case when $m_2 = m_3 = n_2 = n_3 = 0$.
(ii) System (4.266) actually has only 3 independent parameters. One can apply a proper scaling to remove one parameter. For example, if $a_{21} \neq 0$, then substitute the scalings

$$x \to \frac{x}{\sqrt{|a_{21}|}}, \qquad y \to \frac{y}{\sqrt{|a_{21}|}}, \tag{4.275}$$

$$a_{03} \to m_3 a_{21}, \qquad b_{30} \to n_2 a_{21}, \qquad b_{12} \to n_3 a_{21},$$

into (4.254) to obtain

$$\dot{x} = y - \text{sign}(a_{21})x^2 y - m_3 y^3,$$
$$\dot{y} = -x - n_2 x^3 - n_3 x y^2, \tag{4.276}$$

which has only 3 independent parameters. So it is not surprising that the maximal number of local critical periods for this case is 3. But note that we need to deal with the case $a_{21} = 0$ separately.

(B) Case (2): $a_{20} = a_{02} = 0$, $b_{11} \neq 0$ For this case, the system is given by (4.267). Again like Case (1), we have only 4 independent parameters. However, compared to system (4.266), we can see that this case has an extra term xy in the second equation. Similarly, applying the Maple program results in

$$b_1 = \frac{1}{8}(n_3 - m_2 - 3m_3 + 3n_2) - \frac{1}{24}. \tag{4.277}$$

Letting

$$n_3 = m_2 + 3m_3 - 3n_2 + \frac{1}{3} \tag{4.278}$$

yields $b_1 = 0$, and

$$b_2 = -\frac{1}{8}(m_3 + n_2)m_2 - \frac{3}{16}(3m_3^2 + n_2^2) - \frac{1}{48}(m_3 - n_2). \tag{4.279}$$

Further, setting $b_2 = 0$ gives

$$m_2 = -\frac{9(3m_3^2 + n_2^2) + m_3 - n_2}{6(m_3 + n_2)}, \tag{4.280}$$

and then we obtain

```
b3:=-1/622080/(m3+n2)^2*(10*m3*n2-450*m3^2*n2+2430*m3*n2^2
    +218700*m3^3*n2+224370*m3^2*n2^2-21060*m3*n2^3-25*n2^2
    -m3^2+594*m3^3+450*n2^3+58320*m3*n2^4-58320*m3^4*n2
    +641520*m3^3*n2^2-641520*m3^2*n2^3-2025*n2^4+119151*m3^4):
b4:= 1/14929920/(m3+n2)^3*(15*m3^2*n2-75*m3*n2^2-3924*m3^3*n2
    +486*m3^2*n2^2-12132*m3*n2^3-m3^3+125*n2^3+200853*m3*n2^4
    -1851741*m3^4*n2-3293622*m3^3*n2^2-2475306*m3^2*n2^3
    -3015*n2^4+729*m3^4+23895*n2^5-20655*m3^5*n2+20111409*m3^2*n2^4
    -7725942*m3^5*n2-14508315*m3^4*n2^2-5360580*m3^3*n2^3
    -1120230*m3*n2^5+4186647*m3^6-61965*n2^6-1049760*m3^6*n2
    +7698240*m3^5*n2^2-41290560*m3^4*n2^3-33242400*m3^2*n2^5
    +65784960*m3^3*n2^4+2099520*m3*n2^6):
b5:= ...
```

Now, we cannot simply solve for m_2 or n_2 explicitly from the equation $b_3 = 0$. So, eliminating m_3 from the two equations, $b_3 = b_4 = 0$, yields the solution

```
m3:=-5*n2*(56924780201528555520000000*n2^17
     -103113441440789193008640000000*n2^16
     +466529487811481854805391360000*n2^15
     +729111368647971303747081984000*n2^14
     -344019091962565312974780606720*n2^13
     +343889545264614390656135782944*n2^12
     -137841104039681609548659321235*n2^11
     +285964218564988124360928779568*n2^10
     -341240436357877880297190581040*n2^9
     +243221430171134993966016822*n2^8
     -102825774441224160041266596*n2^7+243543978454673314458368*n2^6
     -269206840066291302909600*n2^5+28477324117008592587*n2^4
     -24412806206412462*n2^3+227374081948134*n2^2+10829164128*n2-89476)
     /(93237097492083621086208000000000*n2^17
     -445172671516124668821672960000000*n2^16
     -614124511813088929545454080000000*n2^15
     +321743370081378172281237529344000*n2^14
     -355843063477576132735248304473600*n2^13
     +179370522907753892786267929778880*n2^12
     -523017341879354233422577878752644*n2^11
     +888637590476064387466140884457*n2^10
     -868968351982197524387386778472*n2^9
     +480555478736057781376723789544*n2^8
     -140101644794602585509718599*n2^7+171155229262466428048567128*n2^6
     -19139109085867805531199*n2^5+541907819900619509166*n2^4
     -205550961979095693*n2^3-232683432511500*n2^2
     -16584226269*n2-246059):
```

$$(4.281)$$

and the resultant

```
F1:= n2*(120*n2+1)*(18*n2-1)*(9*n2-1)
     *(24987557838805401600000000*n2^14
     -177656908908380700067200000*n2^13
     +132033355862506632806400000*n2^12
     +116972435487395393766604800*n2^11
     -312815713413038729734287360*n2^10
     +326210648416092258308527104*n2^9
     -164073549847183061353286400*n2^8+396624873034635307560842440*n2^7
     -435402408957642380821377*n2^6+167083114350796340077368*n2^5
     +7636557097500298985166*n2^4-3114647882885854395*n2^3
     -245131814155149*n2^2-14233762746*n2-559225):
```

A simple numerical scheme can be employed to show that the polynomial equation $F_1(n_2) = 0$ has 14 real solutions for n_2. The first four solutions are $n_2 = 0$, $-\frac{1}{120}, \frac{1}{18}, \frac{1}{9}$. It is easy to verify that the first three of them are not solutions, while the last one results in $m_3 = m_2 = n_3 = 0$, leading to $b_3 = b_4 = b_5 = \cdots = 0$. This indicates that for this solution, the origin is an isochronous center.

For the remaining 10 roots of the equation $F_1(n_2) = 0$, we have used the built-in Maple solver fsolve to numerically compute these real solutions up to 1000 digits, guaranteeing the accuracy of computation. (The 10 solutions are not listed here for brevity.) Note that for a univariate polynomial, fsolve can be used to find all real roots of the polynomial up to a very high accuracy. In fact, the Maple command solve can be employed to find all (real and complex) roots of a univariate polynomial. It can be shown that for all these 10 solutions, $b_1 = b_2 = b_3 = b_4 = 0$ but $b_5 \neq 0$. This implies that (n_2, m_3, m_2, n_3) has 10 sets of real solutions for which system (4.267) has 4 local critical periods bifurcating from the weak center, the origin.

The results obtained above for Case (2) are summarized in the following theorem.

Theorem 4.36 *For the reversible system (4.267), there are* 10 *sets of solutions for the critical point* $(n_{2c}, m_{3c}, m_{2c}, n_{3c})$ *which can be perturbed to generate* 4 *local critical periods. Moreover, when* $n_2 = \frac{1}{9}, m_3 = m_2 = n_3 = 0$, *the origin is an isochronous center.*

Remark 4.37 It should be pointed out that although Theorem 4.36 states that there are only 10 sets of solutions which generate 4 local critical periods, there is actually an infinite number of solutions since $b_{11}(\neq 0)$ can be chosen arbitrarily.

(C) Case (3): $a_{20} = 0, a_{02} \neq 0$ Now we consider Case (3) described by (4.268), which has 5 independent parameters. So it is possible to have 5 local critical periods. Similarly, we apply the Maple program to obtain

$$b_1 = \frac{1}{24}(3n_3 - 3m_2 - 9m_3 + 9n_2 + n_1 - n_1^2) - \frac{5}{12}, \qquad (4.282)$$

which yields

$$n_3 = m_2 + 3m_3 - 3n_2 + \frac{1}{3}n_1(n_1 - 1) + \frac{10}{3}, \qquad (4.283)$$

such that $b_1 = 0$. Then,

$$b_2 = -\frac{1}{48}\left[(6m_3 + 6n_2 - n_1 + 11)m_2 + m_3\left(n_1 + \frac{23}{2}\right)^2\right.$$

$$\left. + 27\left(m_3 - \frac{17}{24}\right)^2 + (n_1 + 7)^2 + 9\left(n_2 + \frac{1}{18}\right)^2 - n_1^2 n_2 - \frac{13003}{576}\right]. \quad (4.284)$$

Setting $b_2 = 0$ we obtain

$$m_2 = \frac{n_1^2 n_2 - m_3 n_1(n_1 + 23) - m_3(27m_3 + 94) - n_2(9n_2 + 1) - n_1(n_1 + 14) - 40}{6(m_3 + n_2) - n_1 + 11}.$$

$$(4.285)$$

Then b_3, b_4 and b_5 become

$$b_3 = \frac{1}{155520(6m_3 + 6n_2 - n_1 + 11)^2}F_1(m_3, n_2, n_1),$$

$$b_4 = -\frac{1}{1866240(6m_3 + 6n_2 - n_1 + 11)^2}F_2(m_3, n_2, n_1), \qquad (4.286)$$

$$b_5 = \frac{1}{1881169920(6m_3 + 6n_2 - n_1 + 11)^2}F_3(m_3, n_2, n_1),$$

where F_1, F_2 and F_3 are polynomials in m_3, n_2 and n_1. Eliminating m_3 from the three polynomials equations, $F_1 = F_2 = F_3 = 0$, yields a solution $m_3 = m_3(n_2, n_1)$, and two resultant polynomial equations:

$$P_1 = F F_4(n_2, n_1) \quad \text{and} \quad P_2 = F F_5(n_2, n_1), \tag{4.287}$$

where

$$F = 432n_2^2 - 24(n_1^2 + 16n_1 - 3)n_2 + 2n_1^3 + 45n_1^2 - 348n_1 - 499, \tag{4.288}$$

and F_4 and F_5 are respectively 25th- and 26th-degree polynomials with respect to n_2.

We now want to solve the two equations, $P_1 = P_2 = 0$. It can be shown that the roots of $F = 0$ (e.g., solving n_2 in terms of n_1) are not solutions of the original equations $b_3 = b_4 = b_5 = 0$, since they yield $6(m_3 + n_2) - n_1 + 11 = 0$, giving rise to a zero divisor (see (4.285)). Thus, the only possible solutions come from the two equations, $F_4 = F_5 = 0$. However, it is very difficult to follow the above procedure to eliminate one parameter from these two equations, since their degrees are too high. Therefore, we apply the built-in Maple solver fsolve here to find the solutions of $F_4 = F_5 = 0$. But Maple has a limit on solving multivariate polynomials, which gives only one possible real solution. (For univariate polynomials, fsolve can be used to find all real solutions.) Nevertheless, if the solution is a true solution of the system, it is enough for our purpose, since we mainly want to prove the existence of critical periods, rather than finding all solutions. Certainly, if one could find all solutions, it would be better.

By applying fsolve to equations $F_4 = F_5 = 0$, we obtain a solution as follows (up to 1000 digits):

```
n2 :=-1.0374657711409184091779706577299995254461315397204415031414369263370 ...
     ... 79336402835255809692665089727046900731772309088198085488679011713048:
n1 := 8.9323170715252135817419113353613426091223557177893498009791622790995 ...
     ... 8352299418026740387584254236414897562285624783014046210153921292369I:
```

Then the values of m_3, m_2 and n_3 directly follow the formulas given in (4.283), (4.285), and $m_3(n_2, n_1)$ (which is not listed here). By verifying the original equations, we can show that the above solution yields $b_1 = b_2 = b_3 = b_4 = 0$ but $b_5 \neq 0$. Thus, this solution only gives at most 4 local critical periods, not 5 as we are expecting.

The problem is caused by numerically solving the roots of the resultant equations, rather than the original equations. We may apply fsolve directly to the original equations, with the risk that we may not be able to obtain any solutions at all due to too many equations and variables involved. The Maple command

```
with(linalg):
Mysolution := fsolve({B1,b2,b3,b4,b5}, {m2,m3,n1,n2,n3}):
```

yields a solution (up to 1000 digits):

```
m2 :=-21.090600484437152798842381398933514966653977502834642843072384945 ...
     ... 0192340434705916841204595597378248611629849758969906955111700114751:
```

```
m3  := 5.125109394169587039145091618355152678331363487291313143329652082 ...
    ... 708996820915283214563810999105222159218993365237562500814877898150:
n1  := .221144481852042476362097983437931319595147515673132301352980558 5 ...
    ... 380110623946971134688104347855383954946414228889480850825415792946:
n2  := 4.536766340054913651507181574123480691222115160017318145023788504 ...
    ... 774024708155221034216788445170902581275389079050267890102308702990:
n3  :=-16.04965118875927734665732451579703873331519269547524914368562229 ...
    ... 113889180570162284796205552039147067145683154187478411837014672177:
```

Substituting the above solution (referred to as a critical point C) into the b_i's we obtain

$$b_1 = 0, \qquad b_2 = -0.128 \times 10^{-997}, \qquad b_3 = 0.48 \times 10^{-997},$$

$$b_4 = 0.5042 \times 10^{-995}, \qquad b_5 = -0.219 \times 10^{-994},$$

$$b_6 = 63.2614003037798228307321439 80343\dots.$$

Further calculating the Jacobian given in (4.263) at the above critical point shows that

$$\det\left[\frac{\partial(b_1, b_2, b_3, b_4, b_5)}{\partial(n_1, n_2, m_3, m_2, n_3)}\right]_C = -788.59440734553592526150\dots \neq 0,$$

implying that for Case (3) there exist 5 local critical periods bifurcating from the weak center (the origin). The above results are summarized as a theorem below.

Theorem 4.38 *For the reversible system* (4.268), *there exist values of the parameters* n_1, n_2, m_3, m_2, n_3 *such that* 5 *local critical periods are obtained, which bifurcate from the weak center (the origin).*

(D) Case (4): $a_{20} \neq 0$ The last of the cases, $a_{20} \neq 0$, is considered the most general and difficult. The system, described by (4.265), has 6 independent parameters. So it is expected that the system may have 6 local critical periods bifurcating from the weak center (the origin). If all the parameters are chosen freely, then purely symbolic computation becomes intractable. What we will show below includes three cases:

(i) $m_1 = n_1 = 0$: 4 local critical periods (using symbolic computation only);
(ii) $m_1 = 0$: 5 local critical periods (using both symbolic and numerical computations);
(iii) No parameter equals zero: 6 local critical periods (using numerical computation only).

Note here that the cases of 4 and 5 local critical periods are different from those presented in Cases (2) and (3), respectively, since this case contains the term x^2 in the first equation of (4.265).

Subcase (i): $m_1 = n_1 = 0$. For this subcase, we have

$$b_1 = \frac{1}{8}(n_3 - m_2 - 3m_3 + 3n_2) - \frac{1}{6}. \tag{4.289}$$

Thus,

$$n_3 = m_2 + 3m_2 - 3n_2 + \frac{4}{3}, \qquad (4.290)$$

in order to have $b_1 = 0$. Then,

$$b_2 = -\frac{1}{48}\left[(6m_3 + 6n_2 + 1)m_2 + 27m_3^2 + 4m_3 + 9n_2^2 + 17n_2 - \frac{28}{3}\right], \quad (4.291)$$

which, in turn, gives

$$m_2 = -\frac{3(27m_3^2 + 4m_3 + 9n_2^2 + 17n_2) - 28}{(36m_3 + 6n_2 + 1)}. \qquad (4.292)$$

Having determined n_3, m_2, further calculation on the b_i yields

```
b3  := 1/24186470400/(1+6*n2+6*m3)^4
     *(-166336+4299237*m3^2*n2-6138180*m3*n2^3-524880*m3^4*n2
     -860016*n2+8118774*m3*n2+167748*m3+6689538*m3^2+5799564*n2^2
     -5457861*n2^3-13858047*m3*n2^2+7283439*m3^3+790236*m3^4
     +2055780*m3^3*n2-12629196*m3^2*n2^2-3020976*n2^4
     -5773680*m3^2*n2^3+5773680*m3^3*n2^2+524880*m3*n2^4):
b4  := 1/3482851737600/(1+6*n2+6*m3)^6
     *(23079424+1436655204*m3^2*n2+2741028336*m3^3*n2^3
     -3587043042*m3^4*n2^2-1236483144*m3*n2^3-1899088011*m3^4*n2
     +45558144*n2+222969024*m3^5*n2+111996270*m3^6-1152699903*n2^5
     -1335699936*m3*n2-96135264*m3-1533692286*m3^2-816192288*n2^2
     +1653973560*n2^3+5023865106*m3*n2^2-1559217870*m3^3
     +1445706144*m3^4+1161666360*m3^3*n2+3218307552*m3^2*n2^2
     -168315894*m3^2*n2^4-28343520*m3^6*n2-133898832*n2^4
     -897544800*m3^2*n2^5+923597802*m3^2*n2^3-1114845120*m3^4*n2^3
     -1265905584*m3*n2^5-1445873814*m3^3*n2^2-4577964723*m3*n2^4
     +207852480*m3^5*n2^2+1712007657*m3^5+56687040*m3*n2^6
     -271822230*n2^6+1776193920*m3^3*n2^4):
```

Now eliminating n_2 from the two equations $b_3 = b_4 = 0$ (ignoring the constant factors and the denominator) results in a solution $n_2 = n_2(m_3)$ and the following resultant:

```
F  := m3*(432*m3^2-120*m3-143)*(250923216733026906789642240000000000*m3^17
     -126054277488251166385108942848000000000*m3^16+ ...
     +178488837097252527009160287244355814400*m3
     +126869487069973102400323268975096320000):
```

The solution $m_3 = 0$ gives $n_2 = \frac{4}{9}, m_2 = n_3 = 0$, leading to $b_1 = b_2 = \cdots = 0$, implying that the origin is an isochronous center. The two solutions from the second factor are actually not solutions of the original equations. So other possible solutions come from the 17th-degree polynomial, which has 13 real solutions. By verifying the original b_i equations, there are only 11 solutions satisfying the original equations. Further, by checking the determinant (nonzero) of the Jacobian in Theorem 4.2, we know that perturbing each of these 11 solutions results in 4 local critical periods. Thus, we have the following theorem.

Theorem 4.39 *For the reversible system* (4.265) *when* $m_1 = n_1 = 0$, *there are* 11 *sets of solutions* (m_3, n_2, m_2, n_3) *leading to 4 local critical periods. Moreover, when* $n_2 = \frac{4}{9}, m_3 = m_2 = n_3 = m_1 = n_1 = 0$, *the origin is an isochronous center.*

Subcase (ii): $m_1 = 0$. For this subcase, if we use an elimination procedure, it will lead to very high-degree polynomials and it is difficult to obtain the final resultant with only one variable. Thus, we try to use the Maple command, `fsolve`, to find a possible solution, since one solution is enough for proving the existence of certain orders of critical periods. To do this, let $m_1 = 0$ in b_i, $i = 1, 2, \ldots, 5$. Then use the command

```
with(linalg):
m1 := 0:
Mysolution := fsolve({B1,b2,b3,b4,b5}, {m2,m3,n1,n2,n3}):
```

to obtain a solution (up to 1000 digits):

```
m1 := 0:
m2 := 6.035400017219239604088365819876376821283045937559186370150833917 ...
     ... 521247539246439636893372513430973055913423563721427540938133657870:
m3 :=-2.780809904103370988424442531627905476289418239384590526626481008 ...
     ... 834576020348013746716149424390335117837648597349089950519527455229:
n1 := 0.621849114501545687930704699633765375651291643986255358128944707 ...
     ... 257901313693726483172377467569487002179968213262136823553358496491:
n2 :=-2.047466873502703059425954703630994770898922153574473647213673030 ...
     ... 666859689375467940323918533739308056608575305705993922947509110621:
n3 := 4.261187841650111830405730002765259103184527609789400252719948417 ...
     ... 683595939560330109243994695209408865769445515993036954490477716439:
```

Now, it is very important to verify whether this approximate solution indeed implies the existence of a true solution. To do this, we substitute the numerical solution into the explicit expressions of the b_i's to obtain

$$b_1 = -0.1 \times 10^{-999}, \qquad b_2 = -0.127 \times 10^{-998}, \qquad b_3 = 0.836 \times 10^{-998},$$

$$b_4 = 0.121 \times 10^{-996}, \qquad b_5 = -0.46863 \times 10^{-995},$$

$$b_6 = -7.421658867638726722085244758030\ldots.$$

Because the symbolic expressions of the b_i's are exact before the substitution, the above verification scheme indeed shows that there exists a solution such that $b_i = 0, i = 1, 2, \ldots, 5$, but $b_6 \neq 0$.

Moreover, the Jacobian given in (4.263) for this case evaluated at the above critical point yields

$$\det\left[\frac{\partial(b_1, b_2, b_3, b_4, b_5)}{\partial(m_2, m_3, n_1, n_2, n_3)}\right]_C = 364.78657557777209224666\ldots \neq 0.$$

Thus, based on Theorem 4.2, we know that Subcase (ii) has 5 local critical periods bifurcating from the weak center (the origin). A theorem summarizing the above results is given below.

Theorem 4.40 *For the reversible system* (4.265) *when* $m_1 = 0$, *a solution* $(m_2, m_3, n_1, n_2, n_3)$ *exists such that the system has 5 local critical periods.*

Subcase (iii): *no parameter equals zero.* For this case, the computation is more involved than any of the other cases discussed above. Unless done with a very powerful computer system, purely symbolic computation is almost impossible for finding the solutions for a possible 6 local critical periods. The Maple command

```
with(linalg):
Mysolution := fsolve({B1,b2,b3,b4,b5,b6}, {m1,m2,m3,n1,n2,n3}):
```

has been used to obtain the following solution (up to 500 digits):

```
m1 :=-1.9112713114122488943183767403133372917996393865562255381775990088 ...
    ... 551695141943450774237846747710516252168805098710405990834877601960:
m2 :=-5.5061408929743706996878298788459440636088120761563092662384462197 ...
    ... 4539257965884497132209263913702839404991917774657813626935750555392:
m3 :=-0.9646247375962511793182213859915933558992414217937891567460303895 ...
    ... 4887887458334775633330575954945388557737349274887010532840115877824:
n1 := 0.0109531615574163384904551402432056612710930615278298506363227350 ...
    ... 817276254120361520768053706683747606003767341822911703779322849287:
n2 := 0.0009319940685681757280685531003689111451453621486822581628144240 ...
    ... 387802846258161211724575939719403782034564144996916537003441939052:
n3 :=-1.2750924974463412684941367121996445298187025226622867449018543590 ...
    ... 410836309921145454838123983140640614491450117156176212929362565911:
```

for which the verification scheme shows that

$$b_1 = -0.9 \times 10^{-499}, \qquad b_2 = -0.68 \times 10^{-499}, \qquad b_3 = 0.69269 \times 10^{-497},$$

$$b_4 = 0.753 \times 10^{-496}, \qquad b_5 = -0.405311585 \times 10^{-494},$$

$$b_6 = 0.174948224057419 \times 10^{-492},$$

$$b_7 = 0.000285510949123486875739425029\ldots,$$

indicating that there exists a solution $(m_1, m_2, m_3, n_1, n_2, n_3)$ such that $b_i = 0, i = 1, 2, \ldots, 6$, but $b_7 \neq 0$.

Further, substituting the above critical values into the Jacobian results in

$$\det \left[\frac{\partial(b_1, b_2, b_3, b_4, b_5)}{\partial(m_2, m_3, n_1, n_2, n_3)} \right]_C = 0.00003708074926875589\ldots \neq 0.$$

Therefore, based on Theorem 4.2, we can conclude that Subcase (iii) can have 6 local critical periods bifurcating from the weak center (the origin), as summarized in the following theorem.

Theorem 4.41 *For the reversible system* (4.265) *there exists a solution* $(m_1, m_2, m_3, n_1, n_2, n_3)$ *for the critical point such that* 6 *local critical periods bifurcate from the weak center.*

We note from the analysis above that if we follow the classification given at the beginning of this section, we can have more cases, and combining the case studies with the results obtained above leads to the following result.

Theorem 4.42 *For the general reversible system* (4.254), *the maximal number of local critical periods bifurcating from the weak center is equal to the number of the independent parameters contained in the system.*

(E) Numerical Examples We have established several theorems for the properties of local critical periods and the isochronous center of cubic reversible systems. In this section, we present two numerical examples to demonstrate how to perturb parameters from a critical point to obtain the exact number of local critical periods given in the theorems.

Remark 4.43 We have established Theorem 4.32 which theoretically guarantees the existence of k local critical periods if the conditions given in the theorem are satisfied. However, in practice it is not easy to find a particular set of perturbations to obtain a numerical realization. If the parameters can be perturbed, one by one, separately for each of the b_i's, the process is straightforward. When the perturbation parameters are coupled, such as those cases considered in Cases (2), (3) and (4), it is very difficult to find such a set of perturbations. In particular, when more parameters are coupled, like the case of 6 local critical periods (Theorem 4.42), it is extremely difficult to obtain a numerical set of perturbations.

In the following, we give two examples, one for the 3 local critical periods considered in (A) and the other for the 4 local critical periods discussed in (B).

Example 4.44 Consider the 3 local critical periods given in Theorem 4.34. For this example, $T'(h)$ is given by

$$T'(h) = \frac{-2\pi p'(h)}{(1 + p(h))^2},$$

where

$$p_4'(h) = b_1 + 2b_2 h + 3b_3 h^2 + 4b_4 h^3, \tag{4.293}$$

in which the subscript 4 denotes that $p(h)$ is a fourth-degree polynomial in h.

Taking $n_2 = 0.01$, and applying the perturbations

$$m_3 = (5 + 2\sqrt{6})n_2 + (0.1 \times 10^{-3}),$$

$$m_2 = -\frac{3(3m_3^2 + n_2^2)}{2(m_3 + n_2)} + (0.1 \times 10^{-8}),$$

$$n_3 = m_2 + 3m_3 - 3n_2 + (0.1 \times 10^{-13}),$$

yields the third-degree polynomial

$$p_4'(h) = (0.125 \times 10^{-14}) - (0.27272820328796 \times 10^{-10})h$$
$$+ (0.2046433012965192 \times 10^{-7})h^2$$
$$- (0.1229176145646577605 \times 10^{-5})h^3.$$

The roots of $p_4'(h) = 0$ are

$$h_1 = 0.47522968629422680728 \times 10^{-4},$$

$$h_2 = 0.14084921112674735809 \times 10^{-2}, \qquad (4.294)$$

$$h_3 = 0.15192803075362020703 \times 10^{-1},$$

as expected. Therefore, $T'(h_i) = 0$, $i = 1, 2, 3$, and

$$T'(h) > 0 \quad \forall h \in (0, h_1) \cup (h_2, h_3) \quad \text{and} \quad T'(h) < 0 \quad \forall h \in (h_1, h_2).$$

In terms of the amplitude of the periodic solution, $r = \sqrt{h}$ (see (4.258)), the amplitudes corresponding to the three critical points (see (4.294)) are

$$r_1 = 0.00689369049417093288, \qquad r_2 = 0.03752988291038853708,$$

$$r_3 = 0.12325908922007342952.$$

In order to show that higher-order terms added to $p_4'(h)$ do not affect the number of real roots of $p_4'(h)$ for $0 < h \ll 1$, we expand $p'(h)$ up to b_7 using the above perturbed parameter values to obtain

$$\begin{aligned}
p_7'(h) = {} & (0.125 \times 10^{-14}) - (0.27272820328796 \times 10^{-10})h \\
& + (0.2046433012965192 \times 10^{-7})h^2 \\
& - (0.122917614564657764 \times 10^{-5})h^3 \\
& + (0.30977476026963404769 \times 10^{-7})h^4 \\
& - (0.13132778484607037717 \times 10^{-7})h^5 \\
& + (0.90644113252474127528 \times 10^{-9})h^6,
\end{aligned}$$

which has the following 4 real roots:

$$h_1 = 0.47522968635658762712 \times 10^{-4},$$

$$h_2 = 0.14084868274006266263 \times 10^{-2}, \qquad (4.295)$$

$$h_3 = 0.15199198405041823566 \times 10^{-1},$$

$$h_4 = 17.115958097383951681.$$

Compared with the roots of $p_4'(h)$, the first three roots of $p_7'(h)$ are almost the same as those of $p_4'(h)$ (see (4.294)). The extra root of $p_7'(h)$, 17.115958097383951681, is obviously not in the interval $0 < h \ll 1$. This clearly shows that adding higher-order terms to $p_4'(h)$ does not change the number of local critical periods for small values of h.

Example 4.45 Consider the case of 4 local critical periods as discussed in Part (B). For this case,

$$p_5'(h) = b_1 + 2b_2 h + 3b_3 h^2 + 4b_4 h^3 + 5b_5 h^4. \qquad (4.296)$$

Note that here we cannot follow the procedure of Example 4.44, since for this case the parameters m_3 and n_2 are coupled in the two equations, $b_3(m_3, n_2) = b_4(m_3, n_2) = 0$. Although we obtain the exact expression $m_3 = m_3(n_2)$, given in (4.281), we cannot treat these two parameters independently. Thus, we have to find the perturbations simultaneously for b_3 and b_4, by using m_3 and n_2. Having determined perturbations on m_3 and n_2, we can determine the perturbations on m_2 and n_3, one by one, since they are separated.

It was shown in Part (B) that we have 10 sets of real solutions of n_2 for the 4 local critical periods. The critical values of n_3, m_2 and m_3 are given by (4.278), (4.280) and (4.281), respectively. The 10 sets of solutions of n_2 are given below (computed with up to 1000 digits, but here we only list the first 30 digits for brevity):

$$n_{2c} = -2.98855639079543384724018446 5716,$$

$$- 0.63538286307505131070801453 7214 \times 10^{-2},$$

$$- 0.54546674100569958464135398 2592 \times 10^{-4},$$

$$0.27069130345735445244582162 4251 \times 10^{-2},$$

$$0.10533244528224431048647324 2133,$$

$$0.14676833315855338716345475 7140,$$

$$0.22972579437866785131334370 7244,$$

$$1.26245327024229244121803353 0515,$$

$$1.82330015790697964733987013 6974,$$

$$4.98613392130385918196636597 2699.$$

Theoretically, for all the above 10 sets of critical solutions, we should be able to find perturbations which yield exactly 4 local critical periods. However, we have found that except for the ninth solution, $n_2 = 1.82330015790697964734$, it is very difficult to use other 9 solutions to obtain proper perturbations.

For the ninth set of solutions,

$$n_{2c} = 1.82330015790697964733987013 6974,$$

$$m_{3c} = 0.05474757744936263141191770 8322,$$

$$m_{2c} = -2.50545509802902096030149096 1062,$$

$$n_{3c} = -7.47777950606853867475201491 3683,$$

we have

$$b_1 = b_2 = b_3 = b_4 = 0, \qquad b_5 = -0.42487704474573231015 \times 10^{-2} < 0.$$

Thus, we need perturbations such that

$$b_4 > 0, \qquad b_3 < 0, \qquad b_2 > 0, \qquad b_1 < 0 \quad \text{and}$$

$$|b_i| \ll |b_{i+1}| \ll 1, \quad i = 1, 2, 3, 4.$$

First, consider perturbations simultaneously on n_{2c} and m_{3c} for b_4 and b_3. Following the procedure given in [239], we obtain

$$n_2 = n_{2c} + \varepsilon_1 = n_{2c} + 0.001 = 1.8243001579069796473398701369 74,$$

$$m_3 = m_{3c} + \varepsilon_2 = m_{3c} - 0.000025572$$

$$= 0.0547220054493626314119177083 22,$$

for which (4.296) has 2 real solutions for h. Then take

$$\varepsilon_3 = -0.1 \times 10^{-15} \quad \text{and} \quad \varepsilon_4 = -0.1 \times 10^{-22},$$

for m_2 and n_3, respectively, to obtain

$$m_2 = m_{2c} + \varepsilon_3 = -2.5069708382294280467639904 14400,$$

$$n_3 = n_{3c} + \varepsilon_2 = -7.4823745929393911995540432 92633.$$

Under the above perturbed parameter values, we have

$$b_1 = -0.1250000000001 \times 10^{-23},$$

$$b_2 = 0.23487766293640493222454 \times 10^{-16},$$

$$b_3 = -0.42058141423438673184346818 4691 \times 10^{-10},$$

$$b_4 = 0.33913761594472503735992422 8352 \times 10^{-5},$$

$$b_5 = -0.42714695334058353236697110 2420 \times 10^{-2},$$

for which (4.296) has 4 real roots:

$$h_1 = 0.169811971816428230300926230 677 \times 10^{-3},$$

$$h_2 = 0.598730983112402851118019026 925 \times 10^{-3},$$

$$h_3 = 0.300806322819703128399192749 467 \times 10^{-2},$$

$$h_4 = 0.250333629057619075954658809 755 \times 10^{-1},$$

$$(4.297)$$

as expected. If we add two more terms, $6b_6 h^5$ and $7b_7 h^6$, to (4.296), it still gives only 4 real roots, which are almost exactly the same as those given in (4.297).

4.6.3 Cubic-Order Planar Hamiltonian Systems

In this subsection, we consider the bifurcation of local critical periods from the weak center of a cubic-order Hamiltonian system with the following Hamiltonian function:

$$H(x, y) = \frac{1}{2}(x^2 + y^2) + \sum_{i+j=2}^{4} h_{ij}x^i y^j, \qquad (4.298)$$

which contains 9 coefficients h_{ij}. We will show that a system with Hamiltonian (4.298) can have a maximum of 7 local critical periods.

(A) General Formulation   The Hamiltonian system considered can be described by

$$\frac{dx}{dt'} = \frac{\partial H}{\partial y}, \qquad \frac{dy}{dt'} = -\frac{\partial H}{\partial x}, \qquad (4.299)$$

where the Hamiltonian $H(x, y)$ is given in (4.298). It is noted that the Hamiltonian has 9 coefficients. But in fact, one can reduce them to 7 coefficients. To show this, first note that a general third-order polynomial planar system with a Hopf singularity at the origin can be written in the form (4.26), i.e.,

$$\dot{x} = y + ax^2 + (b + 2d)xy + cy^2 + fx^3 + gx^2y + (h - 3p)xy^2 + ky^3,$$

$$\dot{y} = -x + dx^2 + (e - 2a)xy - dy^2 \qquad (4.300)$$

$$+ lx^3 + (m - h - 3f)x^2y + (n - g)xy^2 + py^3,$$

where α has been set to zero (see (4.26)). System (4.300) has 13 parameters.

Now, assume that system (4.300) is a Hamiltonian system. Then, it is easy to find that the Hamiltonian of the system is given by

$$H(x, y) = \frac{1}{2}(x^2 + y^2) - \frac{1}{3}dx^3 + \left(a - \frac{1}{2}e\right)x^2y + dxy^2 + \frac{1}{3}cy^3$$

$$- \frac{1}{4}lx^4 - \frac{1}{3}(m - h - 3f)x^3y$$

$$+ \frac{1}{2}(g - n)x^2y^2 - pxy^3 + \frac{1}{4}ky^4. \qquad (4.301)$$

It is clear to see that the Hamiltonian function given in (4.301) has only 8 parameters, because the coefficients of the term x^3 and xy^2 are not independent. Comparing the Hamiltonian (4.301) with the original Hamiltonian (4.298) shows that $h_{12} = -3h_{30}$. This reduction of one more parameter is due to rotation (since no rotation is considered in the Hamiltonian given by (4.298)). In other words, one can directly apply a rotation to system (4.299) to reduce by one more parameter. Thus, for convenience we still use (4.298) in the following analysis, but assume

$$h_{12} = -3h_{30}. \qquad (4.302)$$

More specifically, we consider the following Hamiltonian system:

$$\dot{x} = y + h_{21}x^2 - 6h_{30}xy + 3h_{03}y^2 + h_{31}x^3$$
$$+ 2h_{22}x^2y + 3h_{13}xy^2 + 4h_{04}y^3,$$
$$\dot{y} = -x - 3h_{30}x^2 - 2h_{21}xy + 3h_{30}y^2$$
$$- 4h_{40}x^3 - 3h_{31}x^2y - 2h_{22}xy^2 - h_{13}y^3,$$
$$\text{(4.303)}$$

which has 8 parameters (coefficients). However, in general, we can further reduce by one more parameter. To achieve this, assume that $h_{21} \neq 0$ (in the case $h_{21} = 0$, the system has only 7 parameters), then we can use the scaling

$$x \to \frac{x}{h_{21}}, \qquad y \to \frac{y}{h_{21}}, \qquad h_{30} \to a_1 h_{21}, \qquad h_{03} \to a_2 h_{21},$$

$$h_{40} \to a_3 h_{21}^2, \qquad h_{31} \to a_4 h_{21}^2, \qquad h_{22} \to a_5 h_{21}^2, \qquad \text{(4.304)}$$

$$h_{13} \to a_6 h_{21}^2, \qquad h_{04} \to a_7 h_{21}^2,$$

to obtain the following new system (when $h_{21} \neq 0$):

$$\dot{x} = y + x^2 - 6a_1xy + 3a_2y^2 + a_4x^3 + 2a_5x^2y + 3a_6xy^2 + 4a_7y^3,$$
$$\dot{y} = -x - 3a_1x^2 - 2xy + 3a_1y^2 - 4a_3x^3 - 3a_4x^2y - 2a_5xy^2 - a_6y^3. \qquad \text{(4.305)}$$

System (4.305) has only 7 independent parameters. In other words, h_{21} can be chosen arbitrarily (except $h_{21} = 0$) if we use the original Hamiltonian system (4.299). This implies that for the cubic Hamiltonian polynomial system (4.305) (or for the original Hamiltonian (4.299)), in general the maximal number of local critical periods that the system can have is 7.

Note that the advantage of the above scaling is that it reduces the number of system parameters by one, making computation simpler. However, we then need to consider one more possibility: $h_{21} = 0$. When $h_{21} = 0$, there are only 7 parameters. We may assume $h_{30} \neq 0$, and apply a similar scaling to obtain a system like (4.305) with only 6 independent parameters. This clearly shows that such a "degenerate" system has fewer independent parameters and so in general has fewer local critical periods. By doing this, we may have four different cases.

Case (i) $h_{21} = h_{30} = h_{03} = 0$: the corresponding system (no scaling) is given by

$$\dot{x} = y + a_4x^3 + 2a_5x^2y + 3a_6xy^2 + 4a_7y^3,$$
$$\dot{y} = -x - 4a_3x^3 - 3a_4x^2y - 2a_5xy^2 - a_6y^3, \qquad \text{(4.306)}$$

where $a_3 = h_{40}, a_4 = h_{31}, a_5 = h_{22}, a_6 = h_{13}, a_7 = h_{04}$. This system is actually a special Hamiltonian system with only cubic homogeneous polynomials. Note that the advantage without using scaling in (4.306) is that one does not necessarily need to specify one of the 5 parameters to be nonzero, and 5 parameters can be handled computationally.

Case (ii) $h_{21} = h_{30} = 0$, $h_{03} \neq 0$: the system is described by

$$\dot{x} = y + 3y^2 + a_4x^3 + 2a_5x^2y + 3a_6xy^2 + 4a_7y^3,$$
$$\dot{y} = -x - 4a_3x^3 - 3a_4x^2y - 2a_5xy^2 - a_6y^3. \tag{4.307}$$

Case (iii) $h_{21} = 0$, $h_{30} \neq 0$: the system is given by

$$\dot{x} = y - 6xy + 3a_2y^2 + a_4x^3 + 2a_5x^2y + 3a_6xy^2 + 4a_7y^3,$$
$$\dot{y} = -x - 3x^2 + 3y^2 - 4a_3x^3 - 3a_4x^2y - 2a_5xy^2 - a_6y^3. \tag{4.308}$$

Case (iv) $h_{21} \neq 0$: the system is given by (4.305).

In order to compare with Case (i) which only has cubic terms, we consider one more special case which only has quadratic terms.

Case (v) $h_{ij} = 0$, $i + j = 3$: quadratic Hamiltonian system, described by the following general form (a simpler system can be directly obtained from (4.303) by neglecting the cubic terms):

$$\dot{x} = y + h_{21}x^2 + 2h_{12}xy + 3h_{03}y^2,$$
$$\dot{y} = -x - 3h_{30}x^2 - 2h_{21}xy - h_{12}y^2. \tag{4.309}$$

In the following, we consider the above 5 cases, one by one.

(B) Case (i): $h_{21} = h_{30} = h_{03} = 0$ (No Scaling)   The system describing this case is given by (4.306) which has 5 parameters. Applying the Maple program [218] we easily obtain the exact expressions of the coefficients b_i. In particular,

$$b_1 = \frac{1}{2}(3a_7 + 3a_3 + a_5). \tag{4.310}$$

Setting $b_1 = 0$ yields

$$a_7 = -a_3 - \frac{1}{3}a_5, \tag{4.311}$$

and further computation gives

$$b_2 = -\frac{1}{240}[1440a_3^2 + 144a_4^2 + 160(a_5 + 3a_3)^2 + 9(5a_6 + 3a_4)^2] \leq 0. \tag{4.312}$$

This clearly shows that the only solution satisfying $b_2 = 0$ is $a_3 = a_4 = a_5 = a_6 = 0$, and thus $a_7 = 0$, leading to a linear system. There exists an infinite number of nontrivial solutions such that $b_1 = 0$ but $b_2 \neq 0$. For example, let $a_3 = a_4 = a_6 = 0$, $a_5 \neq 0$ and choose $a_7 = -\frac{1}{3}a_5$, then $b_1 = 0$, and $b_2 = -\frac{2}{3}a_5^2 < 0$. Then, giving a small perturbation to a_5 such that $a_5 \Longrightarrow a_5 - \varepsilon$ ($0 < \varepsilon \ll 1$), we obtain

$$b_1 = \frac{1}{2}\varepsilon, \qquad b_2 = -\frac{2}{3}(a_5 - \varepsilon)^2.$$

Thus, we may choose a_5 and $\varepsilon > 0$ such that $b_1 b_2 < 0$ and $0 < b_1 \ll -b_2$. In summary, we have the following theorem for Case (i).

Theorem 4.46 *For the Hamiltonian system* (4.306), *with only cubic homogeneous polynomials, there exists a maximum of 1 local critical period bifurcating from the weak center* (*the origin*).

(C) Case (ii): $h_{21} = h_{30} = 0, h_{03} \neq 0$ The system for this case is described by (4.307), which, like Case (i), has only 5 independent parameters. However, comparing (4.307) with (4.306), there is an extra term, $3y^2$, in the first equation, and thus system (4.307) may exhibit more critical periods. In fact, applying the Maple program results in

$$b_1 = \frac{1}{4}(6a_7 + 6a_3 + 2a_5 - 15). \tag{4.313}$$

Setting $b_1 = 0$ we obtain

$$a_7 = -a_3 - \frac{1}{3}a_5 + \frac{5}{2}, \tag{4.314}$$

and then,

$$b_2 = -\frac{1}{48}\big(45a_6^2 + 54a_4a_6 + 45a_4^2 + 576a_3^2 + 32a_5^2 + 192a_5a_3$$

$$+ 1260a_3 + 330a_5 + 2520\big). \tag{4.315}$$

Letting $b_2 = 0$ yields

$$a_6 = -\frac{1}{15}(3a_4 \pm \sqrt{Q}),$$
$$\tag{4.316}$$
$$Q = 12600 - 144a_4^2 - 2880a_3^2 - 160a_5^2 - 960a_5a_3 - 6300a_3 - 1650a_5.$$

Having determined a_7 and a_6, b_3, b_4 and b_5 are expressed in terms of a_3, a_4, a_5 and Q:

$$b_3 = b_3(a_3, a_4, a_5, Q), \qquad b_4 = b_4(a_3, a_4, a_5, Q), \qquad b_5 = b_5(a_3, a_4, a_5, Q).$$

Eliminating Q from the equations, $b_3 = b_4 = b_5 = 0$, results in $F_1 = F_2 = F_3 = 0$, where

```
F1:= 4722*a5*a3^2*a4^2+863*a5^2*a3*a4^2+132573/8*a5*a3*a4^2
    +48*a5*a3*a4^4+1344*a5*a3^3*a4^2+144*a3^2*a4^4+136/3*a5^3*a3*a4^2
    +328*a5^2*a3^2*a4^2+2880*a3^4*a4^2-166005/2*a5^2*a3
    +849009/128*a5*a4^2-242655/8*a5*a3^2-526617/64*a3*a4^2
    -64860075/32*a5-390913425/64*a3+104056155/64*a5*a3
    +368757585/256*a3^2-2735271/32*a4^2+18832905/64*a5^2+91035*a3^3
    -45115/4*a5^3+8145/16*a4^4+309204*a3^4-37177/24*a5^3*a3
    +223797/2*a5*a3^3+1407*a5^2*a4^2+183267/4*a3^2*a4^2
    +10899/2*a5^2*a3^2+826/9*a5^4*a3-4312*a5^2*a3^3+491/6*a5^3*a4^2
    -5376*a5*a3^4-392*a5^3*a3^2+99*a5*a4^4+522*a3*a4^4
    +13284*a3^3*a4^2-4123/12*a5^4+140/9*a5^5+32256*a3^5+9216*a3^6
```

```
    +5*a5^2*a4^4-104/9*a5^4*a3^2-112*a5^2*a3^4+10/3*a5^4*a4^2
    -416/3*a5^3*a3^3+40/27*a5^5*a3+2304*a5*a3^5+25/81*a5^6
    +278723025/64:
F2:= ...
F3:= ...
```

Next, eliminating a_4 from the three equations, $F_i = 0, i = 1, 2, 3$, we obtain the equation:

$$\begin{aligned}
&\big(10555920 + 995328a_5a_3 + 103680a_5^2 + 2052864a_5 + 10824192a_3 \\
&\quad + 2985984a_3^2\big)a_4^4 + \big(69120a_5^4 + 940032a_5^3a_3 + 6801408a_5^2a_3^2 + 27869184a_5a_3^3 \\
&\quad + 59719680a_3^4 + 1696896a_5^3 + 17895168a_5^2a_3 + 97915392a_5a_3^2 + 275457024a_3^3 \\
&\quad + 29175552a_5^2 + 343629216a_5a_3 + 950056128a_3^2 + 137539458a_5 \\
&\quad - 170623908a_3 - 1772455608\big)a_4^2 + 6400a_5^6 + 30720a_5^5a_3 - 239616a_5^4a_3^2 \\
&\quad - 2875392a_5^3a_3^3 - 2322432a_5^2a_3^4 + 47775744a_5a_3^5 + 191102976a_3^6 + 322560a_5^5 \\
&\quad + 1903104a_5^4a_3 - 8128512a_5^3a_3^2 - 89413632a_5^2a_3^3 - 111476736a_5a_3^4 \\
&\quad + 668860416a_3^5 - 7124544a_5^4 - 32120928a_5^3a_3 + 113000832a_5^2a_3^2 \\
&\quad + 2320327296a_5a_3^3 + 6411654144a_3^4 - 233876160a_5^3 - 1721139840a_5^2a_3 \\
&\quad - 628961760a_5a_3^2 + 1887701760a_3^3 + 6101861220a_5^2 + 33714194220a_5a_3 \\
&\quad + 29869364385a_3^2 - 42029328600a_5 - 126655949700a_3 + 90306260100 = 0,
\end{aligned}$$

$$\tag{4.317}$$

for determining a_4, as well as two resultant equations:

$$F_4(a_3, a_5) = 0, \qquad F_5(a_3, a_5) = 0.$$

Further, eliminating a_5 from the above two equations yields the solution

$$a_5 = a_5(a_3),$$

and the final resultant equation

$$F_6(a_3) = 0,$$

from which we obtain four real solutions, satisfying

$$b_i = 0, \quad i = 1, 2, 3, 4, 5, \quad \text{but} \quad b_6 \neq 0.$$

These four solutions are (the numerical computation using Maple command `fsolve` is up to 100 digits):

$$S_{1,2} = \begin{cases} a_3 = -0.5466124782428906770790767846808235329850828793614 5 \ldots, \\ a_4 = \pm 4.2037413584938534243669343543896195240738731993665 8 \ldots, \\ a_5 = 6.00564660833476047493659993345483363613137207845867 \ldots, \\ a_6 = \mp 1.82381133671814562881757388912123684061481533432877 \ldots, \\ a_7 = 1.04473027546463718543354347352921232094129218654189 \ldots, \end{cases}$$

$$(4.318)$$

and

$$S_{3,4} = \begin{cases} a_3 = 0.74628402190433633862153786736666691360178359116002 \ldots, \\ a_4 = \pm 0.9549488358606783788449107498560439168233040184333 4 \ldots, \\ a_5 = 1.77677679276518287286832995496947484184665887640751 \ldots, \\ a_6 = \mp 1.96857356633865127739656172962377003142054778517880 \ldots, \\ a_7 = 1.16145704717393603708901881431017480578266345003747 \ldots . \end{cases}$$

$$(4.319)$$

Further calculating the Jacobian given in (4.263) at the above four critical points shows that

$$\det \left[\frac{\partial(b_1, b_2, b_3, b_4, b_5)}{\partial(n_1, n_2, m_3, m_2, n_3)} \right]_{S_{1,2}} = -94787249180.04587320676089268 \ldots \neq 0,$$

and

$$\det \left[\frac{\partial(b_1, b_2, b_3, b_4, b_5)}{\partial(n_1, n_2, m_3, m_2, n_3)} \right]_{S_{3,4}} = 26168947720.95200915227587033 34 \ldots \neq 0.$$

Therefore, there exist four solutions for Case (ii), described by system (4.307), to have 5 local critical periods. Moreover, through the above solution procedure, we did not find solutions such that the origin is an isochronous center (except for the elementary center under the conditions $a_i = 0, i = 3, 4, 5, 6, 7$). The above results are summarized in the following theorem.

Theorem 4.47 *For the Hamiltonian system* (4.307), *there are four solutions $S_i, i = 1, 2, 3, 4$, for the critical point* $(a_{3c}, a_{4c}, a_{5c}, a_{6c}, a_{7c})$, *which can be perturbed to generate* 5 *local critical periods. There are no solutions for the origin to be an isochronous center.*

Remark 4.48 It should be pointed out that although Theorem 4.47 states that there are only four solutions which give 5 local critical periods, there is actually an infinite number of solutions since $h_{03}(\neq 0)$ can be chosen arbitrarily.

(D) Case (iii): $h_{21} = 0, h_{30} \neq 0$ The system for this case is described by (4.308), which has 6 independent parameters. So it is possible to have 6 local critical periods. Computation here is more involved. So after determining a_7 from the equation

$$b_1 = \frac{1}{4}(6a_7 + 6a_3 + 2a_5 - 15a_2^2 - 24),$$

$$(4.320)$$

as

$$a_7 = -a_3 - \frac{1}{3}a_5 + \frac{5}{2}a_2^2 + 4, \tag{4.321}$$

we apply a numerical computation scheme, built in Maple, to find a solution $(a_2, a_3, a_4, a_5, a_6)$ such that $b_i = 0, i = 2, 3, \ldots, 6$, but $b_7 \neq 0$. Hence, this case may have a maximum of 6 local critical periods.

With the built-in Maple solver `fsolve`:

```
with(linalg):
Mysolution := fsolve({B2,b3,b4,b5,b6}, {a2,a3,a4,a5,a6}):
```

we obtain the following solution (up to 100 digits):

$a_2 = -1.9530167042546119939292193098056889089603955329148127555 7 \ldots,$

$a_3 = 7.4400917989078701656061658837180529593568608754342112909 4 \ldots,$

$a_4 = 25.6508104516616801211605290313920150504198447482606572416 6 \ldots,$

$a_5 = -15.1161039071423049568380249683362083741611998941678266727 1 \ldots,$

$a_6 = -4.8231644116499007349385547526299334038450962109238948871 3 \ldots,$

and thus

$a_7 = 11.1342951212167645794144701363415223591766536339678646872 00 \ldots,$

for which it can be shown that

$b_1 = 0.2 \times 10^{-97},$

$b_2 = 0.7 \times 10^{-96},$

$b_4 = 0.6 \times 10^{-92},$

$b_5 = -0.2061606 \times 10^{-88},$

$b_6 = -0.2903 \times 10^{-86},$

$b_7 = -455046268530.8138980917633685338370108821312913008510386104 \ldots.$

Theoretically speaking, the above $b_i, i = 1, 2, \ldots, 6$, should be exactly equal to zero. However, due to numerical computation error, they are only very close to zero, which does not affect the conclusion. The above result indicates that we can have at most 6 local critical periods.

Further, substituting the above critical values into the Jacobian results in

$$\det \left[\frac{\partial(b_1, b_2, b_3, b_4, b_5, b_6)}{\partial(a_2, a_3, a_4, a_5, a_6, a_7)} \right]_C = 330430731142097473798555592786.512 \ldots \neq 0,$$

showing that Case (iii) can indeed have 6 local critical periods bifurcating from the weak center (the origin). Thus, we have the following theorem.

Theorem 4.49 *For the Hamiltonian system* (4.308), *there exists a solution* (a_2, a_3, a_4, a_5, a_6, a_7) *for the critical point such that 6 local critical periods bifurcate from the weak center.*

(E) Case (iv): $h_{21} \neq 0$ Now, we consider the most general case $h_{21} \neq 0$, described by (4.305). Since the system has 7 independent parameters, we thus expect to have 7 possible local critical periods bifurcating from the weak center (the origin). If all the parameters are chosen freely, then purely symbolic computation becomes intractable.

First, a_7 can be easily determined from $b_1 = 0$, where

$$b_1 = \frac{1}{4}\left(6a_7 + 6a_3 + 2a_5 - 24a_1^2 - 15a_2^2 - 6a_2 - 3\right), \tag{4.322}$$

as

$$a_7 = -a_3 - \frac{1}{3}a_5 + 4a_1^2 + \frac{5}{2}a_2^2 + a_2 + \frac{2}{3}. \tag{4.323}$$

Then, we apply the built-in Maple solver `fsolve`:

```
with(linalg):
Mysolution := fsolve({B2,b3,b4,b5,b6,b7}, {a1,a2,a3,a4,a5,a6}):
```

to obtain (up to 100 digits):

$a_1 = -1.236424673487758780263249822702619892588728027087886088596\ldots,$

$a_2 = 0.980600925073309753862696720801098646345508800409507885853\ldots,$

$a_3 = 1.167491399446400255941918209739435263092288905381214877376\ldots,$

$a_4 = -1.999491941039390717531984861774052854884533896233252487913\ldots,$

$a_5 = -2.486819742823199313215345865393248642243769821937870829847\ldots,$

$a_6 = 1.575346933954787469584942930641557171866800529638307984076\ldots,$

and so

$a_7 = 9.660978768375130033822984541796120688683095489596055252 1122\ldots.$

With the above solution, we have

$b_1 = -0.13 \times 10^{-97},$

$b_2 = -0.4 \times 10^{-98},$

$b_3 = 0.1719 \times 10^{-93},$

$b_4 = -0.1419 \times 10^{-91},$

$b_5 = 0.1787 \times 10^{-89},$

$b_6 = -0.26853633 \times 10^{-87},$

$b_7 = 0.9740407672794 \times 10^{-85},$

$b_8 = 1816997149.14429824809906913569542197686144472921335\ldots.$

Further, substituting the above critical values into the Jacobian results in

$$\det\left[\frac{\partial(b_1, b_2, b_3, b_4, b_5, b_6, b_7)}{\partial(a_1, a_2, a_3, a_4, a_5, a_6, a_7)}\right]$$
$$= -471971241749135752126311666636803.2562879\ldots \neq 0,$$

implying that Case (iv) can indeed have 7 local critical periods bifurcating from the weak center (the origin). Therefore, we have the following result.

Theorem 4.50 *For the Hamiltonian system* (4.305), *there exists a solution for the critical point such that 7 local critical periods bifurcate from the weak center. This is the maximal number of local critical periods which can be obtained from cubic Hamiltonian systems.*

(F) Case (v): Quadratic Hamiltonian System   Finally, we consider the quadratic system, given by (4.309). For this case, it is easy to show that

$$b_1 = -\frac{3}{4}\left[4\left(h_{30}^2 + h_{03}^2\right) + (h_{21} + h_{03})^2 + (h_{12} + h_{30})^2\right]. \tag{4.324}$$

This clearly indicates that $b_1 \leq 0$, and $b_1 = 0$ only if all the second-order terms equal zero. This implies that the period $T(h)$ monotonically increases for $0 < h \ll 1$ and $\sum_{i+j=2} h_{ij}^2 \neq 0$. Thus, we have the following result.

Theorem 4.51 *For the quadratic Hamiltonian system* (4.309), *there is no critical period near the origin. More specifically, the period $T(h)$ monotonically increases for $0 < h \ll 1$.*

Note that the result given in the above theorem is only a partial result of the general conclusion: a quadratic Hamiltonian system does not have a critical period and the period function monotonically increases for $h > 0$ (e.g., see [127]).

(G) A Numerical Example   Now for an illustration, we present an example chosen from Case (ii) which has 5 local critical periods (see Theorem 4.47). For this case, $h_{21} = h_{30} = 0$, while $h_{03} \neq 0$. The period $T'(h)$ for this example is given by (4.260) in which

$$p_6'(h) = b_1 + 2b_2 h + 3b_3 h^2 + 4b_4 h^3 + 5b_5 h^4 + 6b_6 h^5, \tag{4.325}$$

in which the subscript 6 denotes that $p(h)$ is a sixth-degree polynomial in h.

Note that for this example the parameters a_3, a_4 and a_5 are coupled in the three equations:

$$F_1(a_3, a_4, a_5) = F_2(a_3, a_4, a_5) = F_3(a_3, a_4, a_5) = 0.$$

Although we obtain the exact expressions, $a_4(a_3, a_5)$ and $a_5 = a_5(a_3)$, we cannot treat these three parameters independently. Thus, we have to find the perturbations simultaneously for b_3, b_4 and b_5, by using a_3, a_4 and a_5. Having determined perturbations on a_3, a_4 and a_5, we can determine the perturbations on a_6 and a_7, one by one, since they are separated.

It was shown in (C) that we have 4 real solutions of a_3 for the 4 local critical periods. The complete set of critical values of $(a_{3c}, a_{4c}, a_{5c}, a_{6c}, a_{7c})$ is given in (4.318) and (4.319). We choose the solution S_3 for this example, under which

$$b_1 = b_2 = b_3 = b_4 = b_5 = 0, \qquad b_6 = 4911.5370692734385786465\cdots > 0.$$

Thus, we need perturbations such that $-b_1 \ll b_2 \ll -b_3 \ll b_4 \ll -b_5 \ll 1$.

First, consider perturbations simultaneously on a_{3c}, a_{4c} and a_{5c} for b_5, b_4 and b_3. Following the procedure given in [239], we obtain (computed with up to 100 digits, but here only list the first 30 digits for brevity):

$$a_3 = a_{3c} - \varepsilon_1 = a_{3c} - 0.00070804685 = 0.745575975054336338621537867367,$$

$$a_4 = a_{4c} + \varepsilon_2 = a_{4c} + 0.013925185 = 0.968874020860678378844910749856,$$

$$a_5 = a_{5c} - \varepsilon_3 = a_{5c} - 0.0019420513 = 1.774834741465182872868329954969,$$

for which (4.325) has 3 real solutions for h. Then take

$$\varepsilon_4 = -0.1 \times 10^{-13} \quad \text{and} \quad \varepsilon_5 = -0.5 \times 10^{-20},$$

for a_6 and a_7, respectively, to obtain

$$a_6 = a_{6c} - \varepsilon_4 = a_{6c} - 0.00000000000001$$
$$= 1.969427032656687745526560799043,$$
$$a_7 = a_{7c} - \varepsilon_5 = a_{7c} - 0.00000000000000000000005$$
$$= 1.162812444457269370417352147643.$$

Under the above perturbed parameter values, we have

$$b_1 = -0.75 \times 10^{-20},$$
$$b_2 = 0.4782636049591312994866 \times 10^{-13},$$
$$b_3 = -0.41403486321711356253114862 \times 10^{-7},$$
$$b_4 = 0.43122529716447837933681258104 \times 10^{-2},$$
$$b_5 = -42.344161874069191402765521051235,$$
$$b_6 = 4269.680505038880075940579579767262,$$

for which (4.325) has 5 real roots:

$$h_1 = 0.88441121458464352512707024710 9 \times 10^{-7},$$

$$h_2 = 0.77469356861907450566288753460 5 \times 10^{-6},$$

$$h_3 = 0.70259489705672490170685661193 0 \times 10^{-5}, \qquad (4.326)$$

$$h_4 = 0.74327685122234213846260553602 3 \times 10^{-4},$$

$$h_5 = 0.81822895613934585215102507459 7 \times 10^{-2},$$

as expected.

In terms of the amplitude of periodic solution, $r = \sqrt{h}$ (see (4.258)), the amplitudes corresponding to the 5 critical points (see (4.326)) are

$$r_1 = 0.0002973905, \qquad r_2 = 0.0008801668, \qquad r_3 = 0.0026506507,$$

$$r_4 = 0.0086213505, \qquad r_5 = 0.0904560090.$$

In order to show that higher-order terms added to $p_6'(h)$ do not affect the number of real roots of $p_6'(h)$ for $0 < h \ll 1$, we expand $p'(h)$ up to b_9 using the above perturbed parameter values to obtain

$$p_9'(h) = -\left(0.75 \times 10^{-20}\right) + \left(0.47826360495913129949 \times 10^{-13}\right)h$$

$$- \left(0.41403486321711356253 \times 10^{-7}\right)h^2$$

$$+ \left(0.43122529716447837934 \times 10^{-2}\right)h^3$$

$$- 42.344161874069191402 77h^4$$

$$+ 4269.68050503888007594058h^5$$

$$+ 79515.61266347462865146140h^6$$

$$+ 776520.33802945848451878955h^7$$

$$+ 3929787.82501761055435369607h^8,$$

which has the following 6 real roots:

$$h_1 = 0.88441121458464347567603682647 0 \times 10^{-7},$$

$$h_2 = 0.77469356862089379772978720219 5 \times 10^{-6},$$

$$h_3 = 0.70259488610281832696990105542 9 \times 10^{-5}, \qquad (4.327)$$

$$h_4 = 0.74328915721628212422416151033 5 \times 10^{-4},$$

$$h_5 = 0.70130197637715879754786538835 7 \times 10^{-2},$$

$$h_6 = -0.96952281163357789049159317415 8 \times 10^{-1}.$$

Compared to the roots of $p_6'(h)$, the 5 positive roots of $p_9'(h)$ are almost the same as those of $p_6'(h)$ (see (4.326)). The extra real root of $p_9'(h)$ is negative, which obviously does not belong to the interval $0 < h \ll 1$. This clearly shows that adding higher-order terms to $p_6'(h)$ does not change the number of local critical periods for small values of h.

Remark 4.52 It has been observed from the analysis and results of this chapter that there are two main tasks in determining the limit cycles bifurcating from an isolated Hopf critical point: (1) computing the focus values (with exact symbolic computation); and (2) solving a group of multivariate polynomials. Both tasks can be very computationally demanding, especially for higher-order focus values. As we have seen, the second task usually involves trying to find as many analytical (closed-form) solutions as possible, leaving the last polynomial equation (the last resultant) with only one variable (parameter). Thus, with the help of a computer algebra system such as Maple, Mathematica, etc., one can solve this polynomial equation to find all real (and complex) solutions to as high an accuracy as one wishes. Having found all solutions from the last polynomial, backward substitution yields the solutions for all other variables (parameters). Here, the accuracy of solution is usually not a problem, and the process can be considered as symbolic since the last step (solving a univariate polynomial) can be performed using interval computation to an arbitrary accuracy.

Recently, software packages have been developed with the help of computer algebra systems such as AUTO, Content and Symcon, etc., to analyze the bifurcation of dynamical systems (e.g., see [18, 120–122]). This approach uses the (generalized) critical eigenvectors and adjoint eigenvectors, as well as the multilinear forms, computed in the original basis (and thus it does not need the linear coordinate transformation which is usually used in other methods with Jordan canonical form). This leads to compact explicit expressions of the critical normal form coefficients for all codimension 1 and 2 equilibrium bifurcations (i.e., fold, Hopf, cusp, Bogdanov–Takens, generalized Hopf, Hopf-zero, and double-Hopf bifurcations). These expressions have been implemented in the standard MATLAB bifurcation software MATCONT, and proved to be numerically robust [56, 57]. It should be pointed out that such computer software packages are very useful for plotting bifurcation diagrams associated with not very degenerate singular points, but it is very difficult (if not impossible) to determine multiple limit cycles around a very highly degenerate (weak) focus point, such as the 12 small limit cycles discussed in Sect. 4.2.2.

Chapter 5
Application (II)—Practical Problems

This chapter is focused on the application of Hopf bifurcation theory and normal form computation to practical problems, including those from engineering and biological systems, as well as problems arising from the area of Hopf bifurcation control. In particular, we shall show how to deal with Hopf bifurcation in higher-dimensional systems.

5.1 An Electrical Circuit

The nonlinear electrical circuit, shown in Fig. 5.1, is similar to the one depicted in Fig. 2.5 where there is an electronic source (see Example 2.11 for the 1:1 resonant Hopf bifurcation). The circuit consists of two capacitors C_1, C_2, two inductors L_1, L_2, a resistor R, and a conductance. L_1 and C_1 are connected in parallel, while L_2, C_2 and R in series. All the five elements, L_1, L_2, C_1, C_2 and R are assumed to be linear time-invariant elements, but C_1 and R may be varied. The conductance, however, is a nonlinear element with the characteristic [53, 104]

$$I_G = -\frac{1}{2}V_G + V_G^3, \tag{5.1}$$

where I_G and V_G represent the current and voltage of the conductance, respectively.

The voltages across the capacitors and the currents in the inductors are chosen as the state variables (as shown in Fig. 5.2), denoted by $(z_1, z_2, z_3, z_4) = (V_{C_1}, I_{L_1}, V_{C_2}, I_{L_2})$. Then the state equations of the circuit can be found as follows:

$$\dot{z}_1 = \frac{1}{2}\eta_1 z_1 + \eta_2 z_2 - \eta_1 z_4 - \eta_1 z_1^3,$$

$$\dot{z}_2 = -\frac{\sqrt{2}}{2}z_1, \tag{5.2}$$

$$\dot{z}_3 = (\sqrt{2}+1)z_4,$$

$$\dot{z}_4 = (2 - \sqrt{2})(z_1 - z_3 - \eta_2 z_4),$$

M. Han, P. Yu, *Normal Forms, Melnikov Functions and Bifurcations of Limit Cycles*,
Applied Mathematical Sciences 181,
DOI 10.1007/978-1-4471-2918-9_5, © Springer-Verlag London Limited 2012

Fig. 5.1 A nonlinear
electrical circuit

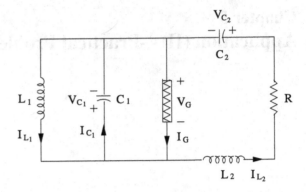

where $\eta_1 = 1/C_1$ and $\eta_2 = R$ are treated as two independent control (perturbation) parameters.

It is easy to show that the Jacobian of (5.2) evaluated on the equilibria $z_i = 0$ at the critical point, defined by

$$\eta_{1c} = 2, \qquad \eta_{2c} = 1 + \frac{\sqrt{2}}{2}, \qquad (5.3)$$

has two pairs of purely imaginary eigenvalues: $\lambda_{1,2} = \pm i$ and $\lambda_{1,2} = \pm\sqrt{2}i$. Further, introducing a linear transformation into system (5.2) yields the system

$$\dot{x}_1 = x_2 - (\sqrt{2} - 1)\left[x_1 + (\sqrt{2} + 1)x_2 - \sqrt{2}(2 + \sqrt{2})x_4\right]^3$$
$$\qquad - \frac{1}{2}\mu_1 + \frac{1}{2}\left[\mu_1 + 4(2 - \sqrt{2})\mu_2\right] - \mu_1 x_3 + 2(\sqrt{2} - 1)\mu_2 x_4$$
$$\equiv F_1(x, \mu),$$
$$\dot{x}_2 = -x_1 - (3\sqrt{2} - 4)\mu_2 x_2 - 2(3 - 2\sqrt{2})\mu_2 x_4$$
$$\equiv F_2(x, \mu),$$
$$\dot{x}_3 = \sqrt{2}x_4 + (\sqrt{2} - 1)\left[x_1 + (\sqrt{2} + 1)x_2 - \sqrt{2}(2 + \sqrt{2})x_4\right]^3 \qquad (5.4)$$
$$\qquad + \frac{1}{2}\mu_1 - \frac{1}{2}\left[\mu_1 + 4(2 - \sqrt{2})\mu_2\right] + \mu_1 x_3 - 2(\sqrt{2} - 1)\mu_2 x_4$$
$$\equiv F_3(x, \mu),$$
$$\dot{x}_4 = -\sqrt{2}x_3 - (3\sqrt{2} - 4)\mu_2 x_2 - 2(3 - 2\sqrt{2})\mu_2 x_4$$
$$\equiv F_4(x, \mu),$$

where $\mu = (\mu_1, \mu_2) = (\eta_1 - \eta_{1c}, \eta_2 - \eta_{2c})$.

Now executing the Maple program (with $\mu = 0$) yields the following normal form, in which the unfolding terms have been added:

Fig. 5.2 Bifurcation diagram for the electrical circuit

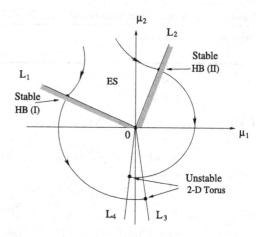

$$\dot{r}_1 = r_1\left[-\frac{1}{4}\mu_1 - \frac{1}{2}(3\sqrt{2}-4)\mu_2 - \frac{3}{4}\sqrt{2}r_1^2 - 3\sqrt{2}r_2^2\right],$$

$$\dot{r}_2 = r_2\left[\frac{1}{2}\mu_1 - (3-2\sqrt{2})\mu_2 - 3r_1^2 - 3r_2^2\right],$$

(5.5)

$$\dot{\theta}_1 = 1 + \frac{1}{4}\mu_1 + (2-\sqrt{2})\mu_2 - \frac{3}{4}(2+\sqrt{2})(r_1^2 + 4r_2^2),$$

$$\dot{\theta}_2 = \sqrt{2} - (\sqrt{2}-1)\mu_2 + 3(\sqrt{2}+1)(r_1^2 + r_2^2).$$

(5.6)

Here, the unfolding terms can be found from (5.4) using the following formulas [104]:

$$\alpha_{11} = \frac{1}{2}\left(\frac{\partial^2 F_1}{\partial x_1 \partial \mu_1} + \frac{\partial^2 F_2}{\partial x_2 \partial \mu_1}\right), \qquad \alpha_{12} = \frac{1}{2}\left(\frac{\partial^2 F_1}{\partial x_1 \partial \mu_2} + \frac{\partial^2 F_2}{\partial x_2 \partial \mu_2}\right),$$

$$\alpha_{21} = \frac{1}{2}\left(\frac{\partial^2 F_3}{\partial x_3 \partial \mu_1} + \frac{\partial^2 F_4}{\partial x_4 \partial \mu_1}\right), \qquad \alpha_{22} = \frac{1}{2}\left(\frac{\partial^2 F_3}{\partial x_3 \partial \mu_2} + \frac{\partial^2 F_4}{\partial x_4 \partial \mu_2}\right),$$

$$\beta_{11} = \frac{1}{2}\left(\frac{\partial^2 F_1}{\partial x_2 \partial \mu_1} - \frac{\partial^2 F_2}{\partial x_1 \partial \mu_1}\right), \qquad \beta_{12} = \frac{1}{2}\left(\frac{\partial^2 F_1}{\partial x_2 \partial \mu_2} - \frac{\partial^2 F_2}{\partial x_1 \partial \mu_2}\right),$$

$$\beta_{21} = \frac{1}{2}\left(\frac{\partial^2 F_3}{\partial x_4 \partial \mu_1} - \frac{\partial^2 F_4}{\partial x_3 \partial \mu_1}\right), \qquad \beta_{22} = \frac{1}{2}\left(\frac{\partial^2 F_3}{\partial x_4 \partial \mu_2} - \frac{\partial^2 F_4}{\partial x_3 \partial \mu_2}\right).$$

(5.7)

Finally, based on (5.5), applying the formulas and the Maple programs given in [223] for bifurcation analysis associated with double-Hopf bifurcation gives the results summarized in Table 5.1, where

Table 5.1 Bifurcation solutions and stability conditions for the electrical circuit

Bifurcation	Solution	Stability	Slope of critical line
ES	$r_1 = 0$	$\mu_2 - L_1\mu_1 > 0$	L_1
	$r_2 = 0$	$\mu_2 - L_2\mu_1 < 0$	L_2
HB (I)	$r_1^2 = -\frac{\sqrt{2}}{6}[\mu_1 + 2(3\sqrt{2} - 4)\mu_2]$	Stable	L_3
	$r_2 = 0$		
	$\omega_1 = 1 - \frac{1}{4}(2 + \sqrt{2})\mu_1 + \frac{3}{2}(2 - \sqrt{2})\mu_2$		
HB (II)	$r_1 = 0$	Stable	L_4
	$r_2^2 = \frac{1}{6}[\mu_1 + 2(2\sqrt{2} - 3)\mu_2]$		
	$\omega_2 = \sqrt{2} + \frac{1}{2}(\sqrt{2} + 1)\mu_1 + \frac{3}{2}(2 - \sqrt{2})\mu_2$		
2D torus	$r_1^2 = \frac{\sqrt{2}}{18}[(2\sqrt{2} + 1)\mu_1 + 2(4 - 3\sqrt{2})\mu_2]$	Unstable	
	$r_2^2 = \frac{-\sqrt{2}}{36}[(2 + \sqrt{2})\mu_1 - 2(4 - 3\sqrt{2})\mu_2]$		
	$\omega_1 = 1 + \frac{1}{4}(2 + \sqrt{2})\mu_1 + \frac{3}{2}(2 - \sqrt{2})\mu_2$		
	$\omega_2 = \sqrt{2} + \frac{1}{2}(\sqrt{2} + 1)\mu_1 + 2(1 - \sqrt{2})\mu_2$		

$$L_1 = -\left(\frac{3}{4}\sqrt{2} + 1\right), \qquad L_2 = \sqrt{2} + \frac{3}{2},$$

$$L_3 = -\frac{1}{2}(7 + 5\sqrt{2}), \qquad L_4 = 4 + \frac{11}{4}\sqrt{2},$$

and the bifurcation diagram is shown in Fig. 5.2. The normal form and the bifurcation results obtained here using the Maple programs indeed agree with the results presented earlier [104], showing the correctness, convenience and efficiency of using a computer algebra system.

5.2 A Double Pendulum System

In this section, we consider a mechanical system—a double pendulum—shown in Fig. 5.3 which consists of two rigid weightless links of equal length l one carrying a concentrated mass $2m$ and the other carrying a concentrated mass m. A follower force, P_1, and a constant directional force (vertical), P_2, are applied to the system.

The system energy for the three linear springs h_1, h_2 and h_3 is assumed to be given by [160]

$$V = \frac{1}{2}\left[(h_1 + h_2 + h_3l^2)\theta_1^2 + 2(h_3l^2 - h_2)\theta_1\theta_2 + (h_2 + h_3l^2)\theta_2^2\right]$$
$$- \frac{1}{6}h_3l^2(\theta_1 + \theta_2)(\theta_1^3 + \theta_2^3), \tag{5.8}$$

Fig. 5.3 A double pendulum system

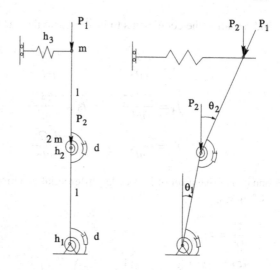

where θ_1 and θ_2 are generalized coordinates which specify the configuration of the system completely.

The kinetic energy T of the system is expressed by

$$T = \frac{ml^2}{2\Omega^2}\left[3\dot{\theta}_1^2 + \dot{\theta}_2^2 + 2\dot{\theta}_1\dot{\theta}_2\cos(\theta_1 - \theta_2)\right], \tag{5.9}$$

where Ω is an arbitrary value rendering the time variable nondimensional, and the dot denotes differentiation with respect to the nondimensional time variable τ and $\tau = \Omega t$.

The components of the generalized forces corresponding to the generalized coordinates θ_1 and θ_2 may be written as

$$\begin{aligned}
Q_1 &= P_1 l \sin(\theta_1 - \theta_2) + 2P_2 l \sin\theta_2, \\
Q_2 &= P_2 l \sin\theta_2,
\end{aligned} \tag{5.10}$$

and the damping can be expressed by

$$D = \frac{1}{2}\left[d_1\dot{\theta}_1^2 + d_2(\dot{\theta}_1 - \dot{\theta}_2)^2\right] + \frac{1}{4}\left[d_3\dot{\theta}_1^4 + d_4(\dot{\theta}_1 - \dot{\theta}_2)^4\right], \tag{5.11}$$

where d_1, d_2, represent the linear parts and d_3, d_4, describe the nonlinear parts, respectively.

With the aid of the *Lagrangian equations*, choosing the state variables

$$z_1 = \theta_1, \qquad z_2 = \dot{\theta}_1, \qquad z_3 = \theta_2 \quad \text{and} \quad z_4 = \dot{\theta}_2, \tag{5.12}$$

and rescaling the coefficients to be dimensionless as

$$f_1 = \frac{h_1\Omega^2}{ml^2}, \qquad f_2 = \frac{h_2\Omega^2}{ml^2}, \qquad f_3 = \frac{h_3\Omega^2}{m},$$

$$f_4 = \frac{P_1\Omega^2}{ml}, \qquad f_5 = \frac{P_2\Omega^2}{ml}, \qquad f_6 = \frac{d_3\Omega^4}{ml^2}, \tag{5.13}$$

$$f_7 = \frac{d_4\Omega^4}{ml^2}, \qquad \eta_1 = \frac{d_1\Omega^2}{ml^2}, \qquad \eta_2 = \frac{d_2\Omega^2}{ml^2},$$

one can derive a set of first-order differential equations with up to third-order terms as follows:

$$\frac{dz_1}{d\tau} = z_2,$$

$$\frac{dz_2}{d\tau} = \left(-\frac{1}{2}f_1 - f_2 + \frac{1}{2}f_4 + f_5\right)z_1 - \left(\frac{1}{2}\eta_1 + \eta_2\right)z_2$$

$$+ \left(f_2 - \frac{1}{2}f_4 - \frac{1}{2}f_5\right)z_3 + \eta_2 z_4 + \left(\frac{1}{4}f_1 + \frac{3}{4}f_2 - \frac{1}{3}f_4 - \frac{2}{3}f_5\right)z_1^3$$

$$+ \left(-\frac{3}{4}f_2 - \frac{1}{2}f_3 + \frac{1}{3}f_4 + \frac{7}{12}f_5\right)z_3^3$$

$$+ f_7 z_4^3 - \left(\frac{1}{2}f_6 + f_7\right)z_2^3 + \left(\frac{3}{4}\eta_2 + \frac{1}{4}\eta_1\right)z_1^2 z_2 - \frac{3}{4}\eta_2 z_1^2 z_4$$

$$- \frac{1}{2}z_1 z_2^2 - \left(\frac{1}{2}f_1 + \frac{9}{4}f_2 - \frac{1}{2}f_3 - f_4 - \frac{3}{2}f_5\right)z_1^2 z_3$$

$$+ \frac{1}{2}z_2^2 z_3 + 3f_7 z_2^2 z_4 + \left(\frac{1}{4}f_1 + \frac{9}{4}f_2 - f_4 - \frac{3}{2}f_5\right)z_1 z_3^2$$

$$+ \left(\frac{1}{4}\eta_1 + \frac{3}{4}\eta_2\right)z_2 z_3^2 - \frac{3}{4}\eta_2 z_3^2 z_4 - \frac{1}{2}z_1 z_4^2 - 3f_7 z_2 z_4^2$$

$$+ \frac{1}{2}z_3 z_4^2 - \left(\frac{1}{2}\eta_1 + \frac{3}{2}\eta_2\right)z_1 z_2 z_3 + \frac{3}{2}\eta_2 z_1 z_3 z_4,$$

$$\frac{dz_3}{d\tau} = z_4, \tag{5.14}$$

$$\frac{dz_4}{d\tau} = \left(\frac{1}{2}f_1 + 2f_2 - f_3 - \frac{1}{2}f_4 - f_5\right)z_1 + \left(\frac{1}{2}\eta_1 + 2\eta_2\right)z_2 - 2\eta_2 z_4$$

$$+ \left(-2f_2 - f_3 + \frac{1}{2}f_4 + \frac{3}{2}f_5\right)z_3 - 2\eta_2 z_4 + \frac{5}{4}\eta_2 z_1^2 z_4 + \frac{3}{2}z_1 z_2^2$$

$$- \left(\frac{1}{2}f_1 + \frac{5}{4}f_2 - \frac{1}{6}f_3 - \frac{7}{12}f_4 - \frac{7}{6}f_5\right)z_1^3 - \frac{3}{2}z_2^2 z_3 - 6f_7 z_2^2 z_4$$

$$+ \left(\frac{5}{4}f_2 + \frac{7}{6}f_3 - \frac{7}{12}f_4 - f_5 \right) z_3{}^3 - \left(\frac{1}{2}\eta_1 + \frac{5}{4}\eta_2 \right) z_2 z_3{}^2$$

$$+ \left(\frac{1}{2}f_6 + 2f_7 \right) z_2{}^3 - 2f_7 z_4{}^3 - \left(\frac{1}{2}\eta_1 + \frac{5}{4}\eta_2 \right) z_1{}^2 z_2 + \frac{5}{4}\eta_2 z_3{}^2 z_4$$

$$+ \left(f_1 + \frac{15}{4}f_2 - \frac{1}{2}f_3 - \frac{7}{4}f_4 - \frac{11}{4}f_5 \right) z_1{}^2 z_3 + \frac{1}{2}z_1 z_4{}^2 + 6f_7 z_2 z_4{}^2$$

$$- \left(\frac{1}{2}f_1 + \frac{15}{4}f_2 - \frac{1}{2}f_3 - \frac{7}{4}f_4 - \frac{5}{2}f_5 \right) z_1 z_3{}^2 - \frac{1}{2}z_3 z_4{}^2$$

$$+ \left(\eta_1 + \frac{5}{2}\eta_2 \right) z_1 z_2 z_3 - \frac{5}{2}\eta_2 z_1 z_3 z_4,$$

where $f_i \geq 0$, $i = 1, 2, 3$, due to physical conditions. (For more detailed discussion, see [231]).

The Jacobian matrix of (5.14) evaluated at an arbitrary point on the initial equilibrium solution $z_i = 0$ takes the form

$$J = \begin{bmatrix} 0 & 1 & 0 & 0 \\ -\frac{1}{2}f_1 - f_2 + \frac{1}{2}f_4 + f_5 & -\eta_2 - \frac{1}{2}\eta_1 & f_2 - \frac{1}{2}f_4 - \frac{1}{2}f_5 & \eta_2 \\ 0 & 0 & 0 & 1 \\ \frac{1}{2}f_1 + 2f_2 - f_3 - \frac{1}{2}f_4 - f_5 & 2\eta_2 + \frac{1}{2}\eta_1 & -2f_2 - f_3 + \frac{1}{2}f_4 + \frac{3}{2}f_5 & -2\eta_2 \end{bmatrix}$$

(5.15)

from which one may obtain the characteristic polynomial

$$P(\lambda) = \lambda^4 + a_1 \lambda^3 + a_2 \lambda^2 + a_3 \lambda + a_4, \tag{5.16}$$

where

$$a_1 = \frac{1}{2}\eta_1 + 3\eta_2,$$

$$a_2 = 3f_2 + \frac{1}{2}\eta_1 \eta_2 + \frac{1}{2}f_1 - f_4 - \frac{5}{2}f_5 + f_3,$$

$$a_3 = 2\eta_2 f_3 - \frac{1}{2}\eta_1 f_5 + \frac{1}{2}f_1 \eta_2 + \frac{1}{2}\eta_1 f_3 + \frac{1}{2}\eta_1 f_2 - \frac{3}{2}\eta_2 f_5, \tag{5.17}$$

$$a_4 = 2f_2 f_3 - \frac{3}{2}f_2 f_5 + \frac{1}{2}f_1 f_2 + \frac{1}{2}f_1 f_3 - \frac{1}{2}f_1 f_5 + \frac{1}{2}f_4 f_5$$

$$- \frac{3}{2}f_5 f_3 + f_5{}^2 - f_4 f_3.$$

Using the Hurwitz criterion [173], one may show that when

$$a_1 > 0, \qquad a_2 > 0, \qquad a_4 > 0 \quad \text{and} \quad a_3(a_1 a_2 - a_3) - a_4 a_1{}^2 > 0, \tag{5.18}$$

the initial equilibrium solution, $z_i = 0$, is stable. It should be noted that the conditions given in (5.18) imply $a_3 > 0$, which is of course as expected. Now let's consider the violations of the four conditions given in (5.18) leading to codimension 1 bifurcations occurring from the stable equilibrium, $z_i = 0$. It is not difficult to find from (5.18) that the condition which may be first violated is either $a_4 > 0$ or $a_3(a_1a_2 - a_3) - a_4a_1^2 > 0$, since if a_1 crosses zero first (i.e., positive a_1 becomes zero) while a_2, a_3 and a_4 are still positive, then the fourth condition becomes $-a_3^2 < 0$, which has already crossed zero. Similarly, if a_2 crosses zero first, but a_1, a_2 and a_4 are still positive, then the fourth condition becomes $-a_3^2 - a_4a_1^2 < 0$, which again already violates the stability conditions. Thus, there exist only two types of codimension 1 bifurcations which may occur from the initial equilibrium: one is a static bifurcation when $a_4 = 0$, which gives a zero eigenvalue at the critical point; the other is a Hopf bifurcation when $a_3(a_1a_2 - a_3) - a_4a_1^2 = 0$ under which the characteristic polynomial (5.16) has one pair of purely imaginary eigenvalues.

5.2.1 Hopf Bifurcation

To have a Hopf bifurcation, we need $a_5(a_1a_2 - a_3) - a_4a_1^2 = 0$ while the other three conditions: $a_1 > 0, a_2 > 0, a_4 > 0$ still hold. To achieve this, take

$$f_1 = 5, \qquad f_2 = f_3 = f_5 = 1, \qquad f_4 = 2 + \mu, \qquad f_6 = f_7 = 0,$$

$$\eta_1 = 2, \qquad \eta_2 = \frac{1}{2},$$

at which the Jacobian has a purely imaginary pair $\pm i$ and two negative real roots, -1 and $-\frac{3}{2}$. By a linear transformation:

$$
\begin{pmatrix} z_1 \\ z_2 \\ z_3 \\ z_4 \end{pmatrix} =
\begin{bmatrix}
\frac{1}{5} & \frac{2}{5} & 1 & 1 \\
-\frac{2}{5} & \frac{1}{5} & -1 & -\frac{3}{2} \\
1 & 0 & -1 & -\frac{6}{5} \\
0 & 1 & 1 & \frac{9}{5}
\end{bmatrix}
\begin{pmatrix} x_1 \\ x_2 \\ x_3 \\ x_4 \end{pmatrix},
\tag{5.19}
$$

system (5.14) is transformed to $\dot{x}_i = F_i(x_1, x_2, x_3, x_4; \mu)$, i.e.,

$$
\frac{dx_1}{d\tau} = x_2 + \left(\frac{9}{65}x_1 - \frac{9}{130}x_2 - \frac{9}{26}x_3 - \frac{99}{260}x_4 \right)\mu + \frac{253}{7800}x_1^3 + \frac{67}{975}x_2^3 + \frac{503}{312}x_3^3
$$

$$
+ \frac{48919}{12000}x_4^3 + \frac{3}{650}x_1^2x_2 - \frac{89}{2600}x_1^2x_3 + \frac{87}{1625}x_1^2x_4 - \frac{53}{325}x_2^2x_1 + \frac{168}{325}x_2^2x_3
$$

$$
+ \frac{9019}{13000}x_2^2x_4 - \frac{439}{520}x_1x_3^2 + \frac{153}{130}x_2x_3^2 + \frac{3503}{520}x_3^2x_4 - \frac{12723}{6500}x_1x_4^2
$$

$$
+ \frac{3996}{1625}x_2x_4^2 + \frac{2383}{260}x_3x_4^2 - \frac{157}{325}x_1x_2x_3 - \frac{2659}{3250}x_1x_2x_4 - \frac{1713}{650}x_1x_3x_4
$$

$$
+ \frac{4543}{1300}x_2x_3x_4 + \cdots,
$$

$$\frac{dx_2}{d\tau} = -x_1 + \left(-\frac{7}{65}x_1 + \frac{7}{130}x_2 + \frac{7}{26}x_3 + \frac{77}{260}x_4\right)\mu + \frac{1577}{7800}x_1^3 - \frac{11}{195}x_2^3$$

$$+ \frac{331}{312}x_3^3 + \frac{31403}{12000}x_4^3 + \frac{119}{650}x_1^2x_2 - \frac{1021}{1625}x_1^2x_4 - \frac{1173}{2600}x_1^2x_3 + \frac{47}{325}x_2^2x_1$$

$$- \frac{174}{325}x_2^2x_3 - \frac{8777}{13000}x_2^2x_4 + \frac{197}{520}x_1x_3^2 - \frac{119}{130}x_2x_3^2 + \frac{2331}{520}x_3^2x_4$$

$$+ \frac{1229}{6500}x_1x_4^2 - \frac{2133}{1625}x_2x_4^2 + \frac{1567}{260}x_3x_4^2 + \frac{151}{325}x_1x_2x_3 + \frac{2097}{3250}x_1x_2x_4$$

$$+ \frac{379}{650}x_1x_3x_4 - \frac{2869}{1300}x_2x_3x_4 + \cdots,$$

$$\tag{5.20}$$

$$\frac{dx_3}{d\tau} = -x_3 + \left(-\frac{3}{5}x_1 + \frac{3}{10}x_2 + \frac{3}{2}x_3 + \frac{33}{20}x_4\right)\mu - \frac{2059}{3000}x_1^3 - \frac{109}{375}x_2^3 - \frac{301}{24}x_3^3$$

$$- \frac{75733}{2400}x_4^3 - \frac{117}{250}x_1^2x_2 + \frac{259}{200}x_1^2x_3 + \frac{147}{125}x_1^2x_4 + \frac{83}{125}x_2^2x_1 - \frac{48}{25}x_2^2x_3$$

$$- \frac{2681}{1000}x_2^2x_4 + \frac{173}{40}x_1x_3^2 - \frac{51}{10}x_2x_3^2 - \frac{2101}{40}x_3^2x_4 + \frac{5841}{500}x_1x_4^2$$

$$- \frac{1512}{125}x_2x_4^2 - \frac{7129}{100}x_3x_4^2 + \frac{47}{25}x_1x_2x_3 + \frac{881}{250}x_1x_2x_4$$

$$+ \frac{747}{50}x_1x_3x_4 - \frac{1637}{100}x_2x_3x_4 + \cdots,$$

$$\frac{dx_4}{d\tau} = -\frac{3}{2}x_4 + \left(-\frac{3}{5}x_1 + \frac{3}{10}x_2 + \frac{3}{2}x_3 + \frac{33}{20}x_4\right)\mu + \frac{584}{975}x_1^3 + \frac{292}{975}x_2^3 + \frac{460}{39}x_3^3$$

$$+ \frac{2227}{75}x_4^3 + \frac{128}{325}x_1^2x_2 - \frac{72}{65}x_1^2x_3 - \frac{304}{325}x_1^2x_4 - \frac{224}{325}x_2^2x_1 + \frac{132}{65}x_2^2x_3$$

$$+ \frac{914}{325}x_2^2x_4 - \frac{56}{13}x_1x_3^2 + \frac{68}{13}x_2x_3^2 + \frac{642}{13}x_3^2x_4 - \frac{3694}{325}x_1x_4^2 + \frac{3942}{325}x_2x_4^2$$

$$+ \frac{4358}{65}x_3x_4^2 - \frac{128}{65}x_1x_2x_3 - \frac{1176}{325}x_1x_2x_4 - \frac{952}{65}x_1x_3x_4$$

$$+ \frac{1076}{65}x_2x_3x_4 + \cdots.$$

Now, by using the formulas given in [245], we obtain the coefficients v_0 and τ_0 as

$$v_0 = \frac{1}{2}\left(\frac{\partial^2 F_1}{\partial x_1 \partial \mu} + \frac{\partial^2 F_2}{\partial x_2 \partial \mu}\right)\bigg|_{\mu=0} = \frac{5}{52},$$

$$\tag{5.21}$$

$$\tau_0 = \frac{1}{2}\left(\frac{\partial^2 F_1}{\partial x_2 \partial \mu} - \frac{\partial^2 F_2}{\partial x_1 \partial \mu}\right)\bigg|_{\mu=0} = \frac{1}{52},$$

and v_1 and τ_1 are obtained by executing the Maple program, yielding

$$v_1 = -\frac{27}{4160}, \qquad \tau_1 = -\frac{281}{4160}.$$

Therefore, the normal form for this case with unfolding is given by

$$\frac{dr}{d\tau} = r\left(\frac{5}{52}\mu - \frac{27}{4160}r^2 + \cdots\right),$$

$$\frac{d\theta}{d\tau} = 1 + \frac{1}{52}\mu - \frac{281}{4160}r^2 + \cdots. \tag{5.22}$$

Based on (5.22), it is easy to carry out the bifurcation and stability analysis. The steady-state solutions are given as

$$\bar{r} = 0 \quad \text{and} \quad \bar{r} = \frac{20}{3}\sqrt{\frac{\mu}{3}}. \tag{5.23}$$

The zero solution, $\bar{r} = 0$, is the initial equilibrium, $x_i = 0$ (or $z_i = 0$), while the nonzero solution, $\bar{r} = \frac{20}{3}\sqrt{\frac{\mu}{3}}$, represents a solution with a family of limit cycles bifurcating from a Hopf critical point at $\mu = \mu_c = 0$. The Hopf bifurcation is super-critical, i.e., when $\mu < 0$, the equilibrium, $r = 0$, is stable. It becomes unstable at $\mu = 0$ and bifurcates into stable limit cycles. The bifurcation diagram can be seen in Fig. 2.2.

A numerical simulation based on (5.14) for $\mu = 0.002$ is depicted in Fig. 5.4. For this particular value of μ, the estimate of the amplitude of the periodic solution is equal to

$$\bar{r} = \frac{2}{3\sqrt{15}} \approx 0.17213.$$

However, the amplitude of the limit cycle shown in Fig. 5.4 is only about 0.01. The big difference is due to the fact that the above estimation is obtained from the normal form (5.22) which is based on normalized system (5.20). From the transformation (5.19) we see that the amplitude for the variables z_i is about 0.03. This indicates that

Fig. 5.4 Simulated phase portrait projected on the z_1–z_2 plane, with the initial condition $(z_1, z_2, z_3, z_4) = (0.02, 0.01, -0.02, 0.01)$, showing a stable limit cycle

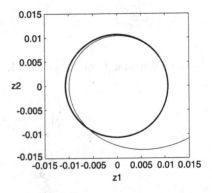

the normal form gives a reasonable qualitative prediction, at least for the stability analysis.

5.2.2 A Double-Zero Singularity

For system (5.14) to have a double-zero eigenvalue, let

$$f_1 = 8, \qquad f_2 = 10, \qquad f_3 = \frac{32}{11}, \qquad f_4 = \frac{134}{11}, \qquad f_5 = \frac{100}{11},$$

$$f_6 = f_7 = 0, \qquad \eta_1 = 2, \qquad \eta_2 = 1, \tag{5.24}$$

then the characteristic polynomial (5.14) has a double-zero and two negative real eigenvalues, $-1, -3$.

Further, choosing f_4 and f_5 (initially, P_1 and P_2) as the parameters, and then using the parametric transformation

$$f_4 = \frac{134}{11} + \mu_1, \qquad f_5 = \frac{100}{11} + \mu_2, \tag{5.25}$$

and the state variable transformation

$$\begin{pmatrix} z_1 \\ z_2 \\ z_3 \\ z_4 \end{pmatrix} = \begin{bmatrix} 1 & 0 & -1 & -2 \\ 0 & 1 & 1 & 6 \\ \frac{13}{7} & -\frac{11}{49} & -\frac{4}{3} & 1 \\ 0 & \frac{13}{7} & \frac{4}{3} & -3 \end{bmatrix} \begin{pmatrix} x_1 \\ x_2 \\ x_3 \\ x_4 \end{pmatrix}, \tag{5.26}$$

one may transform equation (5.14) into a new system:

$$\frac{dx_1}{d\tau} = x_2 + \left(\frac{45}{77}\mu_1 + \frac{1451}{1386}\mu_2 \right)x_1 - \left(\frac{15}{98}\mu_1 + \frac{257}{882}\mu_2 \right)x_2$$

$$- \left(\frac{5}{22}\mu_1 + \frac{109}{297}\mu_2 \right)x_3 + \left(\frac{45}{22}\mu_1 + \frac{797}{198}\mu_2 \right)x_4 + Nf_1,$$

$$\frac{dx_2}{d\tau} = - \left(\frac{6}{11}\mu_1 + \frac{10}{33}\mu_2 \right)x_1 + \left(\frac{1}{7}\mu_1 + \frac{4}{21}\mu_2 \right)x_2$$

$$+ \left(\frac{7}{33}\mu_1 - \frac{14}{99}\mu_2 \right)x_3 - \left(\frac{21}{11}\mu_1 + \frac{112}{33}\mu_2 \right)x_4 + Nf_2,$$

$$\frac{dx_3}{d\tau} = -x_3 + \left(\frac{9}{11}\mu_1 + \frac{213}{154}\mu_2 \right)x_1 - \left(\frac{3}{14}\mu_1 + \frac{39}{98}\mu_2 \right)x_2$$

$$- \left(\frac{7}{22}\mu_1 + \frac{5}{11}\mu_2 \right)x_3 + \left(\frac{63}{22}\mu_1 + \frac{123}{22}\mu_2 \right)x_4 + Nf_3, \tag{5.27}$$

$$\frac{dx_4}{d\tau} = -3x_4 - \left(\frac{9}{77}\mu_1 + \frac{233}{1386}\mu_2\right)x_1 + \left(\frac{3}{98}\mu_1 + \frac{47}{882}\mu_2\right)x_2$$

$$+ \left(\frac{1}{22}\mu_1 + \frac{13}{297}\mu_2\right)x_3 - \left(\frac{9}{22}\mu_1 + \frac{155}{198}\mu_2\right)x_4 + Nf_4,$$

where the Nf_i represent nonlinear terms. The Jacobian matrix of (5.27), evaluated for the initial equilibrium solution, $x_i = 0$, at the critical point, $\mu_{1c} = \mu_{2c} = 0$, is now in Jordan canonical form:

$$J_{(x_i=0,\mu_i=0)} = \begin{bmatrix} 0 & 1 & 0 & 0 \\ 0 & 0 & 0 & 0 \\ 0 & 0 & -1 & 0 \\ 0 & 0 & 0 & -3 \end{bmatrix}. \tag{5.28}$$

The local dynamic behavior of system (5.27) associated with the double-zero singularity can be studied using the normal form of the system. Now apply the Maple program for computing the normal form associated with a double-zero singularity [231] (see Sect. 2.2.2) to obtain the normal form with unfolding as follows:

$$\frac{du_1}{d\tau} = u_2,$$

$$\frac{du_2}{d\tau} = -\left(\frac{6}{11}\mu_1 + \frac{10}{33}\mu_2\right)u_1 + \left(\frac{1}{7}\mu_1 + \frac{4}{21}\mu_2\right)u_2 - \frac{5108}{1617}u_1^3 + \frac{32411}{1617}u_1^2 u_2. \tag{5.29}$$

Now, based on (5.29), we can perform analysis on the local dynamical behavior of the original system (5.27). First, it is easy to find two equilibrium solutions:

$$\text{(I)} \quad u_1 = u_2 = 0, \tag{5.30}$$

$$\text{(II)} \quad u_1 = 7\left[-\frac{1}{2544}(9\mu_1 + 5\mu_2)\right]^{1/2}, \tag{5.31}$$

$$u_2 = 0, \quad \text{for} \quad 9\mu_1 + 5\mu_2 < 0.$$

The stability of the equilibrium solutions is based on the Jacobian matrix of system (5.29), given by

$$J = \begin{bmatrix} 0 & 1 \\ -\left(\frac{6}{11}\mu_1 + \frac{10}{33}\mu_2\right) - \frac{5108}{539}u_1^2 + \frac{64822}{1617}u_1u_2 & \frac{1}{7}\mu_1 + \frac{4}{21}\mu_2 + \frac{32411}{1617}u_1^2 \end{bmatrix}. \tag{5.32}$$

Substituting equilibrium solution (I) into the Jacobian (5.32) results in the following stability condition: the equilibrium solution (I) is asymptotically stable if

$$9\mu_1 + 5\mu_2 > 0 \quad \text{and} \quad 3\mu_1 + 4\mu_2 < 0, \tag{5.33}$$

which leads to two critical bifurcation lines. One of these is

$$L_1: \quad 9\mu_1 + 5\mu_2 = 0 \quad (3\mu_1 + 4\mu_2 < 0), \tag{5.34}$$

along which the equilibrium solution (I) becomes unstable, leading to bifurcation of
the nonzero equilibrium solution (II). It is noted that this special case is a so-called
pitchfork bifurcation (see the solution form (5.31)).

On the other hand, the second critical line,

$$L_2: \quad 3\mu_1 + 4\mu_2 = 0 \quad (9\mu_1 + 5\mu_2 > 0), \tag{5.35}$$

describes a dynamic boundary where the equilibrium solution (I) loses its stability
and bifurcates into a family of limit cycles. By using Hopf bifurcation theory and
the associated normal form, we obtain the normal form for this case up to third order
as

$$\frac{dr}{d\tau} = \frac{1}{42}(3\mu_1 + 4\mu_2)r + \frac{32411}{12936}r^3,$$

$$\frac{d\theta}{d\tau} = \omega_c + \frac{1}{33}(9\mu_1 + 5\mu_2) + \frac{1277}{1078\omega_c}r^2, \quad \omega_c = \sqrt{\frac{7}{22}\mu_1}, \quad \mu_1 > 0, \tag{5.36}$$

where μ_1 and μ_2 are taken around the critical line L_2 such that $3\mu_1 + 4\mu_2 > 0$. The
first equation in (5.36) indicates that the family of limit cycles bifurcating from the
critical line L_2 is unstable since the coefficient of r^3 is positive.

Now substituting the equilibrium solution (II) into the Jacobian (5.32) yields the
following result: the equilibrium solution (II) is stable if

$$9\mu_1 + 5\mu_2 < 0 \quad \text{and} \quad 1957611\mu_1 + 1022009\mu_2 > 0, \tag{5.37}$$

which yields the third critical line:

$$L_3: \quad 1957611\mu_1 + 1022009\mu_2 = 0 \quad (9\mu_1 + 5\mu_2 < 0), \tag{5.38}$$

along which the equilibrium solution (II) becomes unstable, leading to the bifurcation of another family of limit cycles, usually called *secondary* Hopf bifurcations.
Similarly, we can execute the Maple program to obtain the normal form for this case
up to third order as

$$\frac{dr}{d\tau} = -\frac{1}{589974}(1957611\mu_1 + 1022099\mu_2)r - \frac{1}{6468}\left(32411 + \frac{7662}{\omega_c}\right)r^3,$$

$$\frac{d\theta}{d\tau} = \omega_c - \frac{2}{33}(9\mu_1 + 5\mu_2)$$

$$- \frac{1}{99115632}\left(\frac{587062440}{\omega_c} + 1050472921\omega_c\right)r^2 \tag{5.39}$$

$$\left(\omega_c = \sqrt{\frac{71512}{1022009}\mu_1}, \quad \mu_1 > 0\right),$$

Fig. 5.5 Bifurcation diagram for a double-zero singularity: ES (initial equilibrium solution (I)); SB (static bifurcation solution (II)); HB (Hopf bifurcation solution); 2nd HB (secondary Hopf bifurcation solution)

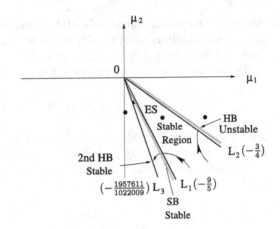

where μ_1 and μ_2 are taken around the critical line L_3 such that $1957611\mu_1 + 1022009\mu_2 < 0$. It follows from the first equation in (5.39) that the family of limit cycles bifurcating from the critical line L_3 is stable since the coefficient of r^3 is negative.

The critical bifurcation lines are illustrated in the parameter space shown in Fig. 5.5, where the numbers given in the brackets denote the slopes of the critical lines. The stable region for the ES is bounded by the critical lines L_1 and L_2. The SB solution branching off the critical line L_1 is stable in the region bounded by the critical lines L_1 and L_3, while the periodic solution bifurcating from the critical line L_2 is unstable. The 2nd HB solution is stable.

The numerical simulation results are shown in Fig. 5.6, where Fig. 5.6(a) shows a trajectory converging to the zero equilibrium solution (I), Fig 5.6(b) shows a trajectory converging to the nonzero equilibrium solution (II), and Figs 5.6(c) and (d) show trajectories converging to stable limit cycles. The physical explanation of the steady-state solutions obtained above is as follows. First, it can be seen from Fig. 5.3 that the initial equilibrium (I) $z_i = 0$ (i.e., $\theta_1 = \theta_2 = 0$, indicating that the two rigid links are both in the vertical direction) exists for all parameter values since both P_1 and P_2 are in the vertical direction. This equilibrium solution (I) is stable if the parameter values of (μ_1, μ_2) are chosen from the stable region bounded by the critical lines L_1 and L_2 (see Fig. 5.5). When the parameter values are chosen from the region bounded by the critical lines L_1 and L_3, a small perturbation will cause the initial equilibrium to move to a nontrivial stable equilibrium (II), which is either on the left or right side of the vertical position, depending upon the perturbation. Further, when the parameters are varied so that the critical boundary L_3 is crossed, the SB solution (II) becomes unstable and a periodic vibration is initiated. The center of the vibration is located either on the left or on the right side of the vertical position, again depending on the initial perturbation.

To end the section, we briefly present together two more singularities: Hopf-zero and double-Hopf, showing quasiperiodic solutions on two- and three-dimensional tori. The detailed analysis can be found in [231].

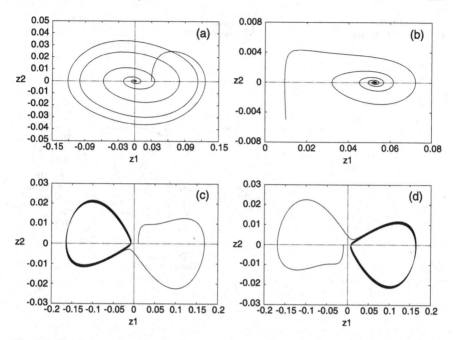

Fig. 5.6 Simulation trajectories for the case of a double-zero singularity: (**a**) converging to the zero equilibrium solution (I) for $(\mu_1, \mu_2) = (0.18, -0.18)$; (**b**) converging to the nonzero equilibrium solution (II) for $(\mu_1, \mu_2) = (0.03, -0.09)$; (**c**) and (**d**) converging to stable limit cycles for $(\mu_1, \mu_2) = (0.05, -0.18)$

5.2.3 Hopf-Zero and Double-Hopf Singularities

Hopf-Zero Singularity   The critical point is determined by

$$f_1 = f_2 = \frac{35}{12}, \qquad f_3 = \frac{35}{6}, \qquad f_4 = -\frac{37}{12} + \mu_1, \qquad f_5 = \frac{91}{12} + \mu_2,$$

$$f_6 = f_7 = 0, \qquad \eta_1 = \eta_2 = 1, \qquad \mu_1 = \mu_2 = 0, \tag{5.40}$$

and the associated eigenvalues are $0, \pm\frac{1}{3}\sqrt{6}i, -\frac{7}{2}$. The normal form for the Hopf-zero singularity can be obtained using the Maple program in [21] as

$$\frac{dy}{d\tau} = y\left[\left(\frac{7}{8}\mu_1 + \frac{23}{56}\mu_2\right) - \frac{635}{1152}y^2 - \frac{1891}{192}\rho^2\right],$$

$$\frac{d\rho}{d\tau} = \rho\left[\left(-\frac{693}{2480}\mu_1 + \frac{33}{496}\mu_2\right) + \frac{175371}{39680}y^2 - \frac{230477}{39680}\rho^2\right], \tag{5.41}$$

$$\frac{d\theta}{d\tau} = \frac{1}{3}\sqrt{6}\left[1 + \frac{99}{1240}\mu_1 - \frac{111}{248}\mu_2 - \frac{222523}{19840}y^2 - \frac{373219}{19840}\rho^2\right].$$

The bifurcation diagram is shown in Fig 5.7, and numerical simulation results for this case are depicted in Fig. 5.8.

Fig. 5.7 Bifurcation diagram
for Hopf-zero singularity

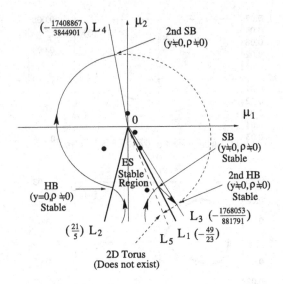

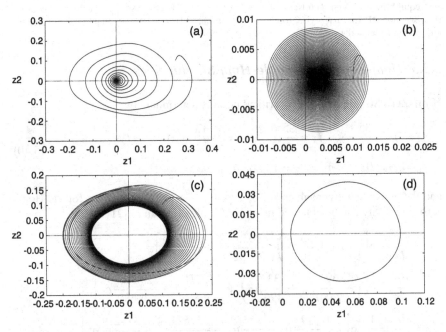

Fig. 5.8 Simulation trajectories for the case of a Hopf-zero singularity: (**a**) converging to the zero ES for $(\mu_1, \mu_2) = (0.1, -0.3)$; (**b**) converging to the nonzero SB for $(\mu_1, \mu_2) = (0.01, -0.0205)$; (**c**) converging to a stable limit cycle from HB for $(\mu_1, \mu_2) = (-0.1, -0.1)$; and (**d**) converging to a stable limit cycle from 2nd HB for $(\mu_1, \mu_2) = (0.02, -0.03)$

Double-Hopf Singularity The critical point is given by

$$f_1 = \frac{4}{7}, \qquad f_2 = \frac{407}{56}, \qquad f_3 = \frac{1}{56}, \qquad f_4 = \frac{535}{28},$$

$$f_5 = f_6 = 0, \qquad f_7 = -1\eta_1 = \mu_1, \qquad \eta_2 = \mu_2, \qquad \mu_1 = \mu_2 = 0,$$

(5.42)

and the eigenvalues of the Jacobian of the system evaluated at the critical point are $\pm i, \pm\sqrt{2}i$. The normal form for this case is obtained as (e.g., using the Maple program in [231])

$$\frac{d\rho_1}{d\tau} = \rho_1\left[-\frac{11}{7}\mu_1 + \frac{75}{56}\mu_2 - \frac{625}{10976}\rho_1^2 - \frac{2025}{5488}\rho_2^2\right],$$

$$\frac{d\rho_2}{d\tau} = \rho_2\left[\frac{37}{28}\mu_1 - \frac{159}{56}\mu_2 + \frac{1325}{5488}\rho_1^2 + \frac{4293}{10976}\rho_2^2\right],$$

$$\frac{d\theta_1}{d\tau} = 1 - \frac{6493901}{22127616}\rho_1^2 - \frac{1423069}{2458624}\rho_2^2,$$

$$\frac{d\theta_2}{d\tau} = \sqrt{2}\left[1 + \frac{14046397}{22127616}\rho_1^2 + \frac{2975501}{9834496}\rho_2^2\right].$$

(5.43)

The bifurcation diagram for this case is shown in Fig 5.9 and some numerical simulation results are shown in Fig. 5.10.

5.3 Induction Machine Model

A model of an induction machine is used to demonstrate the application of the theorems obtained in the previous section. The model is based on the one discussed in [119] and the same notation is adopted here. Since here we are mainly interested in the application of focus value computation, we will not give a detailed derivation of the model.

Induction machines (or asynchronous machines) are widely used in industrial applications. The behavior of induction machines has been studied for years, but the main attention has been focused on steady-state solutions, due to the complexity of the model (even with simplifying assumptions). In order to study the dynamical behavior of the model, such as instability and bifurcations, it is necessary to determine the conditions of the bifurcation (critical) points.

The model is described by a system of seven ordinary differential equations, given as follows:

$$\dot{\phi}_{qs} = \omega_b\left\{u_q - \phi_{ds} + \frac{r_s}{X_{1s}}\left[X_{aq}\left(\frac{\phi_{qs}}{X_{1s}} + \frac{\phi_{qr}'}{X_{1r}'}\right) - \phi_{qs}\right]\right\},$$

$$\dot{\phi}_{ds} = \omega_b\left\{u_d + \phi_{qs} + \frac{r_s}{X_{1s}}\left[X_{aq}\left(\frac{\phi_{ds}}{X_{1s}} + \frac{\phi_{dr}'}{X_{1r}'}\right) - \phi_{ds}\right]\right\},$$

Fig. 5.9 Bifurcation diagram
for double-Hopf singularity

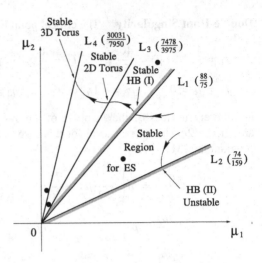

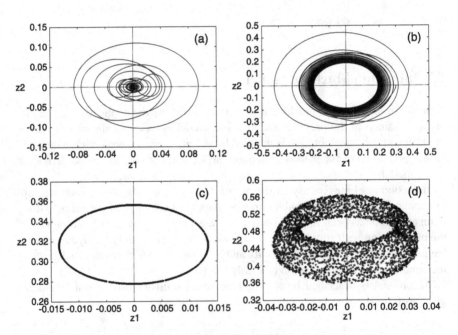

Fig. 5.10 Simulation trajectories for the case of a double-Hopf singularity: (**a**) converging to
the zero ES for $(\mu_1, \mu_2) = (0.05, 0.04)$; (**b**) converging to a stable limit cycle from HB for
$(\mu_1, \mu_2) = (0.05, 0.095)$; (**c**) converging to a quasiperiodic solution on a 2D torus (with a projected
1D Poincaré map as shown) for $(\mu_1, \mu_2) = (0.005, 0.015)$; and (**d**) converging to a quasiperiodic
solution on a 3D torus (with a projected 2D Poincaré map as shown) for $(\mu_1, \mu_2) = (0.005, 0.03)$

$$\dot{\phi}_{0s} = \omega_b \left\{ \frac{r_s}{X_{1s}} (-\phi_{0s}) \right\},$$

$$\dot{\phi}'_{qr} = \omega_b \left\{ -(1 - \omega_r)\phi'_{dr} + \frac{r'_r}{X'_{1r}} \left[X_{aq} \left(\frac{\phi_{qs}}{X_{1s}} + \frac{\phi'_{qr}}{X'_{1r}} \right) - \phi'_{qr} \right] \right\}, \qquad (5.44)$$

$$\dot{\phi}'_{dr} = \omega_b \left\{ (1 - \omega_r)\phi'_{qr} + \frac{r'_r}{X'_{1r}} \left[X_{aq} \left(\frac{\phi_{ds}}{X_{1s}} + \frac{\phi'_{dr}}{X'_{1r}} \right) - \phi'_{dr} \right] \right\},$$

$$\dot{\phi}'_{0r} = \omega_b \left\{ \frac{r'_r}{X'_{1r}} (-\phi'_{0r}) \right\},$$

$$\dot{\omega}'_r = \frac{1}{2H} \left\{ \frac{X_{ad}}{X_{1s} X'_{1r}} (\phi_{qs}\phi'_{dr} - \phi_{ds}\phi'_{qr}) - T_L \right\},$$

where, apart from the state variables, all the variables are system parameters. Letting

$$w_1 = \phi_{qs}, \qquad w_2 = \phi_{ds}, \qquad w_3 = \phi_{0s}, \qquad w_4 = \phi'_{qr},$$

$$w_5 = \phi'_{dr}, \qquad w_6 = \phi'_{0r}, \qquad w_7 = \omega_r,$$

and substituting proper parameter values into (5.44) yields a model of a 3 hp induction machine, $\frac{d\mathbf{w}}{dt} = \mathbf{f}(\mathbf{w}, V)$; namely,

$$\dot{w}_1 = -\frac{3}{10}w_1 - w_2 + \frac{3}{10}w_4 + V,$$

$$\dot{w}_2 = w_1 - \frac{3}{10}w_2 + \frac{3}{10}w_5,$$

$$\dot{w}_3 = -\frac{3}{5}w_3,$$

$$\dot{w}_4 = \frac{1}{2}w_1 - \frac{1}{2}w_4 - w_5 + w_5 w_7, \qquad (5.45)$$

$$\dot{w}_5 = \frac{1}{2}w_2 + w_4 - \frac{1}{2}w_5 - w_4 w_7,$$

$$\dot{w}_6 = -w_6,$$

$$\dot{w}_7 = \frac{7}{120\pi^3} (14 w_1 w_5 - 14 w_2 w_4 - 1),$$

where $V > 0$ is a bifurcation parameter representing the input voltage of the motor.

Setting $\frac{dw_i}{dt} = 0, i = 1, 2, \ldots, 7$, results in two equilibrium solutions (fixed points):

$$w_1^{\pm} = -\frac{3(-350V^2 + 15 \pm 10S)}{7630V}, \qquad w_2^{\pm} = \frac{7315V^2 - 150 \pm 9S}{7360V},$$

$$w_3^{\pm} = 0, \qquad w_4^{\pm} = -\frac{1}{14V}, \qquad w_5^{\pm} = \frac{35V^2 \pm S}{70V}, \qquad (5.46)$$

$$w_6^{\pm} = 0, \qquad w_7^{\pm} = \frac{-350V^2 + 124 \pm 10S}{109},$$

where $S = \sqrt{1225V^4 - 105V^2 - 25}$, indicating that the equilibrium solutions exist when $V^2 \geq \frac{3+\sqrt{109}}{70}$. The Jacobian of (5.45) is given by

$$J(\boldsymbol{w}) = \begin{bmatrix} -\frac{3}{10} & -1 & 0 & \frac{3}{10} & 0 & 0 & 0 \\ 1 & -\frac{3}{10} & 0 & 0 & \frac{3}{10} & 0 & 0 \\ 0 & 0 & -\frac{3}{5} & 0 & 0 & 0 & 0 \\ \frac{1}{2} & 0 & 0 & -\frac{1}{2} & -1+w_7 & 0 & w_5 \\ 0 & \frac{1}{2} & 0 & 1-w_7 & -\frac{1}{2} & 0 & -w_4 \\ 0 & 0 & 0 & 0 & 0 & -1 & 0 \\ \frac{49w_5}{60\pi^3} & -\frac{49w_4}{60\pi^3} & 0 & -\frac{49w_2}{60\pi^3} & \frac{49w_1}{60\pi^3} & 0 & 0 \end{bmatrix}. \qquad (5.47)$$

The conditions for the various singularities of the system have been obtained in [227], but the stability of the bifurcating limit cycles was not discussed. Here, we consider the stability of the limit cycles generated from a Hopf bifurcation. To achieve this, the characteristic polynomial of $J(\boldsymbol{w})$ is obtained as

$$P_7(\lambda) = \lambda^7 + \sum_{i=1}^{7} a_i \lambda^{7-i},$$

where

$$a_1 = \frac{16}{5},$$

$$a_2 = \frac{1}{60\pi^3}\left[49(w_1w_4 + w_2w_5) + 12(29 - 10w_7 + 5w_7^2)\pi^3\right],$$

$$a_3 = -\frac{1}{3000\pi^3}\Big[735\left(w_4^2 + w_5^2\right) - 6615(w_1w_4 + w_2w_5)$$

$$- 2450(w_1w_5 - w_2w_4)(w_7 - 1) - (20352 - 13200w_7 + 6600w_7^2)\pi^3\Big],$$

$$a_4 = -\frac{1}{1500\pi^3}\Big[882\left(w_4^2 + w_5^2\right) - 4410(w_1w_4 + w_2w_5)$$

$$- 2695(w_1w_5 - w_2w_4)(w_7 - 1) - (8676 - 8400w_7 + 3975w_7^2)\pi^3\Big],$$

$$a_5 = -\frac{1}{30000\pi^3}\big[(6468 + 7350w_7)(w_4^2 + w_5^2)$$
$$+ (68600 - 64925w_7)(w_1w_5 - w_2w_4) - 77028(w_1w_4 + w_2w_5)$$
$$- (107520 - 140640w_7 + 63120w_7^2)\pi^3\big],$$

$$a_6 = -\frac{1}{7500\pi^3}\big[-(2058 - 2940w_7)(w_4^2 + w_5^2)$$
$$+ (14357 - 12887w_7)(w_1w_5 - w_2w_4) - 9457(w_1w_4 + w_2w_5)$$
$$- (7380 - 11160w_7 + 4905w_7^2)\pi^3\big],$$

$$a_7 = -\frac{49}{10000\pi^3}\big[30(w_4^2 + w_5^2)(w_7 - 1) + (124 - 109w_7)(w_1w_5 - w_2w_4)$$
$$- 50(w_1w_4 + w_2w_5)\big].$$

$$(5.48)$$

For Hopf bifurcation, the Jacobian (5.47) needs one pair of purely imaginary eigenvalues, which requires [227] that $V \geq V_0 = ((3 + \sqrt{109})/70)^{1/2} \approx 0.4381830425$. Since w^- is always unstable when $V > V_0$, we consider w^+. It can be shown that w^+ is stable when $0.4381830425 < V < 6.2395593195$ or $V > 7.75369242394$; unstable when $6.2395593195 < V < 7.35369242394$. The point $V_0 = 0.4381830425$ is a static critical point. Furthermore, we employ the criterion given in [227] to show that

$$V_h^1 = 6.2395593195 \quad \text{and} \quad V_h^2 = 7.35369242394, \qquad (5.49)$$

are two solutions at which Hopf bifurcations occur. To achieve this, we first state the criterion [227] for determining the Hopf critical point of a general n-dimensional system. Suppose the characteristic polynomial of the system evaluated at a fixed point is given by

$$P_n(\lambda) = \lambda^n + a_1\lambda^{n-1} + a_2\lambda^{n-2} + \cdots + a_{n-2}\lambda^2 + a_{n-1}\lambda + a_n, \qquad (5.50)$$

where the coefficients a_i are expressed in terms of the original system. Then the so-called "Hurwitz principal minors" are:

$$\Delta_1 = \det[a_1] = a_1,$$

$$\Delta_2 = \det\begin{bmatrix} a_1 & 1 \\ a_3 & a_2 \end{bmatrix} = a_1a_2 - a_3,$$

$$\Delta_3 = \det\begin{bmatrix} a_1 & 1 & 0 \\ a_3 & a_2 & a_1 \\ a_5 & a_4 & a_3 \end{bmatrix} = (a_1a_2 - a_3)a_3 - (a_1a_4 - a_5)a_1,$$

$$\vdots$$

$$\Delta_{n-1} = \det \begin{bmatrix} a_1 & 1 & 0 & 0 & \cdots & 0 & 0 & 0 & 0 \\ a_3 & a_2 & a_1 & 1 & \cdots & 0 & 0 & 0 & 0 \\ a_5 & a_4 & a_3 & a_2 & \cdots & 0 & 0 & 0 & 0 \\ \vdots & \vdots & \vdots & \vdots & \ddots & \vdots & \vdots & \vdots & \vdots \\ 0 & 0 & 0 & 0 & \cdots & a_{n-2} & a_{n-3} & a_{n-4} & a_{n-5} \\ 0 & 0 & 0 & 0 & \cdots & a_n & a_{n-1} & a_{n-2} & a_{n-3} \\ 0 & 0 & 0 & 0 & \cdots & 0 & 0 & a_n & a_{n-1} \end{bmatrix},$$

$$\Delta_n = \det \begin{bmatrix} a_1 & 1 & 0 & 0 & \cdots & 0 & 0 & 0 & 0 & 0 \\ a_3 & a_2 & a_1 & 1 & \cdots & 0 & 0 & 0 & 0 & 0 \\ a_5 & a_4 & a_3 & a_2 & \cdots & 0 & 0 & 0 & 0 & 0 \\ \vdots & \vdots & \vdots & \vdots & \ddots & \vdots & \vdots & \vdots & \vdots & \vdots \\ 0 & 0 & 0 & 0 & \cdots & a_n & a_{n-1} & a_{n-2} & a_{n-3} & a_{n-4} \\ 0 & 0 & 0 & 0 & \cdots & 0 & 0 & a_n & a_{n-1} & a_{n-2} \\ 0 & 0 & 0 & 0 & \cdots & 0 & 0 & 0 & 0 & a_n \end{bmatrix}$$

$$= a_n \Delta_{n-1}.$$

Then we have the following result (the proof can be found in [227]).

Theorem 5.1 *The necessary and sufficient condition for system (2.1) to have a Hopf bifurcation at a fixed point of the system is*

$$\Delta_{n-1} = 0. \tag{5.51}$$

For the induction machine example, the Hopf critical point is given by the condition $\Delta_6 = 0$, where

$$\Delta_6 = Q_1(V^2) - Q_2(V^2)\sqrt{1225V^4 - 105V^2 - 25}, \tag{5.52}$$

in which $Q_1(V^2)$ and $Q_2(V^2)$ are polynomials in V^2. Thus we have a polynomial equation in the form

$$Q_1^2(V^2) - Q_2^2(V^2)(1225V^4 - 105V^2 - 25) = 0,$$

which is a 21st-degree polynomial in V^2. One may apply the built-in Maple solver `fsolve` in Maple to solve this polynomial to find all three real solutions:

$$V^2 = 38.9321005020, \quad 54.0767922650 \quad \text{and} \quad -26.6127343788.$$

Taking positive values yields

$$V_h^1 = 6.2395593195 \quad \text{or} \quad V_h^2 = 7.3536924239. \tag{5.53}$$

These two solutions satisfy (5.52) and thus are true Hopf critical points, and are the only Hopf critical points. It can be shown [227] that at these two critical points, except for $\Delta_6 = 0$, all other Hurwitz equalities still remain positive. Thus, at these two Hopf critical points, the characteristic polynomial has a purely imaginary pair and the remaining five eigenvalues still have negative parts.

When the critical point is chosen as $V = V_h^1 = 6.2395593195$, the eigenvalues of $J(q)$ are

$$\pm 0.7905733366i, \qquad -1, \qquad -0.6, \qquad -0.5630004665,$$
$$-0.5184997667 \pm 1.0893171380i,$$

at which the equilibrium solution w^+ becomes (see (5.46))

$$w_1^+ = 0.0000063079, \qquad w_2^+ = 6.2361231187, \qquad w_4^+ = -0.0114476949,$$
$$w_5^+ = 6.2361020924, \qquad w_7^+ = 0.9990816376, \qquad w_3^+ = w_6^+ = 0.$$
$$(5.54)$$

When the critical point is taken as $V = V_h^2 = 7.3536924239$, w^+ is given by

$$w_1^+ = 0.0000038521, \qquad w_2^+ = 7.3507772803, \qquad w_4^+ = -0.0097132933,$$
$$w_5^+ = 7.3507644401, \qquad w_7^+ = 0.9993390384, \qquad w_3^+ = w_6^+ = 0.$$
$$(5.55)$$

In the following, we choose the Hopf critical point $V_h^1 = 6.2395593195$ and carry out a Hopf bifurcation analysis. First, introduce the transformations

$$w = w^+ + Tx \quad \text{and} \quad V = V_h^1 + \mu, \tag{5.56}$$

where w^+ is given in (5.54), μ is a bifurcation parameter, and T is given by

$$T = \begin{bmatrix} 0.108625 & 0.467959 & 0 & 0 & -0.343510 & -0.060050 & 0.418545 \\ 0.505844 & -0.000735 & 0 & 0 & -0.162471 & 0.339091 & -0.010442 \\ 0 & 0 & 0 & 1 & 0 & 0 & 0 \\ 0.561587 & 0.751762 & 0 & 0 & -0.240425 & -0.345722 & -0.557693 \\ 0.145698 & -0.227574 & 0 & 0 & 1.287467 & -0.008888 & -0.156286 \\ 0 & 0 & 1 & 0 & 0 & 0 & 0 \\ -0.058964 & 0.093916 & 0 & 0 & 0.030161 & 0.103256 & -0.092317 \end{bmatrix}$$

under which system (5.45) is transformed into a new system such that its linear part is in Jordan canonical form:

$$\frac{dx}{dt} = Jx + f_2(x, \mu), \tag{5.57}$$

where J is

$$
J = \begin{bmatrix}
0 & 0.790573 & 0 & 0 & 0 & 0 & 0 \\
-0.790573 & 0 & 0 & 0 & 0 & 0 & 0 \\
0 & 0 & -1 & 0 & 0 & 0 & 0 \\
0 & 0 & 0 & -0.6 & 0 & 0 & 0 \\
0 & 0 & 0 & 0 & -0.563000 & 0 & 0 \\
0 & 0 & 0 & 0 & 0 & -0.518500 & 1.089317 \\
0 & 0 & 0 & 0 & 0 & -1.089317 & -0.518500
\end{bmatrix},
$$

and $f_2(x, \mu)$ is given below:

$$
\begin{aligned}
\dot{x}_1 = {} & 0.7905733368x_2 + 0.0088303403x_1\mu + 0.0535254447x_2\mu + \cdots \\
& + 0.0246142803x_1^2 - 0.0215006434x_2^2 + 0.0487837274x_5^2 \\
& + 0.0022408473x_6^2 - 0.0050411213x_7^2 + 0.0301172995x_1x_2 \\
& - 0.0558143814x_1x_5 - 0.0156196278x_1x_6 - 0.0183055538x_1x_7 \\
& - 0.0120371644x_2x_5 - 0.0057118446x_2x_6 + 0.0489610685x_2x_7 \\
& + 0.0624660631x_5x_6 - 0.0806260388x_5x_7 - 0.0109414070x_6x_7,
\end{aligned}
$$

$$
\begin{aligned}
\dot{x}_2 = {} & -0.7905733368x_1 - 0.0538321687x_1\mu + 0.0085082244x_2\mu + \cdots \\
& - 0.0147573718x_1^2 - 0.0332545864x_2^2 - 0.0224230083x_5^2 \\
& + 0.0179395818x_6^2 - 0.0140895136x_7^2 - 0.0191665298x_1x_2 \\
& - 0.0079820592x_1x_5 - 0.0171471036x_1x_6 + 0.0289417583x_1x_7 \\
& + 0.0982747938x_2x_5 - 0.0383337242x_2x_6 + 0.0211820356x_2x_7 \\
& + 0.0482387594x_5x_6 - 0.0030914700x_5x_7 + 0.0163142126x_6x_7,
\end{aligned}
$$

$$
\dot{x}_3 = -x_3,
$$

$$
\dot{x}_4 = -0.6x_4,
$$

$$
\begin{aligned}
\dot{x}_5 = {} & -0.5630004665x_5 + \cdots \\
& + 0.0232681695x_1^2 - 0.0588688619x_2^2 - 0.0046338145x_5^2 \\
& + 0.0315782579x_6^2 - 0.0446396137x_7^2 - 0.0143771854x_1x_2 \\
& - 0.0161802581x_1x_5 - 0.0661356681x_1x_6 + 0.0216321813x_1x_7 \\
& + 0.0068297420x_2x_5 - 0.0400658632x_2x_6 + 0.0974741740x_2x_7 \\
& + 0.0202663375x_5x_6 + 0.0116194327x_5x_7 + 0.0254244420x_6x_7,
\end{aligned}
$$

$$
\begin{aligned}
\dot{x}_6 = {} & -0.5184997668x_6 + 1.0893171378x_7 + \cdots \\
& - 0.0248122565x_1^2 + 0.0036406838x_2^2 - 0.0751095676x_5^2 \\
& + 0.0120420948x_6^2 - 0.0145660213x_7^2 - 0.0520321981x_1x_2
\end{aligned}
$$

(5.58)

$$+ 0.0761405472x_1x_5 - 0.0094222846x_1x_6 + 0.0375977666x_1x_7$$
$$+ 0.0184084401x_2x_5 - 0.0104526601x_2x_6 - 0.0250568650x_2x_7$$
$$- 0.0853951307x_5x_6 + 0.1274433979x_5x_7 + 0.0288348791x_6x_7,$$
$$\dot{x}_7 = -1.0893171378x_6 - 0.5184997668x_7 + \cdots$$
$$+ 0.0256483680x_1^2 - 0.0050320634x_2^2 - 0.0021698205x_5^2$$
$$+ 0.0070056516x_6^2 - 0.0216653968x_7^2 - 0.0056519111x_1x_2$$
$$+ 0.0210545013x_1x_5 - 0.0324057372x_1x_6 - 0.0044594882x_1x_7$$
$$- 0.0985067508x_2x_5 + 0.0099591246x_2x_6 + 0.0400147914x_2x_7$$
$$- 0.0657644904x_5x_6 + 0.0522023474x_5x_7 + 0.0096028248x_6x_7,$$

where "$\cdots$" denotes the other terms $x_i \mu$ which are not used for the normal form.
Now, applying the Maple program [218] results in the following normal form:

$$\dot{r} = r\left(0.0086692824\mu - 0.2245633437r^2 + \cdots\right),$$
$$\dot{\theta} = 0.7905733366 + 0.0536788067\mu - 0.0002514156r^2 + \cdots, \tag{5.59}$$

which implies that the family of limit cycles bifurcating from the critical point V_h^1 in the neighborhood of w^+ are stable, and the estimates of the amplitude and frequency of the limit cycles are given by

$$r = 0.1964817158\sqrt{\mu},$$
$$\omega = 0.7905733366 + 0.0536691008\mu. \tag{5.60}$$

Simulation results for this example using system (5.45) with $V = 6.0$ and $V = 6.5$ are depicted in Fig. 5.11. The initial point is chosen as $w_0 = (0.5, 5.5, 2.0,$

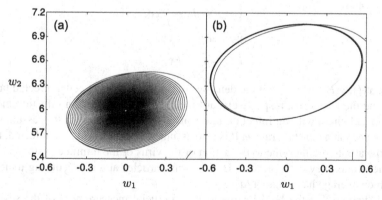

Fig. 5.11 Simulated trajectories of the induction machine model (5.45) projected on the w_1–w_2 plane with initial point $w_0 = (0.5, 5.5, 2.0, -3.0, 1.0, 4.0, -5.0)^T$: (a) convergent to the stable equilibrium point $w^+ = (0.000007, 5.996426, 0, -0.011905, 5.996403, 0, 0.999007)^T$ when $V = 6.0$; and (b) convergent to a stable limit cycle when $V = 6.5$

$-3.0, 1.0, 4.0, -5.0)^T$. It can be seen from this figure, as expected, that when $V = 6.0 < V_h^1$ the trajectory converges to the stable equilibrium point:

$$\boldsymbol{w}^+ = (0.000007, 5.996426, 0, -0.011905, 5.996403, 0, 0.999007)^T,$$

as shown in Fig. 5.11(a). When $V = 6.5 > V_h^1$, i.e., $\mu = 0.5$, the equilibrium point becomes unstable, and a supercritical Hopf bifurcation occurs, giving rise to a stable limit cycle (see Fig. 5.11(b)). Note that the estimation given in (5.60) is based on the normal form of normalized system (5.58) while the simulation is based on the original system (5.45). There is a transformation (5.56) between these two systems.

5.4 An HIV-1 Model

In this section, we consider an HIV-1 model, with attention focused on the study of Hopf bifurcation. The results presented in this section are based on the paper by Jiang, Yuan, Yu and Zou [113].

In recent years, mathematical modeling has contributed greatly to the understanding of HIV-1 infection in a host and has provided valuable insight into HIV-1 pathogenesis. Among various models is the class using differential equations which quantitatively describe the dynamics of the HIV-1 virus, healthy and infected cells, and possibly even the immune response. By studying these models, researchers have gained much knowledge about the mechanism of the interactions of these components within a host, and have thereby enhanced the progress in understanding the HIV-1 infection (e.g., see [172, 175–177]). Such understanding may in turn offer guidance for developing new drugs and for designing an optimal combination of existing therapies (e.g., see [117, 171, 181] and the references therein).

A standard and classical differential equation model for HIV infection is the following system of ordinary differential equations (e.g., see [171, 172, 176]):

$$\dot{x} = \lambda - dx - \beta x v,$$
$$\dot{y} = -ay + \beta x v, \qquad\qquad (5.61)$$
$$\dot{v} = -pv + ky,$$

where $x(t)$, $y(t)$ and $v(t)$ are the densities of uninfected target cells, infected target cells and the free virus, respectively, at time t. Here a mass action infection mechanism is adopted, with an infection rate constant β. The healthy cell is assumed to be produced at a constant rate λ. It is also assumed that once cells are infected, they may die at rate a either due to the action of the virus or the immune system, and in the meantime, they each produce HIV-1 virus particles at a rate k during their life which on average has length $1/a$.

It is known that the HIV-1 virus load is a crucial measurement of the severity of an HIV-1 carrier. When the load exceeds a certain level after a clinically latent phase, the CD^+ T cell count declines drastically, indicating a transition from HIV to AIDS (e.g., see [110, 192]). Most of the existing therapies for HIV and/or AIDS use inhibitors of the enzymes required for replication of the HIV-1 virus to reduce the load (e.g., see [73, 126]). Recent progress in genetic engineering has offered an alternative approach: modification of a viral genome can produce recombinants capable of controlling infections by other viruses. Indeed, this method has been used to modify rhabdovirus, including the rabies and the vesicular stomatitis viruses, making them capable of infecting and killing cells previously attacked by HIV-1; for details see, e.g., [163, 170, 188, 198]. To understand this approach of fighting a virus with a genetically modified virus, Revilla and Garcia-Ramos [181] proposed a mathematical model which is a result of incorporating into (5.61) two more variables: the density w of the recombinant (genetically modified) virus and the density z of doubly-infected cells.

$$
\begin{aligned}
\dot{x} &= \lambda - dx - \beta x v, \\
\dot{y} &= -ay + \beta x v - \alpha w y, \\
\dot{z} &= -bz + \alpha w y, \\
\dot{v} &= -pv + ky, \\
\dot{w} &= -qw + cz.
\end{aligned}
\tag{5.62}
$$

Here it is assumed that: (i) the recombinant infects cells previously infected by the pathogen, turning them at a rate $\alpha w y$ into doubly-infected cells, and in the meantime, recombinants are removed at a rate qw; and (ii) the doubly-infected cells die at a rate of bz and release recombinants at a rate cz.

In [181], the authors only analyzed the structure of equilibria of system (5.62), and performed some numerical simulations. System (5.62) has a *dimension higher than 2*, and it is well known that for systems with higher dimensions, equilibria cannot determine the long-term behavior of solutions and complicated dynamics (periodic solutions and even chaos) may occur which would make the system unpredictable. Therefore, theoretically determining the global dynamics of (5.62) is an important yet challenging problem.

5.4.1 Equilibrium and Stability Analysis

Here, we focus on Hopf bifurcation. For convenience in the analysis, we first do the following rescalings to reduce the number of parameters:

$$
\begin{aligned}
x &\to \mu_1 x, & y &\to \mu_2 y, & z &\to \mu_3 z, \\
v &\to \mu_4 v, & w &\to \mu_5 w, & \tau &= vt
\end{aligned}
\tag{5.63}
$$

$$
\begin{aligned}
\frac{d}{v} &\to d, & \frac{a}{v} &\to a, & \frac{b}{v} &\to b, \\
\frac{p}{v} &\to p, & \frac{q}{v} &\to q, & \frac{\alpha c}{k\beta} &\to c,
\end{aligned}
\tag{5.64}
$$

where

$$v = (\lambda k \beta)^{1/3}, \qquad \mu_1 = \mu_2 = \mu_3 = \frac{v^2}{k\beta}, \qquad \mu_4 = \frac{v}{\beta}, \qquad \mu_5 = \frac{v}{\alpha}. \qquad (5.65)$$

By the above, system (5.62) is transformed to

$$\frac{dx}{d\tau} = 1 - dx - xv,$$

$$\frac{dy}{d\tau} = -ay + xv - yw,$$

$$\frac{dz}{d\tau} = -bz + yw, \qquad\qquad\qquad (5.66)$$

$$\frac{dv}{d\tau} = -pv + y,$$

$$\frac{dw}{d\tau} = -qw + cz.$$

It should be noted that all state variables are non-negative, and all parameters are positive.

It is easy to find that the system has three equilibrium solutions (ES), described as follows:

infection-free ES: $E_0 = (\frac{1}{d}, 0, 0, 0, 0)$;
single-infection ES: $E_s = (ap, \frac{1}{a} - dp, 0, \frac{1}{ap} - d,\ 0)$;
double-infection ES: $E_d = (\frac{cp}{bq+cdp}, \frac{bq}{c}, \frac{q}{bq+cdp} - \frac{a}{c}, \frac{bq}{cp}, \frac{c}{bq+cdp} - a)$.

Using linear system theory, one can show that E_0 is (locally) asymptotically stable for $\mathcal{R}_0 > 1$, and E_s is stable for $1 < \mathcal{R}_0 < R_1$. Recently, in [113] the Fluctuation Lemma is applied to show that the infection-free equilibrium is globally asymptotically stable when $\mathcal{R}_0 < 1$; and a properly chosen Lyapunov function is used to prove that the single-infection equilibrium is globally asymptotically stable when $1 < \mathcal{R}_0 < R_1$.

Here $\mathcal{R}_0$ is the basic reproduction number, defined as

$$\mathcal{R}_0 = \frac{1}{adp}, \qquad\qquad\qquad (5.67)$$

and the critical value R_1 is given by

$$R_1 = 1 + \frac{bq}{adp}. \qquad\qquad\qquad (5.68)$$

In the following, we particularly investigate the stability of E_d for (5.66) and possible bifurcations from this equilibrium. In order to examine the stability of E_d, we compute the Jacobian matrix of system (5.66) as

$$J = \begin{bmatrix} -d-v & 0 & 0 & -x & 0 \\ v & -a-w & 0 & x & -y \\ 0 & w & -b & 0 & y \\ 0 & 1 & 0 & -p & 0 \\ 0 & 0 & c & 0 & -q \end{bmatrix}. \tag{5.69}$$

By straightforward but tedious computation, the characteristic polynomial of J at E_d is obtained as

$$P_d(\xi) = \xi^2(\xi + dR_1)(\xi + b + q)\left(\xi + p + a\frac{\mathcal{R}_0}{\mathcal{R}_1}\right)$$

$$+ \xi(\xi + b + q)\left(\frac{R_1 - 1}{R_1}\right) + abq(\xi + dR_1)(\xi + p)\left(\frac{\mathcal{R}_0 - \mathcal{R}_1}{R_1}\right)$$

$$\equiv \xi^5 + a_1\xi^4 + a_2\xi^3 + a_3\xi^2 + a_4\xi + a_5, \tag{5.70}$$

where

$$a_1 = a\frac{\mathcal{R}_0}{\mathcal{R}_1} + b + dR_1 + p + q,$$

$$a_2 = (b+q)\left(p + a\frac{\mathcal{R}_0}{\mathcal{R}_1}\right) + \left(b + q + p + a\frac{\mathcal{R}_0}{\mathcal{R}_1}\right)dR_1,$$

$$a_3 = (b+q)\left(p + a\frac{\mathcal{R}_0}{\mathcal{R}_1}\right)dR_1 + \frac{R_1 - 1}{R_1} + adq\left(\frac{\mathcal{R}_0 - \mathcal{R}_1}{R_1}\right), \tag{5.71}$$

$$a_4 = (b+q)\left(\frac{R_1 - 1}{R_1}\right) + abq(p + dR_1)\left(\frac{\mathcal{R}_0 - \mathcal{R}_1}{R_1}\right),$$

$$a_5 = \frac{bq}{\mathcal{R}_0}(\mathcal{R}_0 - \mathcal{R}_1).$$

It is obvious that $a_1 > 0, a_2 > 0$ for any positive parameter values. Here we apply the Hurwitz criterion to find the stability of the equilibrium solution E_d. The necessary and sufficient conditions for E_d to be stable are given by

$$\Delta_i > 0, \quad i = 1, 2, 3, 4, 5, \tag{5.72}$$

where

$$\Delta_1 = a_1,$$

$$\Delta_2 = a_1 a_2 - a_3,$$

$$\Delta_3 = a_3 \Delta_2 - a_1(a_1 a_4 - a_5), \tag{5.73}$$

$$\Delta_4 = a_4 \Delta_3 - a_5[a_2 \Delta_2 - (a_1 a_4 - a_5)],$$

$$\Delta_5 = a_5 \Delta_4.$$

It is easy to see that $a_i > 0, i = 1, 2, 3, 4, 5$, since $R_1 > 1$ and we have assumed $\mathcal{R}_0 > R_1$. Now, we need to check the signs of $\Delta_i, i = 2, 3, 4$. First, a straightforward calculation shows that

$$
\Delta_2 = \left(a\frac{\mathcal{R}_0}{R_1} + b + p + q \right)\left[(b+q)p + qa\frac{\mathcal{R}_0}{R_1} \right] + ab\left(a\frac{\mathcal{R}_0}{R_1} + b + p \right)\frac{\mathcal{R}_0}{R_1}
$$
$$
+ \left[\left(a\frac{\mathcal{R}_0}{R_1} + b + dR_1 + q \right)\left(b + q + p + a\frac{\mathcal{R}_0}{R_1} \right) + p(b + p + q) \right]dR_1
$$
$$
+ abq + \frac{1}{R_1}, \tag{5.74}
$$

indicating that $\Delta_2 > 0$ for all positive parameter values.

For Δ_3 and Δ_4, the signs are not easy to determine for general $\mathcal{R}_0$, and hence we use a continuity argument below. At $\mathcal{R}_0 = R_1$, using (5.71) and (5.73) and by direct calculations, we have

$$
\Delta_4|_{\mathcal{R}_0=R_1} = \frac{bq(b+q)}{a^2d^2cR_1^2}\left[1 + a^2d + a^2dR_1 + ad^2(1 + a^3)R_1^2 + a^3d^3R_1^3 \right]
$$
$$
\times \left\{ (b+q)^2 + \left[ad(b+q)^3 + a^2d(b+q)^2 + d(b+q) \right. \right.
$$
$$
\left. \left. + \frac{a^2bdq}{c} \right]R_1 + ad^2(b+q)(a+b+q)R_1^2 \right\} > 0, \tag{5.75}
$$

and

$$
\Delta_3|_{\mathcal{R}_0=R_1} = \frac{c}{abq(b+q)}\Delta_4|_{\mathcal{R}_0=R_1} > 0. \tag{5.76}
$$

Note that Δ_3, Δ_4 and Δ_5 depend continuously on $\mathcal{R}_0$. From (5.71), (5.73)–(5.76) and the continuity, we know there is a $R_2 > R_1$ such that (5.72) holds when $\mathcal{R}_0 \in (R_1, R_2)$, leading to the following conclusion.

Theorem 5.2 *There is an $R_2 > R_1$ such that when $\mathcal{R}_0 \in (R_1, R_2)$, the double-infection equilibrium E_d is asymptotically stable.*

5.4.2 Hopf Bifurcation

When $\mathcal{R}_0$ is further increased, Δ_1 and Δ_2 remain positive, but Δ_3 and Δ_4 may become negative (so may Δ_5 by (5.73) and $a_5 > 0$ as $\mathcal{R}_0 > R_1$). The following lemma identifies the order of possible sign switches for Δ_3 and Δ_4.

Lemma 5.3 *If Δ_3 and Δ_4 can change signs from positive to negative as $\mathcal{R}_0$ is further increased from the value R_2 in Theorem 5.2, then Δ_4 will change before Δ_3 does.*

Proof Assume, for the sake of contradiction, that Δ_3 will change sign no later than Δ_4 does. Then there exists an $R_3 > R_2$ such that

$$
\Delta_3 = \begin{cases} > 0 & \text{when } \mathcal{R}_0 \in (R_1, R_3), \\ = 0 & \text{when } \mathcal{R}_0 = R_3, \end{cases}
$$

$$
\Delta_4 = \begin{cases} > 0 & \text{when } \mathcal{R}_0 \in (R_1, R_3), \\ \geq 0 & \text{when } \mathcal{R}_0 = R_3. \end{cases}
\tag{5.77}
$$

Then, at $\mathcal{R}_0 = R_3$,

$$
a_3 \Delta_2 - a_1(a_1 a_4 - a_5) = 0,
$$

from which we obtain

$$
a_1 a_4 - a_5 = \frac{a_3}{a_1} \Delta_2.
$$

Thus,

$$
\Delta_4 = -a_5 \left[a_2 \Delta_2 - \frac{a_3}{a_1} \Delta_2 \right] = -\frac{a_5}{a_1} \Delta_2^2 < 0,
$$

leading by a contradiction to $\Delta_4 \geq 0$. This completes the proof. $\square$

The next lemma shows that when Δ_4 crosses zero, Δ_3 must remain positive.

Lemma 5.4 *For any $\mathcal{R}_0 > R_1$, if $\Delta_4 = 0$, then $\Delta_3 > 0$.*

Proof Suppose $\Delta_4 = 0$ at $\mathcal{R}_0 = R_4 > R_1$. Then,

$$
a_4 \Delta_3 - a_5 \left[a_2 \Delta_2 - (a_1 a_4 - a_5) \right] = 0,
$$

and hence

$$
a_1 a_4 \Delta_3 = a_1 a_5 \left[a_2 \Delta_2 - (a_1 a_4 - a_5) \right].
\tag{5.78}
$$

On the other hand, the third equation in (5.73) leads to

$$
a_5 \Delta_3 = a_3 a_5 \Delta_2 - a_1 a_5 (a_1 a_4 - a_5).
\tag{5.79}
$$

Subtracting (5.79) from (5.78) results in

$$
\Delta_3 = \frac{a_5(a_1 a_2 - a_3)\Delta_2}{a_1 a_4 - a_5} = \frac{a_5 \Delta_2^2}{a_1 a_4 - a_5}.
\tag{5.80}
$$

Note that $a_5 > 0$ and $\Delta_2 > 0$. Also, a careful calculation gives

$$a_1 a_4 - a_5 = (b+q)\left(a\frac{R_0}{R_1} + b + dR_1 + p + q\right)\left(\frac{R_1 - 1}{R_1}\right)$$
$$+ abq\left[\left(a\frac{R_0}{R_1} + b + dR_1 + q\right)(p + dR_1) + p^2\right]$$
$$\times \left(\frac{R_0 - R_1}{R_1}\right), \tag{5.81}$$

which clearly shows that $a_1 a_4 - a_5 > 0$ when $R_0 > R_1$. This, together with (5.80), confirms $\Delta_3 > 0$, completing the proof. □

The above discussion and the results in Yu [227] imply that there is no static bifurcation, Hopf-zero bifurcation, double-Hopf bifurcation, or double-zero Hopf bifurcation emerging from the equilibrium solution E_d; and the only possibility for E_d to lose stability is the occurrence of Hopf bifurcation when Δ_4 crosses zero from positive to negative as R_0 further increases from R_2 (see Theorem 5.2).

In order to show that Hopf bifurcation can occur, we need to show that Δ_4 can change sign from positive to negative as R_0 further increases from R_1. To this end, we note that $R_0 = 1/(adp)$ and $R_1 = 1 + bq/(cdp)$, implying that as $a \to 0^+$, $R_0 \to +\infty$ while $R_0 > R_1$ remains valid (actually, as long as $\frac{1}{a} > dp + \frac{bq}{c}$). This observation suggests considering small values of a and large values of c. Indeed, by (5.71), (5.73) and tedious but straightforward expansion, we may obtain

$$\Delta_4 = \frac{-bq}{(cdp + bq)^4}\left[c_4 c^4 + c_3 c^3 + c_2 c^2 + c_1 c + c_0 + c_{-1}c^{-1} + c_{-2}c^{-2} + c_{-3}c^{-3}\right]$$
$$+ O(a), \tag{5.82}$$

where

$$c_4 = \left[pdq + pbd^3(d + p) + d(d + q)\left(pd^3 + p^2 d^2 + b + d\right)\right]f(d),$$

in which

$$f(d) = p^4(b + p + q)d^2 - p\left(b^2 + bq + q^2 - 2p^2\right)d - b + p - q. \tag{5.83}$$

Thus, for small a and large c, the sign of Δ_4 is determined by the leading coefficient c_4, i.e., the sign of $f(d)$. In order to have $c_4 > 0$, we need $f(d) > 0$, which holds for appropriate values of d. For example, $f(d) > 0$ when $d < d_1$ or $d > d_2$, where d_1 and d_2 are the two roots of $f(d) = 0$ as a quadratic function of d:

$$d_1 = \frac{(b^2 + bq + q^2 - 2p^2) - \sqrt{(b^2 + q^2)(b + q)^2 + bq(bq + 4p^2)}}{2p^3(b + p + q)},$$
$$\tag{5.84}$$
$$d_2 = \frac{(b^2 + bq + q^2 - 2p^2) + \sqrt{(b^2 + q^2)(b + q)^2 + bq(bq + 4p^2)}}{2p^3(b + p + q)}.$$

Combining the above and the results in [245], we have proved the following.

Theorem 5.5 *For some large values of c and small values of a, together with $d <$* *d_1 or $d > d_2$ (d_1, d_2 as given in (5.84)) (hence $\mathcal{R}_0 \gg \mathcal{R}_1$), the double-infection* *equilibrium E_d loses its stability through Hopf bifurcation, giving rise to a family* *of periodic solutions.*

By the above theorem, E_d can lose its stability through Hopf bifurcation when $\mathcal{R}_0$ is further increased from $\mathcal{R}_1$ in some way. Thus, the numerical results in [181], claiming that all solutions converge to E_d when $\mathcal{R}_0 > \mathcal{R}_1$, are incomplete. It should be pointed out that the conditions obtained above for Δ_4 to change sign (i.e., for large values of c and small values of a) are only sufficient conditions for the requirement $\Delta_4 < 0$, which may be quite conservative. There may be many other choices of the parameters that satisfy this requirement. For some particular choice, we may even find the critical value R_h for $\mathcal{R}_0$, at precisely which Hopf bifurcation occurs. This will be illustrated numerically in the next section.

5.4.3 Numerical Illustrations

In this section, we use a numerical example and some simulations to demonstrate the theoretical results obtained in the previous sections. Due to the larger number of parameters, there are many choices for this purpose. For convenience, we will work on the scaled model (5.66) instead of the original model (5.62). Throughout this section, we fix

$$c = 40, \qquad a = \frac{93}{100}, \qquad b = p = q = \frac{28}{5}, \tag{5.85}$$

but choose d as the bifurcation parameter. Then,

$$\mathcal{R}_0 = \frac{1}{adp} = \frac{125}{651d} \quad \text{and} \quad R_1 = 1 + \frac{bq}{cdp} = 1 + \frac{7}{50d}. \tag{5.86}$$

The infection-free equilibrium becomes

$$E_0 = \left(\frac{1}{d}, 0, 0, 0, 0\right),$$

which is stable when $d > \frac{125}{651}$ (i.e., $\mathcal{R}_0 < 1$). When d decreases below the critical value $\frac{125}{651}$, $\mathcal{R}_0$ increases above the threshold value 1, and E_0 becomes unstable and there occurs the single-infection equilibrium

$$E_s = \left(\frac{651}{125}, \frac{100}{93} - \frac{28}{5}d, 0, \frac{125}{651} - d, 0\right),$$

which is stable for $\frac{1693}{32550} < d < \frac{125}{651}$ (corresponding to $1 < \mathcal{R}_0 < R_1$). When d further decreases below the critical value $\frac{1693}{32550}$, $\mathcal{R}_0$ increases above R_1 and E_s loses its stability to the double-infection equilibrium

$$E_d = \left(1 + \frac{7}{50}d, \frac{98}{125}, \frac{1693 - 32550d}{125(50d + 7)}, \frac{7}{50}, \frac{1693 - 32550d}{700(50d + 7)}\right),$$

which is stable when

$$1 + \frac{7}{50d} < \mathcal{R}_0 < \mathcal{R}_h, \quad \text{or} \quad d_h < d < \frac{1693}{32550} = 0.05201228879,$$

where d_h or $\mathcal{R}_h$ is determined as follows.

For the given parameter values, the coefficients of the characteristic polynomial for E_d become

$$a_1 = \frac{2}{175(50d + 7)}(4375d^2 + 74725d + 11157),$$

$$a_2 = \frac{1}{3500(50d + 7)}(2940000d^2 + 11830450d + 1948639),$$

$$a_3 = \frac{1}{125(50d + 7)}(392000d^2 - 60020d + 19789), \tag{5.87}$$

$$a_4 = \frac{14}{15625(50d + 7)}(573391 - 9257200d - 1627500d^2),$$

$$a_5 = \frac{392}{78125}(1693 - 32550d),$$

and thus Δ_i can be readily obtained. In particular,

$$\Delta_4 = \frac{2}{1068115234375(50d + 7)^4}(484484753320312500000000d^8$$

$$+ 85834882638750000000000d^7 + 370216771606681835937500 0d^6$$

$$+ 92493314786535113281250 0d^5 - 733313450837538235546875 0d^4$$

$$- 77556742983386341889062 5d^3 + 175027848868353875686250 d^2$$

$$+ 15328850593840359524200 d - 46289058675447144169 9). \tag{5.88}$$

A numerical scheme for solving for the roots of the polynomial can be applied here to find three positive real solutions of $\Delta_4 = 0$, given by

$$d = 0.02433284924, \qquad 0.1439442394, \qquad 1.1875365473.$$

It is seen that only the first solution satisfies the requirement, and thus

$$d_h = 0.02433284924,$$

giving a corresponding value

$$\mathcal{R}_h = 7.8910729629$$

for $\mathcal{R}_0$ by the formula for $\mathcal{R}_0$ in (5.86) in terms of d. Hence, when

$$0.02433284924 < d < 0.05201228879, \quad \text{or}$$

$$3.6916715889 < \mathcal{R}_0 < 7.8910729629,$$

the equilibrium solution E_d is stable. At the critical point, $d = d_h$ ($\mathcal{R}_0 = R_h$), the equilibrium solution E_d becomes unstable and a Hopf bifurcation occurs, leading to a family of periodic solutions. In fact, at the critical point $\mathcal{R}_0 = R_h$, other Hurwitz conditions are satisfied. Indeed, at the Hopf critical point, the characteristic polynomial $P_d(\xi)$ includes a pair of purely imaginary roots and three negative real roots:

$$\xi = \pm 0.6996439883\mathrm{i}, \quad -0.1229130660,$$

$$-6.6799164524, \quad -11.2481480850,$$

where i is the imaginary unit, $\mathrm{i}^2 = -1$.

In what follows, we show, via this numerical example, how to obtain more information about a Hopf bifurcation, such as the bifurcation direction and stability, and the magnitudes and periods of the bifurcating periodic solutions. To this end, we apply normal form theory and the program using the computer algebra system Maple [227], and the formulas given in [245] to analyze the Hopf bifurcation of system (5.66) from the critical point $d = d_h$ ($\mathcal{R}_0 = R_h$) (with other parameters given by (5.85).

Suppose the general normal form of the system is given in (4.2)–(4.3). Let $d = d_h - \mu$, where μ is small perturbation (bifurcation) parameter, and

$$T = \begin{bmatrix} -0.980731 & 1.228946 & 7.713205 & -0.573862 & 0.009961 \\ 0.883910 & 0.562986 & -0.287552 & 0.663553 & -0.102481 \\ 2.777049 & -3.816526 & 7.248006 & 0.148704 & 6.696766 \\ 0.167782 & 0.079571 & -0.525009 & -0.614448 & 0.018144 \\ 0.404442 & -0.732052 & 1.323332 & -0.137700 & -1.185657 \end{bmatrix}.$$

By the linear transformation

$$\begin{pmatrix} x \\ y \\ z \\ v \\ w \end{pmatrix} = \begin{pmatrix} 6.0852106234 \\ 0.7840000000 \\ 0.8772106234 \\ 0.1400000000 \\ 0.1566447542 \end{pmatrix} + T \begin{pmatrix} x_1 \\ x_2 \\ x_3 \\ x_4 \\ x_5 \end{pmatrix}, \tag{5.89}$$

system (5.66) is transformed to

$$\frac{dx_i}{d\tau} = F_i(x_1, x_2, x_3, x_4, x_5; \mu), \quad i = 1, 2, \ldots, 5, \tag{5.90}$$

in which the F_i's are polynomials in x_i and μ, given by

$$F_1 = 0.6996439883x_2 + (12.3096075793x_1 + 7.0040169156x_2$$
$$- 4.3728592211x_3 - 10.5713612773x_4 - 0.4156611553x_5)\mu + \cdots ,$$

$$F_2 = -0.6996439883x_1 + (-15.7111809897x_1 - 10.0506358184x_2$$
$$+ 6.2186196979x_3 - 20.3255688833x_4 + 2.2407734339x_5)\mu + \cdots ,$$

$$F_3 = -0.1229130660x_3 + (3.2296425652x_1 + 2.3100610142x_2$$
$$- 0.3267210453x_3 + 4.3446396524x_4 - 0.4813686500x_5)\mu + \cdots ,$$

$$F_4 = -6.6799164524x_4 + (1.5622567928x_1 + 0.7409792295x_2$$
$$- 0.5272185195x_3 - 5.4639965627x_4 + 0.1563670175x_5)\mu + \cdots ,$$

$$F_5 = -11.2481480850x_5 + (17.3226213739x_1 + 11.0868910462x_2$$
$$- 5.6345823041x_3 + 14.4271665998x_4 - 2.0807163549x_5)\mu + \cdots .$$

Now, the Jacobian of system (5.90) evaluated at the trivial equilibrium solution $x_i = 0, i = 1, 2, \ldots, 5$ (corresponding to E_d for (5.66)), is in Jordan canonical form. By the formulas given in [246], the coefficients v_0 and τ_0 are given by

$$v_0 = \frac{1}{2}\left(\frac{\partial^2 F_1}{\partial x_1 \partial \mu} + \frac{\partial^2 F_2}{\partial x_2 \partial \mu}\right)\bigg|_{\mu=0} = 1.1294858805,$$

$$\tau_0 = \frac{1}{2}\left(\frac{\partial^2 F_1}{\partial x_2 \partial \mu} - \frac{\partial^2 F_2}{\partial x_1 \partial \mu}\right)\bigg|_{\mu=0} = 11.3575989526. \tag{5.91}$$

Applying the Maple program [227] to system (5.90) (setting $\mu = 0$) results in

$$v_1 = -0.0607814981, \qquad \tau_1 = -1.0336310494. \tag{5.92}$$

Therefore, the third-order normal form of the system becomes

$$\frac{dr}{d\tau} = r\left(1.1294858805\mu - 0.0607814981r^2\right),$$

$$\frac{d\theta}{d\tau} = 0.6996439883 + 11.3575989526\mu - 1.0336310494r^2. \tag{5.93}$$

The steady-state solutions of (5.93) are determined by setting $\frac{dr}{d\tau} = \frac{d\theta}{d\tau} = 0$, yielding

$$\bar{r} = 0 \quad \text{and} \quad \bar{r}^2 = 18.5827252624\mu. \tag{5.94}$$

The solution $\bar{r} = 0$ actually corresponds to the equilibrium solution E_d of (5.66). A simple linearization of the first equation of (5.93) indicates that $\bar{r} = 0$ (E_d) is stable for $\mu < 0$, as expected. When μ increases from negative to cross zero, a Hopf

Fig. 5.12 Simulated time history of system (5.66) for $d = 0.21$, $a = 0.93$, $c = 40$, $b = p = q = 5.6$, with the initial condition $x(0) = 5.0$, $y(0) = 1.0$, $z(0) = 2.0$, $v(0) = 0.5$, $w(0) = 4.0$, converging to the stable equilibrium solution E_0

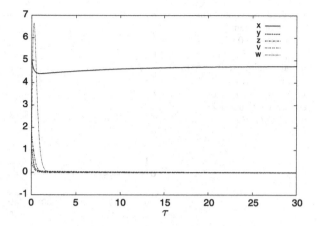

bifurcation occurs and the amplitude of the bifurcation periodic solutions is given by the nonzero steady-state solution

$$\bar{r} = 4.3107685234\sqrt{\mu}, \quad \mu > 0. \tag{5.95}$$

Since $v_1 < 0$, the Hopf bifurcation is supercritical and the bifurcating limit cycle is stable. The amplitude of the bifurcating limit cycle is given by (5.95), and the frequency is determined from the following equation:

$$\omega = 0.6996439883 + 6.9018547601\mu. \tag{5.96}$$

We have performed some numerical simulations for (5.66) by using a fourth-order Runge–Kutta method. We take the parameter values given in (5.85), giving $d_h = 0.02433284924$ and $R_h = 7.8910729629$. We choose four different values for d (and so for $\mathcal{R}_0$), and so based on (5.86) we have

$$d = 0.21 \quad \Longrightarrow \quad \mathcal{R}_0 = 0.9143442323 < 1,$$

$$d = 0.10 \quad \Longrightarrow \quad \mathcal{R}_0 = 1.9201228879 \in (1, 2.4) = (1, R_1),$$

$$d = 0.04 \quad \Longrightarrow \quad \mathcal{R}_0 = 4.8003072197 \in (4.5, 7.8910729629) = (R_1, R_h),$$

$$d = 0.022 \quad \Longrightarrow \quad \mathcal{R}_0 = 8.7278313085 > 7.8910729629 = R_h.$$

According to the above theoretical analysis, the simulation results are expected to have the stable equilibrium solution E_0 for $d = 0.21$, the stable equilibrium solution E_s for $d = 0.10$, the stable equilibrium solution E_d for $d = 0.04$, and a stable limit cycle for $d = 0.022$ (for which $\mu = 0.0023328492$), with an approximate amplitude for the periodic motion, $\bar{r} = 0.2082083010$.

The simulated time history and phase portraits for the above four cases are shown in Figs. 5.12, 5.13, 5.14, 5.15, respectively, where the initial conditions are taken as

$$x(0) = 5.0, \qquad y(0) = 1.0, \qquad z(0) = 2.0,$$
$$v(0) = 0.5, \qquad w(0) = 4.0. \tag{5.97}$$

Fig. 5.13 Simulated time history of system (5.66) for $d = 0.10$, $a = 0.93$, $c = 40$, $b = p = q = 5.6$, with the initial condition $x(0) = 5.0$, $y(0) = 1.0$, $z(0) = 2.0$, $v(0) = 0.5$, $w(0) = 4.0$, converging to the stable equilibrium solution E_s

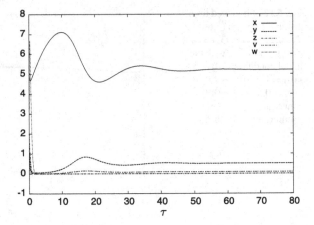

Fig. 5.14 Simulated time history of system (5.66) for $d = 0.04$, $a = 0.93$, $c = 40$, $b = p = q = 5.6$, with the initial condition $x(0) = 5.0$, $y(0) = 1.0$, $z(0) = 2.0$, $v(0) = 0.5$, $w(0) = 4.0$, converging to the stable equilibrium solution E_d

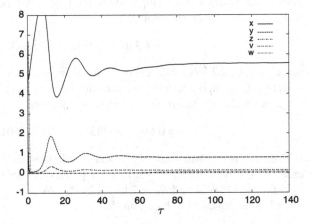

It can be seen from these figures that the numerical simulation results agree with the analytical predictions. The solutions for the first three cases converge to the equilibrium points, E_0, E_s and E_d, respectively. For the last case, the simulated amplitude of the limit cycle (see Fig. 5.15) is close to the predicted value, $\bar{r} = 0.2082$, showing good agreement, not only qualitatively, but also quantitatively, between the theoretical prediction and the numerical simulation. Also, it is seen that the period of motion, $T = \frac{2\pi}{\omega}$ (ω is given in (5.96)), decreases as μ increases. In other words, T decreases as d decreases. However, since μ is quite small, the change of the period due to μ is not significant (hardly observed, see Figs. 5.15 and 5.16). Nevertheless, a small change in μ can cause a large variation in the amplitude. The simulation results shown in Fig. 5.16 use $d = 0.012$, which gives $\mu = 0.0123328492$, and thus the approximation of the amplitude of periodic motion is $\bar{r} = 0.4787253380$, which is almost 2.3 times that when $d = 0.022$. This can be observed from Figs. 5.15(b) and 5.16(b).

Fig. 5.15 Simulation results
of system (5.66) for
$d = 0.022$, $a = 0.93$, $c = 40$,
$b = p = q = 5.6$, with the
initial condition $x(0) = 5.0$,
$y(0) = 1.0$, $z(0) = 2.0$,
$v(0) = 0.5$, $w(0) = 4.0$:
(**a**) time history showing
convergence to a stable
periodic solution; (**b**) phase
portrait projected on x–y
plane indicating a stable limit
cycle

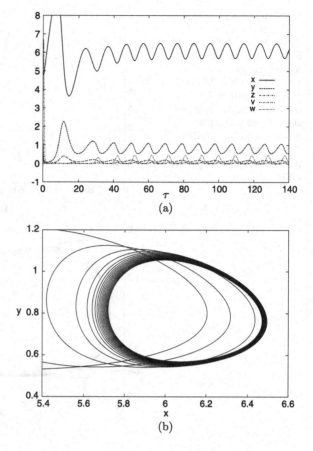

5.5 Hopf Bifurcation Control

Finally, we show the application of normal form theory to Hopf bifurcation control.
In the past two decades, there has been rapidly growing interest in the bifurcation
dynamics of control systems, including controlling and anticontrolling bifurcations
and chaos. Such bifurcation and chaos control techniques have been widely applied
to solve physical and engineering problems (e.g., see [1, 17, 33, 39, 43, 71, 116, 167,
174, 245]). The general goal of bifurcation control is to design a controller such that
the bifurcation characteristics of a nonlinear system undergoing bifurcation can be
modified to achieve a certain desirable dynamical behavior, such as changing a Hopf
bifurcation from subcritical to supercritical, eliminating chaotic motions, etc.

In this section, we consider bifurcation control using nonlinear state feedback.
A general explicit formula is presented for the control strategy, given in the form
of simple homogeneous polynomials [41, 42, 232]. The formula keeps the equilibria of the original system unchanged. The linear part of the formula can be used to
modify the system's linear stability, in order to eliminate or delay an existing bifur-

Fig. 5.16 Simulation results of system (5.66) for $d = 0.012$, $a = 0.93$, $c = 40$, $b = p = q = 5.6$, with the initial condition $x(0) = 5.0$, $y(0) = 1.0$, $z(0) = 2.0$, $v(0) = 0.5$, $w(0) = 4.0$: (**a**) time history showing convergence to a stable periodic solution; (**b**) phase portrait projected on x–y plane indicating a stable limit cycle

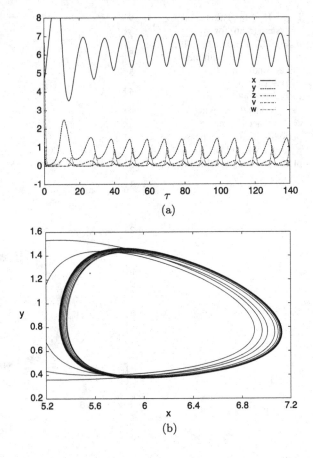

cation. The nonlinear part, on the other hand, can change the stability of bifurcation solutions, for example, by converting a subcritical Hopf bifurcation to supercritical.

Here we particularly want to study Hopf bifurcation since the limit cycles generated by Hopf bifurcation are the most popular phenomenon exhibited in nonlinear dynamical systems.

For convenience, we will use the well-known Lorenz system as an example to illustrate the theory and methodology of Hopf bifurcation control for continuous-time systems, described by ordinary differential equations. Consider the general nonlinear system (2.1):

$$\dot{x} = f(x, \mu), \quad x \in R^n, \mu \in R, \qquad f : R^{n+1} \to R^n, \tag{5.98}$$

where x is an n-dimensional state vector while μ is a scalar parameter called a *bifurcation parameter*. Suppose that at the critical point, $\mu = \mu^*$, on an equilibrium solution, $x = x^*$, the Jacobian of the system has a complex pair of eigenvalues first crossing the imaginary axis. Then Hopf bifurcation occurs at the critical point and a family of limit cycles bifurcate from the equilibrium solution x^*.

Suppose system (5.98) has k equilibria, given by

$$x_i^*(\mu) = \left(x_{1i}^*, x_{2i}^*, \ldots, x_{ni}^*\right), \quad i = 1, 2, \ldots, k, \tag{5.99}$$

satisfying that $f(x^*, \mu) = 0$ for $i = 1, 2, \ldots, k$.

5.5.1 Feedback Control Using a Polynomial Function

A general nonlinear state feedback control is applied so that system (5.98) becomes

$$\dot{x} = f(x, \mu) + u(x, \mu). \tag{5.100}$$

In order for the controlled system (5.100) to keep all the original k equilibria unchanged under the control u, it requires that the following conditions be satisfied:

$$u\left(x_i^*, \mu\right) \equiv (u_1, u_2, \ldots, u_n)^T = 0, \quad i = 1, 2, \ldots, k. \tag{5.101}$$

Then we have the following result [232].

Theorem 5.6 *For system* (5.100), *the feedback control can take the following polynomial function*:

$$
\begin{aligned}
u_q\left(x, x_1^*, x_2^*, \ldots, x_k^*, \mu\right) = {} & \sum_{i=1}^{n} A_{qi} \prod_{j=1}^{k} \left(x_i - x_{ij}^*\right) \\
& + \sum_{i=1}^{n}\sum_{j=1}^{k} B_{qij}\left(x_i - x_{ij}^*\right) \prod_{p=1}^{k}\left(x_i - x_{ip}^*\right) \\
& + \sum_{i=1}^{n}\sum_{j=1}^{k} C_{qij}\left(x_i - x_{ij}^*\right)^2 \prod_{p=1}^{k}\left(x_i - x_{ip}^*\right) \\
& + \sum_{i=1}^{n}\sum_{j=1}^{k} D_{qij}\left(x_i - x_{ij}^*\right)^2 \prod_{p=1}^{k}\left(x_i - x_{ip}^*\right)^2 + \cdots,
\end{aligned}
$$

$$q = 1, 2, \ldots, n. \tag{5.102}$$

It is easy to verify that $u_q(x_i^*, x_1^*, x_2^*, \ldots, x_k^*, \mu) = 0$ for $i = 1, 2, \ldots, k$. Usually, the terms given in (5.102) up to D_{qij} are enough for controlling a bifurcation if the singularity of the system is not highly degenerate. The coefficients A_{qi}, B_{qij}, C_{qij} and D_{qij}, which may be functions of μ, are determined from the stability of an equilibrium under consideration and that of the associated bifurcation solutions. More precisely, linear terms are determined by the requirement of shifting an existing bifurcation (e.g., delaying an existing Hopf bifurcation). The nonlinear terms, on the other hand, can be used to change the stability of an existing bifurcation or create a new bifurcation (e.g., changing an existing subcritical Hopf bifurcation to supercritical). Note that it is not only the A_{qi} terms but also the B_{qij} terms, etc. that may contain linear terms.

It is not necessary to take all the components $u_q, i = 1, 2, \ldots, n$, in the controller. In most cases, using fewer components or just one component may be enough to satisfy the predesigned control objectives. It is preferable to have the simplest possible design for engineering applications. If $x_{i1}^* = x_{i2}^* = \cdots = x_{ik}^*$ for some i, then one only needs to use these terms and omit the remaining terms in the control law. Moreover, lower-order terms related to these equilibrium components can be added. This greatly simplifies the control formula. For example, if $i = 1$, then the general controller can be taken as

$$
u_q = \sum_{i=1}^{k-1} a_{qi} \left(x_1 - x_{11}^* \right)^i + A_{q1} \left(x_1 - x_{11}^* \right)^k
$$

$$
+ B_{q11} \left(x_1 - x_{11}^* \right)^{k+1} + C_{q11} \left(x_1 - x_{11}^* \right)^{k+2},
$$

where the a_{qi}'s denote the added lower-order terms.

The goals of Hopf bifurcation control are:

(i) to move the critical point (x^*, μ^*) to a designated position $(\tilde{x}, \tilde{\mu})$;
(ii) to stabilize all possible Hopf bifurcations.

Goal (i) only requires linear analysis, while goal (ii) must apply nonlinear systems theory. In general, if the purpose of the control is to avoid bifurcations, one should employ linear analysis to maximize the stable interval for the equilibrium. The best result is to completely eliminate possible bifurcations using a feedback control. If this is not feasible, then one may have to consider stabilizing the limit cycles by using a nonlinear state feedback.

5.5.2 The Lorenz System

It is well known that the Lorenz system [154] can exhibit very rich periodic and chaotic motions. In this subsection, we will use a different version of the Lorenz equation, which contains only two parameters, given below [37, 201]:

$$
\begin{aligned}
\dot{x} &= -p(x - y), \\
\dot{y} &= -xz - y, \\
\dot{z} &= xy - z - r,
\end{aligned}
\tag{5.103}
$$

where p and r are positive constants, which are considered control parameters. One can easily show that system (5.103) is a special case of the general system [232].

System (5.103) has three equilibrium solutions, C_0, C_+ and C_-, given by

$$C_0: \quad x_e^0 = y_e^0 = 0, \qquad z_e^0 = -r,$$
$$C_\pm: \quad x_e^\pm = y_e^\pm = \pm\sqrt{r-1}, \qquad z_e^\pm = -1. \tag{5.104}$$

Suppose the parameters p and r are positive. Then C_0 is stable for $0 \le r < 1$, and a pitchfork bifurcation occurs at $r = 1$, where the equilibrium C_0 loses its stability and bifurcates into either C_+ or C_-. The two equilibria C_+ and C_- are stable for $1 < r < r_H$, where

$$r_H = \frac{p(p+4)}{p-2}, \qquad p > 2, \tag{5.105}$$

and at this critical point C_+ and C_- lose their stability, giving rise to a Hopf bifurcation. We fix $p = 4$, which was used in [37, 201]. Then $r_H = 16$, and the Lorenz system (5.103) exhibits chaotic motion when $r > 16$. In fact, one can use numerical simulation to show the coexistence of locally stable equilibria $C_\pm$ and (global) chaotic attractors at the same value of r, with different initial conditions [232].

(A) **Without Control**   We first consider system (5.103) without control. The critical point is $p = 4, r_H = 16$, at which the Jacobian of system (5.103) evaluated at C_+ and C_- has a real eigenvalue, -6, and a purely imaginary pair, $\pm 2\sqrt{5}i$. Using the shift given by

$$x = \pm\sqrt{r-1} + \tilde{x}, \qquad y = \pm\sqrt{r-1} + \tilde{y}, \qquad z = -1 + \tilde{z}, \tag{5.106}$$

to move $C_\pm$ to the origin and then applying an appropriate linear transformation to system (5.103), we obtain the new system

$$\dot{\tilde{x}} = 2\sqrt{5}\tilde{y} + \frac{1}{84}(\tilde{x} + 4\sqrt{5}\tilde{y} - 6\tilde{z})\mu - \frac{\sqrt{15}}{21}(\tilde{x} - 2\sqrt{5}\tilde{y})(\tilde{x} - 2\tilde{z}) + \cdots,$$

$$\dot{\tilde{y}} = -2\sqrt{5}\tilde{x} - \frac{\sqrt{5}}{2100}(155\tilde{x} - 10\sqrt{5}\tilde{y} - 6\tilde{z})\mu$$
$$\qquad - \frac{\sqrt{3}}{105}(55\tilde{x} - 5\sqrt{5}\tilde{y} + 42\tilde{z})(\tilde{x} - 2\tilde{z}) + \cdots,$$

$$\dot{\tilde{z}} = -6\tilde{z} + \frac{1}{168}(\tilde{x} + 4\sqrt{5}\tilde{y} - 6\tilde{z})\mu - \frac{\sqrt{15}}{42}(\tilde{x} - 2\sqrt{5}\tilde{y})(\tilde{x} - 2\tilde{z}) + \cdots,$$

where $\mu = r - 16$ is a bifurcation parameter.

Applying the Maple programs given on "Springer Extras" (to find the programs, visit extras.springer.com and search for this book using its ISBN) for computing the normal forms of Hopf and generalized Hopf bifurcations yields the normal form

$$\dot{\rho} = r\left(\frac{1}{56}\mu + \frac{31}{3248}r^2\right) + \cdots,$$

$$\dot{\theta} = 2\sqrt{5}\left(1 + \frac{17}{560}\mu - \frac{851}{48720}r^2\right) + \cdots,$$

(5.107)

where r and θ represent the amplitude and phase of the motion, respectively. The first equation of (5.107) clearly shows that the Hopf bifurcation is subcritical since the coefficient of r^3 is $\frac{31}{3248} > 0$.

(B) With Control   Now, we apply a feedback control to stabilize system (5.13). A washout filter control was used by Wang and Abed [201] for the Lorenz system (5.103). The disadvantage of this method is that it increases the dimension of the original system by one, unnecessarily increasing the complexity of the controlled system and the difficulty of analysis. Here we apply the control formula (5.105) to control the Hopf bifurcation. Due to the symmetry of the system and $z_{\pm} = -1$, we may use a control law with one variable only:

$$u_3 = -k_{31}(z+1) - k_{33}(z+1)^3. \tag{5.108}$$

The closed-loop system is now given by

$$\dot{x} = -p(x-y),$$

$$\dot{y} = -xz - y, \tag{5.109}$$

$$\dot{z} = xy - z - r - k_{31}(z+1) - k_{33}(z+1)^3,$$

where the negative signs are used for the k_{ij}'s for consistency with that of the controller based on the washout filter. Introducing the transformation (5.106) into (5.109) results in

$$\dot{x} = -p(\tilde{x} - \tilde{y}),$$

$$\dot{y} = -\tilde{x}\tilde{z} + \tilde{x} - \tilde{y} \mp \sqrt{r-1}\tilde{z}, \tag{5.110}$$

$$\dot{z} = \tilde{x}\tilde{y} \pm \sqrt{r-1}(\tilde{x} + \tilde{y}) - \tilde{z} - k_{31}\tilde{z} - k_{33}\tilde{z}^3.$$

Then $O_e = (\tilde{x}, \tilde{y}, \tilde{z}) = (0, 0, 0)$ is an equilibrium of system (5.110), corresponding to the equilibria C_+ and C_- of the original system (5.103). The characteristic polynomial of system (5.110) for the equilibrium point O_e is

$$P(\lambda) = \lambda^3 + (p + 2 + k_{31})\lambda^2 + (p + r + k_{31} + pk_{31})\lambda + 2p(r-1),$$

which shows that only the linear term of the controller u_3 affects the linear stability. The stability conditions for O_e (under the assumption that $p, r > 0$) can be obtained as

$$p + 2 + k_{31} > 0,$$

$$p + r + k_{31}(p + 1) > 0,$$

$$2p(r - 1) > 0,$$

$$p(p + 4) - r(p - 2 - k_{31}) + k_{31}^2(p + 1) + k_{31}(p^2 + 4p + 2) > 0.$$

If choosing $k_{31} > 0$, then only $r > 1$ is required to satisfy the first 3 inequalities above. The last condition implies a critical point at which the controlled system has a Hopf bifurcation emerging from the equilibrium O_e, defined by

$$r_H = \frac{p(p + 4) + k_{31}^2(p + 1) + k_{31}(p^2 + 4p + 2)}{(p - 2 - k_{31})}, \tag{5.111}$$

for $0 < k_{31} < p - 2$.

Setting $k_{31} = 0$ yields $r_H = \frac{p(p+4)}{p-2}$, $p > 2$, which is the condition given in (5.105) for the system without control. It can be seen from (5.111) that the parameter r_H for the controlled system can reach very large values as long as k_{31} is chosen close to $(p - 2)$. For example, when $p = 4$, choosing $k_{31} = 1.5$ gives $r_H = 188.5$ (and $r_H = 71$ if $k_{31} = 1$). These values of r_H are much larger than $r_H = 16$ for the uncontrolled system. If we choose $r > 1$ and $0 < p - 2 < k_{31}$, then the equilibria C_+ and C_- are always stable, and no Hopf bifurcations can occur from the two equilibria.

Next, we perform a nonlinear analysis to determine the stability of the Hopf bifurcation. If $p = 4$, then $k_{31} \in (0, 2)$, and for determination we choose $k_{31} = \frac{2\sqrt{1006} - 58}{5} \approx 1.087$, thus $r_H = 82$. Let $r = r_H + \mu = 82 + \mu$, where μ is a perturbation from the critical point. Then, we have the closed-loop system

$$\dot{\tilde{x}} = -8(\tilde{x} - \tilde{y}),$$

$$\dot{\tilde{y}} = -\tilde{x}\tilde{z} + \tilde{x} - \tilde{y} \mp \sqrt{81 + \mu}\tilde{z}, \tag{5.112}$$

$$\dot{\tilde{z}} = \tilde{x}\tilde{y} \pm \sqrt{81 + \mu}(\tilde{x} + \tilde{y}) - \frac{2\sqrt{1006} - 53}{5}\tilde{z} - k_{33}\tilde{z}^3.$$

The eigenvalues of the Jacobian of system (5.112), when evaluated at the equilibrium O_e, are: $\lambda_{1,2} = \pm\sqrt{2\sqrt{1006} + 28}\,i \approx 9.5621i$ and $\lambda_3 = -\frac{2\sqrt{1006} - 28}{5} \approx -7.0870$. Now introduce the transformation

$$\tilde{x} = u - \frac{24 + \sqrt{1006}}{43}w,$$

$$\tilde{y} = u + \frac{2\sqrt{1006} + 28}{4}v + w,$$

$$\tilde{z} = \pm\frac{\sqrt{1006} + 14}{18}u \mp \frac{5(2\sqrt{1006} + 28)}{36}v \pm \frac{9\sqrt{1006} - 171}{215}w;$$

Fig. 5.17 Stable equilibria, $C_{\pm}$, of the controlled Lorenz system (5.116) with the control law (5.115) and initial conditions $(x_0, y_0, z_0) = (\pm 3.0, \pm 12.0, -2.5)$ when (**a**) and (**b**) $r = 20$; (**c**) and (**d**) $r = 55$; and (**e**) and (**f**) $r = 81$

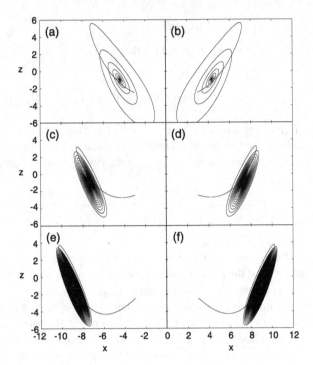

to (5.112), and then apply the Maple program to obtain an identical normal form for the system associated with the two equilibria C_+ and C_-, given in polar coordinates as

$$\dot{r} = \rho \left[\frac{1249 - 34\sqrt{1006}}{52942} \mu \right.$$
$$+ \left(\frac{4646315818 - 102399253\sqrt{1006}}{358010321904} - \frac{5746272 + 187233\sqrt{1006}}{4235360} k_{33} \right) \rho^2 \right]$$
$$+ \cdots, \tag{5.113}$$

$$\dot{\theta} = \sqrt{2\sqrt{1006} + 28} \left[1 + \frac{122602 - 773\sqrt{1006}}{17153208} \mu \right.$$
$$- \left(\frac{21706679417 + 211691192\sqrt{1006}}{6444185794272} + \frac{34871 + 1594\sqrt{1006}}{16941440} k_{33} \right) \rho^2 \right]$$
$$+ \cdots. \tag{5.114}$$

Approximations up to third order for the steady-state solutions and their stability can be found from (5.113): the solution $r = 0$ represents the initial equilibrium solution O_e (or $C_{\pm}$ for the original system (5.103)), which is stable when $\mu < 0$ (i.e., $r < r_H = 82$) and unstable when $\mu > 0$ ($r > 82$). The supercritical Hopf bifurcation

Fig. 5.18 Stable limit cycles around $C_\pm$ of the controlled Lorenz system (5.116) with the control law (5.115) and initial conditions $(x_0, y_0, z_0) = (\pm 3.0, \pm 12.0, -2.5)$ when (a) and (b) $r = 83$; (c) and (d) $r = 90$; and (e) and (f) $r = 101$

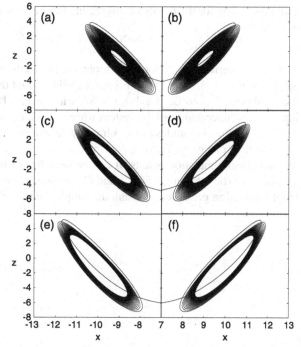

solution can be obtained if

$$\frac{4646315818 - 102399253\sqrt{1006}}{358010321904} - \frac{5746272 + 187233\sqrt{1006}}{4235360}k_{33} < 0,$$

i.e.

$$k_{33} > \frac{3672843514\sqrt{1006} - 115816173526}{478327912875} \approx 0.001416.$$

Choosing $k_{33} = 0.01$, we have the controller

$$u = -1.087(z + 1) - 0.01(z + 1)^3. \tag{5.115}$$

So the controlled system described in the original state is given by

$$\dot{x} = -4(x - y),$$
$$\dot{y} = -xz - y, \tag{5.116}$$
$$\dot{z} = xy - z - r - 1.087(z + 1) - 0.01(z + 1)^3.$$

The corresponding normal form then becomes

$$\dot{r} = r\left(0.003222\mu - 0.023683r^2\right) + \cdots,$$
$$\dot{\theta} = 9.562165 + 0.054678\mu - 0.046512r^2 + \cdots,$$

and the solution for the family of bifurcating limit cycles is obtained as

$$r = 0.136070\sqrt{\mu} = 0.136070\sqrt{r - 82}. \tag{5.117}$$

Some numerical simulation results, obtained from the controlled system (5.116), are given in Figs. 5.17 and 5.18. Figure 5.17 shows that the trajectories converge to the equilibria C_+ and C_- for $1 < r < 82$, while Fig. 5.18 demonstrates the stable limit cycles bifurcated from the system when $r > 82$. By using equation (5.117) one can estimate the amplitudes of the 3 limit cycles shown in Fig. 5.18 as 0.136, 0.385 and 0.593, respectively. These approximations are a good prediction, confirmed by the numerical simulation results. It can be seen from Figs. 5.17 and 5.18 that the symmetry of the two equilibria C_+ and C_- remains unchanged before and after the Hopf bifurcation generated by using the simple control (5.115).

Chapter 6
Fundamental Theory of the Melnikov Function Method

Another perturbation problem that has a much longer tradition than the singular one (discussed in previous chapters) is perturbation from a period annulus. In perturbating from a period annulus, limit cycles can be created from the boundary or from the interior. The latter is a purely analytical question; the former is much more irregular, dealing with the limit behavior of analytic functions. An exception has to be made for an inner boundary consisting of a nondegenerate singularity which is also an analytical problem. In any case, all limit cycles under consideration coincide with fixed points of the Poincaré return map which is an analytic function near any periodic orbit in the period annulus. The perturbation problem can be reduced to the perturbation from an analytic Hamiltonian system with Hamiltonian H. Using 1-forms the system can be written as the equivalent Pfaffian equation

$$dH + \varepsilon\omega = 0 \quad \text{with } \omega \equiv q\,dx - p\,dy. \tag{6.1}$$

Then to first-order approximation the limit cycles are given by the isolated zeros of the so-called *Melnikov function*

$$M(h) = \int_{H(x,y)=h} q\,dx - p\,dy. \tag{6.2}$$

This integral depends on the same parameters that the equation is subject to and the study of the zero set of the integral is an important question. A great deal of attention has already been given to the case in which both H and ω are polynomials. The integral is then called an *Abelian integral* and the search for the maximum number, $Z(m, n)$, of isolated zeros of such integrals, depending only on the degrees m and n of H and ω, respectively, is called the *weakened* (or *infinitesimal* or *tangential*) *Hilbert 16th problem*. Varchenko and Khovanskii [197] have proved that the asymptotic growth of $Z(m, n)$ with respect to n is linear for a fixed Hamiltonian. More detailed information can be found in [106]. However, except for a very limited number of cases, the precise value of $Z(m, n)$ is not known. It is known for H of degree ≤ 3. Also, $Z(n) = Z(n + 1, n)$ is known only for $n = 2$. Recently, the upper bound $Z(n) \leq 2^{2^{P(n)}}$, for some explicit polynomial $P(n)$, has been provided in [24].

M. Han, P. Yu, *Normal Forms, Melnikov Functions and Bifurcations of Limit Cycles*, Applied Mathematical Sciences 181, DOI 10.1007/978-1-4471-2918-9_6, © Springer-Verlag London Limited 2012

There are many papers concerning the weakened Hilbert 16th problem: using the Picard–Fuchs equations, the Argument Principle, the averaging method, the Picard–Lefschets formula; for example, see [49] for a detailed introduction to the basics of these methods and its references to related papers.

It is necessary to observe that merely studying the integrals in (6.2), even when they are quite generic, is not sufficient to obtain all limit cycles that can be perturbed from the boundary of a period annulus. This has been discussed in [32] and first been shown in [62] for a 2-saddle cycle as boundary, giving rise to so-called "alien" limit cycles (see also [155] for a concrete example). In perturbing a saddle loop (homoclinic loop), all limit cycles that bifurcate from the saddle loop can be controlled by the zeros of the related Melnikov function (see [183]). But perturbing from the heteroclinic k-saddle cycle Γ, where $k \geq 2$, the lower bound determined by the number of zeros of the Melnikov function is in general a strict lower bound for the number of limit cycles bifurcating from Γ (see [62]).

For the purpose of following chapters, in this chapter we introduce the fundamental theory of the Melnikov function method. Basic definitions and fundamental lemmas are presented. A major theory on the number of limit cycles is given.

Consider a C^∞ system of the form

$$\dot{x} = H_y + \varepsilon p(x, y, \varepsilon, \delta), \qquad \dot{y} = -H_x + \varepsilon q(x, y, \varepsilon, \delta), \tag{6.3}$$

where $H(x, y), p(x, y, \varepsilon, \delta), q(x, y, \varepsilon, \delta)$ are C^∞ functions, $\varepsilon \geq 0$ is small and $\delta \in D \subset R^m$ is a vector parameter with D compact. For $\varepsilon = 0$, (6.3) becomes

$$\dot{x} = H_y, \qquad \dot{y} = -H_x, \tag{6.4}$$

which is a Hamiltonian system. Hence, (6.3) is called a *near-Hamiltonian system*.

For (6.4) we suppose there exists a family of periodic orbits given by

$$L_h : H(x, y) = h, \quad h \in (\alpha, \beta),$$

such that L_h approaches an elementary center point, denoted by L_α, as $h \to \alpha$, and an invariant curve, denoted by L_β, as $h \to \beta$. If L_β is bounded, it is usually a homoclinic loop consisting of a saddle and a connection or a heteroclinic loop consisting of at least two saddles and connections between them. In the homoclinic case, the phase portrait of the family $\{L_h : \alpha \leq h \leq \beta\}$ is shown in Fig. 6.1 for the case of clockwise orientation.

Fig. 6.1 The phase portrait of (6.4) with a clockwise-oriented homoclinic loop

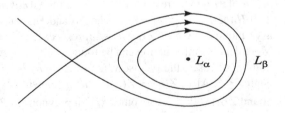

Next, introduce an open set G:

$$G = \bigcup_{\alpha < h < \beta} L_h.$$

Our main task is to study the number of limit cycles of (6.3) in a neighborhood of the closure $\bar{G}$ of G for $\varepsilon > 0$, small, and $\delta \in D$.

Note that if (6.3) has a limit cycle $L(\varepsilon, \delta)$ for $\varepsilon > 0$, small, and $\delta \in D_0 \subset D$, then the limit of the cycle as $\varepsilon \to 0$ is either the center L_α, a periodic orbit L_h with $h \in (\alpha, \beta)$ or the boundary L_β; i.e.,

$$\lim_{\varepsilon \to 0} L(\varepsilon, \delta) = L_h, \quad h \in [\alpha, \beta].$$

In this case, it is said that the limit cycle $L(\varepsilon, \delta)$ is generated from L_h. Thus, in order to study the number of limit cycles, we first need to study the number of limit cycles generated from each L_h.

6.1 Definition and Lemmas

First, we introduce notation below.

Definition 6.1 We say that (6.3) has cyclicity k at a given $L_h, h \in [\alpha, \beta]$, if there exist $\varepsilon_0 > 0$ and a neighborhood V of L_h such that (6.3) has at most k limit cycles in V for $0 < \varepsilon < \varepsilon_0, \delta \in D$ and if k limit cycles can appear in an arbitrary neighborhood of L_h for some (ε, δ) with $\varepsilon > 0$, sufficiently small. More specifically, k is said to be the Hopf (Poincaré or homoclinic) cyclicity when $h = \alpha$ ($h \in (\alpha, \beta)$ or $h = \beta$), respectively, with L_β a homoclinic loop.

Take $h = h_0 \in (\alpha, \beta)$ and $A_0 \in L_{h_0}$. Let l be a cross section of (6.4) passing through A_0. Let

$$\mathbf{n}_1 = \big(H_y(A_0), -H_x(A_0)\big), \qquad \mathbf{n}_0 = \big(H_x(A_0), H_y(A_0)\big).$$

Then $\mathbf{n}_1$ is tangent to L_{h_0}, and $\mathbf{n}_0$ is normal to L_{h_0} at A_0. Let $\mathbf{n}_l$ denote a unit vector parallel to l, as shown in Fig. 6.2. Then $\mathbf{n}_0 \cdot \mathbf{n}_l \neq 0$ since l is a cross section, and we

Fig. 6.2 The cross section, l

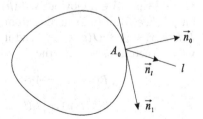

can write

$$l = \{A_0 + u\mathbf{n}_l, \ u \in \mathbf{R}, \ \text{for } |u| \text{ small}\}.$$

Consider the equation

$$G(h, u) \equiv H(A_0 + u\mathbf{n}_l) - h = 0.$$

We have

$$G(h_0, 0) = H(A_0) - h_0 = 0$$

$$\frac{\partial G}{\partial u}(h_0, 0) = DH(A_0) \cdot \mathbf{n}_l = \mathbf{n}_0 \cdot \mathbf{n}_l \neq 0.$$

Hence, by the implicit function theorem, the equation $G(h, u) = 0$ defines a unique function $u = a(h)$ with $a(h) \in C^\infty$ and $a(h_0) = 0$. Let

$$A(h) = A_0 + a(h)\mathbf{n}_l \quad \text{for } |h - h_0| \text{ small}.$$

Then $A(h) = L_h \cap l \in C^\infty$ with $A(h_0) = A_0$.

For $|\varepsilon|$ small, let $\varphi(t, A, \varepsilon, \delta)$ denote the solution of (6.3) satisfying $\varphi(0, A, \varepsilon, \delta) = A$. Then the solution is C^∞ in its variables. Let $T = T(h)$ denote the period of L_h. Then

$$\varphi(0, A, 0, \delta) = \varphi(T, A, 0, \delta) = A.$$

Consider the following equation

$$G_1(t, h, \varepsilon, \delta) \equiv \big(\varphi(t, A(t), \varepsilon, \delta) - A(t)\big) \cdot \mathbf{n}_l^\perp = 0,$$

where $\mathbf{n}_l^\perp \neq 0$ is a vector normal to l. Then

$$G_1\big(T(h_0), h_0, 0, \delta\big) = 0,$$

$$\frac{\partial G_1}{\partial t}\big(T(h_0), h_0, 0, \delta\big) = \big(H_y(A_0), -H_x(A_0)\big) \cdot \mathbf{n}_l^\perp = \mathbf{n}_1 \cdot \mathbf{n}_l^\perp \neq 0.$$

Using the implicit function theorem again, we know that there exists $t = \tau(h, \varepsilon, \delta) = T(h_0) + O(|\varepsilon| + |h - h_0|) \in C^\infty$, such that

$$G_1(\tau, h, \varepsilon, \delta) = 0, \quad \text{or} \quad \big(\varphi(\tau, A, \varepsilon, \delta) - A\big) \cdot \mathbf{n}_l^\perp = 0.$$

This shows that the vector $\varphi(\tau, A, \varepsilon, \delta) - A$ is parallel to $\mathbf{n}_l$. Thus, $\varphi(\tau, A, \varepsilon, \delta) \in l$ since $A \in l$. Obviously, $T(h) = \tau(h, 0, \delta) \in C^\infty$. Consider the positive orbit $\gamma^+(h, \varepsilon, \delta)$ of (6.3) starting from $A(h)$. Let $B(h, \varepsilon, \delta) = \varphi(\tau, A(t), \varepsilon, \delta)$. Then $B(h, \varepsilon, \delta) = A + O(\varepsilon) \in C^\infty$, and it is the first intersection point of the orbit γ^+ with l. Therefore, we can write

$$B(h, \varepsilon, \delta) = A_0 + b(h, \varepsilon, \delta)\mathbf{n}_l,$$

$$b(h, \varepsilon, \delta) = \big[B(h, \varepsilon, \delta) - A_0\big] \cdot \mathbf{n}_l = a(h) + O(\varepsilon).$$

Fig. 6.3 Poincaré map

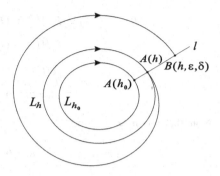

Then we have

$$H(B) - H(A) = \int_{\widehat{AB}} dH = \int_{\widehat{AB}} H_x\, dx + H_y\, dy$$

$$= \int_0^\tau \left[H_x(H_y + \varepsilon p) + H_y(-H_x + \varepsilon q) \right] dt$$

$$= \varepsilon \int_0^\tau (H_x p + H_y q)\, dt \equiv \varepsilon F(h, \varepsilon, \delta). \tag{6.5}$$

Obviously,

$$F(h, 0, \delta) = \oint_{L_h} (H_y q + H_x p)|_{\varepsilon=0}\, dt$$

$$= \oint_{L_h} (q\, dx - p\, dy)|_{\varepsilon=0} = \int\int_{H \leq h} (p_x + q_y)|_{\varepsilon=0}\, dx\, dy$$

$$\equiv M(h, \delta). \tag{6.6}$$

The function $F(h, \varepsilon, \delta)$ in (6.5) is called a *bifurcation function* of (6.3), and the function $M(h, \delta)$ is called the *Melnikov function*. The resulting map from $A(h)$ to $B(h, \varepsilon, \delta)$ is called a *Poincaré map* of (6.3); see Fig. 6.3.

For the properties of F we have the following lemma.

Lemma 6.2 *Let $h_0 \in (\alpha, \beta)$. Then*

(i) *the function $F(h, \varepsilon, \delta)$ is C^∞ in (h, ε, δ) for $|\varepsilon| + |h - h_0|$ small and $\delta \in D$; in particular, $M \in C^\infty$ for $h \in (\alpha, \beta)$;*

(ii) *for $|\varepsilon| + |h - h_0|$ small and $\delta \in D$, (6.3) has a periodic orbit near L_{h_0} if and only if the equation $F(h, \varepsilon, \delta) = 0$ has a zero for h near h_0.*

Proof Conclusion (i) is clear from the discussion above. We only need to prove conclusion (ii). Since the orbit $\gamma^+(h, \varepsilon, \delta)$ starting from $A(h)$ is closed if and only if $A = B$, we need only to show that $A = B$ if and only if $H(A) = H(B)$ by (6.5). In fact, by the differential mean value theorem we obtain

$$H(B) - H(A) = DH\big(A + O(B - A)\big) \cdot (B - A)$$
$$= DH\big(A + O(\varepsilon)\big) \cdot (B - A)$$
$$= \big[DH(A_0) + O\big(|h - h_0| + |\varepsilon|\big)\big] \cdot (b - a)\boldsymbol{n}_l$$
$$= \big[\boldsymbol{n}_0 \cdot \boldsymbol{n}_l + O\big(|h - h_0| + |\varepsilon|\big)\big](b - a).$$

Note that $\boldsymbol{n}_0 \cdot \boldsymbol{n}_l \neq 0$. Hence,

$$A = B \quad \Longleftrightarrow \quad a = b \quad \Longleftrightarrow \quad H(B) = H(A).$$

The proof is complete.   □

In many cases we can use the Melnikov function $M(h, \delta)$ to determine the cyclicity at a periodic orbit. Based on the above lemma we can prove the results given in next section.

Remark 6.3 The Melnikov function $M(h, \delta)$ is, more precisely, called the *first-order Melnikov function*, since it is the ε-order approximation of the bifurcation function $F(h, \varepsilon, \delta)$. If the first-order Melnikov function is identically zero one computes the so-called "higher-order Melnikov functions", and without more effort it is seen that Lemma 6.2 can be generalized in a straightforward way when the first-order Melnikov function is replaced by the first nonzero higher-order Melnikov function. Computation of higher-order Melnikov functions may be found, for example, in [65]. However, the discussions and computations presented in this book are restricted to the first-order Melnikov function.

6.2 Main Theorem and Corollaries

Theorem 6.4 *Let $h_0 \in (\alpha, \beta)$, $\delta_0 \in D$.*

(i) *There is no limit cycle near L_{h_0} for $\varepsilon + |\delta - \delta_0|$ small, if $M(h_0, \delta_0) \neq 0$.*
(ii) *There is exactly one (at least one) limit cycle $L(\varepsilon, \delta)$ for $\varepsilon + |\delta - \delta_0|$ small, which approaches L_{h_0} as $(\varepsilon, \delta) \to (0, \delta_0)$ if $M(h_0, \delta_0) = 0$, $M_h(h_0, \delta_0) \neq 0$ (h_0 is a zero of $M(h, \delta_0)$ with odd multiplicity), respectively.*
(iii) *The cyclicity of (6.3) at L_{h_0} is at most k if for any $\delta \in D$ there exists $0 \leq j \leq k$ such that*

$$M_h^{(j)}(h_0, \delta) \neq 0.$$

Proof By (6.5) and Lemma 6.2, conclusion (i) is obvious. Conclusion (ii) follows from the implicit function theorem if h_0 is a simple zero of $M(h, \delta_0)$. Let h_0 be a multiple zero of $M(h, \delta_0)$ with odd multiplicity. Then for $\varepsilon_0 > 0$, small, we have

$$M(h_0 - \varepsilon_0, \delta_0) \cdot M(h_0 + \varepsilon_0, \delta_0) < 0.$$

Hence, by (6.5) we have

$$F(h_0 - \varepsilon_0, \varepsilon, \delta) \cdot F(h_0 + \varepsilon_0, \varepsilon, \delta) < 0$$

for $0 < \varepsilon < \varepsilon_0$, $|\delta - \delta_0| < \varepsilon_0$ as long as ε_0 is sufficiently small. Thus, the function $F(h, \varepsilon, \delta)$ has a zero $h^* \in (h_0 - \varepsilon_0, h_0 + \varepsilon_0)$.

For conclusion (iii), if (6.3) has cyclicity at least $(k + 1)$ at L_{h_0}, then there exist some $\varepsilon_n \to 0$, $\delta^{(n)} \in D$ such that for $(\varepsilon, \delta) = (\varepsilon_n, \delta^{(n)})$, (6.3) has at least $k + 1$ limit cycles which approach L_{h_0} as $n \to \infty$. We can suppose $\delta^{(n)} \to \delta_0$ as $n \to \infty$. On the other hand, by our assumption, $M(h, \delta_0)$ has a zero h_0 with multiplicity at most k. By (6.5) and Rolle's theorem, for $\varepsilon + |\delta - \delta_0|$ small, the function $F(h, \varepsilon, \delta)$ has at most k zeros for h near h_0, which follows from the fact that at most k limit cycles exist near L_{h_0} for $\varepsilon + |\delta - \delta_0|$ small. Thus a contradiction appears if $(\varepsilon, \delta) = (\varepsilon_n, \delta^{(n)})$ with n sufficiently large.

The proof is finished. $\qquad\square$

It should be mentioned that an early work related to part (ii) of Theorem 6.4 is due to Pontryagin [179] where the correspondence between regular zeros of the Melnikov function and limit cycles in the perturbed system was first established.

By the above proof we immediately have the following corollaries.

Corollary 6.5 *Suppose for some $\delta_0 \in D$, $M(h, \delta_0)$ has k zeros in $h \in (\alpha, \beta)$, each having an odd multiplicity. Then for $\varepsilon + |\delta - \delta_0|$ small, (6.3) has k limit cycles in a compact subset of the open set G.*

Corollary 6.6 *If there exist $h_0 \in (\alpha, \beta)$, $\delta_0 \in D$ such that for an arbitrary neighborhood of L_{h_0}, (6.3) has k limit cycles in the neighborhood for some (ε, δ) with $\varepsilon + |\delta - \delta_0|$ sufficiently small, then $M(h, \delta_0) = O(|h - h_0|^k)$.*

Corollary 6.7 *If the C^∞ system (6.4) has a periodic orbit, then all orbits near it are also periodic.*

We remark that in many cases the Melnikov function M can be written in the form [133]

$$M(h, \delta) = I(h)[\lambda - P(h)].$$

Then one can find limit cycles by studying the geometrical property of the function $\lambda = P(h)$. This function was introduced by Li [133] and called the *detection function*. Many interesting applications to polynomial systems can be found in [128, 132, 134–136] and [138, 139] and the references therein.

Let $L(\varepsilon, \delta)$ be the limit cycle defined in Theorem 6.4. To determine its stability we need to consider the sign of the integral

$$\varepsilon \oint_{L(\varepsilon, \delta)} (p_x + q_y) \, dt = \varepsilon \left[\oint_{L_{h_0}} (p_x + q_y)|_{\varepsilon=0} \, dt + O(\varepsilon) \right].$$

Obviously, if

$$\sigma(h_0, \delta_0) = \oint_{L_{h_0}} (p_x + q_y)|_{\varepsilon=0, \delta=\delta_0} \, dt \neq 0,$$

the stability can be easily determined. The following lemma shows that $\sigma(h, \delta)$ is the derivative of $M(h, \delta)$ in h.

Lemma 6.8 ([75])

(i) *The closed curve L_h expands (shrinks) with h increasing if L_h is oriented clockwise (counterclockwise), respectively.*

(ii) *For the function M given by (6.6), we have*

$$M_h(h, \delta) = \oint_{L_h} (p_x + q_y)|_{\varepsilon=0} \, dt = \sigma(h, \delta).$$

Proof Fix $h_0 \in (\alpha, \beta)$ and let $A(h_0)$ denote the rightmost point on L_{h_0}. Then, by (6.4) we have

$$H_y(A(h_0)) = 0, \qquad H_x(A(h_0)) \neq 0.$$

Note that L_{h_0} is oriented clockwise (counterclockwise) if $H_x(A(h_0)) > 0 \, (< 0)$, respectively. Introduce a cross section l of (6.4) passing through $A(h_0)$, then for h near h_0 we can write

$$L_h \cap l = A(h) = (a_1(h), a_2(h)), \qquad H(A(h)) = h,$$

which yields

$$H_x(A(h))a_1'(h) + H_y(A(h))a_2'(h) = 1,$$

and hence

$$H_x(A(h_0))a_1'(h_0) = 1, \quad a_1'(h_0) \neq 0.$$

Thus, L_{h_0} is oriented clockwise (counterclockwise) if $a_1'(h_0) > 0 \, (a_1'(h_0) < 0)$, respectively. On the other hand, it is easy to see that L_h expands (shrinks) with h increasing near h_0 if $a_1'(h_0) > 0 \, (a_1'(h_0) < 0)$, respectively. Therefore, L_h expands (shrinks) with h increasing if L_h is oriented clockwise (counterclockwise), respectively.

To be specific, we suppose L_h is oriented clockwise, which yields that L_h expands with h increasing near h_0. Thus for $h > h_0$, applying Green's formula we have

$$M(h, \delta) - M(h_0, \delta) = \int\int_{\Delta(h)} (p_x + q_y)|_{\varepsilon=0} \, dx \, dy, \qquad (6.7)$$

where $\Delta(h)$ denotes the annulus bounded by L_h and L_{h_0}. Let $u(t, h)$ denote a representation of L_h satisfying

$$H(u(t, h)) = h, \qquad 0 \leq t \leq T(h), h \in (\alpha, \beta),$$

where $T(h)$ denotes the period of L_h. Then it follows that

$$DH\big(u(t,h)\big) \cdot D_h u(t,h) = 1.$$

Next, consider the integral transformation of variables given by

$$(x, y) = u(t, r), \quad 0 \le t \le T(r), h_0 < r < h.$$

Noting that $u(t, h)$ satisfies (6.4), we have

$$\det \frac{\partial u(t, r)}{\partial(t, r)} = DH\big(u(t, r)\big) \cdot D_h u(t, r) = 1,$$

which, together with (6.7) results in

$$M(h, \delta) - M(h_0, \delta) = \int_{h_0}^{h} dr \int_{0}^{T(r)} (p_x + q_y)\big(u(t, r), 0, \delta\big) dt.$$

Then differentiating the above equation with respect to h yields

$$M_h(h, \delta) = \int_{0}^{T(h)} (p_x + q_y)\big(u(t, h), 0, \delta\big) dt.$$

This ends the proof.   □

6.3 An Illustrative Example

Example 6.9 Consider van der Pol's equation

$$\ddot{x} + \varepsilon\big(x^2 - 1\big)\dot{x} + x = 0,$$

which can be rewritten as

$$\dot{x} = y, \qquad \dot{y} = -x - \varepsilon\big(x^2 - 1\big)y. \tag{6.8}$$

For $\varepsilon = 0$, (6.8) has periodic orbits $L_h : \frac{1}{2}(x^2 + y^2) = h, h > 0$. By (6.6) we have

$$M(h) = \iint_{x^2+y^2 \le 2h} \big(1 - x^2\big) dx\, dy = \pi h(2 - h).$$

The function M has a unique positive zero, $h = 2$. By Lemma 6.8 we have

$$\sigma_0 = \oint_{L_2} \big(1 - x^2\big) dt = M'(2) = -2\pi.$$

Hence, by Theorem 6.4, (6.8) has a unique limit cycle $L(\varepsilon)$ for $\varepsilon > 0$ which is stable, simple and approaches the circle $x^2 + y^2 = 4$ as $\varepsilon \to 0$.

We remark that one often introduces a suitable linear change of variables or rescaling of the time to transform the linear part of the Hamiltonian system at a center or saddle point into a normal form. This procedure may cause a change of the Melnikov function. More precisely, making a linear change of the form

$$u = a(x - x_0) + b(y - y_0), \qquad v = c(x - x_0) + d(y - y_0),$$

and $\tau = kt$, where $D = ad - bc \neq 0$, we can obtain from (6.3)

$$\frac{du}{d\tau} = \tilde{H}_v + \varepsilon \tilde{p}, \qquad \frac{dv}{d\tau} = -\tilde{H}_u + \varepsilon \tilde{q}, \tag{6.9}$$

where

$$\tilde{H}(u, v) = \frac{D}{k} H(x, y),$$

$$\tilde{p}(u, v, \varepsilon, \delta) = \frac{1}{k} \big[ap(x, y, \varepsilon, \delta) + bq(x, y, \varepsilon, \delta) \big],$$

$$\tilde{q}(u, v, \varepsilon, \delta) = \frac{1}{k} \big[cp(x, y, \varepsilon, \delta) + dq(x, y, \varepsilon, \delta) \big],$$

$$x = x_0 + \frac{1}{D}(du - bv), \qquad y = y_0 + \frac{1}{D}(-cu + av).$$

Let $\tilde{M}$ denote the Melnikov function of (6.9) which is given by

$$\tilde{M}(h, \delta) = \oint_{\tilde{H}(u,v)=h} \tilde{q}\, du - \tilde{p}\, dv|_{\varepsilon=0}.$$

Then it is easy to show that

$$\tilde{M}(h, \delta) = \oint_{\tilde{H}(u,v)=h} \left(\tilde{q}\frac{du}{d\tau} - \tilde{p}\frac{dv}{d\tau} \right) d\tau|_{\varepsilon=0}$$

$$= \operatorname{sgn}(k) \oint_{H(x,y)=kh/D} \left[(a\tilde{q} - c\tilde{p})\frac{dx}{dt} + (b\tilde{q} - d\tilde{p})\frac{dy}{dt} \right] dt|_{\varepsilon=0}$$

$$= \operatorname{sgn}(k)\frac{D}{k} \oint_{H(x,y)=kh/D} (q\, dx - p\, dy)|_{\varepsilon=0} = \frac{D}{|k|} M\left(\frac{k}{D}h, \delta \right).$$

Therefore,

$$M(h, \delta) = \frac{|k|}{D} \tilde{M}\left(\frac{D}{k}h, \delta \right). \tag{6.10}$$

Chapter 7
Limit Cycle Bifurcations Near a Center

In this chapter, particular attention is given to the bifurcation of limit cycles near a center. After normalizing the Hamiltonian function, detailed steps for computing the Melnikov function are described, and formulas are explicitly given. Maple programs for computing the coefficients of the Melnikov function are developed and illustrative examples are presented.

We again consider the C^∞ near-Hamiltonian system (6.1)

$$\dot{x} = H_y + \varepsilon p(x, y, \varepsilon, \delta),$$
$$\dot{y} = -H_x + \varepsilon q(x, y, \varepsilon, \delta),$$

with the unperturbed system (6.4).

7.1 Normalized Hamiltonian Function

Suppose that (6.4) has an elementary center at the origin, i.e., the function H satisfies $H_x(0, 0) = H_y(0, 0) = 0$, and

$$\det \frac{\partial(H_y, -H_x)}{\partial(x, y)}(0, 0) > 0.$$

Then, without loss of generality, we may suppose

$$H_{yy}(0, 0) = \omega, \qquad H_{xx}(0, 0) = \omega, \qquad H_{xy}(0, 0) = 0, \quad \omega > 0.$$

It follows that the expansion of H at the origin has the form

$$H(x, y) = \frac{\omega}{2}(x^2 + y^2) + \sum_{i+j \geq 3} h_{ij} x^i y^j, \quad \omega > 0. \tag{7.1}$$

M. Han, P. Yu, *Normal Forms, Melnikov Functions and Bifurcations of Limit Cycles*, Applied Mathematical Sciences 181, DOI 10.1007/978-1-4471-2918-9_7, © Springer-Verlag London Limited 2012

The implicit function theorem implies that a unique C^∞ function $\varphi(x)$ exists such that $H_y(x, \varphi(x)) = 0$ for $|x|$ small. Thus, let

$$\varphi(x) = \sum_{j \geq 2} e_j x^j. \tag{7.2}$$

We can find (e.g., using Maple) that

$$e_2 = -\frac{1}{\omega} h_{21},$$

$$e_3 = -\frac{1}{\omega^2}(h_{31}\omega - 2h_{12}h_{21}),$$

$$e_4 = -\frac{1}{\omega^3}\left(h_{41}\omega^2 - 2h_{12}h_{31}\omega - 2h_{22}h_{21}\omega + 3h_{03}h_{21}^2 + 4h_{12}^2 h_{21}\right),$$

$$e_5 = \frac{1}{\omega^4}\left(-h_{51}\omega^3 + 2h_{12}h_{41}\omega^2 + 2h_{32}h_{21}\omega^2 + 2h_{22}h_{31}\omega^2 - 8h_{12}h_{22}h_{21}\omega\right.$$
$$\left. - 4h_{12}^2 h_{31}\omega - 6h_{03}h_{21}h_{31}\omega - 3h_{13}h_{21}^2\omega + 18h_{12}h_{03}h_{21}^2 + 8h_{12}^3 h_{21}\right),$$

$$e_6 = -\frac{1}{\omega^5}\left(h_{61}\omega^4 - 2h_{42}h_{21}\omega^3 - 2h_{22}h_{41}\omega^3 - 2h_{12}h_{51}\omega^3 - 2h_{32}h_{31}\omega^3\right.$$
$$+ 8h_{12}h_{32}h_{21}\omega^2 + 3h_{03}h_{31}^2\omega^2 + 4h_{22}^2 h_{21}\omega^2 + 3h_{23}h_{21}^2\omega^2 + 4h_{12}^2 h_{41}\omega^2$$
$$+ 6h_{03}h_{21}h_{41}\omega^2 + 8h_{22}h_{12}h_{31}\omega^2 + 6h_{13}h_{21}h_{31}\omega^2 - 4h_{04}h_{21}^3\omega$$
$$- 18h_{03}h_{21}^2 h_{22}\omega - 24h_{22}h_{12}^2 h_{21}\omega - 8h_{12}^3 h_{31}\omega - 18h_{12}h_{13}h_{21}^2\omega$$
$$\left. - 36h_{03}h_{21}h_{12}h_{31}\omega + 18h_{03}^2 h_{21}^3 + 16h_{12}^4 h_{21} + 72h_{03}h_{21}^2 h_{12}^2\right).$$

By introducing a new variable, $v = y - \varphi(x)$, (6.1) becomes

$$\dot{x} = H_v^*(x, v) + \varepsilon p^*(x, v, \varepsilon, \delta),$$
$$\dot{v} = -H_x^*(x, v) + \varepsilon q^*(x, v, \varepsilon, \delta), \tag{7.3}$$

where

$$H^*(x, v) = H\left(x, v + \varphi(x)\right), \qquad p^*(x, v, \varepsilon, \delta) = p\left(x, v + \varphi(x), \varepsilon, \delta\right),$$
$$q^*(x, v, \varepsilon, \delta) = q\left(x, v + \varphi(x), \varepsilon, \delta\right) - \varphi'(x)p^*(x, v, \varepsilon, \delta). \tag{7.4}$$

By (7.1) and noting $H_y(x, \varphi(x)) = 0$, we have

$$H^*(x, v) = H_0^*(x) + \sum_{j \geq 1} H_j^*(x)v^{j+1} \equiv H_0^*(x) + v^2 \tilde{H}(x, v), \tag{7.5}$$

where

$$H_0^*(x) = H\big(x, \varphi(x)\big) = \sum_{j \geq 2} h_j x^j,$$

$$H_j^*(x) = \frac{1}{(j+1)!} \frac{\partial^{j+1} H}{\partial y^{j+1}} \big(x, \varphi(x)\big), \quad j \geq 1, \tag{7.6}$$

$$\tilde{H}(0, 0) = \frac{\omega}{2},$$

and

$$h_2 = \frac{\omega}{2}, \qquad h_3 = h_{30}, \qquad h_4 = \frac{1}{2\omega}\big(2h_{40}\omega - h_{21}^2\big),$$

$$h_5 = \frac{1}{\omega^2}\big(h_{50}\omega^2 - h_{31}h_{21}\omega + h_{12}h_{21}^2\big),$$

$$\begin{aligned}
h_6 = \frac{1}{2\omega^3}\big(&2h_{60}\omega^3 - h_{31}^2\omega^2 - 2h_{41}h_{21}\omega^2 + 4h_{31}h_{12}h_{21}\omega + 2h_{22}h_{21}^2\omega \\
&- 4h_{12}^2 h_{21}^2 - 2h_{03}h_{21}^3\big),
\end{aligned} \tag{7.7}$$

$$\begin{aligned}
h_7 = \frac{1}{\omega^4}\big(&h_{70}\omega^4 - h_{31}h_{41}\omega^3 - h_{51}h_{21}\omega^3 + h_{31}^2 h_{12}\omega^2 + 2h_{31}h_{22}h_{21}\omega^2 \\
&+ h_{32}h_{21}^2\omega^2 + 2h_{12}h_{21}h_{41}\omega^2 - 4h_{12}h_{21}^2 h_{22}\omega - 4h_{31}h_{12}^2 h_{21}\omega \\
&- h_{13}h_{21}^3\omega - 3h_{31}h_{03}h_{21}^2\omega + 6h_{12}h_{21}^3 h_{03} + 4h_{12}^3 h_{21}^2\big).
\end{aligned}$$

7.2 Computation of the Melnikov Function, M

By (7.1), surrounding the origin there exists a family of periodic orbits defined by
the equation $H(x, y) = h$ or $H^*(x, v) = h$ for $h > 0$, small. Then it is obvious that

$$\begin{aligned}
M(h, \delta) &= \oint_{H^*(x,v)=h} q^* \, dx - p^* \, dv|_{\varepsilon=0} \\
&= \iint_{H^* \leq h} \big(p_x^* + q_v^*\big) \, dx \, dv \\
&= \oint_{H^*(x,v)=h} \bar{q}(x, v, \delta) \, dx, \tag{7.8}
\end{aligned}$$

where

$$\begin{aligned}
\bar{q}(x, v, \delta) &= q^*(x, v, 0, \delta) - q^*(x, 0, 0, \delta) + \int_0^v p_x^*(x, u, 0, \delta) \, du \\
&= \int_0^v \big[p_x^*(x, u, 0, \delta) + q_v^*(x, u, 0, \delta) \big] \, du \tag{7.9}
\end{aligned}$$

satisfying

$$\bar{q}_v = \left(p_x^* + q_v^*\right)\big|_{\varepsilon=0} \quad \text{and} \quad \bar{q}(x, 0, \delta) = 0.$$

By (7.2), (7.4) and (7.9), we obtain that when

$$p_x(x, y, 0, \delta) + q_y(x, y, 0, \delta) = \sum_{i+j \geq 0} c_{ij} x^i y^j, \tag{7.10}$$

$$\bar{q}(x, v, \delta) = v \sum_{i+j \geq 0} \bar{b}_{ij} x^i v^j = \sum_{j \geq 1} q_j(x) v^j, \tag{7.11}$$

where

$$q_{j+1}(x) = \frac{1}{(j+1)!} \frac{\partial^j}{\partial v^j} \left(p_x^* + q_v^*\right)\big|_{\varepsilon=v=0}$$

$$= \frac{1}{(j+1)!} \frac{\partial^j}{\partial y^j} (p_x + q_y)\left(x, \varphi(x), 0, \delta\right)$$

$$= \sum_{i \geq 0} \bar{b}_{ij} x^i, \quad j \geq 0.$$

The coefficients $\bar{b}_{ij}$ (obtained using Maple) are given by

$$\bar{b}_{00} = c_{00}, \qquad \bar{b}_{10} = c_{10}, \qquad \bar{b}_{20} = c_{01}e_2 + c_{20}, \qquad \bar{b}_{30} = c_{01}e_3 + c_{11}e_2 + c_{30},$$

$$\bar{b}_{40} = c_{11}e_3 + c_{21}e_2 + c_{02}e_2^2 + c_{40} + c_{01}e_4,$$

$$\bar{b}_{50} = c_{31}e_2 + c_{11}e_4 + 2c_{02}e_2e_3 + c_{21}e_3 + c_{50} + c_{12}e_2^2 + c_{01}e_5,$$

$$\bar{b}_{60} = 2c_{12}e_2e_3 + 2c_{02}e_2e_4 + c_{03}e_2^3 + c_{41}e_2 + c_{60} + c_{31}e_3 + c_{22}e_2^2 + c_{11}e_5$$

$$\qquad + c_{01}e_6 + c_{21}e_4 + c_{02}e_3^2,$$

$$\bar{b}_{01} = \frac{1}{2}c_{01}, \qquad \bar{b}_{11} = \frac{1}{2}c_{11}, \qquad \bar{b}_{21} = \frac{1}{2}c_{21} + c_{02}e_2,$$

$$\bar{b}_{31} = \frac{1}{2}c_{31} + c_{02}e_3 + c_{12}e_2, \qquad \bar{b}_{41} = \frac{1}{2}c_{41} + c_{02}e_4 + c_{12}e_3 + c_{22}e_2 + \frac{3}{2}c_{03}e_2^2,$$

$$\bar{b}_{02} = \frac{1}{3}c_{02}, \qquad \bar{b}_{12} = \frac{1}{3}c_{12}, \qquad \bar{b}_{22} = \frac{1}{3}c_{22} + c_{03}e_2,$$

$$\bar{b}_{32} = \frac{1}{3}c_{32} + c_{03}e_3 + c_{13}e_2, \qquad \bar{b}_{42} = \frac{1}{3}c_{42} + c_{03}e_4 + c_{13}e_3 + 2c_{04}e_2^2 + c_{23}e_2,$$

$$\bar{b}_{03} = \frac{1}{4}c_{03}, \qquad \bar{b}_{13} = \frac{1}{4}c_{13}, \qquad \bar{b}_{23} = \frac{1}{4}c_{23} + c_{04}e_2, \qquad \bar{b}_{04} = \frac{1}{5}c_{04},$$

$$\bar{b}_{14} = \frac{1}{4}c_{14}, \qquad \bar{b}_{24} = \frac{1}{4}c_{24} + c_{05}e_2, \qquad \bar{b}_{05} = \frac{1}{6}c_{05}, \qquad \bar{b}_{06} = \frac{1}{7}c_{06}.$$

$$\tag{7.12}$$

Consider the equation $H^*(x, v) = h$; we have the following lemma.

Lemma 7.1 *The equation $H^*(x, v) = h$ has exactly two C^∞ solutions $v_1(x, w)$ and $v_2(x, w)$ in v satisfying*

$$v_1(x, w) = \sqrt{\frac{2}{\omega}} w \left(1 + O(|x, w|)\right), \qquad v_2(x, w) = v_1(x, -w),$$

where $w = \sqrt{h - H_0^*(x)}$.

Proof By (7.5) the equation $H^*(x, v) = h$ is equivalent to

$$w = |v| \left(\tilde{H}(x, v)\right)^{\frac{1}{2}}. \tag{7.13}$$

Introduce an equation of the form

$$u = v \left(\tilde{H}(x, v)\right)^{\frac{1}{2}}.$$

Noting that $[\tilde{H}(x, v)]^{\frac{1}{2}} \in C^\infty$ for $|x| + |v|$ small, the implicit function theorem implies that the above equation has a unique solution in v, denoted by $v^*(x, u)$. Obviously, the solution is C^∞ with $v^*(x, u) = \sqrt{\frac{2}{\omega}} u (1 + O(|x, u|))$. Let

$$v_1(x, w) = v^*(x, w), \qquad v_2(x, w) = v^*(x, -w).$$

Then, $v_1(x, w)$ and $v_2(x, w)$ are the only solutions of (7.13) for $v > 0$ and $v < 0$, respectively, with

$$v_1(x, w) = v_2(x, -w) \quad \text{for } |x| + |w| \text{ small.}$$

The proof is complete. $\square$

Further, we can write

$$v_1(x, w) = \sum_{j \geq 1} a_j(x) w^j. \tag{7.14}$$

Then by (7.13) and (7.5) we obtain

$$w^2 = \left(v_1(x, w)\right)^2 \sum_{j \geq 1} H_j^*(x) \left[v_1(x, w)\right]^{j-1}$$

$$= H_1^* a_1^2 w^2 + \left(H_2^* a_1^3 + 2 H_1^* a_1 a_2\right) w^3$$

$$+ \left(2 H_1^* a_1 a_3 + H_3^* a_1^4 + H_1^* a_2^2 + 3 H_2^* a_1^2 a_2\right) w^4$$

$$+ \left(3 H_2^* a_1 a_2^2 + 2 H_1^* a_2 a_3 + H_4^* a_1^5 + 2 H_1^* a_1 a_4 + 4 H_3^* a_1^3 a_2 + 3 H_2^* a_1^2 a_3\right) w^5$$

$$+ \left(3H_2^* a_1^2 a_4 + H_2^* a_2^3 + H_5^* a_1^6 + 6H_3^* a_1^2 a_2^2 + 2H_1^* a_1 a_5 + 4H_3^* a_1^3 a_3\right.$$

$$+ 6H_2^* a_1 a_2 a_3 + 5H_4^* a_1^4 a_2 + 2H_1^* a_2 a_4 + \left. H_1^* a_3^2\right)w^6$$

$$+ \left(5H_4^* a_1^4 a_3 + 3H_2^* a_1 a_3^2 + 2H_1^* a_1 a_6 + 2H_1^* a_3 a_4 + 2H_1^* a_2 a_5\right.$$

$$+ 3H_2^* a_1^2 a_5 + 4H_3^* a_2^3 a_1 + 3H_2^* a_2^2 a_3 + 10H_4^* a_1^3 a_2^2$$

$$+ 12H_3^* a_1^2 a_2 a_3 + H_6^* a_1^7 + 6H_2^* a_1 a_2 a_4 + 4H_3^* a_1^3 a_4 + 6H_5^* a_1^5 a_2\right)w^7$$

$$+ \left(6H_2^* a_1 a_3 a_4 + H_1^* a_4^2 + 20H_4^* a_1^3 a_2 a_3 + 2H_1^* a_3 a_5 + 10H_4^* a_1^2 a_2^3\right.$$

$$+ 12H_3^* a_1^2 a_2 a_4 + 6H_5^* a_1^5 a_3 + 3H_2^* a_2^2 a_4 + 2H_1^* a_1 a_7 + 3H_2^* a_1^2 a_6$$

$$+ H_7^* a_1^8 + 6H_3^* a_1^2 a_3^2 + 7H_6^* a_1^6 a_2 + 15H_5^* a_1^4 a_2^2 + 5H_4^* a_4 a_1^4$$

$$+ 6H_2^* a_1 a_2 a_5 + 4H_3^* a_1^3 a_5 + 2H_1^* a_2 a_6 + 3H_2^* a_2 a_3^2$$

$$+ 12H_3^* a_2^2 a_1 a_3 + H_3^* a_2^4\right)w^8 + \cdots .$$

Thus, comparing the terms on both sides of the above equation yields

$$a_1(x) = \frac{1}{\sqrt{H_1^*(x)}}, \qquad a_2(x) = -\frac{H_2^*(x)}{2(H_1^*(x))^2},$$

$$a_3(x) = -\frac{1}{8(H_1^*)^{\frac{7}{2}}}\left[4H_1^* H_3^* - 5\left(H_2^*\right)^2\right],$$

$$a_4(x) = -\frac{1}{2H_1^{*5}}\left(2H_2^{*3} + H_4^* H_1^{*2} - 3H_2^* H_3^* H_1^*\right),$$

$$a_5(x) = -\frac{1}{128(H_1^*)^{\frac{13}{2}}}\left(-231H_2^{*4} + 64H_5^* H_1^{*3} + 504H_3^* H_2^{*2} H_1^*\right.$$

$$\left. - 112H_3^{*2} H_1^{*2} - 224H_4^* H_2^* H_1^{*2}\right),$$

$$a_6(x) = -\frac{1}{2H_1^{*8}}\left(-20H_3^* H_2^{*3} H_1^* + H_6^* H_1^{*4} + 10H_2^* H_3^{*2} H_1^{*2} - 4H_3^* H_1^{*3} H_4^*\right.$$

$$\left. + 10H_2^{*2} H_4^* H_1^{*2} - 4H_2^* H_5^* H_1^{*3} + 7H_2^{*5}\right),$$

$$a_7(x) = -\frac{1}{1024H_1^{*19/2}}\left(512H_7^* H_1^{*5} - 20592H_2^{*2} H_3^{*2} H_1^{*2} - 2304H_5^* H_1^{*4} H_3^*\right.$$

$$+ 25740H_3^* H_2^{*4} H_1^* - 2304H_6^* H_2^* H_1^{*4} - 13728H_4^* H_2^{*3} H_1^{*2}$$

$$+ 6336H_5^* H_2^{*2} H_1^{*3} + 12672H_4^* H_1^{*3} H_2^* H_3^* - 7293H_2^{*6}$$

$$\left. - 1152H_4^{*2} H_1^{*4} + 2112H_3^{*3} H_1^{*3}\right).$$

On the other hand, by (7.1), (7.2) and (7.6) we have

$$H_1^* = \frac{\omega}{2} + h_{12}x + (h_{22} + 3h_{03}e_2)x^2 + (3h_{03}e_3 + 3h_{13}e_2 + h_{32})x^3$$
$$+ (3h_{13}e_3 + h_{42} + 6h_{04}e_2^2 + 3h_{03}e_4 + 3h_{23}e_2)x^4$$
$$+ (12h_{04}e_2e_3 + 3h_{13}e_4 + 3h_{23}e_3 + 3h_{33}e_2 + 6h_{14}e_2^2 + 3h_{03}e_5 + h_{52})x^5$$
$$+ (3h_{23}e_4 + 12h_{04}e_2e_4 + 3h_{43}e_2 + 3h_{33}e_3 + 6h_{04}e_3^2 + 6h_{24}e_2^2$$
$$+ 12h_{14}e_2e_3 + 3h_{03}e_6 + 3h_{13}e_5 + 10h_{05}e_2^3 + h_{62})x^6 + O(x)^7,$$

$$H_2^* = h_{03} + h_{13}x + (4e_2h_{04} + h_{23})x^2 + (4e_3h_{04} + 4e_2h_{14} + h_{33})x^3$$
$$+ (10h_{05}e_2^2 + 4h_{04}e_4 + h_{43} + 4h_{14}e_3 + 4h_{24}e_2)x^4 + O(x)^5,$$

$$H_3^* = h_{04} + h_{14}x + (5e_2h_{05} + h_{24})x^2 + (5e_3h_{05} + 5e_2h_{15} + h_{34})x^3$$
$$+ (15h_{06}e_2^2 + 5h_{25}e_2 + h_{44} + 5h_{15}e_3 + 5h_{05}e_4)x^4 + O(x)^5,$$

$$H_4^* = h_{05} + h_{15}x + (6h_{06}e_2 + h_{25})x^2 + O(x^3),$$

$$H_5^* = h_{06} + h_{16}x + (7h_{07}e_2 + h_{26})x^2 + O(x^3),$$

$$H_6^* = h_{07} + O(x), \qquad H_7^* = h_{08} + O(x).$$

Therefore, by writing

$$a_j(x) = \sum_{i \geq 0} \bar{a}_{ij}x^i, \quad j = 1, 2, 3, 4, 5, 6, 7, \tag{7.15}$$

we obtain

$$\bar{a}_{01} = \sqrt{2\omega^{-1}}, \qquad \bar{a}_{11} = -\frac{\sqrt{2\omega^{-1}}}{\omega}h_{12},$$

$$\bar{a}_{21} = -\frac{\sqrt{2\omega^{-1}}}{2\omega^2}(2h_{22}\omega + 6h_{03}e_2\omega - 3h_{12}^2),$$

$$\bar{a}_{31} = -\frac{\sqrt{2\omega^{-1}}}{2\omega^3}\big[(6h_{13}e_2 + 2h_{32} + 6h_{03}e_3)\omega^2$$
$$- (18h_{12}h_{03}e_2 + 6h_{12}h_{22})\omega + 5h_{12}^3\big],$$

$$\bar{a}_{41} = -\frac{\sqrt{2\omega^{-1}}}{8\omega^4}\big[(24h_{03}e_4 + 24h_{13}e_3 + 8h_{42} + 48h_{04}e_2^2 + 24h_{23}e_2)\omega^3$$
$$- (24h_{12}h_{32} + 108h_{03}^2e_2^2 + 72h_{12}h_{03}e_3 + 12h_{22}^2 + 72h_{22}h_{03}e_2$$
$$+ 72h_{12}h_{13}e_2)\omega^2 + (60h_{12}^2h_{22} + 180h_{12}^2h_{03}e_2)\omega - 35h_{12}^4\big],$$

$$\bar{a}_{51} = -\frac{\sqrt{2\omega^{-1}}}{8\omega^5}\big[(24h_{23}e_3 + 24h_{33}e_2 + 48h_{14}e_2^2 + 24h_{13}e_4 + 96h_{04}e_2e_3$$
$$+ 8h_{52} + 24h_{03}e_5)\omega^4 - (24h_{12}h_{42} + 72h_{12}h_{03}e_4 + 24h_{22}h_{32}$$

$$+ 72h_{12}h_{23}e_2 + 72h_{12}h_{13}e_3 + 72h_{22}h_{13}e_2 + 216h_{03}^2e_2e_3 + 216h_{03}e_2^2h_{13}$$
$$+ 144h_{12}h_{04}e_2^2 + 72h_{22}h_{03}e_3 + 72h_{03}e_2h_{32})\omega^3 + (60h_{12}^2h_{32}$$
$$+ 540h_{12}h_{03}^2e_2^2 + 180h_{12}^2h_{03}e_3 + 180h_{12}^2h_{13}e_2 + 60h_{12}h_{22}^2$$
$$+ 360h_{12}h_{22}h_{03}e_2)\omega^2 - (420h_{12}^3h_{03}e_2 + 140h_{12}^3h_{22})\omega + 63h_{12}^5 \big],$$

$$\bar{a}_{61} = -\frac{\sqrt{2\omega^{-1}}}{16\omega^6}\Big[(192h_{14}e_2e_3 + 96h_{04}e_3^2 + 48h_{23}e_4 + 16h_{62} + 48h_{33}e_3$$

$$+ 160h_{05}e_2^3 + 96h_{24}e_2^2 + 48h_{03}e_6 + 192h_{04}e_2e_4 + 48h_{13}e_5 + 48h_{43}e_2)\omega^5$$

$$- (144h_{03}e_3h_{32} + 144h_{13}e_2h_{32} + 864h_{03}e_3h_{13}e_2 + 144h_{12}h_{13}e_4$$
$$+ 144h_{12}h_{23}e_3 + 144h_{22}h_{13}e_3 + 288h_{22}h_{04}e_2^2 + 144h_{12}h_{03}e_5$$
$$+ 144h_{22}h_{03}e_4 + 144h_{12}h_{33}e_2 + 288h_{12}h_{14}e_2^2 + 864h_{03}e_2^3h_{04}$$
$$+ 432h_{03}^2e_2e_4 + 432h_{03}e_2^2h_{23} + 144h_{03}e_2h_{42} + 48h_{12}h_{52} + 48h_{22}h_{42}$$
$$+ 576h_{12}h_{04}e_2e_3 + 216h_{03}^2e_3^2 + 216h_{13}^2e_2^2 + 24h_{32}^2 + 144h_{22}h_{23}e_2)\omega^4$$
$$+ (720h_{12}h_{03}e_2h_{32} + 40h_{22}^3 + 2160h_{12}h_{03}e_2^2h_{13} + 720h_{12}h_{22}h_{03}e_3$$
$$+ 720h_{12}h_{22}h_{13}e_2 + 2160h_{12}h_{03}^2e_2e_3 + 720h_{12}^2h_{04}e_2^2 + 360h_{12}^2h_{23}e_2$$
$$+ 360h_{22}^2h_{03}e_2 + 1080h_{03}^3e_2^3 + 1080h_{22}h_{03}^2e_2^2 + 360h_{12}^2h_{03}e_4$$
$$+ 120h_{12}^2h_{42} + 360h_{12}^2h_{13}e_3 + 240h_{12}h_{22}h_{32})\omega^3 - (840h_{12}^3h_{13}e_2$$
$$+ 280h_{12}^3h_{32} + 3780h_{12}^2h_{03}^2e_2^2 + 420h_{12}^2h_{22}^2 + 2520h_{12}^2h_{22}h_{03}e_2$$
$$+ 840h_{12}^3h_{03}e_3)\omega^2 + (1890h_{12}^4h_{03}e_2 + 630h_{12}^4h_{22})\omega - 231h_{12}^6 \big],$$

$$\bar{a}_{02} = -\frac{2}{\omega^2}h_{03},$$

$$\bar{a}_{12} = -\frac{2}{\omega^3}(h_{13}\omega - 4h_{12}h_{03}),$$

$$\bar{a}_{22} = -\frac{2}{\omega^4}\big[(h_{23} + 4h_{04}e_2)\omega^2 - (4h_{12}h_{13} + 4h_{03}h_{22} + 12h_{03}^2e_2)\omega + 12h_{03}h_{12}^2\big],$$

$$\bar{a}_{32} = -\frac{2}{\omega^5}\big[(h_{33} + 4h_{14}e_2 + 4h_{04}e_3)\omega^3 - (4h_{12}h_{23} + 16h_{12}h_{04}e_2 + 4h_{13}h_{22}$$
$$+ 24h_{13}h_{03}e_2 + 12h_{03}^2e_3 + 4h_{03}h_{32})\omega^2 + (12h_{13}h_{12}^2 + 24h_{03}h_{12}h_{22}$$
$$+ 72h_{03}^2h_{12}e_2)\omega - 32h_{03}h_{12}^3\big],$$

$$\bar{a}_{42} = -\frac{2}{\omega^6}\big[(4h_{04}e_4 + 4h_{24}e_2 + 4h_{14}e_3 + 10h_{05}e_2^2 + h_{43})\omega^4 - (24h_{23}h_{03}e_2$$
$$+ 16h_{04}e_2h_{22} + 4h_{12}h_{33} + 4h_{23}h_{22} + 4h_{03}h_{42} + 4h_{13}h_{32} + 12h_{03}^2e_4$$

$$+ 12h_{13}^2 e_2 + 16h_{12}h_{04}e_3 + 24h_{13}h_{03}e_3 + 72h_{04}e_2^2 h_{03} + 16h_{12}h_{14}e_2)\omega^3$$

$$+ (12h_{12}^2 h_{23} + 108h_{03}^3 e_2^2 + 12h_{03}h_{22}^2 + 24h_{13}h_{12}h_{22} + 72h_{03}^2 h_{22}e_2$$

$$+ 48h_{12}^2 h_{04}e_2 + 24h_{03}h_{12}h_{32} + 72h_{03}^2 h_{12}e_3 + 144h_{13}h_{12}h_{03}e_2)\omega^2$$

$$- (288h_{03}^2 h_{12}^2 e_2 + 96h_{03}h_{12}^2 h_{22} + 32h_{13}h_{12}^3)\omega + 80h_{03}h_{12}^4],$$

$$\bar{a}_{03} = -\frac{\sqrt{2\omega^{-1}}}{\omega^3}(2h_{04}\omega - 5h_{03}^2),$$

$$\bar{a}_{13} = -\frac{\sqrt{2\omega^{-1}}}{\omega^4}\left[2h_{14}\omega^2 - (10h_{04}h_{12} + 10h_{03}h_{13})\omega + 35h_{12}h_{03}^2\right],$$

$$\bar{a}_{23} = -\frac{\sqrt{2\omega^{-1}}}{2\omega^5}\left[(20h_{05}e_2 + 4h_{24})\omega^3 - (140h_{04}h_{03}e_2 + 20h_{14}h_{12} + 20h_{04}h_{22}\right.$$

$$+ 20h_{03}h_{23} + 10h_{13}^2)\omega^2 + (70h_{12}^2 h_{04} + 140h_{12}h_{03}h_{13} + 70h_{03}^2 h_{22}$$

$$\left. + 210h_{03}^3 e_2)\omega - 315h_{12}^2 h_{03}^2\right],$$

$$\bar{a}_{33} = -\frac{\sqrt{2\omega^{-1}}}{2\omega^6}\left[(20h_{15}e_2 + 20h_{05}e_3 + 4h_{34})\omega^4 - (140h_{04}h_{13}e_2 + 20h_{14}h_{22}\right.$$

$$+ 140h_{14}h_{03}e_2 + 100h_{12}h_{05}e_2 + 20h_{13}h_{23} + 140h_{04}h_{03}e_3 + 20h_{04}h_{32}$$

$$+ 20h_{12}h_{24} + 20h_{03}h_{33})\omega^3 + (210h_{03}^3 e_3 + 70h_{03}^2 h_{32} + 140h_{03}h_{13}h_{22}$$

$$+ 980h_{12}h_{04}h_{03}e_2 + 630h_{03}^2 h_{13}e_2 + 140h_{12}h_{04}h_{22} + 140h_{12}h_{03}h_{23}$$

$$+ 70h_{12}h_{13}^2 + 70h_{12}^2 h_{14})\omega^2 - (210h_{12}^3 h_{04} + 1890h_{12}h_{03}^3 e_2$$

$$\left. + 630h_{12}h_{03}^2 h_{22} + 630h_{12}^2 h_{03}h_{13})\omega + 1155h_{12}^3 h_{03}^2\right],$$

$$\bar{a}_{43} = -\frac{\sqrt{2\omega^{-1}}}{8\omega^7}\left[(80h_{15}e_3 + 240h_{06}e_2^2 + 16h_{44} + 80h_{25}e_2 + 80h_{05}e_4)\omega^5\right.$$

$$- (560h_{03}h_{24}e_2 + 560h_{03}h_{14}e_3 + 560h_{04}h_{03}e_4 + 400h_{22}h_{05}e_2$$

$$+ 2000h_{03}h_{05}e_2^2 + 400h_{12}h_{05}e_3 + 40h_{23}^2 + 80h_{03}h_{43} + 560h_{13}h_{14}e_2$$

$$+ 560h_{23}h_{04}e_2 + 80h_{12}h_{34} + 560h_{04}h_{13}e_3 + 400h_{12}h_{15}e_2 + 80h_{14}h_{32}$$

$$+ 1120h_{04}^2 e_2^2 + 80h_{22}h_{24} + 80h_{13}h_{33} + 80h_{04}h_{42})\omega^4$$

$$+ (3920h_{12}h_{04}h_{03}e_3 + 560h_{03}h_{22}h_{23} + 2520h_{03}^2 e_2h_{23} + 560h_{14}h_{12}h_{22}$$

$$+ 280h_{03}^2 h_{42} + 560h_{12}h_{04}h_{32} + 1400h_{12}^2 h_{05}e_2 + 3920h_{12}h_{04}h_{13}e_2$$

$$+ 3920h_{03}h_{22}h_{04}e_2 + 2520h_{03}^2 h_{13}e_3 + 280h_{04}h_{22}^2 + 10920h_{03}^2 e_2^2 h_{04}$$

$$+ 560h_{12}h_{03}h_{33} + 3920h_{12}h_{03}h_{14}e_2 + 560h_{12}h_{13}h_{23} + 2520h_{03}h_{13}^2 e_2$$

$$+ 280h_{13}^2 h_{22} + 840h_{03}^3 e_4 + 280h_{12}^2 h_{24} + 560h_{03}h_{13}h_{32})\omega^3$$

$$- \left(840h_{12}^3h_{14} + 1260h_{12}^2h_{13}^2 + 22680h_{12}h_{03}^2h_{13}e_2 + 1260h_{03}^2h_{22}^2 \right.$$

$$+ 17640h_{12}^2h_{04}h_{03}e_2 + 2520h_{12}h_{03}^2h_{32} + 2520h_{12}^2h_{03}h_{23} + 7560h_{12}h_{03}^3e_3$$

$$+ 11340h_{03}^4e_2^2 + 5040h_{12}h_{03}h_{13}h_{22} + 2520h_{12}^2h_{04}h_{22} + 7560h_{03}^3h_{22}e_2 \big)\omega^2$$

$$+ \left(2310h_{12}^4h_{04} + 41580h_{12}^2h_{03}^3e_2 + 13860h_{12}^2h_{03}^2h_{22} + 9240h_{12}^3h_{03}h_{13}\right)\omega$$

$$- 15015h_{12}^4h_{03}^2\big],$$

$$\bar{a}_{04} = -\frac{4}{\omega^5}\left(h_{05}\omega^2 - 6h_{03}h_{04}\omega + 8h_{03}^3\right),$$

$$\bar{a}_{14} = -\frac{4}{\omega^6}\left[h_{15}\omega^3 - (6h_{05}h_{12} + 6h_{03}h_{14} + 6h_{13}h_{04})\omega^2 + \left(24h_{03}^2h_{13}\right.\right.$$

$$+ 48h_{03}h_{04}h_{12})\omega - 80h_{12}h_{03}^3\big],$$

$$\bar{a}_{24} = -\frac{4}{\omega^7}\left[(6h_{06}e_2 + h_{25})\omega^4 - \left(6h_{13}h_{14} + 6h_{04}h_{23} + 24h_{04}^2e_2 + 6h_{12}h_{15}\right.\right.$$

$$+ 6h_{05}h_{22} + 48h_{05}h_{03}e_2 + 6h_{03}h_{24})\omega^3 + \left(24h_{12}^2h_{05} + 48h_{12}h_{13}h_{04}\right.$$

$$+ 240h_{03}^2h_{04}e_2 + 24h_{13}^2h_{03} + 48h_{03}h_{04}h_{22} + 24h_{03}^2h_{23} + 48h_{12}h_{03}h_{14})\omega^2$$

$$- \left(240h_{12}h_{03}^2h_{13} + 240h_{12}^2h_{03}h_{04} + 80h_{03}^3h_{22} + 240h_{03}^4e_2\right)\omega + 480h_{12}^2h_{03}^3\big],$$

$$\bar{a}_{05} = -\frac{\sqrt{2\omega^{-1}}}{2\omega^6}\left[8h_{06}\omega^3 - \left(28h_{04}^2 + 56h_{05}h_{03}\right)\omega^2 + 252h_{04}h_{03}^2\omega - 231h_{03}^4\right],$$

$$\bar{a}_{15} = -\frac{\sqrt{2\omega^{-1}}}{2\omega^7}\left[8h_{16}\omega^4 - (56h_{06}h_{12} + 56h_{04}h_{14} + 56h_{05}h_{13} + 56h_{15}h_{03})\omega^3\right.$$

$$+ \left(504h_{05}h_{03}h_{12} + 504h_{04}h_{03}h_{13} + 252h_{14}h_{03}^2 + 252h_{04}^2h_{12}\right)\omega^2$$

$$- \left(924h_{03}^3h_{13} + 2772h_{04}h_{03}^2h_{12}\right)\omega + 3003h_{12}h_{03}^4\big],$$

$$\bar{a}_{25} = -\frac{\sqrt{2\omega^{-1}}}{4\omega^8}\left[(112h_{07}e_2 + 16h_{26})\omega^5 - \left(112h_{06}h_{22} + 112h_{16}h_{12} + 112h_{15}h_{13}\right.\right.$$

$$+ 56h_{14}^2 + 112h_{04}h_{24} + 1008h_{03}h_{06}e_2 + 112h_{05}h_{23} + 1008h_{05}h_{04}e_2$$

$$+ 112h_{03}h_{25})\omega^4 + \left(5544h_{04}^2h_{03}e_2 + 1008h_{12}h_{05}h_{13} + 504h_{04}^2h_{22}\right.$$

$$+ 504h_{03}^2h_{24} + 1008h_{04}h_{14}h_{12} + 1008h_{05}h_{03}h_{22} + 1008h_{04}h_{03}h_{23}$$

$$+ 5544h_{03}^2h_{05}e_2 + 504h_{12}^2h_{06} + 1008h_{12}h_{15}h_{03} + 504h_{04}h_{13}^2$$

$$+ 1008h_{14}h_{03}h_{13})\omega^3 - \left(1848h_{03}^3h_{23} + 5544h_{04}h_{03}^2h_{22} + 2772h_{12}^2h_{04}^2\right.$$

$$+ 2772h_{03}^2h_{13}^2 + 24024h_{04}h_{03}^3e_2 - 11088h_{12}h_{04}h_{03}h_{13} + 5544h_{12}^2h_{05}h_{03}$$

$$+ 5544h_{12}h_{14}h_{03}^2)\omega^2 + \left(18018h_{03}^5e_2 + 24024h_{12}h_{03}^3h_{13} + 6006h_{03}^4h_{22}\right.$$

$$+ 36036h_{12}^2h_{04}h_{03}^2)\omega - 45045h_{12}^2h_{03}^4\big],$$

$$a_{06} = -\frac{8}{\omega^8}\left[h_{07}\omega^4 - (8h_{04}h_{05} + 8h_{03}h_{06})\omega^3\right.$$
$$\left. + (40h_{03}^2h_{05} + 40h_{04}^2h_{03})\omega^2 - 160h_{03}^3h_{04}\omega + 112h_{03}^5\right],$$

$$\bar{a}_{07} = -\frac{\sqrt{2\omega^{-1}}}{2\omega^9}\left[16h_{08}\omega^5 - (144h_{07}h_{03} + 144h_{06}h_{04} + 72h_{05}^2)\omega^4\right.$$
$$+ (792h_{06}h_{03}^2 + 1584h_{05}h_{03}h_{04} + 264h_{04}^3)\omega^3 - (5148h_{03}^2h_{04}^2$$
$$\left. + 3432h_{05}h_{03}^3)\omega^2 + 12870h_{04}h_{03}^4\omega - 7293h_{03}^6\right].$$

$$(7.16)$$

Let $x_1(h) > 0$ and $x_2(h) < 0$ be the solutions of the equation $H_0^*(x) = h$. Then, it follows from (7.8) that

$$M(h, \delta) = \int_{x_2(h)}^{x_1(h)} \left[\bar{q}(x, v_1(x, w)) - \bar{q}(x, v_2(x, w))\right] dx.$$

Hence, by Lemma 7.1, the function $\bar{q}(x, v_1) - \bar{q}(x, v_2)$ is odd in w. Thus, we can write

$$\bar{q}(x, v_1) - \bar{q}(x, v_2) = \sum_{j\geq 0} \bar{q}_j(x)w^{2j+1}, \tag{7.17}$$

where, by (7.11), (7.14), (7.15) and Lemma 7.1,

$$\bar{q}_0(x) = 2q_1(x)a_1(x) = \sum_{i\geq 0}\alpha_{i0}x^i,$$

$$\bar{q}_1(x) = 2\left[q_1(x)a_3(x) + 2q_2(x)a_1(x)a_2(x) + q_3(x)a_1^3(x)\right] = \sum_{i\geq 0}\alpha_{i1}x^i,$$

$$\bar{q}_2(x) = 2\left(q_5a_1^5 + 2q_2a_1a_4 + q_1a_5 + 2q_2a_2a_3 + 3q_3a_1a_2^2 + 4q_4a_1^3a_2 + 3q_3a_1^2a_3\right)$$
$$= \sum_{i\geq 0}\alpha_{i2}x^i, \tag{7.18}$$

$$\bar{q}_3(x) = 2\left(4q_4a_1^3a_4 + 6q_6a_1^5a_2 + 2q_2a_1a_6 + 10q_5a_1^3a_2^2 + 3q_3a_1a_3^2\right.$$
$$+ 2q_2a_3a_4 + 2q_2a_2a_5 + 5q_5a_1^4a_3 + 4q_4a_1a_2^3 + 3q_3a_1^2a_5$$
$$\left. + 6q_3a_1a_2a_4 + 12q_4a_1^2a_2a_3 + q_1a_7 + q_7a_1^7 + 3q_3a_2^2a_3\right)$$
$$= \sum_{i\geq 0}\alpha_{i3}x^i,$$

with

$$\alpha_{00} = 2\bar{a}_{01}\bar{b}_{00} = 2\sqrt{2\omega^{-1}}(a_{10} + b_{01}),$$
$$\alpha_{10} = 2(\bar{a}_{11}\bar{b}_{00} + \bar{a}_{01}\bar{b}_{10}),$$

$$\alpha_{20} = 2(\bar{a}_{21}\bar{b}_{00} + \bar{a}_{11}\bar{b}_{10} + \bar{a}_{01}\bar{b}_{20}),$$

$$\alpha_{30} = 2(\bar{a}_{31}\bar{b}_{00} + \bar{a}_{21}\bar{b}_{10} + \bar{a}_{11}\bar{b}_{20} + \bar{a}_{01}\bar{b}_{30}),$$

$$\alpha_{40} = 2(\bar{a}_{41}\bar{b}_{00} + \bar{a}_{31}\bar{b}_{10} + \bar{a}_{21}\bar{b}_{20} + \bar{a}_{11}\bar{b}_{30} + \bar{a}_{01}\bar{b}_{40}),$$

$$\alpha_{50} = 2(\bar{a}_{51}\bar{b}_{00} + \bar{a}_{41}\bar{b}_{10} + \bar{a}_{31}\bar{b}_{20} + \bar{a}_{21}\bar{b}_{30} + \bar{a}_{11}\bar{b}_{40} + \bar{a}_{01}\bar{b}_{50}),$$

$$\alpha_{60} = 2(\bar{a}_{61}\bar{b}_{00} + \bar{a}_{51}\bar{b}_{10} + \bar{a}_{41}\bar{b}_{20} + \bar{a}_{31}\bar{b}_{30} + \bar{a}_{21}\bar{b}_{40} + \bar{a}_{11}\bar{b}_{50} + \bar{a}_{01}\bar{b}_{60}),$$

$$\alpha_{01} = 2(\bar{a}_{03}\bar{b}_{00} + 2\bar{a}_{01}\bar{a}_{02}\bar{b}_{01} + \bar{a}_{01}^3\bar{b}_{02}),$$

$$\alpha_{11} = 2\big[\bar{a}_{13}\bar{b}_{00} + 2\bar{a}_{01}\bar{a}_{12}\bar{b}_{01} + 3\bar{a}_{01}^2\bar{a}_{11}\bar{b}_{02} + \bar{a}_{03}\bar{b}_{10} \\ + 2\bar{a}_{02}(\bar{a}_{11}\bar{b}_{01} + \bar{a}_{01}\bar{b}_{11}) + \bar{a}_{01}^3\bar{b}_{12}\big],$$

$$\alpha_{21} = 2\big[\bar{a}_{23}\bar{b}_{00} + 2\bar{a}_{02}\bar{a}_{21}\bar{b}_{01} + 2\bar{a}_{01}\bar{a}_{22}\bar{b}_{01} + 3\bar{a}_{01}\bar{a}_{11}^2\bar{b}_{02} + 3\bar{a}_{01}^2\bar{a}_{21}\bar{b}_{02} \\ + \bar{a}_{13}\bar{b}_{10} + 2\bar{a}_{01}\bar{a}_{12}\bar{b}_{11} + \bar{a}_{11}(2\bar{a}_{12}\bar{b}_{01} + 2\bar{a}_{02}\bar{b}_{11} + 3\bar{a}_{01}^2\bar{b}_{12}) \\ + \bar{a}_{03}\bar{b}_{20} + 2\bar{a}_{01}\bar{a}_{02}\bar{b}_{21} + \bar{a}_{01}^3\bar{b}_{22}\big],$$

$$\alpha_{31} = 2\big(3\bar{b}_{12}\bar{a}_{01}^2\bar{a}_{21} + \bar{b}_{30}\bar{a}_{03} + 2\bar{b}_{01}\bar{a}_{01}\bar{a}_{32} + 3\bar{b}_{12}\bar{a}_{01}\bar{a}_{11}^2 + 2\bar{b}_{01}\bar{a}_{31}\bar{a}_{02} \\ + 6\bar{b}_{02}\bar{a}_{01}\bar{a}_{11}\bar{a}_{21} + 2\bar{b}_{01}\bar{a}_{21}\bar{a}_{12} + \bar{b}_{00}\bar{a}_{33} + 3\bar{b}_{22}\bar{a}_{01}^2\bar{a}_{11} + 2\bar{b}_{11}\bar{a}_{11}\bar{a}_{12} \\ + 2\bar{b}_{21}\bar{a}_{01}\bar{a}_{12} + \bar{b}_{10}\bar{a}_{23} + 2\bar{b}_{11}\bar{a}_{21}\bar{a}_{02} + \bar{b}_{02}\bar{a}_{11}^3 + 2\bar{b}_{21}\bar{a}_{11}\bar{a}_{02} + \bar{b}_{32}\bar{a}_{01}^3 \\ + \bar{b}_{20}\bar{a}_{13} + 2\bar{b}_{11}\bar{a}_{01}\bar{a}_{22} + 2\bar{b}_{01}\bar{a}_{11}\bar{a}_{22} + 2\bar{b}_{31}\bar{a}_{01}\bar{a}_{02} + 3\bar{b}_{02}\bar{a}_{01}^2\bar{a}_{31}\big),$$

$$\alpha_{41} = 2\big(2\bar{b}_{21}\bar{a}_{21}\bar{a}_{02} + 2\bar{b}_{01}\bar{a}_{41}\bar{a}_{02} + 3\bar{b}_{32}\bar{a}_{01}^2\bar{a}_{11} + 6\bar{b}_{12}\bar{a}_{01}\bar{a}_{11}\bar{a}_{21} + 2\bar{b}_{31}\bar{a}_{11}\bar{a}_{02} \\ + 3\bar{b}_{02}\bar{a}_{11}^2\bar{a}_{21} + \bar{b}_{00}\bar{a}_{43} + 6\bar{b}_{02}\bar{a}_{01}\bar{a}_{11}\bar{a}_{31} + 2\bar{b}_{11}\bar{a}_{11}\bar{a}_{22} + 2\bar{b}_{01}\bar{a}_{21}\bar{a}_{22} \\ + 3\bar{b}_{22}\bar{a}_{01}\bar{a}_{11}^2 + 2\bar{b}_{11}\bar{a}_{21}\bar{a}_{12} + 2\bar{b}_{41}\bar{a}_{01}\bar{a}_{02} + 2\bar{b}_{21}\bar{a}_{11}\bar{a}_{12} + 2\bar{b}_{01}\bar{a}_{31}\bar{a}_{12} \\ + 3\bar{b}_{12}\bar{a}_{01}^2\bar{a}_{31} + 2\bar{b}_{31}\bar{a}_{01}\bar{a}_{12} + 3\bar{b}_{02}\bar{a}_{01}^2\bar{a}_{41} + 2\bar{b}_{01}\bar{a}_{11}\bar{a}_{32} + 2\bar{b}_{11}\bar{a}_{31}\bar{a}_{02} \\ + 3\bar{b}_{02}\bar{a}_{21}^2\bar{a}_{01} + 3\bar{b}_{22}\bar{a}_{01}^2\bar{a}_{21} + 2\bar{b}_{21}\bar{a}_{01}\bar{a}_{22} + 2\bar{b}_{11}\bar{a}_{01}\bar{a}_{32} + 2\bar{b}_{01}\bar{a}_{01}\bar{a}_{42} \\ + \bar{b}_{12}\bar{a}_{11}^3 + \bar{b}_{42}\bar{a}_{01}^3 + \bar{b}_{10}\bar{a}_{33} + \bar{b}_{20}\bar{a}_{23} + \bar{b}_{30}\bar{a}_{13} + \bar{b}_{40}\bar{a}_{03}\big),$$

$$\alpha_{02} = 2\big(3\bar{b}_{02}\bar{a}_{01}\bar{a}_{02}^2 + \bar{b}_{00}\bar{a}_{05} + 2\bar{b}_{01}\bar{a}_{02}\bar{a}_{03} + 2\bar{b}_{01}\bar{a}_{01}\bar{a}_{04} + 4\bar{b}_{03}\bar{a}_{01}^3\bar{a}_{02} \\ + 3\bar{b}_{02}\bar{a}_{01}^2\bar{a}_{03} + \bar{b}_{04}\bar{a}_{01}^5\big),$$

$$\alpha_{12} = 2\big(3\bar{b}_{02}\bar{a}_{11}\bar{a}_{02}^2 + 2\bar{b}_{01}\bar{a}_{02}\bar{a}_{13} + 12\bar{b}_{03}\bar{a}_{01}^2\bar{a}_{11}\bar{a}_{02} + 3\bar{b}_{02}\bar{a}_{01}^2\bar{a}_{13} \\ + 2\bar{b}_{11}\bar{a}_{01}\bar{a}_{04} + 2\bar{b}_{01}\bar{a}_{12}\bar{a}_{03} + 3\bar{b}_{12}\bar{a}_{01}\bar{a}_{02}^2 + 6\bar{b}_{02}\bar{a}_{01}\bar{a}_{02}\bar{a}_{12} + \bar{b}_{00}\bar{a}_{15} \\ + 6\bar{b}_{02}\bar{a}_{01}\bar{a}_{11}\bar{a}_{03} + 2\bar{b}_{11}\bar{a}_{02}\bar{a}_{03} + \bar{b}_{14}\bar{a}_{01}^5 + 2\bar{b}_{01}\bar{a}_{11}\bar{a}_{04} + 5\bar{b}_{04}\bar{a}_{01}^4\bar{a}_{11} \\ + 3\bar{b}_{12}\bar{a}_{01}^2\bar{a}_{03} + 4\bar{b}_{13}\bar{a}_{01}^3\bar{a}_{02} + 4\bar{b}_{03}\bar{a}_{01}^3\bar{a}_{12} + 2\bar{b}_{01}\bar{a}_{01}\bar{a}_{14} + \bar{b}_{10}\bar{a}_{05}\big),$$

$$
\begin{aligned}
\alpha_{22} = 2\big(&3\bar{b}_{02}\bar{a}_{01}^2\bar{a}_{23} + 3\bar{b}_{12}\bar{a}_{11}\bar{a}_{02}^2 + 2\bar{b}_{01}\bar{a}_{11}\bar{a}_{14} + 2\bar{b}_{21}\bar{a}_{01}\bar{a}_{04} + 4\bar{b}_{03}\bar{a}_{01}^3\bar{a}_{22} \\
&+ 12\bar{b}_{03}\bar{a}_{01}\bar{a}_{11}^2\bar{a}_{02} + 6\bar{b}_{12}\bar{a}_{01}\bar{a}_{11}\bar{a}_{03} + 6\bar{b}_{12}\bar{a}_{01}\bar{a}_{02}\bar{a}_{12} + 12\bar{b}_{03}\bar{a}_{01}^2\bar{a}_{21}\bar{a}_{02} \\
&+ 6\bar{b}_{02}\bar{a}_{01}\bar{a}_{11}\bar{a}_{13} + 12\bar{b}_{03}\bar{a}_{01}^2\bar{a}_{11}\bar{a}_{12} + 12\bar{b}_{13}\bar{a}_{01}^2\bar{a}_{11}\bar{a}_{02} + 6\bar{b}_{02}\bar{a}_{01}\bar{a}_{21}\bar{a}_{03} \\
&+ 6\bar{b}_{02}\bar{a}_{11}\bar{a}_{02}\bar{a}_{12} + 6\bar{b}_{02}\bar{a}_{01}\bar{a}_{02}\bar{a}_{22} + 3\bar{b}_{22}\bar{a}_{01}^2\bar{a}_{03} + 10\bar{b}_{04}\bar{a}_{01}^3\bar{a}_{11}^2 \\
&+ 2\bar{b}_{01}\bar{a}_{01}\bar{a}_{24} + \bar{b}_{10}\bar{a}_{15} + \bar{b}_{24}\bar{a}_{01}^5 + \bar{b}_{20}\bar{a}_{05} + \bar{b}_{00}\bar{a}_{25} + 5\bar{b}_{04}\bar{a}_{01}^4\bar{a}_{21} \\
&+ 4\bar{b}_{13}\bar{a}_{01}^3\bar{a}_{12} + 2\bar{b}_{01}\bar{a}_{22}\bar{a}_{03} + 4\bar{b}_{23}\bar{a}_{01}^3\bar{a}_{02} + 2\bar{b}_{11}\bar{a}_{01}\bar{a}_{1,4} + 3\bar{b}_{02}\bar{a}_{21}\bar{a}_{02}^2 \\
&+ 2\bar{b}_{01}\bar{a}_{12}\bar{a}_{13} + 5\bar{b}_{14}\bar{a}_{01}^4\bar{a}_{11} + 3\bar{b}_{12}\bar{a}_{01}^2\bar{a}_{13} + 3\bar{b}_{02}\bar{a}_{01}\bar{a}_{12}^2 + 3\bar{b}_{02}\bar{a}_{11}^2\bar{a}_{03} \\
&+ 2\bar{b}_{11}\bar{a}_{11}\bar{a}_{04} + 2\bar{b}_{01}\bar{a}_{02}\bar{a}_{23} + 3\bar{b}_{22}\bar{a}_{01}\bar{a}_{02}^2 + 2\bar{b}_{01}\bar{a}_{21}\bar{a}_{04} \\
&+ 2\bar{b}_{21}\bar{a}_{02}\bar{a}_{03} + 2\bar{b}_{11}\bar{a}_{12}\bar{a}_{03} + 2\bar{b}_{11}\bar{a}_{02}\bar{a}_{13}\big), \\
\alpha_{03} = 2\big(&2\bar{b}_{01}\bar{a}_{02}\bar{a}_{05} + 6\bar{a}_{02}\bar{b}_{05}\bar{a}_{01}^5 + 2\bar{b}_{01}\bar{a}_{03}\bar{a}_{04} + 3\bar{b}_{02}\bar{a}_{01}\bar{a}_{03}^2 \\
&+ 3\bar{b}_{02}\bar{a}_{02}^2\bar{a}_{03} + 4\bar{b}_{03}\bar{a}_{01}\bar{a}_{02}^3 + 10\bar{b}_{04}\bar{a}_{01}^3\bar{a}_{02}^2 + \bar{b}_{00}\bar{a}_{07} \\
&+ 6\bar{b}_{02}\bar{a}_{01}\bar{a}_{02}\bar{a}_{04} + 5\bar{a}_{03}\bar{b}_{04}\bar{a}_{01}^4 + 4\bar{a}_{04}\bar{b}_{03}\bar{a}_{01}^3 + 3\bar{b}_{02}\bar{a}_{01}^2\bar{a}_{05} \\
&+ 2\bar{b}_{01}\bar{a}_{01}\bar{a}_{06} + \bar{b}_{06}\bar{a}_{01}^7 + 12\bar{a}_{03}\bar{b}_{03}\bar{a}_{01}^2\bar{a}_{02}\big).
\end{aligned}
$$

$$(7.19)$$

Hence,

$$
M(h,\delta) = \sum_{j\geq0} \int_{x_2(h)}^{x_1(h)} \bar{q}_j(x) w^{2j+1}\, dx.
$$

Let $\psi(x) = \mathrm{sgn}(x)[H_0^*(x)]^{\frac{1}{2}}$. By (7.6), the function ψ is C^∞ for $|x|$ small with $\psi'(0) = h^{\frac{1}{2}}$. Then introduce the new variable $u = \psi(x)$ so that

$$
M(h,\delta) = \sum_{j\geq0} \int_{-h^{\frac{1}{2}}}^{h^{\frac{1}{2}}} \tilde{q}_j(u) w^{2j+1}\, du = \sum_{j\geq0} \int_{0}^{h^{\frac{1}{2}}} [\tilde{q}_j(u) + \tilde{q}_j(-u)] w^{2j+1}\, du,
$$

where $w = \sqrt{h - u^2}$ and

$$
\tilde{q}_j(u) = \frac{\bar{q}_j(x)}{\psi'(x)}\bigg|_{x=\psi^{-1}(u)}.
$$

$$(7.20)$$

We can suppose

$$
\tilde{q}_j(u) + \tilde{q}_j(-u) = \sum_{i\geq0} r_{ij} u^{2i}.
$$

$$(7.21)$$

(We note that the coefficients r_{ij} in (7.21) can be found by using (7.1), (7.2), (7.6)–(7.7), (7.10)–(7.12) and (7.14)–(7.20). We will do this later.) Then

$$M(h, \delta) = \sum_{i+j \geq 0} r_{ij} I_{ij}(h), \qquad (7.22)$$

where

$$I_{ij}(h) = \int_0^{h^{\frac{1}{2}}} u^{2i} w^{2j+1} \, du$$

$$= \int_0^{h^{\frac{1}{2}}} u^{2i} (h - u^2)^j \sqrt{h - u^2} \, du.$$

Lemma 7.2 *Let*

$$\beta_{ij} = \int_0^1 v^{2i} (1 - v^2)^j \sqrt{1 - v^2} \, dv. \qquad (7.23)$$

Then

$$I_{ij}(h) = \beta_{ij} h^{1+i+j}, \quad h > 0, 0 < \beta_{ij} < 1.$$

Proof By introducing $v = u/h^{\frac{1}{2}}$, then a straightforward calculation shows that

$$I_{ij}(h) = h^{\frac{2i+1}{2}+j+\frac{1}{2}} \int_0^1 v^{2i} (1 - v^2)^j \sqrt{1 - v^2} \, dv.$$

$\square$

Remark 7.3 Now, for perturbed Hamiltonian systems, we can either use the method developed above using the Melnikov function, or use the computation of focus values (see Chaps. 2 and 3) to determine the limit cycles bifurcating from a center. In general, formulas in the focus value computation (or in the normal form computation) are less involved than the method using the Melnikov function, since more transformations are used in the latter case. It has been shown that the focus value computation is more efficient, in particular, for higher-order focus values (or higher-order normal forms).

7.3 Determining the Number of Limit Cycles Near a Center

From (7.17), (7.20) and (7.21), we know that if system (6.1) is analytical, then the series $\sum_{i,j \geq 0} r_{ij} u^{2i} w^{2j+1}$ is convergent for (u, w) near the origin. It follows that the series $\sum_{i,j \geq 0} |r_{ij}| \mu^{i+j}$ is convergent for some constant $\mu > 0$ in this case.

By (7.22) and Lemma 7.2, we have

$$M(h, \delta) = h \sum_{i+j \geq 0} r_{ij} \beta_{ij} h^{i+j} = h \sum_{l \geq 0} b_l(\delta) h^l,$$

where

$$b_l(\delta) = \sum_{i+j=l} r_{ij} \beta_{ij}. \tag{7.24}$$

Noting that

$$\sum_{i+j \geq 0} |r_{ij}| \beta_{ij} h^{i+j} \leq \sum_{i+j \geq 0} |r_{ij}| \mu^{i+j} \quad \text{for } 0 < h \leq \mu,$$

we have proved the following theorem which was obtained before [79].

Theorem 7.4 *Let* (7.1) *hold. Then* $M(h, \delta)$ *is* C^∞ *in* $0 \leq h \ll 1$ *with*

$$M(h, \delta) = h \sum_{l \geq 0} b_l(\delta) h^l \tag{7.25}$$

formally for $0 \leq h \ll 1$. *Moreover, if* (6.1) *is analytic, so is* M.

Using expansion (7.25) we have the following result.

Theorem 7.5 (Han [79]) *Under the condition of Theorem* 7.4, *if there exist* $k \geq 1$, $\delta_0 \in D$ *such that* $b_k(\delta_0) \neq 0$ *and*

$$b_j(\delta_0) = 0, \quad j = 0, 1, \ldots, k-1, \qquad \det \frac{\partial(b_0, \ldots, b_{k-1})}{\partial(\delta_1, \ldots, \delta_k)}(\delta_0) \neq 0, \tag{7.26}$$

where $\delta = (\delta_1, \ldots, \delta_m)$, $m \geq k$, *then for any* $\varepsilon_0 > 0$ *and any neighborhood* V *of the origin there exist* $0 < \varepsilon < \varepsilon_0$ *and* $|\delta - \delta_0| < \varepsilon_0$ *such that* (6.1) *has* k *limit cycles in* V.

Proof Fix $\delta_j = \delta_{j0}$ for $j = k+1, \ldots, m$. By (7.26) the change of parameters

$$b_j = b_j(\delta), \quad j = 0, \ldots, k-1,$$

has inverse $\delta_j = \delta_j(b_0, \ldots, b_{k-1})$, $j = 1, \ldots, k$. Then (7.25) becomes

$$M(h, \delta) = h[b_0 + b_1 h + \cdots + b_{k-1} h^{k-1} + b_k h^k + O(h^{k+1})],$$

where $b_k = b_k(\delta_0) \neq 0$ as $b_0 = \cdots = b_{k-1} = 0$. Next, changing the sign of $b_{k-1}, b_{k-2}, \ldots, b_0$, such that

$$b_{j-1} b_j < 0, \quad j = k, k-1, \ldots, 1, \qquad 0 < |b_0| \ll |b_1| \ll \cdots \ll |b_{k-1}| \ll 1,$$

we can find k simple positive zeros, $h_1, h_2, \ldots, h_k$, with $0 < h_k < h_{k-1} < \cdots < h_1 \ll 1$. By (6.5) and the implicit function theorem, the function F has k zeros $h_j + O(\varepsilon)$, $j = 1, \ldots, k$.

This ends the proof.   □

On the maximal number of limit cycles near the origin we have the following theorem.

Theorem 7.6 (Han [79]) *Let (7.1) hold. If there exist $k \geq 1, \delta_0 \in D$ such that*

$$b_k(\delta_0) \neq 0, \qquad b_j(\delta_0) = 0, \quad j = 0, 1, \ldots, k-1, \qquad (7.27)$$

then there exist $\varepsilon_0 > 0$, and a neighborhood V of the origin such that (6.1) has at most k limit cycles in V for $0 < |\varepsilon| < \varepsilon_0, |\delta - \delta_0| < \varepsilon_0$.

Proof With (7.1) we can assume

$$p(0, 0, \varepsilon, \delta) = q(0, 0, \varepsilon, \delta) = 0.$$

Then (6.1) can be written as

$$\begin{pmatrix} \dot{x} \\ \dot{y} \end{pmatrix} = W(\varepsilon, \delta) \begin{pmatrix} x \\ y \end{pmatrix} + R(x, y, \varepsilon, \delta) \qquad (7.28)$$

with $R = O(|x, y|^2)$ for (x, y) near the origin, and $W(0, \delta) = \begin{pmatrix} 0 & \omega \\ -\omega & 0 \end{pmatrix}$.

Let $(x(t), y(t)) = (x(t, x_0, \varepsilon, \delta), y(t, x_0, \varepsilon, \delta))$ denote the solution of (6.1) satisfying $(x(0), y(0)) = (x_0, 0)$. Applying the formula for constant variation to (7.24) yields

$$\begin{pmatrix} x(t) \\ y(t) \end{pmatrix} = e^{W(\varepsilon, \delta)t} \begin{pmatrix} x_0 \\ 0 \end{pmatrix} + R_0(t, x_0, \varepsilon, \delta), \qquad (7.29)$$

where

$$R(0, x_0, \varepsilon, \delta) = 0, \qquad R_0(t, x_0, \varepsilon, \delta) = O(x_0^2).$$

Since

$$e^{W(0, \delta)t} = e^{\begin{pmatrix} 0 & \omega \\ -\omega & 0 \end{pmatrix}t} = \begin{pmatrix} \cos \omega t & \sin \omega t \\ -\sin \omega t & \cos \omega t \end{pmatrix},$$

it follows that

$$e^{W(\varepsilon, \delta)t} \begin{pmatrix} x_0 \\ 0 \end{pmatrix} = \left[\begin{pmatrix} \cos \omega t \\ -\sin \omega t \end{pmatrix} + O(\varepsilon) \right] x_0.$$

Hence, (7.29) becomes

$$\begin{aligned} x(t) &= x_0 \big[\cos \omega t + \bar{R}_1(t, x_0, \varepsilon, \delta) \big], \\ y(t) &= x_0 \big[-\sin \omega t + \bar{R}_2(t, x_0, \varepsilon, \delta) \big], \end{aligned} \qquad (7.30)$$

where $\bar{R}_j(t, x_0, \varepsilon, \delta) = O(|\varepsilon, x_0|)$. It is clear that the solution $(x(t), y(t))$ is periodic with the smallest period T if and only if $(x(T), y(T)) = (x_0, 0) \neq (0, 0)$ and $(x(t), y(t)) \neq (x_0, 0)$ for $0 < t < T$.

By (7.30) and using the implicit function theorem, we conclude that there is a unique $T = \frac{2\pi}{\omega} + O(|\varepsilon, x_0|)$ being C^∞ in (ε, x_0) such that

$$y(T) = 0, \qquad x_0 x(T) > 0 \quad \text{for } x_0 \neq 0.$$

Define

$$G(x_0, \varepsilon, \delta) = x(T) - x_0.$$

Then $G \in C^\infty$ with $G(0, \varepsilon, \delta) = 0$. We call the function G a *bifurcation function* of (6.1). Since all nontrivial orbits near the origin are periodic for $\varepsilon = 0$, it implies that $G(x_0, 0, \delta) = 0$ for $|x_0|$ small. Thus, we can write

$$G(x_0, \varepsilon, \delta) = \varepsilon \bar{G}(x_0, \varepsilon, \delta),$$

where $\bar{G} \in C^\infty$, $G(0, \varepsilon, \delta) = 0$. Obviously, a nontrivial periodic orbit exists near the origin if and only if $G(x_0, \varepsilon, \delta) = G(x_0', \varepsilon, \delta) = 0$ for some $x_0 > 0$ and $x_0' < 0$ with x_0 and $|x_0'|$ small. Thus, by Rolle's theorem, if

$$\bar{G}(x_0, 0, \delta_0) = g_k x_0^{2k+1} + O\left(x_0^{2k+2}\right), \quad g_k \neq 0, \tag{7.31}$$

there exists $\varepsilon_0 > 0$ such that $\bar{G}$ has at most $2k$ zeros on the region $(-\varepsilon_0, \varepsilon_0) - \{0\}$ for $0 < |\varepsilon| < \varepsilon_0$, $|\delta - \delta_0| < \varepsilon_0$.

Now it suffices to prove that (7.27) holds if and only if (7.31) holds. In fact, if we take $x_0 > 0$ satisfying $H(x_0, 0) = h$, then by (6.5) we have

$$H\left(x(T), 0\right) - H(x_0, 0) = \varepsilon M(h, \delta) + O\left(\varepsilon^2\right).$$

On the other hand,

$$H\left(x(T), 0\right) - H(x_0, 0) = H_x(x_0 + \theta \varepsilon \bar{G}) \varepsilon \bar{G}, \quad 0 < \theta < 1.$$

Therefore, we obtain

$$H_x(x_0, 0) \bar{G}(x_0, 0, \delta) = M(h, \delta). \tag{7.32}$$

From $H(x_0, 0) = h$ we can solve for $x_0 = \sqrt{\frac{2h}{\omega}} + O(h)$. Then noting $H_x(x_0, 0) = \omega x_0 (1 + O(x_0))$, if (7.31) holds, then from (7.32) we have

$$M(h, \delta_0) = \omega g_k x_0^{2k+2} + o\left(x_0^{2k+2}\right)$$

$$= \omega g_k \left(\frac{2}{\omega}\right)^{k+1} h^{k+1} + o\left(h^{k+1}\right).$$

Comparing with (7.25), we obtain (7.27). Conversely, (7.31) follows from (7.32) in the same way.

The proof is finished. $\square$

Furthermore, we can prove the following theorem.

Theorem 7.7 (Han [79]) *Let* (7.1) *hold. Suppose there exist* $k \geq 1, \delta_0 \in D$ *such that* (7.26) *holds. If further, there exists a* k-*dimensional vector function* $\varphi(\varepsilon, \delta_{k+1}, \ldots, \delta_m)$ *such that* (6.1) *has a center at the origin when* $(\delta_1, \ldots, \delta_k) = \varphi(\varepsilon, \delta_{k+1}, \ldots, \delta_m)$ *for* $|\varepsilon| + |\delta - \delta_0|$ *small, then there exist* $\varepsilon_0 > 0$ *and a neighborhood* V *of the origin such that* (6.1) *has at most* $(k-1)$ *limit cycles in* V *for* $0 < |\varepsilon| < \varepsilon_0, |\delta - \delta_0| < \varepsilon_0$. *Moreover,* $(k-1)$ *limit cycles can appear in an arbitrary neighborhood of the origin for some* (ε, δ) *sufficiently near* $(0, \delta_0)$. *In other words,* (6.1) *has cyclicity* $(k-1)$ *at the origin for* $|\varepsilon| + |\delta - \delta_0|$ *small.*

Proof Let $h^*(x_0) = H(x_0, 0) = \frac{1}{2}\omega x_0^2 + O(x_0^3)$. Substituting $h = h^*(x_0)$ into $M(h, \delta) = \sum_{j \geq 0} b_j h^{j+1}$ yields

$$M\big(h^*(x_0), \delta\big) = \frac{1}{2}\omega x_0^2 \sum_{j \geq 0} B_j(\delta) x_0^j, \tag{7.33}$$

where

$$B_0 = b_0, \qquad B_1 = L_1(b_0),$$

$$B_{2j} = \left(\frac{\omega}{2}\right)^j b_j + L_{2j}(b_0, b_1, \ldots, b_{j-1}),$$

$$B_{2j+1} = L_{2j+1}(b_0, b_1, \ldots, b_j), \quad j \geq 1,$$

where $L_{2j} = O(|b_0, \ldots, b_{j-1}|)$ and $L_{2j+1} = O(|b_0, \ldots, b_j|)$ are linear in their variables. Hence, by (7.32) and (7.33) we have

$$\bar{G}(x_0, 0, \delta) = \frac{M(h^*(x_0), \delta)}{(h^*)'(x_0)} = \frac{x_0}{2(1 + h_1(x_0))} \sum_{j \geq 1} B_j x_0^j,$$

where $h_1(x_0) = O(x_0) \in C^\infty$. Since $\bar{G} \in C^\infty$, we can rewrite it in the form

$$\bar{G}(x_0, \varepsilon, \delta) = \frac{x_0}{2(1 + h_1(x_0))} \sum_{j \geq 1} c_j(\varepsilon, \delta) x_0^j, \tag{7.34}$$

where

$$c_j(\varepsilon, \delta) = B_j(\delta) + O(\varepsilon), \quad j \geq 0.$$

In particular,

$$c_0 = b_0 + O(\varepsilon), \qquad c_{2j} = \left(\frac{\omega}{2}\right)^j b_j + L_{2j}(b_0, b_1, \ldots, b_{j-1}) + O(\varepsilon), \quad j \geq 1.$$

It then follows that

$$\det \frac{\partial(c_0, c_2, \ldots, c_{2k-2})}{\partial(b_0, b_1, \ldots, b_{k-1})}\bigg|_{\varepsilon=0} \neq 0,$$

which implies

$$\det \frac{\partial(c_0, c_2, \ldots, c_{2k-2})}{\partial(\delta_1, \delta_2, \ldots, \delta_k)}\bigg|_{\varepsilon=0, \delta=\delta_0} \neq 0.$$

Thus, by the implicit function theorem, the equations

$$a_{j+1} = c_{2j}(\varepsilon, \delta), \quad j = 0, \ldots, k-1,$$

have a unique solution $(\delta_1, \ldots, \delta_k) = \tilde{\varphi}(\varepsilon, a_1, \ldots, a_k, \delta_{k+1}, \ldots, \delta_n)$ for (ε, δ) near $(0, \delta_0)$. In particular,

$$c_{2j}(\varepsilon, \delta) = 0, \quad j = 0, \ldots, k-1 \quad \Longleftrightarrow \quad (\delta_1, \ldots, \delta_k) = \varphi^*(\varepsilon, \delta_{k+1}, \ldots, \delta_n),$$

where

$$\varphi^*(\varepsilon, \delta_{k+1}, \ldots, \delta_n) = \tilde{\varphi}(\varepsilon, 0, \ldots, 0, \delta_{k+1}, \ldots, \delta_n).$$

On the other hand, by our assumption, if $(\delta_1, \ldots, \delta_k) = \varphi(\varepsilon, \delta_{k+1}, \ldots, \delta_n)$ then $c_j(\varepsilon, \delta) = 0$ for all $j \geq 0$. Therefore,

$$(\delta_1, \ldots, \delta_k) = \varphi(\varepsilon, \delta_{k+1}, \ldots, \delta_n) \quad \Longrightarrow \quad c_{2j}(\varepsilon, \delta) = 0, \quad j = 0, \ldots, k-1.$$

This implies that the function $(\delta_1, \ldots, \delta_k) = \varphi(\varepsilon, \delta_{k+1}, \ldots, \delta_n)$ is a solution of the equations $c_{2j}(\varepsilon, \delta) = 0$, $j = 0, \ldots, k-1$. Then, by the uniqueness of the solution, one must have $\varphi^* = \varphi$. Hence,

$$(\delta_1, \ldots, \delta_k) = \tilde{\varphi}(\varepsilon, a_1, \ldots, a_k, \delta_{k+1}, \ldots, \delta_n),$$

$$= \varphi(\varepsilon, \delta_{k+1}, \ldots, \delta_n) + O(|a_1, \ldots, a_k|).$$

Inserting the above equation into $c_{2j+1}(\varepsilon, \delta)$ and noting that

$$a_1 = \cdots = a_k = 0 \quad \Longrightarrow \quad (\delta_1, \ldots, \delta_k) = \varphi(\varepsilon, \delta_{k+1}, \ldots, \delta_n)$$

$$\Longrightarrow \quad c_{2j+1}(\varepsilon, \delta) = 0,$$

we can write

$$c_{2j+1}(\varepsilon, \delta) = \sum_{j=1}^{k} a_i d_{ij}(\varepsilon, a_1, \ldots, a_k, \delta_{k+1}, \ldots, \delta_n), \quad j \geq 0. \tag{7.35}$$

Note that if $\bar{G} \neq 0$ for $0 < |x_0| \ll 1$ (i.e., the origin is a focus) then the sign of $x_0 \bar{G}$ remains unchanged (positive or negative) for $0 < |x_0| \ll 1$. This implies that $c_0 = \cdots = c_{2j} = 0 \Longrightarrow c_{2j+1} = 0$. Therefore,

$$c_0(\varepsilon, \delta) = a_1 \quad \text{and} \quad c_1(\varepsilon, \delta) = 0 \quad \text{if } a_1 = 0,$$

$$c_2(\varepsilon, \delta) = a_2 \quad \text{and} \quad c_3(\varepsilon, \delta) = 0 \quad \text{if } a_1 = a_2 = 0,$$

$$\cdots$$

$$c_{2k-2}(\varepsilon, \delta) = a_k \quad \text{and} \quad c_{2k-1}(\varepsilon, \delta) = 0 \quad \text{if } a_1 = \cdots = a_k = 0,$$

which indicates that the functions d_{ij} in (7.35) satisfy

$$d_{ij} = 0 \quad \text{for } i \geq j + 2. \tag{7.36}$$

Then, substituting (7.35) into (7.34) and using (7.36) we obtain

$$\bar{G}(x_0, \varepsilon, \delta) = \frac{x_0}{2(1 + h_1(x_0))} \sum_{j=1}^{k} a_j x_0^{2j-2} P_j(x_0, \varepsilon, a_1, \ldots, a_k, \delta_{k+1}, \ldots, \delta_n)$$

$$\equiv \frac{x_0}{2(1 + h_1(x_0))} \tilde{G}_k(x_0, \varepsilon, a_1, \ldots, a_k, \delta_{k+1}, \ldots, \delta_n), \tag{7.37}$$

where $P_j = 1 + O(x_0) \in C^\infty$. We claim that there exists $\varepsilon_0 > 0$ such that the function $\tilde{G}_k$ has at most $(k - 1)$ positive zeros in the interval $(0, \varepsilon_0)$ for all $0 < |a_1| + \cdots + |a_k| < \varepsilon_0$ and $0 < |\varepsilon| + |\delta - \delta_0| < \varepsilon_0$. We prove this by using the method of mathematical induction on k.

First, for $k = 1$, it is obvious that for any C^∞ function, $P_1(x_0, \varepsilon, a_1, \delta_2, \ldots, \delta_n)$ satisfying $P_1 = 1 + O(x_0)$, the function $\tilde{G}_1 = a_1 P_1$ has no zeros in x_0 for $|a_1| > 0$. Now suppose that for any C^∞ functions $\tilde{P}_1, \ldots, \tilde{P}_{k-1}$ satisfying $\tilde{P}_j = 1 + O(x_0)$, $j = 1, \ldots, k - 1$, the function

$$\tilde{G}_{k-1}(x_0, \varepsilon, a_1, \ldots, a_{k-1}, \delta_k, \ldots, \delta_n)$$

$$= \sum_{j=1}^{k-1} a_j x_0^{2j-2} \tilde{P}_j(x_0, \varepsilon, a_1, \ldots, a_{k-1}, \delta_k, \ldots, \delta_n)$$

has at most $(k - 2)$ positive zeros in x_0 near $x_0 = 0$. Let

$$\tilde{G}_k(x_0, \varepsilon, a_1, \ldots, a_k, \delta_{k+1}, \ldots, \delta_n)$$

$$= \sum_{j=1}^{k} a_j x_0^{2j-2} P_j(x_0, \varepsilon, a_1, \ldots, a_k, \delta_{k+1}, \ldots, \delta_n),$$

where $P_j = 1 + O(x_0) \in C^\infty$. Then,

$$\tilde{G}_k = P_1 \sum_{j=1}^{k} a_j x_0^{2j-2} (P_j/P_1) \equiv P_1 \bar{G}_k.$$

It is easy to see that $(\bar{G}_k)'_{x_0}$ has the following form

$$(\bar{G}_k)'_{x_0} = x_0 \sum_{j=1}^{k-1} \bar{a}_j x_0^{2j-2} \bar{P}_j \equiv x_0 G_{k-1}^*,$$

where $\bar{a}_j = 2ja_{j+1}$, $\bar{P}_j = 1 + O(x_0) \in C^\infty$, $j = 1, \ldots, k-1$. By the assumption, G_{k-1}^* has at most $(k-2)$ positive zeros in x_0 near $x_0 = 0$. The same is true for $(\bar{G}_k)'_{x_0}$. Hence, by Rolle's theorem, $\bar{G}_k$ has at most $(k-1)$ positive zeros in x_0. The same is true for $\tilde{G}_k$. Then the claim follows.

Finally, using (7.37) one can take $a_k \neq 0$ and then change the sign of $a_{k-1}, \ldots, a_1$, in turn to find $(k-1)$ positive simple zeros of $\tilde{G}_k$, which give $(k-1)$ limit cycles of (6.1) near the origin.

This finishes the proof.   □

We remark that in Theorems 7.6 and 7.7, the vector parameter is required to vary near δ_0. This restriction is removed in the following theorem under the condition that the functions p and q depend on δ linearly.

Remark 7.8 An interesting work related to Theorems 7.5, 7.6 and 7.7 is given in [28], where other approaches are presented for studying the asymptotics of a non-vanishing higher-order Melnikov function, by calculating the so-called "reduced focus values" (reduced Lyapunov quantities), reduced coefficients in the normal forms or reduced Bautin Ideal. In particular, this is used to describe necessary and sufficient conditions to guarantee the Hopf bifurcation of limit cycles perturbing from a Hamiltonian center uniformly in terms of the parameter and the phase variable.

Theorem 7.9 (Han [79]) *Let (7.1) hold, and let the functions p and q in (6.1) be linear in δ. Suppose further that for an integer $k \geq 1$*

(i) rank $\frac{\partial(b_0, \ldots, b_{k-1})}{\partial(\delta_1, \ldots, \delta_m)} = k, m \geq k$;
(ii) *Equation (6.1) has a center at the origin as $b_j(\delta) = 0$, $j = 0, 1, \ldots, k-1$.*

Then for any given $N > 0$, there exist $\varepsilon_0 > 0$ and a neighborhood V of the origin such that (6.1) has at most $(k-1)$ limit cycles in V for $0 < |\varepsilon| < \varepsilon_0, |\delta| \leq N$. Moreover, $(k-1)$ limit cycles can be obtained in an arbitrary neighborhood of the origin for some (ε, δ).

Proof As before, let $a_{j+1} = c_{2j}(\varepsilon, \delta)$, $j = 0, \ldots, k-1$, which have the solution

$$(\delta_1, \ldots, \delta_k) = \tilde{\varphi}(\varepsilon, a_1, \ldots, a_k, \delta_{k+1}, \ldots, \delta_n)$$

with

$$\tilde{\varphi}(0, 0, \ldots, 0, \delta_{k+1}, \ldots, \delta_n) = \varphi(\delta_{k+1}, \ldots, \delta_n),$$

where $(\delta_1, \ldots, \delta_k) = \varphi(\delta_{k+1}, \ldots, \delta_n)$ is the solution of equations $b_j = 0, j = 0, \ldots, k-1$. One can check easily that (7.34)–(7.37) hold for $|\varepsilon|$ small and

$(a_1, \ldots, a_k)$ bounded. In the same way, using (7.37) we can prove that the function $\bar{G}(x_0, \varepsilon, \delta)$ has at most $(k-1)$ positive zeros in x_0 near $x_0 = 0$ for all $|\varepsilon|$ small and $(a_1, \ldots, a_k)$ bounded, and that it has exactly $(k-1)$ positive zeros for some $(\varepsilon, a_1, \ldots, a_k)$.

The proof is complete. $\square$

Example 7.10 Consider a Liénard system of the form

$$\dot{x} = y - \varepsilon \sum_{i=1}^{2n+1} a_i x^i, \qquad \dot{y} = -x,$$

where $n \geq 1$, $|a_i| \leq 1$, and $\varepsilon > 0$ is small. We claim that the system has Hopf cyclicity n at the origin.

In fact, we have $H(x, y) = \frac{1}{2}(x^2 + y^2)$. The curve L_h given by $H(x, y) = h$ has the representation $(x, y) = \sqrt{2h}(\cos t, -\sin t)$. Hence, by (6.6) we have

$$M(h) = -\sum_{j=0}^{n} 2^{j+1} N_j a_{2j+1} h^{j+1},$$

where

$$N_j = \int_0^{2\pi} \cos^{2(j+1)} t \, dt > 0.$$

Set $\delta = (a_1, a_2, \ldots, a_{2n+1})$, $b_j = -2^{j+1} N_j a_{2j+1}$, $j = 0, \ldots, n$, and $k = n + 1$. Note that by symmetry the above system has a center at the origin as $b_j = 0$, $j = 0, \ldots, n$. Thus, the claim follows from Theorem 7.9.

7.4 Computation of the b_j Coefficients of M

In this section, we use (7.6), (6.1), (7.12), (7.16), (7.18), (7.20), (7.21), (7.23) and (7.24) to derive formulas for computing b_0, b_1, b_2 and b_3.

Note that $\psi(x) = \text{sgn}(x)[H_0^*(x)]^{\frac{1}{2}}$. If we let

$$\psi(x) = \sum_{i \geq 1} d_i x^i, \qquad \psi^{-1}(u) = \sum_{i \geq 1} \bar{d}_i u^i, \qquad \frac{1}{\psi'(x)} = \sum_{i \geq 0} n_i x^i, \qquad (7.38)$$

then by (7.6) and (6.1) we can find the formulas for the first 6 coefficients for the above functions as follows:

$$d_1 = h_2^{1/2}, \qquad d_2 = \frac{1}{2} h_2^{-1/2} h_3,$$

$$d_3 = -\frac{1}{8} h_2^{-3/2} (h_3^2 - 4h_2 h_4),$$

$$d_4 = \frac{1}{16}h_2^{-5/2}\left(h_3^3 - 4h_2h_3h_4 + 8h_2^2h_5\right),$$

$$d_5 = -\frac{1}{128}h_2^{-7/2}\left(16h_2^2h_4^2 - 64h_2^3h_6 + 32h_2^2h_3h_5 - 24h_2h_3^2h_4 + 5h_3^4\right),$$

$$d_6 = \frac{1}{256}h_2^{-9/2}\left(128h_2^4h_7 + 48h_2^2h_3h_4^2 - 64h_2^3h_3h_6 - 64h_2^3h_4h_5\right.$$
$$\left. + 48h_2^2h_3^2h_5 - 40h_2h_3^3h_4 + 7h_3^5\right),$$

$$\bar{d}_1 = d_1^{-1}, \qquad \bar{d}_2 = -d_1^{-3}d_2,$$

$$\bar{d}_3 = -d_1^{-5}\left(d_1d_3 - 2d_2^2\right),$$

$$\bar{d}_4 = -d_1^{-7}\left(5d_2^3 - 5d_1d_2d_3 + d_1^2d_4\right),$$

$$\bar{d}_5 = -d_1^{-9}\left(d_1^3d_5 - 14d_2^4 + 21d_1d_2^2d_3 - 6d_1^2d_2d_4 - 3d_1^2d_3^2\right),$$

$$\bar{d}_6 = -d_1^{-11}\left(d_1^4d_6 + 42d_2^5 - 84d_1d_2^3d_3 + 28d_1^2d_2^2d_4\right.$$
$$\left. + 28d_1^2d_2d_3^2 - 7d_1^3d_3d_4 - 7d_1^3d_2d_5\right),$$

$$n_0 = d_1^{-1}, \qquad n_1 = -2d_1^{-2}d_2,$$

$$n_2 = -d_1^{-3}\left(3d_1d_3 - 4d_2^2\right),$$

$$n_3 = -4d_1^{-4}\left(2d_2^3 - 3d_1d_2d_3 + d_1^2d_4\right),$$

$$n_4 = -d_1^{-5}\left(36d_1d_2^2d_3 - 16d_1^2d_2d_4 - 16d_2^4 + 5d_1^3d_5 - 9d_1^2d_3^2\right),$$

$$n_5 = -2d_1^{-6}\left(16d_2^5 - 10d_1^3d_2d_5 - 12d_1^3d_3d_4 + 24d_1^2d_2^2d_4\right.$$
$$\left. + 27d_1^2d_2d_3^2 + 3d_1^4d_6 - 48d_1d_2^3d_3\right).$$

By (7.18), (7.20), (7.21) and (7.38) we have $r_{00} = 2n_0\alpha_{00}$. For simplicity, let $\alpha_{00} = 0$, i.e., $a_{10} + b_{01} = 0$. Then, we have

$$r_{i0}|_{\alpha_{00}=0} = 2\sum_{l=1}^{2i} m_{i0}^{(l)}\alpha_{l0}, \quad i = 1, 2, 3, \tag{7.39}$$

where

$$m_{10}^{(1)} = n_0\bar{d}_2 + n_1\bar{d}_1^2, \qquad m_{10}^{(2)} = n_0\bar{d}_1^2,$$

$$m_{20}^{(1)} = n_0\bar{d}_4 + 3n_2\bar{d}_1^2\bar{d}_2 + n_3\bar{d}_1^4 + 2n_1\bar{d}_1\bar{d}_3 + n_1\bar{d}_2^2,$$

$$m_{20}^{(2)} = 3n_1\bar{d}_1^2\bar{d}_2 + n_2\bar{d}_1^4 + 2n_0\bar{d}_1\bar{d}_3 + n_0\bar{d}_2^2,$$

$$m_{20}^{(3)} = 3n_0\bar{d}_1^2\bar{d}_2 + n_1\bar{d}_1^4, \qquad m_{20}^{(4)} = n_0\bar{d}_1^4,$$

$$m_{30}^{(1)} = n_0\bar{d}_6 + n_1\bar{d}_3^2 + n_2\bar{d}_2^3 + n_5\bar{d}_1^6 + 6n_2\bar{d}_1\bar{d}_2\bar{d}_3 + 6n_3\bar{d}_1^2\bar{d}_2^2 + 2n_1\bar{d}_2\bar{d}_4$$
$$+ 3n_2\bar{d}_1^2\bar{d}_4 + 4n_3\bar{d}_1^3\bar{d}_3 + 2n_1\bar{d}_1\bar{d}_5 + 5n_4\bar{d}_1^4\bar{d}_2,$$

$$m_{30}^{(2)} = n_0\bar{d}_3^2 + n_1\bar{d}_2^3 + n_4\bar{d}_1^6 + 6n_1\bar{d}_1\bar{d}_2\bar{d}_3 + 6n_2\bar{d}_1^2\bar{d}_2^2 + 2n_0\bar{d}_2\bar{d}_4 \qquad (7.40)$$
$$+ 3n_1\bar{d}_1^2\bar{d}_4 + 4n_2\bar{d}_1^3\bar{d}_3 + 2n_0\bar{d}_1\bar{d}_5 + 5n_3\bar{d}_1^4\bar{d}_2,$$
$$m_{30}^{(3)} = n_0\bar{d}_2^3 + n_3\bar{d}_1^6 + 6n_0\bar{d}_1\bar{d}_2\bar{d}_3 + 3n_0\bar{d}_1^2\bar{d}_4 + 4n_1\bar{d}_1^3\bar{d}_3 + 5n_2\bar{d}_1^4\bar{d}_2$$
$$+ 6n_1\bar{d}_1^2\bar{d}_2^2,$$
$$m_{30}^{(4)} = n_2\bar{d}_1^6 + 6n_0\bar{d}_1^2\bar{d}_2^2 + 4n_0\bar{d}_1^3\bar{d}_3 + 5n_1\bar{d}_1^4\bar{d}_2,$$
$$m_{30}^{(5)} = n_1\bar{d}_1^6 + 5n_0\bar{d}_1^4\bar{d}_2,$$
$$m_{30}^{(6)} = n_0\bar{d}_1^6.$$

Similarly, we obtain

$$r_{01} = 2n_0\alpha_{01},$$
$$r_{11} = 2(n_2\bar{d}_1^2 + n_1\bar{d}_2)\alpha_{01} + 2(n_0\bar{d}_2 + n_1\bar{d}_1^2)\alpha_{11} + 2n_0\bar{d}_1^2\alpha_{21},$$
$$r_{21} = 2(n_4\bar{d}_1^4 + 2n_2\bar{d}_1\bar{d}_3 + 3n_3\bar{d}_1^2\bar{d}_2 + n_1\bar{d}_4 + n_2\bar{d}_2^2)\alpha_{01}$$
$$+ 2(n_3\bar{d}_1^4 + n_0\bar{d}_4 + n_1\bar{d}_2^2 + 3n_2\bar{d}_1^2\bar{d}_2 + 2n_1\bar{d}_1\bar{d}_3)\alpha_{11}$$
$$+ 2(2n_0\bar{d}_1\bar{d}_3 + n_0\bar{d}_2^2 + n_2\bar{d}_1^4 + 3n_1\bar{d}_1^2\bar{d}_2)\alpha_{21} \qquad (7.41)$$
$$+ 2(3n_0\bar{d}_1^2\bar{d}_2 + n_1\bar{d}_1^4)\alpha_{31} + 2n_0\bar{d}_1^4\alpha_{41},$$
$$r_{02} = 2n_0\alpha_{02},$$
$$r_{12} = 2\big[(n_1\bar{d}_2 + n_2\bar{d}_1^2)\alpha_{02} + (n_0\bar{d}_2 + n_1\bar{d}_1^2)\alpha_{12} + n_0\bar{d}_1^2\alpha_{22}\big],$$
$$r_{03} = 2n_0\alpha_{03}.$$

Next, it follows from (7.23) that

$$\beta_{00} = 14\pi, \qquad \beta_{10} = \frac{1}{16}\pi, \qquad \beta_{01} = \frac{3}{16}\pi, \qquad \beta_{02} = \frac{5}{32}\pi,$$
$$\beta_{20} = \frac{1}{32}\pi, \qquad \beta_{11} = \frac{1}{32}\pi, \qquad \beta_{30} = \frac{5}{256}\pi, \qquad \beta_{12} = \frac{5}{256}\pi,$$
$$\beta_{21} = \frac{3}{256}\pi, \qquad \beta_{03} = \frac{35}{256}\pi.$$

Then, by (7.24) we have

$$b_0(\delta) = r_{00}\beta_{00} = \frac{1}{4}\pi\bar{b}_0,$$
$$b_1(\delta) = r_{10}\beta_{10} + r_{01}\beta_{01} + O(\bar{b}_0) = \frac{1}{16}\pi\bar{b}_1 + O(\bar{b}_0),$$
$$b_2(\delta) = r_{02}\beta_{02} + r_{11}\beta_{11} + r_{20}\beta_{20} + O(\bar{b}_0) = \frac{1}{32}\pi\bar{b}_2 + O(\bar{b}_0), \qquad (7.42)$$
$$b_3(\delta) = r_{03}\beta_{03} + r_{12}\beta_{12} + r_{21}\beta_{21} + r_{30}\beta_{30} + O(\bar{b}_0) = \frac{1}{256}\pi\bar{b}_3 + O(\bar{b}_0),$$

where

$$\bar{b}_0 = r_{00}, \qquad \bar{b}_1 = r_{10} + 3r_{01}, \qquad \bar{b}_2 = 5r_{02} + r_{11} + r_{20},$$

$$\bar{b}_3 = 35r_{03} + 5r_{12} + 3r_{21} + 5r_{30}. \tag{7.43}$$

When (7.1) holds with $\omega = 1$, Hou and Han [101] have obtained the following result: if

$$(p_x + q_y)|_{\varepsilon=0} = \sum_{i+j\geq 0} c_{ij} x^i y^j,$$

then $b_0 = 2\pi c_{00}$, and

$$b_1 = -c_{10}\pi(h_{12} + 3h_{30}) - c_{01}\pi(h_{21} + 3h_{03}) + c_{20}\pi + c_{02}\pi,$$

$$b_2 = c_{10}a_{10} + c_{01}a_{01} + c_{20}a_{20} + c_{11}a_{11} + c_{02}a_{02} - c_{30}\pi(h_{12} + 5h_{30})$$
$$\qquad - c_{12}\pi(h_{30} + h_{12}) - c_{21}\pi(h_{03} + h_{21}) - c_{03}\pi(h_{21} + 5h_{03}) \tag{7.44}$$
$$\qquad + \pi c_{40} + \pi c_{04} + \frac{\pi}{3}c_{22}$$

as $c_{00} = 0$. Here,

$$a_{10} = 35\pi h_{30}h_{40} + 5\pi(h_{12}h_{04} + h_{12}h_{40} + h_{21}h_{31} + h_{03}h_{13} + h_{30}h_{22})$$
$$\qquad + 3\pi(h_{12}h_{22} + h_{13}h_{21} + h_{03}h_{31} + h_{04}h_{30})$$
$$\qquad - \frac{5}{2}\pi\left(h_{12}^3 + 3h_{03}^2h_{30} + 3h_{12}^2h_{30} + 3h_{21}^2h_{12} + 6h_{12}h_{21}h_{03} + 6h_{03}h_{21}h_{30}\right)$$
$$\qquad - \pi(h_{14} + h_{32} + 5h_{50}) - \frac{105}{2}\pi h_{30}^3 - \frac{35}{2}\pi\left(h_{03}^2h_{12} + h_{21}^2h_{30} + h_{30}^2h_{12}\right),$$

$$a_{01} = 35\pi h_{03}h_{04} + 5\pi(h_{21}h_{40} + h_{21}h_{04} + h_{12}h_{13} + h_{30}h_{31} + h_{03}h_{22})$$
$$\qquad + 3\pi(h_{21}h_{22} + h_{12}h_{31} + h_{30}h_{13} + h_{40}h_{03})$$
$$\qquad - \frac{5}{2}\pi\left(h_{21}^3 + 3h_{30}^2h_{03} + 3h_{21}^2h_{03} + 3h_{12}^2h_{21} + 6h_{12}h_{21}h_{30} + 6h_{03}h_{12}h_{30}\right)$$
$$\qquad - \pi(h_{41} + h_{23} + 5h_{05}) - \frac{105}{2}\pi h_{03}^3 - \frac{35}{2}\pi\left(h_{30}^2h_{21} + h_{12}^2h_{03} + h_{03}^2h_{21}\right),$$

$$a_{20} = \frac{35}{2}\pi h_{30}^2 + \frac{5}{2}\pi\left(h_{03}^2 + h_{21}^2 + 2h_{12}h_{30}\right) + \frac{3}{2}\pi\left(h_{12}^2 + 2h_{03}h_{21}\right)$$
$$\qquad - \pi(h_{04} + h_{22} + 5h_{40}),$$

$$a_{11} = 5\pi(h_{03}h_{12} + h_{30}h_{21}) + 3\pi(h_{03}h_{30} + h_{12}h_{21}) - \pi(h_{13} + h_{31}),$$

$$a_{02} = \frac{35}{2}\pi h_{03}^2 + \frac{5}{2}\pi\left(h_{30}^2 + h_{12}^2 + 2h_{21}h_{03}\right) + \frac{3}{2}\pi\left(h_{21}^2 + 2h_{30}h_{12}\right)$$
$$\qquad - \pi(h_{40} + h_{22} + 5h_{04}).$$

If, instead of (7.1), the function H has the form

$$H(x, y) = \frac{1}{2}y^2 + h_{20}x^2 + \sum_{i+j \geq 3} h_{ij}x^i y^j, \quad h_{20} > 0,$$

then letting $u = \sqrt{2h_{20}}x$ yields

$$H(x, y) = \frac{1}{2}(u^2 + y^2) + \sum_{i+j \geq 3} h_{ij}(2h_{20})^{-i/2}u^i y^j,$$

which has the form of (7.1) with $\omega = 1$.

Example 7.11 Consider a quadratic system of the form

$$\dot{x} = H_y + \varepsilon(\delta x + lx^2 + mxy + ny^2),$$
$$\dot{y} = -H_x + \varepsilon(ax^2 + bxy), \tag{7.45}$$

where

$$H(x, y) = \frac{1}{2}(x^2 + y^2) + l_0 x^2 y + \frac{1}{3}n_0 y^3 - \frac{1}{3}a_0 x^3.$$

The divergence of (7.45) is

$$\text{div}_{(7.45)} = \varepsilon(\delta + \bar{b}x + my), \quad \bar{b} = b + 2l.$$

Using (7.42), (7.43) or (7.44), we have

$$b_0 = 2\pi\delta, \qquad b_1 = \pi[a_0\bar{b} - (l_0 + n_0)m] + O(|\delta|),$$
$$b_2 = \pi[K_1\bar{b} - K_2m] + O(|\delta|),$$

where

$$K_1 = \frac{5}{18}a_0[7a_0^2 + 21l_0^2 + n_0^2 + 6l_0n_0],$$

$$K_2 = \frac{5}{18}[7(n_0^3 + l_0 a_0^2 + l_0 n_0^2) + 9l_0^3 + a_0^2 n_0 + 9n_0 l_0^2].$$

Note that

$$\det \frac{\partial(b_0, b_1, b_2)}{\partial(\delta, \bar{b}, m)} = -\frac{10}{3}\pi^3 a_0[(n_0 + l_0)^2(n_0 - 2l_0) - a_0^2 n_0] \equiv \Delta.$$

Hence, by Theorem 7.9, we know that (7.45) has Hopf cyclicity 2 at the origin provided $\Delta \neq 0$.

7.5   A Generalization of Theorem 7.5

In many cases the Hamiltonian system (6.4) contains some constants. If we take them as parameters and change them properly, then we can find more limit cycles. More precisely, suppose $H(x, y, a)$ with $a \in R^n$ satisfies (7.1) where the coefficients h_{ij} depend on a. Then, by Theorem 7.4, we have

$$M(h, \delta, a) = h \sum_{l \geq 0} b_l(\delta, a) h^l. \tag{7.46}$$

For simplicity, suppose the functions p and q in (6.1) are linear in δ. Then, the coefficients $b_l(\delta, a)$ are linear in δ. Suppose there exist an integer $k > 0$, $\delta_0 \in R^m$ and $a_0 \in R^n$ such that

$$b_j(\delta_0, a_0) = 0, \quad j = 0, \dots, k - 1, \quad \det \frac{\partial(b_0, \dots, b_{k-1})}{\partial(\delta_1, \dots, \delta_k)}(a_0) \neq 0. \tag{7.47}$$

Then, the linear equations $b_j = 0$, $j = 0, \dots, k - 1$, of δ have a unique solution of the form

$$(\delta_1, \dots, \delta_k) = \varphi(\delta_{k+1}, \dots, \delta_m, a)$$

for a near a_0. Obviously, φ is linear in $\delta_{k+1}, \dots, \delta_m$. Further, let

$$b_{k+j}|_{(\delta_1, \dots, \delta_k) = \varphi(\delta_{k+1}, \dots, \delta_m, a)} = L_j(\delta_{k+1}, \dots, \delta_m) \Delta_j(a), \quad j = 0, \dots, n. \tag{7.48}$$

We have the following theorem.

Theorem 7.12 ([92]) *Consider the near-Hamiltonian system* (6.1), *where* $H(x, y, a)$ *with* $a \in R^n$ *satisfies* (7.1) *and the functions* p *and* q *are linear in* $\delta \in R^m$. *Suppose there exists an integer* $k > 0$ *and* $\delta_0 = (\delta_{10}, \dots, \delta_{m0}) \in R^m$ *and* $a_0 \in R^n$ *such that* (7.47) *and* (7.48) *hold with*

$$L_j(\delta_{k+1,0}, \dots, \delta_{m0}) \neq 0, \quad j = 0, \dots, n, \tag{7.49}$$
$$\Delta_j(a_0) = 0, \quad j = 0, \dots, n - 1, \quad \Delta_n(a_0) \neq 0,$$

and

$$\det \frac{\partial(\Delta_0, \dots, \Delta_{n-1})}{\partial(a_1, \dots, a_n)}(a_0) \neq 0. \tag{7.50}$$

Then for all (ε, δ, a) *near* $(0, \delta_0, a_0)$, (6.1) *has at most* $(k + n)$ *limit cycles near the origin, and for some* (ε, δ, a) *near* $(0, \delta_0, a_0)$, (6.1) *can have* $(k + n)$ *limit cycles near the origin.*

Proof We fix $(\delta_{k+1}, \dots, \delta_m) = (\delta_{k+1,0}, \dots, \delta_{m0})$ so that

$$L_j(\delta_{k+1}, \dots, \delta_m) = L_j(\delta_{k+1,0}, \dots, \delta_{m0}) \equiv L_{j0} \neq 0.$$

Then noting that $b_j = 0$ for $j = 0, \ldots, k-1$, and $(\delta_1, \ldots, \delta_k) = \varphi(\delta_{k+1}, \ldots, \delta_m, a)$, by (7.46)–(7.48), we have

$$M(h, \delta, a)|_{(\delta_1, \ldots, \delta_k) = \varphi(\delta_{k+1}, \ldots, \delta_m, a)} = h^{k+1} \sum_{j \geq 0} L_{j0} \Delta_j(a) h^j \equiv \widetilde{M}(h, a). \quad (7.51)$$

Further, by (7.50), we can change a near a_0 such that

$$L_{i0} L_{i+1,0} \Delta_i \Delta_{i+1} < 0, \qquad |\Delta_i| \ll |\Delta_{i+1}|, \quad i = 0, \ldots, n-1, \qquad (7.52)$$

which implies that the function $\widetilde{M}$ in (7.51) has n positive simple zeros $h_n^* < \cdots < h_1^*$ near $h = 0$. Now fix a satisfying (7.52). Then by (7.47) we can change $(\delta_1, \ldots, \delta_k)$ near $\varphi(\delta_{k+1,0}, \ldots, \delta_{m0}, a)$ such that

$$b_j b_{j+1} < 0, \qquad |b_j| \ll |b_{j+1}|, \quad j = 0, \ldots, k-1, \qquad (7.53)$$

which implies that the function M given by (7.46) has k simple zeros in the interval $(0, h_n^*)$. Clearly, under (7.53) the zeros $h_n^*, \ldots, h_1^*$, remain in existence. Thus, under (7.52) and (7.53) the function M has $(n+k)$ positive simple zeros altogether. Hence, by (6.5) the function F can have $(n+k)$ positive zeros in h near $h = 0$ which give $(n+k)$ hyperbolic limit cycles.

Finally, by (7.47)–(7.49), we have

$$b_j(\delta_0, a_0) = 0, \quad j = 0, \ldots, n+k-1, \qquad b_{n+k}(\delta_0, a_0) = L_{n0} \Delta_n(a_0) \neq 0.$$

Following the proof of Theorem 7.6, one can show that (6.1) has at most $(n+k)$ limit cycles near the origin for all (ε, δ, a) near $(0, \delta_0, a_0)$.

This completes the proof. □

7.6 Maple Programs

In this section, we list Maple programs for computing the coefficients $\{b_j\}$ in (7.25) or (7.46) based on the proof in Sect. 2.1. The details are given below.

(1) Find e_j given by (7.2).

First, by (7.1) we have

$$H_y(x, y) = \omega y + \sum_{i+j \geq 3} j h_{ij} x^i y^{j-1}.$$

Then using the identity $H_y(x, \varphi(x)) \equiv 0$ and (7.2), we can find e_j. The Maple program for this part follows.

```
Program 1.
> restart; with(LinearAlgebra): n:=6:
> H1:=omega*y[n-1]:
> for i from 3 to trunc(n/2)+1 do
```

```
> for j from 1 to i do
> H1:=H1+j*h[i-j,j]*x^(i-j)*y[n+3-i-j]^(j-1):
> od:
> od:
> H2:=H1:
> for i from trunc(n/2)+2 to n+1 do
> for j from 1 to n+2-i do
> H2:=H2+j*h[i-j,j]*x^(i-j)*y[n+3-i-j]^(j-1):
> od:
> od:
> H:=sort(H2,[x,y],tdeg):
> phi[0]:=0:
> for i from 1 to n-1 do
> phi[i]:=phi[i-1]+e[i+1]*x^(i+1):
> od:
> for i from 1 to n-1 do
> H:=subs(y[i]=phi[i],H):
> od:
> for i from 2 to n do
> temp1:=subs(x=0,diff(H,x$i)/i!):
> e[i]:=solve(temp1,e[i]):
> print(i,e[i]):
> od: save e,"teste20.m":
```

The expressions $e_2, \ldots, e_6$, listed below (7.2) are obtained by executing the above program on a laptop computer. We can obtain higher-order coefficients as necessary.

(2) Find h_j given in (7.6).

Using (7.1) and (7.2) we can easily find h_j in (7.6). The Maple program for this step is listed below.

```
Program 2.
> restart; with(LinearAlgebra):n:=7:
> H1:=(omega/2)*y[n-3]^2+omega/2*x^2:
> for i from 3 to trunc(n/2) do
> for j from 0 to i do
> H1:=H1+h[i-j,j]*x^(i-j)*y[n+1-i-j]^j:
> od:
> od:
> H2:=H1:
> for i from trunc(n/2)+1 to n do
> for j from 0 to n-i do
> H2:=H2+h[i-j,j]*x^(i-j)*y[n+1-i-j]^j:
> od:
> od:
» phi[0]:=0:
> for i from 1 to n-3 do
> phi[i]:=phi[i-1]+e[i+1]*x^(i+1):
> od:
> for i from 1 to n-3 do
> H2:=subs(y[i]=phi[i],H2):
> od:
> read "teste20.m":
> for i from 2 to n do
> h[i]:=simplify(expand(subs(x=0,diff(H2,x$i)/i!))):
> print(i,h[i]):
```

```
> od:
> save h,"testh22.m":
```

The results given in (7.7) are obtained by executing the above program.

(3) Find $\bar{b}_{ij}$ given in (7.11).

Based on (7.2) and (7.10), the corresponding Maple program is coded as follows.

```
Program 3.
> restart;with(LinearAlgebra):n:=6:
> temp1:=0:
> for i from 0 to n do
> for j from 0 to i do
> temp1:=temp1+c[i-j,j]*x^(i-j)*y^j:
> od:
> od:
> phi:=0:
> for i from 2 to n do
> phi:=phi+e[i]*x^i:
> od:
> temp[0]:=sort(expand(subs(y=phi,temp1)),x):
> for j from 1 to n do
> temp[j]:=expand(subs(y=phi,diff(temp1,y$j)/(j+1)!)):
> od:
> for j from 0 to n do
> for i from 0 to n-j do
> barb[i,j]:=coeff(temp[j],x,i):
> print([i,j],barb[i,j]):
> od:
> od:save barb, "barb.m":
```

Executing the above the program yields the explicit expressions given in (7.12).

(4) Find H_j^* in (7.5) together with (7.1), (7.2) and (7.6).

The Maple program is given below.

```
Program 4.
> restart;with(LinearAlgebra):n:=6:
> H:=(omega/2)*y^2+(omega/2)*x^2:
> for i from 3 to n+2 do
> for j from 0 to i do
> H:=H+h[i-j,j]*x^(i-j)*y^j:
> od:
> od:
> phi:=0:
> for i from 2 to n do
> phi:=phi+e[i]*x^i:
> od:
> for j from 1 to n+1 do
> temp[j]:=sort(expand(subs(y=phi,diff(H,y$(j+1))/(j+1)!)),x):
> od:
> for i from 1 to n+1 do
> Hstar[i]:=0:
> od:
> for i from 1 to n+1 do
> for j from 0 to n-i+1 do
> Hstar[i]:=Hstar[i]+coeff(temp[i],x,j)*x^j:
> od:print(i,Hstar[i]):
```

```
> od:
> save Hstar, "Hstar.m":
```

Executing the above program yields the expansions of $H_1^*, \ldots, H_7^*$; for example, see those given above (7.15).

(5) Find $a_j(x)$ given in (7.14) by using (7.5) and (7.13).

This step needs two programs, listed below as Program 5 and Program 6. First, find the formulas for $a_j(x)$ and then obtain their expansions using the expansions of H_j^*.

```
Program 5.
> restart; with(LinearAlgebra): n:=7:Eta:=0:
> for i from 1 to n do
> Eta:=Eta+Hstar[i]*v[n-i+1]^(i+1):
> od:
> V[0]:=0:
> for i from 1 to n do
> V[i]:=V[i-1]+a[i]*w^i:
> od:
> for i from 1 to n do
> Eta:=subs(v[i]=V[i],Eta):
> od:
> temp1:=expand(Eta-w^2):
> a[1]:=1/(sqrt(Hstar[1])):
> for i from 3 to n+1 do
> temp2:=coeff(temp1,w,i):
> a[i-1]:=solve(temp2,a[i-1]):
> print(i-1,a[i-1]):
> od: save a, "a.m":
```

```
Program 6.
> restart;with(LinearAlgebra):n:=6:
> read "Hstar.m";
> read "a.m":
> for i from 1 to n+1 do
> temp1[i]:=taylor(1/(1+s)^((3*i-2)/2),s=0,n+2-i):
> temp[i]:=convert(temp1[i],polynom):
> od:
> s:=(2/omega)*Hstar[1]-1:
> for i from 1 to n+1 do
> temp[i]:=(2/omega)^((3*i-2)/2)*temp[i]:
> od:
> for i from 1 to n+1 do
> temp2[i]:=sort(expand(a[i]*Hstar[1]^((3*i-2)/2)*temp[i]),x):
> od:
> for j from 1 to n+1 do
> for i from 0 to n-j+1 do
> bara[i,j]:=simplify(expand(coeff(temp2[j],x,i))):
> print([i,j],bara[i,j]):
> od:
> od:save bara, "bara.m":
```

Executing Program 5 we obtain the expressions $a_1, \ldots, a_7$, given below (7.14), and executing Program 6 yields the results given in (7.16).

(6) Find $\bar{q}_j$ given in (7.17).

Similarly to the above steps, we need two programs: Program 7 and Program 8. First, we find formulas for $\bar{q}_j$ and then obtain their expansions.

```
Program 7.
> restart; with(LinearAlgebra): n:=3:Q:=0:
> for i from 1 to 2*n+1 do
> Q:=Q+q[i]*((v1)[2*n+2-i]^i-(v2)[2*n+2-i]^i):
> od:
> V1[0]:=0:
> for i from 1 to 2*n+1 do
> V1[i]:=V1[i-1]+a[i]*w^i:
> od:
> V2[0]:=0:
> for i from 1 to 2*n+1 do
> V2[i]:=V2[i-1]+a[i]*(-w)^i:
> od:
> for i from 1 to 2*n+1 do
> Q:=subs([v1[i]=V1[i],v2[i]=V2[i]],Q):
> od:
> temp1:=sort(expand(Q),w):
> for i from 0 to n do
> barq[i]:=coeff(temp1,w,2*i+1):
> print(i,barq[i]):
> od:save barq, "barq.m":
```

```
Program 8.
> restart; with(LinearAlgebra): n:=3:
> for i from 1 to 2*n+1 do
> q[i]:=0:
> a[i]:=0:
> od:
> for j from 1 to 2*n+1 do
> for i from 0 to 2*n+1 do
> q[j]:=q[j]+barb[i,j-1]*x^i:
> a[j]:=a[j]+bara[i,j]*x^i:
> od:
> od:
> read "barq.m";
> for j from 0 to n do
> for i from 0 to 2*(n-j) do
> alpha[i,j]:=simplify(coeff(barq[j],x,i)):
> print([i,j],alpha[i,j]):
> od:
> od:save alpha, "alpha.m":
```

Executing Program 7 yields the expression for $\bar{q}_j$ given in (7.18), and executing Program 8 gives the results in (7.19).

(7) Find $\psi(x)$, $\psi^{-1}(x)$ and $\frac{1}{\psi'(x)}$ given in (7.38).

The following three programs, Programs 9, 10 and 11, are used to find d_i, $\bar{d}_i$ and n_i in (7.38), respectively.

```
Program 9.
> restart; with(LinearAlgebra): n:=6:
> S:=taylor((1+s)^(1/2),s=0,n+1):
> temp1:=1:
```

```
> for i from 1 to n do
> temp1:=temp1+subs(s=0,(diff(S,s$i)/i!))*s[n+1-i]^i:
> od:
> S[0]:=0:
> for i from 1 to n do
> S[i]:=S[i-1]+(h[i+2]/h[2])*x^i:
> od:
> for i from 1 to n do
> temp1:=subs(s[i]=S[i],temp1):
> od:
> temp2:=sort(expand(h[2]^(1/2)*x*temp1),x):
> for i from 1 to n do
> d[i]:=simplify(coeff(temp2,x,i)):
> print([i],d[i]):
> od:save d,"d.m":

Program 10.
> restart; with(LinearAlgebra): n:=6:psi:=0:
> for i from 1 to n do
> psi:=psi+d[i]*x[n+1-i]^i:
> od:
> X[0]:=0:
> for i from 1 to n do
> X[i]:=X[i-1]+bard[i]*u^i:
> od:
> read "d.m":
> for i from 1 to n do
> psi:=subs(x[i]=X[i],psi):
> od:
> temp1:=sort(expand(psi-u),u):
> for i from 1 to n do
> temp2:=coeff(temp1,u,i):
> bard[i]:=solve(temp2,bard[i]):
> print(i,bard[i]):
> od:save bard, "bard.m":

Program 11.
> restart; with(LinearAlgebra): n:=5:
> S:=taylor(1/(1+s),s=0,n+1):
> temp1:=1:
> for i from 1 to n do
> temp1:=temp1+subs(s=0,(diff(S,s$i)/i!))*s[n+1-i]^i:
> od:
> psi[0]:=d[1]*x:
> for i from 1 to n do
> psi[i]:=psi[i-1]+d[i+1]*x^(i+1):
> od:
> read "d.m":
> for i from 1 to n do
> S[i]:=expand(diff(psi[i],x)/d[1])-1:
> od:
> for i from 1 to n do
> temp1:=subs(s[i]=S[i],temp1):
> od:
> temp2:=sort(expand((1/d[1])*temp1),x):
> for i from 0 to n do
> N[i]:=simplify(coeff(temp2,x,i)):
```

```
> print(i,N[i]):
> od:save N,"N.m":
```

Executing the above three programs we obtain the formulas listed below (7.38).

(8) Find r_{ij} given in (7.21).

First, by (7.20), (7.21) and (7.38) we have $r_{00} = 2n_0\alpha_{00}$; to find the remaining r_{ij}, for simplicity we suppose $\alpha_{00} = 0$. The program is listed below.

Program 12.

```
> restart; with(LinearAlgebra): n:=6:
> for i from 0 to n/2 do
> qbar[i]:=0:
> temp[i]:=0:
> od:
> alpha[0,0]:=0:
> for j from 0 to n/2 do
> for i from 0 to n-2*j do
> qbar[j]:=qbar[j]+alpha[i,j]*x^i:
> temp[j]:=temp[j]+N[i]*x^i:
> od:
> od:
> for i from 0 to n/2 do
> temp2[i]:=sort(expand(qbar[i]*temp[i]),x):
> od:
> for i from 0 to n/2 do
> W[i]:=0:
> od:
> for j from 0 to n/2 do
> for i from 0 to n-2*j do
> W[j]:=W[j]+coeff(temp2[j],x,i)*x[n-2*j+1-i]^i:
> od:
> od:
> X[0]:=0:
> for i from 1 to n do
> X[i]:=X[i-1]+bard[i]*u^i:
> od:
> for j from 0 to n/2 do
> for i from 1 to n-2*j do
> W[j]:=sort(subs(x[i]=X[i],W[j]),u):
> od:
> od:
> for j from 0 to n/2 do
> for i from 0 to n-2*j do
> temp3[j,i]:=coeff(W[j],x,i):
> od:
> od:
> for j from 0 to n/2 do
> W1[j]:=0:
> W2[j]:=0:
> od:
> for j from 0 to n/2 do
> for i from 0 to n-2*j do
> W1[j]:=W1[j]+temp3[j,i]*u^i:
> W2[j]:=W2[j]+temp3[j,i]*(-u)^i:
> od:
> od:
> for j from 0 to n/2 do
```

```
> temp5[j]:=W1[j]+W2[j]:
> od:
> for j from 0 to n/2 do
> for i from 0 to (n-2*j)/2 do
> r[i,j]:=expand(coeff(temp5[j],u,2*i)):
> print([i,j],r[i,j]):
> od:
> od:save r, "rij.m":
```

Executing this program we obtain (7.41) with $\alpha_{00} = 0$.

(9) Find b_l using (7.24).

First, by (7.42) we have $b_0 = r_{00}\beta_{00} = 2n_0\alpha_{00}$; then, under the assumption $\alpha_{00} = 0$, we have $b_0 = 0$. The following program can be used to find b_j for $j \geq 1$, where the β_{ij} are first calculated using (7.23) and then substituted into (7.24).

Program 13.
```
> restart; with(LinearAlgebra): n:=3:
> for i from 0 to n do
> for j from 0 to i do
> beta[i-j,j]:=int(x^(2*(i-j))*(1-x^2)^j*sqrt(1-x^2),x=0..1):
> od:
> od:
> for i from 1 to n do
> b[i]:=0:
> od:
> for i from 1 to n do:
> for j from 0 to i do
> b[i]:=b[i]+r[i-j,j]*beta[i-j,j]:
> od:print(i,b[i]):
> od: save beta, "beta.m":
```

Executing this program we finally obtain the formulas for $b_j(\delta)|_{b_0=0}$ for $j = 1, 2, 3$, given in (7.42) and (7.43).

7.7 Application

As an application of the method, the formulas and programs developed in previous sections, we study the Hopf bifurcation of a class of cubic near-Hamiltonian systems, described below.

From [14, 149, 187] we know that a quadratic Hamiltonian system having an elementary center can be transformed into the form

$$\dot{x} = y + bx^2 - 2axy + cy^2,$$
$$\dot{y} = -x - ax^2 - 2bxy + ay^2.$$

Now we perturb this system by cubic polynomials to obtain a near-Hamiltonian system of the form

$$\dot{x} = y + bx^2 - 2axy + cy^2 + \varepsilon p(x, y),$$
$$\dot{y} = -x - ax^2 - 2bxy + ay^2 + \varepsilon q(x, y),$$

(7.54)

where

$$p(x, y) = \sum_{1 \le i+j \le 3} a_{ij} x^i y^j, \qquad q(x, y) = \sum_{1 \le i+j \le 3} b_{ij} x^i y^j.$$

The unperturbed system $(7.54)|_{\varepsilon=0}$ is Hamiltonian with

$$H = \frac{1}{2}(x^2 + y^2) + \frac{a}{3}x^3 + bx^2 y - axy^2 + \frac{c}{3}y^3.$$

For system (7.54) we will consider the cases of $a = 0$ and $a \ne 0$ separately.

7.7.1 Case $a \ne 0$

First, assume $a \ne 0$. Without loss of generality, we let $a = 1$ (otherwise, we can introduce a suitable rescaling of (x, y)). Further, we can assume $b \ge 0$ (otherwise, we need only to change the sign of y and t). Then system (7.54) becomes

$$
\begin{aligned}
\dot{x} &= y + bx^2 - 2xy + cy^2 + \varepsilon p(x, y), \\
\dot{y} &= -x - x^2 - 2bxy + y^2 + \varepsilon q(x, y).
\end{aligned}
\tag{7.55}
$$

Correspondingly,

$$
\begin{aligned}
H(x, y) &= \frac{1}{2}(x^2 + y^2) + \frac{1}{3}x^3 + bx^2 y - xy^2 + \frac{c}{3}y^3 \\
&= \frac{1}{2}(x^2 + y^2) + \sum_{i+j \ge 3} h_{ij} x^i y^j, \quad b > 0,
\end{aligned}
\tag{7.56}
$$

and

$$p_x + q_y = \sum_{0 \le i+j \le 2} c_{ij} x^i y^j,$$

where

$$
\begin{aligned}
c_{00} &= a_{10} + b_{01}, & c_{10} &= 2a_{20} + b_{11}, & c_{01} &= a_{11} + 2b_{02}, \\
c_{20} &= 3a_{30} + b_{21}, & c_{11} &= 2(a_{21} + b_{12}), & c_{02} &= a_{12} + 3b_{03}.
\end{aligned}
\tag{7.57}
$$

Theorem 7.13 ([92]) *For system (7.55), introduce*

$$\delta = (c_{00}, c_{10}, c_{01}, c_{02}, c_{11}, c_{20}), \qquad \sigma = (b, c),$$

and

$$\delta_0 = (c_{00}^*, c_{10}^*, c_{01}^*, c_{02}^*, c_{11}^*, c_{20}^*), \qquad \sigma_0 = (b^*, c^*),$$

where δ_0 and σ_0 satisfy

$$b^* = \frac{2}{13}\sqrt{26}, \quad c^* = 2b^*, \quad c_{10}^* - c^* c_{01}^* + c_{02}^* \neq 0, \quad c_{20}^* = 3b^* c_{01}^* - c_{02}^*,$$

$$c_{00}^* = 0, \quad c_{11}^* = 15b^* c_{10}^*/4 - \left(2 + 7(b^*)^2/2\right)c_{01}^* + 7b^* c_{02}^*/4.$$

Then, for all $(\varepsilon, \delta, \sigma)$ near $(0, \delta_0, \sigma_0)$, system (7.55) has Hopf cyclicity 5 at the origin.

Proof We want to find the formulas for b_i, $i = 0, 1, 2, 3, 4, 5$, and use them to study the limit cycles of (7.55) near the origin by executing a series of Maple programs similar to those listed in the previous section.

By (7.56), the following Program I gives the values of h_{ij}, $3 \leq i + j \leq 11$.

```
Program I:
> restart;with(LinearAlgebra): n:=11:a:=1:
> H1:=1/2*x^2+1/2*y^2+a/3*x^3+b*x^2*y-a*x*y^2+c/3*y^3:
> for i from 0 to n do
> for j from 0 to i do
> h[i-j,j]:=coeftayl(H1,[x,y]=[0,0],[i-j,j]):
> print([i-j,j],h[i-j,j]):
> od:
> od:save h,"hij.m":
```

Similarly to Program I, by (7.57) the following program outputs c_{ij}, $3 \leq i + j \leq 10$.

```
Program II
> restart;with(LinearAlgebra):n:=10:
> P:=0:
> for i from 0 to 2 do
> for j from 0 to i do
> P:=P+c[i-j,j]*x^(i-j)*y^j:
> od:
> od:
> for i from 0 to n do
> for j from 0 to i do
> c[i-j,j]:=coeftayl(P,[x,y]=[0,0],[i-j,j]):
> print([i-j,j],c[i-j,j]):
> od:
> od:save c,"c.m":
```

Executing the above program we obtain

$$c_{ji} = c_{ij} \quad \text{for } 0 \leq i + j \leq 2, \qquad c_{ij} = 0 \quad \text{for } 3 \leq i + j \leq 10.$$

To find the explicit formulas for b_j, $j = 0, 1, 2, 3, 4, 5$, we need to execute Programs 1–13. When we do this we need to choose values for ω and n. It is obvious that $\omega = 1$ for (7.55). The value of n is different in each program. Sometimes we need to read the resulting output from some programs to simplify the calculation; Table 7.1 explains system of programs.

Table 7.1 System of Maple Programs

Program number	n	Reads	Output by program number
1	10	hij.m	I
2	11	hij.m	I
		teste20.m	1
3	10	c.m	II
4	10	hij.m	I
		teste20.m	1
5	11		
6	10	a.m	5
		teste20.m	1
7	5		
8	5	a.m	5
		barq.m	7
		bara.m	6
		barb.m	3
		teste20.m	1
9	10	testh22.m	2
10	10	d.m	9
11	9	d.m	9
12	10	alpha.m	8
		bard.m	10
		N.m	11
13	5	rij.m	12

By Program 9 it is easy to see that $b_0 = 2n_0\alpha_{00}\beta_{00} = 2\pi c_{00}$. Then we find the formulas for b_j, $j = 1, 2, 3, 4, 5$, with $b_0 = 0$ as follows.

$$b_1|_{b_0=0} = \pi\left[c_{20} + c_{02} - (b+c)c_{01}\right],$$

$$b_2|_{b_0=0} = \pi\left[\left(\frac{5}{3}b^2 + \frac{5}{3}c^2 + \frac{10}{3}cb\right)c_{10} + \left(\frac{5}{18}c^2 + cb + \frac{16}{9} + \frac{5}{2}b^2\right)c_{20}\right.$$
$$+ \left(-\frac{40}{9}b - \frac{5}{2}cb^2 - \frac{40}{9}c - \frac{35}{18}c^2b - \frac{5}{2}b^3 - \frac{35}{18}c^3\right)c_{01} \qquad (7.58)$$
$$+ \left(-\frac{4}{3}b - \frac{4}{3}c\right)c_{11} + \left.\left(\frac{16}{9} + \frac{5}{3}cb + \frac{3}{2}b^2 + \frac{35}{18}c^2\right)c_{02}\right],$$

$$b_i|_{b_0=0} = \pi(\delta_{i2}c_{10} + \delta_{i3}c_{01} + \delta_{i4}c_{20} + \delta_{i5}c_{11} + \delta_{i6}c_{02}), \quad i = 3, 4, 5,$$

where

$$\delta_{32} = \frac{105}{8}b^4 + \frac{140}{9}b^2 + \frac{140}{9}c^2 + \frac{385}{72}c^4 + \frac{875}{36}c^2b^2 + \frac{280}{9}cb + \frac{175}{6}cb^3$$
$$+ \frac{245}{18}c^3b,$$

$$\delta_{33} = -\frac{315}{32}b^5 - \frac{1225}{96}b^4c - \frac{175}{16}b^3c^2 - 35b^3 - \frac{385}{48}b^2c^3 - \frac{175}{3}b^2c - \frac{5005}{864}bc^4$$

$$- \frac{1435}{27}bc^2 - \frac{560}{27}b - \frac{5005}{864}c^5 - \frac{805}{27}c^3 - \frac{560}{27}c,$$

$$\delta_{34} = \frac{385}{864}c^4 + \frac{175}{48}c^2b^2 + \frac{35}{2}b^2 + \frac{175}{24}cb^3 + \frac{35}{24}c^3b + \frac{245}{54}c^2 + \frac{140}{27} + \frac{315}{32}b^4$$

$$+ \frac{35}{3}cb,$$

$$\delta_{35} = -\frac{35}{4}b^3 - \frac{175}{12}b^2c - \frac{385}{36}bc^2 - \frac{70}{9}b - \frac{175}{36}c^3 - \frac{70}{9}c,$$

$$\delta_{36} = \frac{175}{32}b^4 + \frac{175}{24}b^3c + \frac{105}{16}b^2c^2 + \frac{35}{2}b^2 + \frac{385}{72}bc^3 + \frac{245}{9}bc + \frac{5005}{864}c^4$$

$$+ \frac{1085}{54}c^2 + \frac{140}{27},$$

$$\delta_{42} = \frac{3080}{27}c^2 + \frac{6160}{27}cb + \frac{3080}{27}b^2 + \frac{29645}{54}c^2b^2 + \frac{5005}{9}cb^3 + \frac{8855}{27}c^3b$$

$$+ \frac{2695}{12}b^4 + \frac{11935}{108}c^4 + \frac{3773}{16}cb^5 + \frac{8085}{32}c^2b^4 + \frac{13475}{72}c^3b^3 + \frac{3003}{32}b^6$$

$$+ \frac{17017}{864}c^6 + \frac{95095}{864}c^4b^2 + \frac{23023}{432}c^5b,$$

$$\delta_{43} = -\frac{3003}{64}b^7 - \frac{4851}{64}b^6c - \frac{4851}{64}b^5c^2 - \frac{2079}{8}b^5 - \frac{35035}{576}b^4c^3 - \frac{13475}{24}b^4c$$

$$- \frac{25025}{576}b^3c^4 - \frac{2695}{4}b^3c^2 - \frac{3080}{9}b^3 - \frac{17017}{576}b^2c^5 - \frac{20405}{36}b^2c^3$$

$$- \frac{6160}{9}b^2c - \frac{323323}{15552}bc^6 - \frac{235235}{648}bc^4 - \frac{52360}{81}bc^2 - \frac{24640}{243}b$$

$$- \frac{323323}{15552}c^7 - \frac{109109}{648}c^5 - \frac{24640}{81}c^3 - \frac{24640}{243}c,$$

$$\delta_{44} = \frac{13475}{648}c^4 + \frac{4480}{243} + \frac{280}{3}cb + \frac{980}{9}b^2 + \frac{3220}{81}c^2 + \frac{1715}{12}c^2b^2 + \frac{1085}{6}cb^3$$

$$+ \frac{1155}{8}b^4 + \frac{1295}{18}c^3b + \frac{2695}{144}c^3b^3 + \frac{1617}{32}cb^5 + \frac{2205}{64}c^2b^4 + \frac{3003}{64}b^6$$

$$+ \frac{17017}{15552}c^6 + \frac{5005}{576}c^4b^2 + \frac{1001}{288}c^5b,$$

$$\delta_{45} = -\frac{231}{4}b^5 - \frac{245}{2}b^4c - \frac{245}{2}b^3c^2 - \frac{350}{3}b^3 - \frac{770}{9}b^2c^3 - \frac{2030}{9}b^2c$$

$$- \frac{5005}{108}bc^4 - \frac{4970}{27}bc^2 - \frac{1120}{27}b - \frac{1001}{54}c^5 - \frac{2030}{27}c^3 - \frac{1120}{27}c,$$

$$\delta_{46} = \frac{1617}{64}b^6 + \frac{1323}{32}b^5c + \frac{2695}{64}b^4c^2 + \frac{1085}{8}b^4 + \frac{5005}{144}b^3c^3 + \frac{1715}{6}b^3c$$

$$+ \frac{5005}{192}b^2c^4 + \frac{1295}{4}b^2c^2 + 140b^2 + \frac{17017}{864}bc^5 + \frac{13475}{54}bc^3 + \frac{6440}{27}bc$$

$$+ \frac{323323}{15552}c^6 + \frac{85085}{648}c^4 + \frac{12460}{81}c^2 + \frac{4480}{243},$$

$$\delta_{52} = \frac{350350}{243}c^4 + \frac{1121120}{729}cb + \frac{560560}{81}cb^3 + \frac{1821820}{243}c^2b^2 + \frac{1310309}{1944}c^6$$

$$+ \frac{560560}{729}b^2 + \frac{560560}{729}c^2 + \frac{70070}{27}b^4 + \frac{1121120}{243}c^3b + \frac{175175}{72}b^6$$

$$+ \frac{1296295}{162}c^3b^3 + \frac{665665}{72}c^2b^4 + \frac{763763}{108}b^5c + \frac{9844835}{1944}c^4b^2$$

$$+ \frac{2277275}{972}c^5b + \frac{77077}{36}c^3b^5 + \frac{875875}{576}c^4b^4 + \frac{85085}{128}b^8 + \frac{15015}{8}b^7c$$

$$+ \frac{77077}{32}c^2b^6 + \frac{595595}{648}c^5b^3 + \frac{11316305}{23328}c^6b^2$$

$$+ \frac{323323}{1458}c^7b + \frac{7436429}{93312}c^8,$$

$$\delta_{53} = -\frac{255255}{1024}b^9 - \frac{495495}{1024}b^8c - \frac{143143}{256}b^7c^2 - \frac{185185}{96}b^7 - \frac{385385}{768}b^6c^3$$

$$- \frac{161161}{32}b^6c - \frac{595595}{1536}b^5c^4 - \frac{231231}{32}b^5c^2 - \frac{49049}{12}b^5 - \frac{11316305}{41472}b^4c^5$$

$$- \frac{6341335}{864}b^4c^3 - \frac{3538535}{324}b^4c - \frac{11316305}{62208}b^3c^6 - \frac{5080075}{864}b^3c^4$$

$$- \frac{805805}{54}b^3c^2 - \frac{700700}{243}b^3 - \frac{7436429}{62208}b^2c^7 - \frac{10125115}{2592}b^2c^5$$

$$- \frac{2137135}{162}b^2c^3 - \frac{1541540}{243}b^2c - \frac{185910725}{2239488}bc^8 - \frac{151638487}{69984}bc^6$$

$$- \frac{23158135}{2916}bc^4 - \frac{13313300}{2187}bc^2 - \frac{1121120}{2187}b - \frac{185910725}{2239488}c^9$$

$$- \frac{64341277}{69984}c^7 - \frac{8275267}{2916}c^5 - \frac{5745740}{2187}c^3 - \frac{1121120}{2187}c,$$

$$\delta_{54} = \frac{160160}{2187} + \frac{255255}{1024}b^8 + \frac{7436429}{2239488}c^8 + \frac{160160}{243}cb + \frac{160160}{243}b^2 + \frac{640640}{2187}c^2$$

$$+ \frac{205205}{81}cb^3 + \frac{385385}{162}c^2b^2 + \frac{35035}{27}c^3b + \frac{1056055}{2916}c^4 + \frac{55055}{36}b^4$$

$$+ \frac{175175}{108}c^3b^3 + \frac{35035}{16}c^2b^4 + \frac{49049}{24}b^5c + \frac{385385}{432}c^4b^2 + \frac{235235}{648}c^5b$$

$$+ \frac{3250247}{34992}c^6 + \frac{55055}{48}b^6 + \frac{77077}{384}c^3b^5 + \frac{175175}{1536}c^4b^4 + \frac{77077}{256}c^2b^6$$

$$+ \frac{595595}{10368}c^5b^3 + \frac{1616615}{62208}c^6b^2 + \frac{323323}{31104}c^7b + \frac{45045}{128}b^7c,$$

$$\delta_{55} = -\frac{25025}{64}b^7 - \frac{63063}{64}b^6c - \frac{77077}{64}b^5c^2 - \frac{23023}{18}b^5 - \frac{595595}{576}b^4c^3$$

$$- \frac{175175}{54}b^4c - \frac{1226225}{1728}b^3c^4 - \frac{35035}{9}b^3c^2 - \frac{10010}{9}b^3$$

$$- \frac{2127125}{5184}b^2c^5 - \frac{245245}{81}b^2c^3 - \frac{190190}{81}b^2c - \frac{9376367}{46656}bc^6 - \frac{785785}{486}bc^4$$

$$- \frac{490490}{243}bc^2 - \frac{160160}{729}b - \frac{3556553}{46656}c^7 - \frac{251251}{486}c^5 - \frac{190190}{243}c^3$$

$$- \frac{160160}{729},$$

$$\delta_{56} = \frac{135135}{1024}b^8 + \frac{33033}{128}b^7c + \frac{77077}{256}b^6c^2 + \frac{49049}{48}b^6 + \frac{35035}{128}b^5c^3$$

$$+ \frac{21021}{8}b^5c + \frac{2977975}{13824}b^4c^4 + \frac{175175}{48}b^4c^2 + \frac{205205}{108}b^4$$

$$+ \frac{1616615}{10368}b^3c^5 + \frac{385385}{108}b^3c^3 + \frac{385385}{81}b^3c + \frac{2263261}{20736}b^2c^6$$

$$+ \frac{1176175}{432}b^2c^4 + \frac{35035}{6}b^2c^2 + \frac{80080}{81}b^2 + \frac{7436429}{93312}bc^7 + \frac{3250247}{1944}bc^5$$

$$+ \frac{1056055}{243}bc^3 + \frac{1281280}{729}bc + \frac{185910725}{2239488}c^8 + \frac{26835809}{34992}c^6$$

$$+ \frac{5260255}{2916}c^4 + \frac{2322320}{2187}c^2 + \frac{160160}{2187}.$$

Let

$$\tilde{b}_j = b_j|_{b_0=0}, \quad j = 1, 2, 3, 4, 5. \tag{7.59}$$

Noting (7.58), we can solve for c_{20} and c_{11} from $\tilde{b}_1 = \tilde{b}_2 = 0$ as follows:

$$c_{20} = (b+c)c_{01} - c_{02},$$
$$c_{11} = \frac{5}{4}(c+b)c_{10} + \left(-\frac{5}{4}c^2 + \frac{3}{4}cb - 2\right)c_{01} + \left(\frac{5}{4}c - \frac{3}{4}b\right)c_{02}. \tag{7.60}$$

Substituting (7.60) into $\tilde{b}_3, \tilde{b}_4, \tilde{b}_5$, we obtain

$$\tilde{b}_{3+i} = \Delta_i \pi(c_{10} - c_{01}c + c_{02}) \equiv \Delta_i L, \quad i = 0, 1, 2, \tag{7.61}$$

where $L = \pi(c_{10} - c_{01}c + c_{02})$, and

$$\Delta_0 = \frac{35}{16}b^4 - \frac{175}{24}b^2c^2 + \frac{35}{6}b^2 - \frac{35}{6}bc^3 + \frac{35}{3}bc - \frac{35}{48}c^4 + \frac{35}{6}c^2$$

$$= -\frac{35}{48}(b+c)^2\left(c + 3b - 2\sqrt{3b^2 + 2}\right)\left(c + 3b + 2\sqrt{3b^2 + 2}\right),$$

$$\Delta_1 = -\frac{7}{288}(b+c)^2\left(-891b^4 + 1350b^3c + 396b^2c^2 - 3240b^2 + 858bc^3\right.$$

$$\left. + 1200bc + 143c^4 - 680c^2 - 2560\right),$$

$$\Delta_2 = -\frac{1001}{20736}(b+c)^2\left(-3645b^6 + 4050b^5c + 2349b^4c^2 - 17280b^4\right.$$

$$+ 4860b^3c^3 + 5184b^3c + 1581b^2c^4 + 192b^2c^2 - 24960b^2 + 1938bc^5$$

$$\left. + 7872bc^3 - 3840bc + 323c^6 - 576c^4 - 9600c^2 - 10240\right).$$

For c near c^* and (c_{10}, c_{01}, c_{02}) near $(c_{10}^*, c_{01}^*, c_{02}^*)$ we have $L \neq 0$. Thus $\tilde{b}_3 = 0$ if and only if $\Delta_0 = 0$. For (b, c) near (b^*, c^*) we have $b + c \neq 0$. Then $\Delta_0 = 0$ if and only if $c = c_1(b)$ or $c = c_2(b)$, where

$$c_i(b) = -3b - (-1)^i 2\sqrt{3b^2 + 2}, \quad i = 1, 2.$$

If $c = c_i(b)$, we have

$$\Delta_1 = -28\left(b + (-1)^i\sqrt{3b^2 + 2}\right)^2$$

$$\times \left(43b^4 + (-1)^i b(25b^2 + 11)\sqrt{3b^2 + 2} + 33b^2 + 4\right)$$

$$\equiv \bar{\Delta}_1(b).$$

Noting $b \geq 0$, one can see that $\Delta_1 < 0$ for $i = 2$, and $\Delta_1 = 0$ if and only if $b = \frac{2}{13}\sqrt{26} \equiv b^*$ for $i = 1$. In this case, it follows that when $\Delta_1 = 0$ we have $c = 2b^*$ and

$$\bar{\Delta}_1'(b^*) \neq 0, \qquad \Delta_2 = -\frac{898128}{2197} \approx -408.7974511.$$

This clearly indicates that the conclusion is true by following Theorem 7.12.
The proof is completed.   □

7.7.2 Case $a = 0$

Next, we consider (7.54) for the case $a = 0$. As before, we have $b_0 = 2\pi c_{00}$. For simplicity, let $b_0 = 0$, i.e., $c_{00} = a_{10} + b_{01} = 0$. Take $a = 0$ in Program I and use the same procedure as for the case $a = 1$; we obtain the following formulas.

$$b_1|_{b_0=0} = \pi\left[-(b+c)c_{01} + c_{20} + c_{02}\right],$$

$$b_2|_{b_0=0} = \pi\left[\left(-\frac{5}{2}cb^2 - \frac{5}{2}b^3 - \frac{35}{18}c^2b - \frac{35}{18}c^3\right)c_{01} + \left(cb + \frac{5}{2}b^2 + \frac{5}{18}c^2\right)c_{20}\right.$$
$$\left. + \left(\frac{3}{2}b^2 + \frac{5}{3}cb + \frac{35}{18}c^2\right)c_{02}\right],$$

$$b_3|_{b_0=0} = \pi\left[\left(-\frac{1225}{96}b^4c - \frac{315}{32}b^5 - \frac{175}{16}c^2b^3 - \frac{385}{48}c^3b^2 - \frac{5005}{864}c^4b\right.\right.$$
$$\left. - \frac{5005}{864}c^5\right)c_{01} + \left(\frac{175}{24}b^3c + \frac{315}{32}b^4 + \frac{175}{48}c^2b^2\right.$$
$$\left. + \frac{35}{24}c^3b + \frac{385}{864}c^4\right)c_{20}$$
$$\left. + \left(\frac{175}{32}b^4 + \frac{175}{24}b^3c + \frac{105}{16}c^2b^2 + \frac{385}{72}c^3b + \frac{5005}{864}c^4\right)c_{02}\right].$$

Also, let

$$\tilde{b}_j = b_j|_{b_0=0}, \quad j = 1, 2, 3.$$

Then $\tilde{b}_1 = 0$ gives

$$-(b+c)c_{01} + c_{20} + c_{02} = 0,$$

or

$$c_{20} = (b+c)c_{01} - c_{02}. \tag{7.62}$$

Substituting (7.62) into $\tilde{b}_2$ and $\tilde{b}_3$ yields

$$\tilde{b}_2 = \Delta_0\tilde{L}, \qquad \tilde{b}_3 = \Delta_1\tilde{L}, \tag{7.63}$$

where $\tilde{L} = \pi(-c_{01}c + c_{02})$ and

$$\Delta_0 = \frac{2}{3}cb - b^2 + \frac{5}{3}c^2 = \frac{1}{3}(b+c)(5c - 3b),$$

$$\Delta_1 = \frac{35}{72}(b+c)\left(11c^3 - 3bc^2 + 9b^2c - 9b^3\right).$$

Let (b, c) and (c_{20}, c_{01}) satisfy $b + c \neq 0$ and $\tilde{L} \neq 0$. Then

$$\tilde{b}_2 = 0 \quad \Longleftrightarrow \quad \Delta_0 = 0 \quad \Longleftrightarrow \quad c = \frac{3}{5}b.$$

In this case we have

$$\Delta_1 = -\frac{224}{125}b^4.$$

Thus, we have proved the following

Theorem 7.14 ([92]) *For system (7.54), let $a = 0$ and introduce*

$$\delta = (c_{00}, c_{10}, c_{01}, c_{02}, c_{11}, c_{20}), \qquad \sigma = (b, c),$$

and

$$\bar{\delta} = (\bar{c}_{00}, \bar{c}_{10}, \bar{c}_{01}, \bar{c}_{02}, \bar{c}_{11}, \bar{c}_{20}), \qquad \bar{\sigma} = (\bar{b}, \bar{c}),$$

where $\bar{\delta}$ and $\bar{\sigma}$ satisfy

$$\bar{b} \neq 0, \qquad \bar{c} = \frac{3}{5}\bar{b}, \qquad \bar{c}_{00} = 0, \qquad \bar{c}_{20} = (\bar{b} + \bar{c})\bar{c}_{01} - \bar{c}_{02}, \qquad \bar{c}_{02} - \bar{c}\bar{c}_{01} \neq 0.$$

Then for all $(\varepsilon, \delta, \sigma)$ near $(0, \bar{\delta}, \bar{\sigma})$ the system (7.54) has Hopf cyclicity 3 at the origin.

Chapter 8
Limit Cycles Near a Homoclinic or Heteroclinic Loop

This chapter is devoted to considering the bifurcation of limit cycles in the near-Hamiltonian system (6.1) near a homoclinic or heteroclinic loop. The method of computing the Melnikov function near a homoclinic or heteroclinic loop is developed and explicit formulas for the coefficients in the expansion of the Melnikov function are derived. Double homoclinic loops are also studied in this chapter. We first consider the case of a homoclinic loop, and then a heteroclinic loop.

8.1 Bifurcation of Limit Cycles Near a Homoclinic Loop

8.1.1 Single Homoclinic Loop

Suppose the unperturbed system (6.4),

$$\dot{x} = H_y, \qquad \dot{y} = -H_x,$$

has a homoclinic loop L_0, given by $H(x, y) = 0$, passing through the origin, which is a hyperbolic saddle; i.e., instead of (7.1), we assume

$$H(x, y) = \frac{\lambda}{2}(y^2 - x^2) + \sum_{i+j\geq 3} h_{ij}x^i y^j, \quad \lambda \neq 0. \tag{8.1}$$

For definiteness, let $\lambda > 0$. In this case, we can suppose that in the right half plane there is a family of periodic orbits of (6.4) near L_0 given by

$$L_h : H(x, y) = h, \quad 0 < -h \ll 1.$$

M. Han, P. Yu, *Normal Forms, Melnikov Functions and Bifurcations of Limit Cycles*, Applied Mathematical Sciences 181, DOI 10.1007/978-1-4471-2918-9_8, © Springer-Verlag London Limited 2012

8.1.2 Computation of the Melnikov Function, M

The corresponding Melnikov function is given by

$$M(h,\delta) = \oint_{L_h} q\,dx - p\,dy|_{\varepsilon=0} = \oint_{L_h} \tilde{q}(x,y,\delta)\,dx, \qquad (8.2)$$

where

$$\tilde{q}(x,y,\delta) = q(x,y,0,\delta) - q(x,0,0,\delta) + \int_0^y p_x(x,v,0,\delta)\,dv$$

$$= \int_0^y \left[p_x(x,v,0,\delta) + q_y(x,v,0,\delta)\right]dv. \qquad (8.3)$$

Suppose L_h lies in the half plane $x > 0$. Take a small constant $x_0 > 0$ so that for $|h| > 0$, small, the line $x = x_0$ divides L_h into two parts $L_{h,1} = L_h|_{x \le x_0}$ and $L_{h,2} = L_h|_{x \ge x_0}$.

Then by (8.2) we have

$$M(h,\delta) = I_1(h,\delta) + I_2(h,\delta), \qquad (8.4)$$

where

$$I_j(h,\delta) = \int_{L_{h,j}} \tilde{q}\,dx, \quad j = 1,2.$$

Obviously, $I_2(h,\delta) \in C^\infty$ in h for $|h| \ll 1$. Thus, in order to understand the analytical properties we need to study the expansion of I_1 near $h = 0$.

As before, let $H^*(x,v) = H(x, v + \varphi(x))$, $H_0^*(x) = H(x, \varphi(x))$, where $H_y(x, \varphi(x)) = 0$. Then (7.5) and (7.6) hold. Under (8.1) the equation $H_0^*(x) = h$ has a unique solution $x = a(h) = (h/h_2)^{\frac{1}{2}} + O(|h|)$ in $x > 0$ with

$$a(h) > 0 \quad \text{for } 0 < -h \ll 1.$$

Similarly to Lemma 7.1, we have the following lemma.

Lemma 8.1 *Let* (8.1) *hold with* $\lambda > 0$. *Then the equation* $H^*(x,v) = h$ *has exactly two* C^∞ *solutions,* $v_1(x,w)$ *and* $v_2(x,w)$ *in* v, *satisfying*

$$v_1(x,w) = \sqrt{\frac{2}{\lambda}}w\bigl(1 + O(|x,w|)\bigr), \qquad v_2(x,w) = v_1(x,-w),$$

where $w = \sqrt{h - H_0^*(x)}$, $a(h) \le x \le x_0$, $0 \le -h \ll 1$.

Let $y_i(x,h) = v_i(x,w) + \varphi(x)$, $i = 1,2$. Then for $a(h) \le x \le x_0$ we have

$$y_2(x,h) \le \varphi(x) \le y_1(x,h), \qquad H\bigl(x, y_i(x,h)\bigr) = h, \quad i = 1,2.$$

It follows that

$$I_1(h, \delta) = \int_{L_{h,1}} \tilde{q}\, dx = \int_{a(h)}^{x_0} \left[\tilde{q}\big(x, y_1(x, h), \delta\big) - \tilde{q}\big(x, y_2(x, h), \delta\big) \right] dx$$

$$= \int_{a(h)}^{x_0} \left[\bar{q}\big(x, v_1(x, w), \delta\big) - \bar{q}\big(x, v_2(x, w), \delta\big) \right] dx,$$

where $\bar{q}(x, v, \delta) = \tilde{q}(x, v + \varphi(x), \delta)$. Note that formulas (7.14) and (7.17) remain true. We obtain

$$I_1(h, \delta) = \sum_{j \geq 0} \int_{a(h)}^{x_0} \bar{q}_j(x) w^{2j+1}\, dx.$$

Further, let $\psi(x) = \mathrm{sgn}(x)[-H_0^*(x)]^{\frac{1}{2}} = |h_2|^{\frac{1}{2}} x + O(x^2)$. Then introducing the change of variables $u = \psi(x)$, we have

$$I_1(h, \delta) = \sum_{j \geq 0} \int_{|h|^{\frac{1}{2}}}^{u_0} \tilde{q}_j(u) w^{2j+1}\, du, \quad u_0 = \psi(x_0) > 0, \tag{8.5}$$

where $w = \sqrt{u^2 + h}$ and $\tilde{q}_j$ satisfies (7.20).

Suppose

$$\tilde{q}_j(u) = \sum_{i \geq 0} \tilde{r}_{ij} u^i, \tag{8.6}$$

and

$$I_{ij}(h) = \int_{|h|^{\frac{1}{2}}}^{u_0} u^i w^{2j+1}\, du$$

$$= \int_{|h|^{\frac{1}{2}}}^{u_0} u^i \big(u^2 + h\big)^j \sqrt{u^2 + h}\, du \quad \text{for } 0 < -h \ll 1. \tag{8.7}$$

Then

$$I_1(h, \delta) = \sum_{i,j \geq 0} \tilde{r}_{ij} I_{ij}(h) \quad \text{for } 0 < -h \ll 1. \tag{8.8}$$

To study the properties of the functions I_{ij} we use the following formulas:

$$\int u^m \big(u^2 + h\big)^{\frac{n}{2}}\, du = \frac{u^{m-1}(u^2 + h)^{\frac{n}{2}+1}}{n + m + 1}$$

$$- \frac{(m-1)h}{n + m + 1} \int u^{m-2}\big(u^2 + h\big)^{\frac{n}{2}}\, du, \tag{8.9}$$

$$\int \big(u^2 + h\big)^{\frac{n}{2}}\, du = \frac{u(u^2 + h)^{\frac{n}{2}}}{n + 1} + \frac{nh}{n + 1} \int \big(u^2 + h\big)^{\frac{n}{2}-1}\, du, \tag{8.10}$$

$$\int u(u^2 + h)^{\frac{n}{2}} \, du = \frac{(u^2 + h)^{\frac{n}{2}+1}}{n+2} + C, \tag{8.11}$$

$$\int (u^2 + h)^{\frac{1}{2}} \, du = \frac{u}{2}(u^2 + h)^{\frac{1}{2}} + \frac{h}{2} \ln(u + \sqrt{u^2 + h}) + C. \tag{8.12}$$

It follows from (8.9) that

$$I_{ij}(h) = \frac{u_0^{i-1}(u_0^2 + h)^{j+\frac{3}{2}}}{2j + i + 2} - \frac{(i-1)h}{2j + i + 2} I_{i-2,j}(h), \tag{8.13}$$

and from (8.10) that

$$I_{0j}(h) = \frac{u_0(u_0^2 + h)^{\frac{2j+1}{2}}}{2(j+1)} + \frac{(2j+1)h}{2(j+1)} I_{0,j-1}. \tag{8.14}$$

By (8.11) and (8.13), we can show that $I_{ij} \in C^\omega$ for i odd. By (8.10) and (8.13), we have for $i \geq 2$, even,

$$I_{ij}(h) = \tilde{I}_{ij}(h) + (-1)^{\frac{i}{2}} \frac{(i-1)!! h^{\frac{i}{2}}}{2^{\frac{i}{2}}(j + \frac{i}{2} + 1)(j + \frac{i}{2}) \cdots (j+2)} I_{0,j}(h),$$

$$\tilde{I}_{ij} \in C^\omega. \tag{8.15}$$

Similarly, by (8.12) and (8.14),

$$I_{0j} = \tilde{I}_{0j}(h) - \frac{(2j+1)!!}{2^{j+2}(j+1)!} h^{j+1} \ln |h|, \quad \tilde{I}_{0j} \in C^\omega. \tag{8.16}$$

Thus, by (8.15) and (8.16), we can write for $i \geq 0$, even,

$$I_{ij}(h) = I_{ij}^*(h) + \alpha_{ij}^* h^{\frac{i}{2}+j+1} \ln |h|,$$

where $I_{ij}^* \in C^\omega$, and the α_{ij}^* are constants with $|\alpha_{ij}^*| < 1$ and in particular,

$$\alpha_{00}^* = -\frac{1}{4}, \qquad \alpha_{01}^* = -\frac{3}{16}, \qquad \alpha_{20}^* = -\frac{1}{4}\alpha_{00}^* = \frac{1}{16}. \tag{8.17}$$

Therefore, we have proved the following lemma.

Lemma 8.2 *For the functions defined by (8.7), we have for $i \geq 1$, odd, $I_{ij} \in C^\omega$, and for $i \geq 0$, even,*

$$I_{ij}(h) = I_{ij}^*(h) + \alpha_{ij}^* h^{\frac{i}{2}+j+1} \ln |h|,$$

where $I_{ij}^* \in C^\omega$, the α_{ij}^* are constants satisfying $|\alpha_{ij}^*| < 1$ and (8.17).

Further, by (8.7), for $i, j \geq 0$ we have

$$\left|I_{ij}(h)\right| \leq \int_{|h|^{1/2}}^{u_0} u^i \left(2u^2\right)^{j+\frac{1}{2}} du \leq \frac{1}{2(j+1)+i}\left(\sqrt{2}u_0\right)^{2j+1}u_0^{i+1}$$

$$\leq \frac{\varepsilon_0^{2j+2+i}}{2(j+1)+i} \leq \frac{1}{2}\varepsilon_0^{2j+2+i} \quad \text{if } \sqrt{|h|} \leq u_0 \leq \frac{\varepsilon_0}{\sqrt{2}}.$$

For i even, we have

$$\left|I_{ij}^*\right| \leq \left|I_{ij}\right| + |h|^{\frac{2j+i}{2}}\left|h \ln|h|\right| \leq \varepsilon_0^{2j+i}\left(\frac{\varepsilon_0^2}{2} + \left|h \ln|h|\right|\right) \quad \text{for } \sqrt{|h|} \leq u_0 \leq \frac{\varepsilon_0}{\sqrt{2}}.$$

By (8.5) and (8.6) we know that if (6.1) is C^ω then the series $\sum_{i,j\geq 0}\tilde{r}_{ij}u^i w^{2j+1}$ is convergent for (u, w) near the origin. Hence, there exists $\varepsilon_0 > 0$ such that $\sum_{i,j\geq 0}|\tilde{r}_{ij}|\varepsilon_0^{2j+i}$ is convergent. Thus, by (8.4), (8.8) and Lemma 8.2 we obtain the following theorem.

Theorem 8.3 *Let* (8.1) *hold. Then*

$$M(h, \delta) = N_1(h, \delta) + N_2(h, \delta)h \ln|h|,$$

where

$$N_1(h, \delta) = \sum_{k\geq 0} c_{2k}h^k, \qquad N_2(h, \delta) = \sum_{k\geq 0} c_{2k+1}h^k,$$

formally for $0 \leq -h \ll 1$ *with*

$$c_{2k+1} = \sum_{l+j=k} \tilde{r}_{2l,j}\alpha_{2l,j}^*. \tag{8.18}$$

Moreover, if (6.1) *is* C^ω, *then* $N_1, N_2 \in C^\omega$.

The form of the expansion of M in Theorem 8.3 was obtained in [183]. The formulas for the coefficients c_0, c_1, c_2, c_3, are given in the next subsection.

8.1.3 Computation of the c_j Coefficients of M

Theorem 8.4 *Let* (8.1) *and* (7.10) *hold. Then for the coefficients* c_0, c_1, c_2, c_3, *in* (8.18) *we have*

$$c_0(\delta) = \oint_{L_0} q \, dx - p \, dy|_{\varepsilon=0} = \oint_{L_0} \tilde{q} \, dx,$$

$$c_1(\delta) = -\frac{a_{10} + b_{01}}{|\lambda|},$$

$$c_2(\delta) = \oint_{L_0} (p_x + q_y - a_{10} - b_{01})|_{\varepsilon=0}\, dt + bc_1(\delta),$$

$$c_3(\delta) = \frac{-1}{2|\lambda|\lambda}\left\{(-3a_{30} - b_{21} + a_{12} + 3b_{03})\right.$$

$$\left. -\frac{1}{\lambda}\big[(2b_{02} + a_{11})(3h_{03} - h_{21}) + (2a_{20} + b_{11})(3h_{30} - h_{12})\big]\right\} + \bar{b}c_1(\delta),$$

for some constants b and $\bar{b}$.

Proof As before, let $\lambda > 0$. The formula for c_0 follows from (6.6) directly.

Next, we deduce the formulas for c_1, c_2 and c_3. Let (7.10) hold. Then by (8.6) and (7.18)–(7.20), we have

$$\tilde{r}_{00} = \tilde{q}_0(0) = |h_2|^{-\frac{1}{2}}\bar{q}_0(0) = \sqrt{\frac{2}{\lambda}}\alpha_{00} = \frac{4}{\lambda}(a_{10} + b_{01}). \qquad (8.19)$$

Further, by (8.17) and (8.18) we obtain

$$c_1 = \tilde{r}_{00}\alpha_{00}^* = -\frac{1}{\lambda}(a_{10} + b_{01}).$$

Thus, $\tilde{r}_{00} = 0$ if and only if $a_{10} + b_{01} = 0$. By Theorem 8.3 we have

$$M'(h, \delta) = c_1 \ln|h| + c_1 + c_2 + O\big(h\ln|h|\big)$$

for $0 < -h \ll 1$. Hence, from Lemma 6.2 we have

$$\oint_{L_h} (p_x + q_y)|_{\varepsilon=0}\, dt = c_1 \ln|h| + c_1 + c_2 + O\big(h\ln|h|\big). \qquad (8.20)$$

In particular, taking $p = x$ and $q = 0$ in (8.20) and using the formula for c_1 gives

$$\oint_{L_h} dt = T(h) = -\frac{1}{\lambda}\ln|h| + T_0 + O\big(h\ln|h|\big),$$

where $T(h)$ denotes the period of L_h, and T_0 is a constant. Thus

$$T(h) + \frac{1}{\lambda}\ln|h| \to T_0 \quad \text{as } h \to 0.$$

Note that

$$\oint_{L_h} (p_x + q_y)|_{\varepsilon=0}\, dt = \oint_{L_h} (p_x + q_y - a_{10} - b_{01})|_{\varepsilon=0}\, dt + (a_{10} + b_{01})T(h).$$

Then with the formula for c_1 it follows from (8.20) that

$$
\begin{aligned}
c_2 &= \lim_{h \to 0} \left[\oint_{L_h} (p_x + q_y)|_{\varepsilon=0} \, dt - c_1 \ln|h| \right] - c_1 \\
&= \oint_{L_0} (p_x + q_y - a_{10} - b_{01})|_{\varepsilon=0} \, dt + (a_{10} + b_{01}) \lim_{h \to 0} \left(T(h) + \frac{1}{\lambda} \ln|h| \right) - c_1 \\
&= \oint_{L_0} (p_x + q_y - a_{10} - b_{01})|_{\varepsilon=0} \, dt + bc_1.
\end{aligned}
$$

As in the previous section, we can derive the formulas for $\tilde{r}_{10}$, $\tilde{r}_{01}$ and $\tilde{r}_{20}$. In fact, noting $\psi(x) = (\operatorname{sgn} x)[-H_0^*(x)]^{\frac{1}{2}}$, we have

$$
\psi(x) = \sum_{i=1}^{4} \mu_i x^i + O(x^5),
$$

$$
\psi^{-1}(u) = \sum_{i=1}^{4} \bar{\mu}_i u^i + O(u^5),
$$

$$
\frac{1}{\psi'(x)} = \sum_{i=0}^{3} \bar{n}_i x^i + O(x^4),
$$

where

$$
\begin{aligned}
\mu_1 &= |h_2|^{\frac{1}{2}}, & \mu_3 &= \frac{1}{8}|h_2|^{-\frac{3}{2}}(4h_2 h_4 - h_3^2), \\
\mu_2 &= -\frac{1}{2}|h_2|^{-\frac{1}{2}} h_3, & \mu_4 &= -\frac{1}{16}|h_2|^{-\frac{5}{2}}(8h_2^2 h_5 - 4h_2 h_3 h_4 + h_3^3), \\
\bar{\mu}_1 &= \mu_1^{-1}, & \bar{\mu}_3 &= -\mu_1^{-5}(\mu_1 \mu_3 - 2\mu_2^2), \\
\bar{\mu}_2 &= -\mu_1^{-3}\mu_2, & \bar{\mu}_4 &= -\mu_1^{-7}(5\mu_2^3 - 5\mu_1 \mu_2 \mu_3 + \mu_1^2 \mu_4), & (8.21) \\
\bar{n}_0 &= \mu_1^{-1}, & \bar{n}_2 &= -\mu_1^{-3}(3\mu_1 \mu_3 - 4\mu_2^2), \\
\bar{n}_1 &= -2\mu_1^{-2}\mu_2, & \bar{n}_3 &= -4\mu_1^{-4}(2\mu_2^3 - 3\mu_1 \mu_2 \mu_3 + \mu_1^2 \mu_4).
\end{aligned}
$$

Hence, by (7.18) and (7.20) we have

$$
\tilde{q}_j(u) = \left. \frac{\bar{q}_j(x)}{\psi'(x)} \right|_{x=\psi^{-1}(u)} = \sum_{i \geq 0} \tilde{r}_{ij} u^i, \quad j = 0, 1,
$$

with

$$\tilde{r}_{10} = \bar{n}_1 \bar{\mu}_1 \alpha_{00} + \bar{n}_0 \bar{\mu}_1 \alpha_{10},$$

$$\tilde{r}_{20} = \left(\bar{n}_1 \bar{\mu}_2 + \bar{n}_2 \bar{\mu}_1^2\right)\alpha_{00} + \left(\bar{n}_0 \bar{\mu}_2 + \bar{n}_1 \bar{\mu}_1^2\right)\alpha_{10} + \bar{n}_0 \bar{\mu}_1^2 \alpha_{20},$$

$$\tilde{r}_{01} = \bar{n}_0 \alpha_{01}.$$

Then by (8.18), (8.17), (8.21), (7.7), (7.12), (7.16) and (7.19) we obtain the formula for c_3 readily. If $\lambda < 0$, we can prove in a similar way that the formulas for c_0, c_1, c_2 and c_3 remain the same.

The proof is completed. $\square$

The formulas for c_1 and c_2 were obtained in [93], and the formula for c_3 was obtained in [91] in the case of $\lambda = 1$. From Theorem 8.4 we see that the quantities $c_0(\delta)$ and $c_2(\delta)$ etc., depend on the values of the functions p and q in a neighborhood of the homoclinic loop L_0, while the quantities $c_1(\delta)$ and $c_3(\delta)$ etc., depend on the values of the functions p and q in a neighborhood of the origin. For the sake of convenience in the following, we call the coefficients $c_1(\delta)$ and $c_3(\delta)$ in (8.18) the first and second local Melnikov coefficients at the saddle and at the origin, respectively.

8.1.4 Alternative Formulas for the c_j Coefficients for Different Hamiltonian Functions

Remark 8.5 If, instead of (8.1), the Hamiltonian function H takes the form

$$H(x, y) = \frac{1}{2}y^2 - h_{20}x^2 + \sum_{i+j \geq 3} h_{ij}x^i y^j, \quad h_{20} > 0,$$

then letting $u = \sqrt{2h_{20}}x$ yields

$$H(x, y) = \frac{1}{2}\left(y^2 - u^2\right) + \sum_{i+j \geq 3} h_{ij}(2h_{20})^{-i/2}u^i y^j,$$

which has the form of (8.1) with $\lambda = 1$. In this case, by Theorem 8.4, we have

$$M(h, \delta) = c_0(\delta) + c_1(\delta)h \ln|h| + c_2(\delta)h + c_3(\delta)h^2 \ln|h| + O\left(h^2\right),$$

where

$$c_0 = \oint_{L_0} q\,dx - p\,dy|_{\varepsilon=0} = \oint_{L_0} \tilde{q}\,dx,$$

$$c_1 = -\frac{\sqrt{2}}{2}|h_2|^{-\frac{1}{2}}(a_{10} + b_{01}),$$

$$c_2 = \oint_{L_0} (p_x + q_y - a_{10} - b_{01})|_{\varepsilon=0}\, dt + bc_1,$$

$$c_3 = \frac{\sqrt{2}}{16}|h_{20}|^{-\frac{5}{2}}\big[-2h_{20}(3a_{30} + b_{21} + 2h_{20}a_{12} + 6h_{20}b_{03})$$
$$+ 2h_{20}(2b_{02} + a_{11})(6h_{20}h_{03} + h_{21})$$
$$+ (2a_{20} + b_{11})(3h_{30} + 2h_{20}h_{12})\big] + \bar{b}c_1,$$

for some constants b and $\bar{b}$.

Remark 8.6 If, instead of (8.1), the function H has the form

$$H(x, y) = \lambda xy + \sum_{i+j\geq 3} h_{ij}x^i y^j,$$

then we have the same expansion of $M(h, \delta)$ with

$$c_0(\delta) = M(0, \delta) = \oint_{L_0} q\, dx - p\, dy|_{\varepsilon=0},$$

$$c_1(\delta) = -\frac{a_{10} + b_{01}}{|\lambda|},$$

$$c_2(\delta) = \oint_{L_0} (p_x + q_y - a_{10} - b_{01})|_{\varepsilon=0}\, dt + bc_1(\delta),$$

$$c_3(\delta) = \frac{-1}{|\lambda|\lambda}\left\{(a_{21} + b_{12}) - \frac{1}{\lambda}\big[h_{12}(2a_{20} + b_{11}) + h_{21}(a_{11} + 2b_{02})\big]\right\} + \bar{b}c_1(\delta),$$

for some constants b and $\bar{b}$. The result can be obtained by first introducing a linear change of variables of the form $u = \frac{1}{\sqrt{2}}(x - y)$, $v = \frac{1}{\sqrt{2}}(x + y)$ to system (6.1) and then using Theorem 8.4. The details are omitted here.

Next, we have the following theorem about the number of limit cycles near L_0 which was proved in [94] for the case $l = 2$ and in [91] for the case $l = 3$.

Theorem 8.7 *Let (8.1) and (7.10) hold. If there exist $\delta_0 \in \mathbf{R}^m$ and $1 \leq l \leq 3$ such that*

$$\bar{c}_l(\delta_0) \neq 0, \qquad \bar{c}_j(\delta_0) = 0, \quad j = 0, \ldots, l - 1,$$

$$\mathrm{rank}\, \frac{\partial(\bar{c}_0, \bar{c}_1, \bar{c}_2, \ldots, \bar{c}_{l-1})}{\partial(\delta_1, \ldots, \delta_m)}(\delta_0) = l,$$

where $\bar{c}_i = c_i$, $i = 0, 1$, and $\bar{c}_j = c_j|_{c_1=0}$, $j = 2, 3$, then (6.1) can have l limit cycles near L_0 for some (ε, δ) near $(0, \delta_0)$.

Proof We present our proof for the case of $l = 3$. Also, for definiteness, we can suppose $m = 3$ and

$$\bar{c}_3(\delta_0) > 0, \qquad \bar{c}_0(\delta_0) = \bar{c}_1(\delta_0) = \bar{c}_2(\delta_0) = 0.$$

Then, we can choose c_0, c_1, c_2, as free parameters so that by Theorem 8.3 we can write

$$M(h, \delta) = c_0 + c_1 h \ln|h| + c_2 h + \tilde{c}_3 h^2 \ln|h| + O(h^2),$$

where $\tilde{c}_3 = \bar{c}_3(\delta_0) + O(|c_0, c_1, c_2|) > 0$. It is easy to see that if

$$0 < c_0 \ll -c_1 \ll -c_2 \ll 1,$$

then M has 3 zeros in $0 < -h \ll 1$. In this case, system (6.1) can have 3 limit cycles for $|\varepsilon| \ll c_0$.

This ends the proof.   □

8.1.5 Double Homoclinic Loops

Now, we suppose that the equation $H(x, y) = 0$ defines a double homoclinic loop $L = L_1 \cup L_2$ consisting of two homoclinic loops L_1 and L_2 with a hyperbolic saddle (see Fig. 8.1) where S_0 represents a saddle point. As before, suppose the saddle is at the origin. Then the equation $H(x, y) = h$ defines a family of periodic orbits $L(h)$ for h on one side of $h = 0$ and two families of periodic orbits, $L_1(h)$ and $L_2(h)$, for h on the other side of $h = 0$. For definiteness, let $L(h)$ exist for $0 < h \ll 1$ and both $L_1(h)$ and $L_2(h)$ exist for $0 < -h \ll 1$. Thus, we have three Melnikov functions as follows:

$$M(h, \delta) = \oint_{L(h)} q\, dx - p\, dy|_{\varepsilon=0}, \quad 0 < h \ll 1,$$

$$M_i(h, \delta) = \oint_{L_i(h)} q\, dx - p\, dy|_{\varepsilon=0}, \quad 0 < -h \ll 1,\ i = 1, 2.$$

Similarly to Theorems 8.3 and 8.4 we have the following theorem.

Theorem 8.8 ([214]) *Suppose* (8.1) *and* (7.10) *hold. Then*

$$M(h, \delta) = c_0(\delta) + 2c_1(\delta)h \ln|h| + c_2(\delta)h + 2c_3(\delta)h^2 \ln|h| + O(h^2),$$

$$M_i(h, \delta) = c_{0i}(\delta) + c_1(\delta)h \ln|h| + c_{2i}(\delta)h + c_3(\delta)h^2 \ln|h| + O(h^2), \quad i = 1, 2,$$

Fig. 8.1 A double homoclinic loop

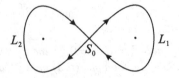

where

$$c_{0i} = M_i(0, \delta) = \oint_{L_i} q\,dx - p\,dy|_{\varepsilon=0}, \qquad c_0 = c_{01} + c_{02},$$

$$c_1 = -\frac{a_{10} + b_{01}}{|\lambda|},$$

$$c_{2i} = \oint_{L_i} (p_x + q_y - a_{10} - b_{01})|_{\varepsilon=0}\,dt + b_i c_1, \quad i = 1, 2, \tag{8.22}$$

$$c_2 = \sum_{i=1}^{2} \oint_{L_i} (p_x + q_y - a_{10} - b_{01})|_{\varepsilon=0}\,dt + b c_1,$$

for some constants b, b_1 *and* b_2, *and with* c_3 *as given in Theorem 8.4.*

Clearly, we have

$$c_2 = c_{21} + c_{22} + (b - b_1 - b_2)c_1.$$

Now we apply Theorem 8.8 to study the number of limit cycles near the double homoclinic loop L. There are two cases: symmetric and nonsymmetric cases.

8.1.6 Symmetric Double Homoclinic Loops

First, for the centrally symmetric case we have the following theorem.

Theorem 8.9 ([80, 214]) *Suppose the functions* H_x, H_y, p *and* q *in* (6.1) *are all odd in* (x, y). *Suppose* (8.1) *holds. Define*

$$\bar{c}_0 = c_{01}, \qquad \bar{c}_1 = c_1, \qquad \bar{c}_3 = c_3|_{c_1=0},$$

$$\bar{c}_2 = \oint_{L_1} (p_x + q_y - a_{10} - b_{01})|_{\varepsilon=0}\,dt. \tag{8.23}$$

If there exist $\delta_0 \in R^m$ *and* $1 \le k \le 3$ *such that*

$$\bar{c}_j(\delta_0) = 0, \quad j = 0, \dots, k-1, \qquad \bar{c}_k(\delta_0) \ne 0,$$

$$\text{rank}\, \frac{\partial(\bar{c}_0, \dots, \bar{c}_{k-1})}{\partial(\delta_1, \dots, \delta_m)}(\delta_0) = k,$$

then (6.1) *can have* $[\frac{5k}{2}]$ *limit cycles near* L *for some* (ε, δ) *near* $(0, \delta_0)$.

Proof As before, we consider the case of $k = 3$ with $\bar{c}_3(\delta_0) > 0$ and $\delta \in R^3$. Since system (6.1) is centrally symmetric we have $M_1(h, \delta) = M_2(h, \delta)$ for $0 < -h \ll 1$.

Then, by the assumption, we can take c_{01}, c_1 and c_{21} as free parameters, and write

$$M_1(h, \delta) = c_{01} + c_1 h \ln|h| + c_{21}h + \tilde{c}_3(c_{01}, c_1, c_{21})h^2 \ln|h| + O(h^2)$$

$$= \tilde{M}_1(h, c_{01}, c_1, c_{21}),$$

$$M(h, \delta) = 2c_{01} + 2c_1 h \ln|h| + (2c_{21} + (b - 2b_1)c_1)h$$

$$+ 2\tilde{c}_3(c_{01}, c_1, c_{21})h^2 \ln|h| + O(h^2)$$

$$= \tilde{M}(h, c_{01}, c_1, c_{21}).$$

First, the functions M_1 and $M_2(= M_1)$ have the same negative zero, $h_1 < 0$ for $c_{01} = c_1 = 0$ and $0 < -c_{21} \ll 1$, and M has no positive zero in this step. Further, let

$$0 < c_{01} \ll -c_1 \ll |c_{21}|,$$

so that M_1 and M_2 have two more negative zeros, h_2 and h_3 with $|h_3| \ll |h_2| \ll |h_1|$, and at the same time the function M has a positive zero, $h^* > 0$. Clearly, the zeros h_1, h_2 and h_3 lead to 6 limit cycles with 3 of them being near L_1 and the other 3 near L_2, and the zero h^* leads to a large limit cycle near $L = L_1 \cup L_2$ surrounding the origin and the 6 limit cycles.

     This ends the proof.   □

Remark 8.10   The number $[\frac{5k}{2}]$ of limit cycles near the double homoclinic loop L is maximal. This fact was proved in [74].

8.1.7 Nonsymmetric Double Homoclinic Loops

Now we turn to the nonsymmetric case. Let c_{01}, c_{02}, c_1 and c_3 be as before, and let

$$\bar{c}_{2i}(\delta) = \oint_{L_i} (p_x + q_y - a_{10} - b_{01})|_{\varepsilon=0} \, dt, \quad i = 1, 2.$$

Theorem 8.11 *Suppose (8.1) holds.*

(1) *If there exist $\delta_0, v \in \mathbf{R}^m$ such that*

$$c_{02}(\delta) = \alpha(\delta)c_{01}(\delta) \quad \text{for } \delta = \delta_0 + \mu v, \mu \in \mathbf{R},$$

$$c_{01}(\delta_0) = 0, \quad c_1(\delta_0) \neq 0, \quad \alpha(\delta_0) > 0, \quad \frac{\partial c_{01}}{\partial \mu}(\delta_0 + \mu v)|_{\mu=0} \neq 0,$$

     then 2 limit cycles can appear near L for some (ε, δ) near $(0, \delta_0)$.

(2) *If there exist $\delta_0, v_1, v_2 \in \mathbf{R}^m$ such that*

$$c_{02}(\delta) = \alpha(\delta)c_{01}(\delta) \quad \text{for } \delta = \delta_0 + \mu_1 v_1 + \mu_2 v_2, \mu_1, \mu_2 \in \mathbf{R},$$

$$c_{01}(\delta_0) = c_1(\delta_0) = 0, \qquad \alpha(\delta_0)\big(\alpha(\delta_0) + 1\big) > 0, \qquad \bar{c}_{21}(\delta_0)\bar{c}_{22}(\delta_0) > 0,$$

$$\text{rank } \frac{\partial(c_{01}, c_1)}{\partial(\mu_1, \mu_2)}(\delta_0 + \mu_1 v_1 + \mu_2 v_2)|_{\mu_1 = \mu_2 = 0} = 2,$$

then 5 limit cycles can appear near L for some (ε, δ) near $(0, \delta_0)$.

(3) *If there exist $\delta_0, v_1, v_2, v_3 \in R^m$ such that*

$$c_{02}(\delta) = \alpha(\delta)c_{01}(\delta), \qquad \bar{c}_{22}(\delta) = \beta(\delta)\bar{c}_{21}(\delta) \quad \text{for } \delta = \delta_0 + \sum_{i=1}^{3} \mu_i v_i, \mu_i \in R,$$

$$c_{01}(\delta_0) = c_1(\delta_0) = \bar{c}_{21}(\delta_0) = 0, \qquad \alpha(\delta_0)\big(\alpha(\delta_0) + 1\big) > 0,$$

$$\beta(\delta_0) > 0, \qquad \bar{c}_3(\delta_0) \neq 0,$$

$$\text{rank } \frac{\partial(c_{01}, c_1, \bar{c}_{21})}{\partial(\mu_1, \mu_2, \mu_3)}\left(\delta_0 + \sum_{i=1}^{3} \mu_i v_i\right)\Bigg|_{\mu_i = 0} = 3,$$

then 7 limit cycles can appear near L for some (ε, δ) near $(0, \delta_0)$.

Proof We only prove the third conclusion. The proof of the first 2 conclusions is similar and simpler.

As before, assume $\bar{c}_3(\delta_0) > 0$. Then under the assumptions, we can uniquely solve for $(\mu_1, \mu_2, \mu_3) = \phi(c_{01}, c_1, c_{21})$ from the equations

$$c_{01} = c_{01}\left(\delta_0 + \sum_{i=1}^{3} \mu_i v_i\right),$$

$$c_1 = c_1\left(\delta_0 + \sum_{i=1}^{3} \mu_i v_i\right),$$

$$c_{21} = c_{21}\left(\delta_0 + \sum_{i=1}^{3} \mu_i v_i\right).$$

Hence we can write

$$M_i(h, \delta) = c_{0i} + c_1 h \ln|h| + c_{2i} h + \tilde{c}_3 h^2 \ln|h| + O(h^2)$$
$$\equiv \tilde{M}_i(h, c_{01}, c_1, c_{21}), \quad h < 0, i = 1, 2,$$

$$M(h, \delta) = c_{01} + c_{02} + 2c_1 h \ln|h| + \big(c_{21} + c_{22} + (b - b_1 - b_2)c_1\big)h$$
$$+ 2\tilde{c}_3 h^2 \ln|h| + O(h^2)$$
$$\equiv \tilde{M}(h, c_{01}, c_1, c_{21}), \quad h > 0,$$

Fig. 8.2 Two distributions of
7 limit cycles

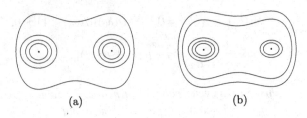

(a)   (b)

where c_{01}, c_1 and c_{21} are taken as free parameters,

$$c_{02} = \alpha^*(c_{01}, c_1, c_{21})c_{01}, \qquad c_{22} = \beta^*(c_{01}, c_1, c_{21})c_{21},$$

$$\alpha^*(0) = \alpha(\delta_0), \qquad \beta^*(0) = \beta(\delta_0),$$

and $\tilde{c}_3 \geq \frac{1}{2}\bar{c}_3(\delta_0) > 0$ for those parameters near zero.

First, let $c_{01} = c_1 = 0$ and $0 < -c_{21} \ll 1$, implying $c_{02} = 0$ and $0 < -c_{22} \ll 1$, so that M_i has a negative zero, $h_{1i} < 0, i = 1, 2$, and M has no positive zero. Further, when $\alpha(\delta_0) > 0$, let

$$0 < c_{01} \ll -c_1 \ll |c_{21}|, \tag{8.24}$$

implying $0 < c_{02} \ll -c_1$, so that M_i has two more negative zeros, h_{2i} and h_{3i} with $|h_{3i}| \ll |h_{2i}| \ll |h_{1i}|$, while M has a positive zero $h^* > 0$ under the same condition. Then, fixing the chosen (c_{01}, c_1, c_{21}) there exist 7 limit cycles of (6.1) for $|\varepsilon| > 0$, sufficiently small. See Fig. 8.2(a) for the distribution of the limit cycles.

When $\alpha(\delta_0) < -1$, it follows from (8.24) that

$$0 < -(c_{01} + c_{02}) \ll -c_1.$$

Thus, in this case, M_1 has 3 negative zeros, h_{31}, h_{21} and h_{11}, with $0 < |h_{31}| \ll |h_{21}| \ll |h_{11}|$, M_2 has 2 negative zeros, h_{22} and h_{12}, with $|h_{22}| \ll |h_{12}|$, and M has 2 positive zeros, h_2^* and h_1^*, with $0 < h_2^* \ll h_1^*$, which also give 7 limit cycles of (6.1). Their distribution is given in Fig. 8.2(b).

The proof is completed.   □

Remark 8.12 From the above proof we see that 7 limit cycles can appear with both distributions near the double homoclinic loop if the coefficients $c_{01}, c_{02}, c_1, \bar{c}_{21}$ and $\bar{c}_{22}$ can be taken as free parameters.

It was proved in [85] that there are at most 2 limit cycles near L for all (ε, δ) near $(0, \delta_0)$ if $c_{01}(\delta_0) = c_{02}(\delta_0) = 0$ and $c_1(\delta_0) \neq 0$, and shown in [80] that there are at most 5 limit cycles near L for all (ε, δ) near $(0, \delta_0)$ if

$$c_{01}(\delta_0) = c_{02}(\delta_0) = c_1(\delta_0) = 0 \quad \text{and} \quad \bar{c}_{21}(\delta_0)\bar{c}_{22}(\delta_0) \neq 0.$$

It was conjectured in [80] that 7 is the maximum number of limit cycles under the condition of conclusion (3) of Theorem 8.11. This problem is still open.

Fig. 8.3 A 2- or 3-polycycle

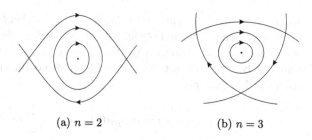

(a) $n = 2$ (b) $n = 3$

8.2 Bifurcation of Limit Cycles Near a Heteroclinic Loop

Now we turn to consider the expansion of the Melnikov function M near a hetero-clinic loop consisting of n hyperbolic saddles with n heteroclinic orbits connecting the saddles. A loop of this type is also called an *n-polycycle*. See Fig. 8.3 for the cases $n = 2$ and 3.

Let the Hamiltonian system (6.4) have an n-polycycle Γ^n:

$$\Gamma^n = \bigcup_{i=1}^{n}(L_i \cup S_i),$$

with n saddles $S_1, \ldots, S_n$, and n heteroclinic orbits $L_1, \ldots, L_n$. Without loss of generality, suppose Γ^n satisfies $H(x, y) = 0$ and a family of periodic orbits L_h of (6.4) exist near Γ^n and are defined by $H(x, y) = h, h \in J$, where J is an interval having an endpoint at $h = 0$.

For a given $1 \le i \le n$, there exists a transformation of the form

$$\begin{pmatrix} x \\ y \end{pmatrix} = T\begin{pmatrix} u \\ v \end{pmatrix} + S_i, \tag{8.25}$$

where T is a 2×2 matrix satisfying $\det T = 1$, so that (6.3) becomes

$$\dot{u} = \tilde{H}_v + \varepsilon\tilde{p}(u, v, \varepsilon, \delta), \qquad \dot{v} = -\tilde{H}_u + \varepsilon\tilde{q}(u, v, \varepsilon, \delta), \tag{8.26}$$

where

$$\tilde{H}(u, v) = \lambda_i uv + \sum_{k+j\ge 3} h_{kj} u^k v^j,$$

or

$$\tilde{H}(u, v) = \frac{1}{2}\lambda_i(v^2 - u^2) + \sum_{k+j\ge 3} h_{kj} u^k v^j,$$

and $\pm\lambda_i$ are the eigenvalues of S_i for system (6.4).

As in (7.10) we can write

$$\tilde{p}(u, v, 0, \delta) = \sum_{k+j\ge 0} a_{kj} u^k v^j, \qquad \tilde{q}(u, v, 0, \delta) = \sum_{k+j\ge 0} b_{kj} u^k v^j.$$

Then we apply the formulas for $c_1(\delta)$ and $c_3(\delta)$ in Theorem 8.4 or Remark 8.6 to the new system (8.26), and obtain the corresponding values for the saddle S_i, and denote them $c_1(S_i, \delta)$ and $c_3(S_i, \delta)$, respectively; i.e., $c_1(S_i, \delta)$ and $c_3(S_i, \delta)$ are, respectively, the first and second local Melnikov coefficients of system (6.1) at the saddle S_i. For example,

$$c_1(S_i, \delta) = -\frac{1}{|\lambda_i|}(\tilde{p}_u + \tilde{q}_v)(0, 0, \delta) = -\frac{1}{|\lambda_i|}(p_x + q_y)(S_i, 0, \delta).$$

8.2.1 Computation of the Melnikov Function, M

Now we are in a position to state the following theorem.

Theorem 8.13 *Let*

$$M(h, \delta) = \oint_{L^h} q\, dx - p\, dy|_{\varepsilon=0},$$

where L^h approaches the n-polycycle Γ^n as $h \to 0$ in J. Then for $h \in J$ and $|h|$ small, we have

$$M(h, \delta) = \sum_{j\geq 0}\left(c_{2j}(\delta) + c_{2j+1}(\delta)h\ln|h|\right)h^j, \tag{8.27}$$

where

$$c_0(\delta) = M(0, \delta) = \sum_{i=1}^{n}\int_{L_i} q\, dx - p\, dy|_{\varepsilon=0},$$

$$c_1(\delta) = \sum_{i=1}^{n} c_1(S_i, \delta), \tag{8.28}$$

$$c_3(\delta) = \sum_{i=1}^{n} c_3(S_i, \delta),$$

and

$$c_2(\delta) = \lim_{h\to 0, h\in J}\oint_{L^h}(p_x + q_y)|_{\varepsilon=0}\, dt \tag{8.29}$$

if $c_1(\delta) = 0$. In particular,

$$c_2(\delta) = \oint_{\Gamma^n}(p_x + q_y)|_{\varepsilon=0}\, dt = \sum_{i=1}^{n}\int_{L_i}(p_x + q_y)|_{\varepsilon=0}\, dt \tag{8.30}$$

if $c_1(S_i, \delta) = 0, i = 1, \ldots, n$.

Proof For simplicity, we take $n = 2$. The proof for the general case is similar. Let U_i denote a closed set with diameter $\varepsilon_0 > 0$ and containing S_i in its interior, $i = 1, 2$. Then for ε_0 sufficiently small, we can write

$$M(h, \delta) = I_1(h, \delta) + I_2(h, \delta) + I_3(h, \delta), \tag{8.31}$$

where

$$I_i(h, \delta) = \int_{L_i^h} q\, dx - p\, dy|_{\varepsilon=0}, \quad i = 1, 2, 3,$$

$$L_i^h = L^h \cap U_i, \quad i = 1, 2, \qquad L_3^h = L^h - \bigcup_{i=1}^{2} L_i^h.$$

Introducing the transformation (8.25) into (6.1) yields system (8.26) with

$$\tilde{H}(u, v) = \lambda_i uv + \sum_{k+j\geq 3} h_{kj} u^k v^j.$$

Further, noting $\det T = 1$, we can write

$$T = \begin{pmatrix} t_{11} & t_{12} \\ t_{21} & t_{22} \end{pmatrix}, \qquad T^{-1} = \begin{pmatrix} t_{22} & -t_{12} \\ -t_{21} & t_{11} \end{pmatrix},$$

and hence,

$$\tilde{H}(u, v) = H(x, y),$$

$$\tilde{p}(u, v, \varepsilon, \delta) = t_{22} p(x, y, \varepsilon, \delta) - t_{12} q(x, y, \varepsilon, \delta),$$

$$\tilde{q}(u, v, \varepsilon, \delta) = -t_{21} p(x, y, \varepsilon, \delta) + t_{11} q(x, y, \varepsilon, \delta)$$

and

$$p\, dy - q\, dx = \tilde{p}\, dv - \tilde{q}\, du,$$

where

$$x = t_{11} u + t_{12} v + x_i, \qquad y = t_{21} u + t_{22} v + y_i,$$

and (x_i, y_i) denotes the coordinates of S_i.

Thus, we obtain

$$I_i(h, \delta) = \int_{\tilde{L}_i^h} \tilde{q}\, du - \tilde{p}\, dv,$$

where $\tilde{L}_i^h, i = 1, 2$, denotes the image of L_i^h under (8.28).

By the proof of Theorem 8.3, we obtain the following expansion for I_i near $h = 0$:

$$I_i(h, \delta) = -\frac{1}{|\lambda_i|} \sum_{j\geq 1} \frac{d_j(S_i)}{\lambda_i^j} h^j \ln|h| + N_i(h, \delta),$$

where

$$d_1(S_i) = -|\lambda_i|\lambda_i c_1(S_i, \delta), \qquad d_2(S_i) = -|\lambda_i|\lambda_i^2 c_3(S_i, \delta),$$

and $N_i \in C^\omega$ at $h = 0$ with $N_i(0, \delta) = O(\varepsilon_0), i = 1, 2$. Therefore, it follows from (8.31) that

$$M(h, \delta) = \sum_{j \geq 0} c_{2j+1}(\delta) h^{j+1} \ln |h| + N(h, \delta),$$

where

$$c_1(\delta) = c_1(S_1, \delta) + c_1(S_2, \delta),$$

$$c_3(\delta) = c_3(S_1, \delta) + c_3(S_2, \delta),$$

$$N(h, \delta) = N_1(h, \delta) + N_2(h, \delta) + I_3(h, \delta).$$

Let

$$N(h, \delta) = \sum_{j \geq 0} c_{2j}(\delta) h^j.$$

Then we have

$$c_0(\delta) = N_1(0, \delta) + N_2(0, \delta) + I_3(0, \delta)$$

$$= \lim_{\varepsilon_0 \to 0} \left[N_1(0, \delta) + N_2(0, \delta) + I_3(0, \delta) \right]$$

$$= \lim_{\varepsilon_0 \to 0} I_3(0, \delta) = \oint_{\Gamma^2} q\,dx - p\,dy|_{\varepsilon=0} = M(0, \delta),$$

$$N_h(h, \delta) = M_h(h, \delta) - c_1\left(1 + \ln |h|\right) + O\left(h \ln |h|\right).$$

Further, by Lemma 6.2 we obtain

$$N_h(h, \delta) = \oint_{L^h} (p_x + q_y)|_{\varepsilon=0}\,dt - c_1\left(1 + \ln |h|\right) + O\left(h \ln |h|\right),$$

which yields

$$c_2(\delta) = N_h(0, \delta) = \lim_{h \to 0} \left[\oint_{L^h} (p_x + q_y)|_{\varepsilon=0}\,dt - c_1\left(1 + \ln |h|\right) \right].$$

In particular,

$$c_2(\delta) = \lim_{h \to 0} \oint_{L^h} (p_x + q_y)|_{\varepsilon=0}\,dt$$

if $c_1 = 0$, and furthermore,

$$c_2(\delta) = \oint_{\Gamma^2} (p_x + q_y)|_{\varepsilon=0}\,dt = \sum_{i=1}^{2} \int_{L_i} (p_x + q_y)|_{\varepsilon=0}\,dt$$

if $c_1(S_1, \delta) = c_1(S_2, \delta) = 0$ by the proof of Theorem 8.4. Then (8.27)–(8.30) follow for the case $n = 2$.

The proof is complete.   □

Remark 8.14 Some other results can be found in the literature (e.g., see [62, 155, 183]), which are related to limit cycles bifurcating from heteroclinic loops, leading to an upper bound for the limit cycles and the presence of alien limit cycles.

Chapter 9
Finding More Limit Cycles Using Melnikov Functions

In Chaps. 7 and 8 we discussed the problem of the bifurcation of limit cycles from a center or a (double) homoclinic loop by using the expansion of the Melnikov functions. In this chapter, we present an idea for finding more limit cycles, which combines the bifurcation of limit cycles from centers, homoclinic and heteroclinic loops. A generalized theorem is presented. In particular, two polynomial systems are studied. It is shown that one system can have 7 limit cycles while the other can have 5 limit cycles.

Again we consider system (6.1),

$$\dot{x} = H_y + \varepsilon p(x, y, \varepsilon, \delta),$$
$$\dot{y} = -H_x + \varepsilon q(x, y, \varepsilon, \delta),$$

where H, p and q are C^∞ functions, ε is a small parameter and $\delta \in D \subset R^m$ with D a compact set. For $\varepsilon = 0$, system (6.1) reduces to the Hamiltonian system (6.4),

$$\dot{x} = H_y, \qquad \dot{y} = -H_x.$$

9.1 Basic Idea and Formulation

First suppose the Hamiltonian system (6.4) has a family of periodic orbits L_h given by $H(x, y) = h$ for $h \in (\alpha, \beta)$ such that the limit of L_h as $h \to \alpha$ is a center, denoted by L_α, while the limit of L_h as $h \to \beta$ is a homoclinic loop, denoted by L_β.

As before, let

$$M(h, \delta) = \oint_{L_h} q\,dx - p\,dy|_{\varepsilon=0}, \quad h \in (\alpha, \beta). \tag{9.1}$$

M. Han, P. Yu, *Normal Forms, Melnikov Functions and Bifurcations of Limit Cycles*,
Applied Mathematical Sciences 181,
DOI 10.1007/978-1-4471-2918-9_9, © Springer-Verlag London Limited 2012

Then, by the results obtained in previous chapters we formally have

$$M(h, \delta) = \sum_{i \geq 0} b_i (h - \alpha)^{i+1}, \quad 0 < h - \alpha \ll 1,$$

$$M(h, \delta) = \sum_{i \geq 0} [c_{2i} + c_{2i+1}(h - \beta) \ln |h - \beta|](h - \beta)^i, \quad 0 < \beta - h \ll 1,$$

(9.2)

where the coefficients c_0, c_1, c_2, and b_0, b_1, b_2, etc., can be obtained by applying Theorem 8.4 and (7.34) to the resulting system obtained from (6.1) after a suitable transformation.

By considering the expansions of Melnikov functions given in (9.2) we discuss the number of limit cycles of system (6.1) which appear in a neighborhood of the set $L_\alpha \cup L_\beta$. It is also possible to have a limit cycle between L_α and L_β. In fact, it is easy to prove the following theorem.

Theorem 9.1 ([87]) *Let (9.1) and (9.2) hold. Suppose there exist $\delta_0 \in D$ and $k \geq 0$ such that*

$$b_j(\delta_0) = 0, \quad j = 0, \ldots, k-1, \quad b_k(\delta_0) \neq 0,$$

$$c_0(\delta_0) = c_1(\delta_0) = \bar{c}_2(\delta_0) = 0, \quad c_3(\delta_0) \neq 0,$$

(9.3)

and

$$\mathrm{rank} \left. \frac{\partial(b_0, \ldots, b_{k-1}, c_0, c_1, \bar{c}_2)}{\partial(\delta_1, \delta_2, \ldots, \delta_m)} \right|_{\delta = \delta_0} = k + 3,$$

(9.4)

where $\bar{c}_2 = c_2|_{c_1 = 0}$. Then there exists some (ε, δ) near $(0, \delta_0)$ such that (6.1) has $k + 4$ (or $k + 3$) limit cycles if $b_k(\delta_0) c_3(\delta_0) > 0$ (or < 0), respectively.

Proof For simplicity, take $k = 2$ and $m = k + 3 = 5$. Then by (9.4) the coefficients b_0, b_1, c_0, c_1, c_2, can be taken as free parameters. First consider $b_2(\delta_0) c_3(\delta_0) > 0$. For definiteness, assume $b_2(\delta_0) > 0$ and $c_3(\delta_0) > 0$. Then, it follows from (9.2) and (9.3) that

$$M(h, \delta_0) = b_2(\delta_0)(h - \alpha)^3 + O(|h - \alpha|^4) > 0, \quad 0 < h - \alpha \ll 1,$$

$$M(h, \delta_0) = c_3(\alpha_0)(h - \beta)^2 \ln |h - \beta| + O(|h - \beta|^2) < 0, \quad 0 < \beta - h \ll 1.$$

Thus, there exists $h_0 \in (\alpha, \beta)$ such that

$$M(h_0, \delta_0) = 0, \quad M(h_0 - \varepsilon_0, \delta_0) M(h_0 + \varepsilon_0, \delta_0) < 0,$$

(9.5)

provided that $\varepsilon_0 > 0$ is sufficiently small.

Now, change $b_0, b_1, c_0, c_1, \bar{c}_2$ in turn such that

$$0 < b_0 \ll -b_1 \ll c_0 \ll -c_1 \ll -\bar{c}_2 \ll 1,$$

(9.6)

which ensures that $M(h, \delta)$ has 3 simple zeros near $h = \beta$ and 2 simple zeros near $h = \alpha$ in the interval (α, β) with respect to h. At the same time, it follows from (9.5) that under (9.6) there exists h^* near h_0 such that

$$M(h^*, \delta) = 0, \qquad M(h^* - \varepsilon_0, \delta) M(h^* + \varepsilon_0, \delta) < 0.$$

Then, for ε satisfying $0 < |\varepsilon| \ll b_0$, the function F in (6.5) has 6 different zeros for $h \in (\alpha, \beta)$, which leads to 6 limit cycles under (9.6).

For $b_2(\delta_0) c_3(\delta_0) < 0$, we may assume $b_2(\delta_0) > 0$ and $c_3(\delta_0) < 0$ for definiteness. For this case, similarly we can prove that (6.1) has 5 limit cycles in a neighborhood of $L_\alpha \cup L_\beta$ for

$$0 < |\varepsilon| \ll b_0 \ll -b_1 \ll -c_0 \ll \bar{c}_2 \ll 1.$$

This completes the proof.   □

Remark 9.2 From the proof of Theorem 9.1, it is seen that if (9.4) is replaced by the weak condition,

$$\text{rank} \left. \frac{\partial(c_0, c_1, \bar{c}_2)}{\partial(\delta_1, \ldots, \delta_m)} \right|_{\delta=\delta_0} = 3,$$

with

$$b_0(\delta) = \eta \alpha_0(\delta) c_0(\delta), \qquad b_1(\delta) = \eta \alpha_1(\delta) c_1(\delta),$$

or

$$b_0(\delta) = -\eta \alpha_0(\delta) c_1(\delta), \qquad b_1(\delta) = \eta \alpha_1(\delta) \bar{c}_2(\delta),$$

where $\eta = \text{sgn}(b_2(\delta_0) c_3(\delta_0))$, $\alpha_0(\delta_0) > 0$, $\alpha_1(\delta_0) > 0$, then the theorem remains true in the case of $k = 2$. A similar weak condition can also be given for the case of $k = 1$.

Remark 9.3 The idea used in Theorem 7.12 can also be used to give a generalization of Theorem 9.1 for finding more limit cycles if the function H contains some variable constants.

9.2 A Generalized Theorem

Now we suppose that (6.4) has three families of periodic orbits given by L^h: $H(x, y) = h$ for $h > \beta$ and $L_{ih} : H(x, y) = h$ for $h \in (\alpha_i, \beta)$, $i = 1, 2$, such that $L^\beta = L_{1\beta} \cup L_{2\beta}$, where $L_{i\beta} = \lim_{h \to \beta} L_{ih}$ is a homoclinic loop with a hyperbolic saddle, and $L_{i\alpha_i} = \lim_{h \to \alpha_i} L_{ih}$ is a center, $i = 1, 2$. In this case, we have the three Melnikov functions

$$M(h, \delta) = \oint_{L^h} q\, dx - p\, dy|_{\varepsilon=0}, \quad h > \beta$$

and

$$M_i(h,\delta) = \oint_{L_{ih}} q\,dx - p\,dy|_{\varepsilon=0}, \quad h \in (\alpha_i, \beta), i = 1, 2.$$

Then, similarly to (9.2) we have

$$M(h,\delta) = c_0 + 2c_1(h-\beta)\ln|h-\beta| + c_2(h-\beta)$$
$$+ 2c_3(h-\beta)^2\ln|h-\beta| + \cdots, \quad 0 < h - \beta \ll 1,$$
$$M_i(h,\delta) = c_{0i} + c_1(h-\beta)\ln|h-\beta| + c_{2i}(h-\beta)$$
$$+ c_3(h-\beta)^2\ln|h-\beta| + \cdots, \quad 0 < \beta - h \ll 1,$$

and

$$M_i(h,\delta) = b_{0i} + b_{1i}(h-\alpha_i) + b_{2i}(h-\alpha_i)^2 + \cdots,$$
$$0 < h - \alpha_i \ll 1, \ i = 1, 2, \tag{9.7}$$

where $c_0 = c_{01} + c_{02}$, $\bar{c}_2 = c_2|_{c_1=0} = (c_{21} + c_{22})|_{c_1=0} = \bar{c}_{21} + \bar{c}_{22}$. Based on (9.7) we can discuss the number of limit cycles which appear near the set $L_{\alpha_1} \cup L_{\alpha_2} \cup L_\beta$. For example, following Theorems 9.1 and 8.11, it is easy to prove the following result.

Theorem 9.4 *Assume that there exist $\delta_0 \in D$ and $k \geq 0$ such that*

$$b_{j2}(\delta) = \alpha_j(\delta)b_{j1}(\delta), \quad \alpha_j(\delta_0) > 0, \quad j = 0, \ldots, k,$$
$$c_{02}(\delta) = \alpha(\delta)c_{01}(\delta), \qquad \bar{c}_{22}(\delta) = \beta(\delta)\bar{c}_{21}(\delta),$$
$$\alpha(\delta_0)\big(\alpha(\delta_0) + 1\big) > 0, \qquad \beta(\delta_0) > 0,$$
$$\mu = c_3(\delta_0)b_{k1}(\delta_0) \neq 0, \qquad c_{01}(\delta_0) = c_1(\delta_0) = \bar{c}_{21}(\delta_0) = 0,$$
$$b_{j1}(\delta_0) = 0, \quad j = 0, \ldots, k-1,$$

and

$$\text{rank}\,\frac{\partial(b_{01}, \ldots, b_{k-1,1}, c_{01}, c_1, \bar{c}_{21})}{\partial(\delta_1, \delta_2, \ldots, \delta_m)}(\delta_0) = k + 3.$$

Then there exists some (ε, δ) near $(0, \delta_0)$ such that (6.1) has $2k+9$ (or $2k+7$) limit cycles if $\mu > 0$ (or $\mu < 0$), respectively.

Remark 9.5 If system (6.1) is centrally symmetric with $\beta = 0$, then automatically $\bar{\beta}(\delta) = \alpha(\delta) = \alpha_j(\delta) = 1$ for $j = 0, \ldots, k$.

Remark 9.6 More conditions on the existence of limit cycles can be given by using the expansions in (9.7). For example, if there exists $\delta_0 \in D$ and $k_1 \geq 0$, $k_2 \geq 0$ such that

$$b_{ji}(\delta_0) = 0, \quad j = 0, \ldots, k_i - 1, \qquad b_{k_i,i}(\delta_0) \neq 0, \quad i = 1, 2,$$

$$c_0(\delta_0) = c_1(\delta_0) = \bar{c}_2(\delta_0) = 0, \qquad \bar{c}_3(\delta_0) \neq 0, \qquad c_{01}(\delta_0) \neq 0,$$

$$\text{rank} \frac{\partial(b_{01}, \ldots, b_{k_1-1,1}, b_{02}, \ldots, b_{k_2-1,2}, c_0, c_1, \bar{c}_2)}{\partial(\delta_1, \delta_2, \delta_3, \ldots, \delta_m)} = k_1 + k_2 + 3,$$

then

(1) if $c_{01}(\delta_0)b_{k_1,1}(\delta_0) < 0, c_{01}(\delta_0)b_{k_2,2}(\delta_0) > 0$, then system (6.1) can have $(k_1 + k_2 + 5)$ limit cycles;
(2) if $c_{01}(\delta_0)b_{k_1,1}(\delta_0) > 0, c_{01}(\delta_0)b_{k_2,2}(\delta_0) < 0$, then system (6.1) can have $(k_1 + k_2 + 3)$ limit cycles;
(3) if $b_{k_1,1}(\delta_0)b_{k_2,2}(\delta_0) > 0$, then system (6.1) can have $(k_1 + k_2 + 4)$ limit cycles.

Remark 9.7 We can consider more complicated cases. As an example, let us suppose that there exists a constant $\gamma > \beta$ such that the family L^h exists for $h \in (\beta, \gamma)$ and the limit L^γ of L^h as $h \to \gamma$ is a homoclinic or heteroclinic loop. Then, apart from (9.7), we also have

$$M(h, \delta) = a_0 + a_1(h - \gamma)\ln|h - \gamma| + a_2(h - \gamma) + a_3(h - \gamma)^2\ln|h - \gamma| + \cdots,$$
$$(9.8)$$

for $0 < \gamma - h \ll 1$. By using (9.7) and (9.8) we can give some sufficient conditions for (6.1) to have more limit cycles. There are some other cases to consider. For example, when two different double homoclinic loops appear simultaneously for (6.4), we can discuss the problem of the number of limit cycles for (6.1) near the two loops, together with Hopf bifurcations.

9.3 Bifurcation of Limit Cycles in Some Polynomial Systems

In this section, we consider the bifurcation of limit cycles in some polynomial systems. Two particular systems are studied, one that can have 7 limit cycles and the other that can have 5 limit cycles.

9.3.1 System 1: 7 Limit Cycles

We first consider a system of the form

$$\dot{x} = y, \qquad \dot{y} = x - x^3 - \varepsilon y[a_0 + a_1 x^2 + a_2 x^4 + a_3 x^6]. \qquad (9.9)$$

For $\varepsilon = 0$, the system is Hamiltonian with $H(x, y) = \frac{1}{2}(y^2 - x^2) + \frac{1}{4}x^4$. There are three families of periodic orbits given by

$$L_h : H(x, y) = h, \quad h > 0,$$

and

$$L_i(h) : H(x, y) = h, \qquad (-1)^i x < 0, \quad i = 1, 2, -\frac{1}{4} < h < 0,$$

where the limit of L_h as $h \to 0$ is a double homoclinic loop $L_0 : H(x, y) = 0$. For system (9.9), we have the following result.

Theorem 9.8 *For $a_3 \neq 0$, system (9.9) can have 8 limit cycles for some $(\varepsilon, a_0, a_1, a_2, a_3)$ near $(0, 0, \frac{32}{15}a_3, -\frac{16}{5}a_3, a_3)$, of which 7 limit cycles are near the loop L_0.*

Proof Let

$$M_1(h) = -\oint_{L_1(h)} y\left[a_0 + a_1 x^2 + a_2 x^4 + a_3 x^6\right] dx$$

$$= -\left(a_0 I_0(h) + a_1 I_1(h) + a_2 I_2(h) + a_3 I_3(h)\right),$$

where

$$I_i(h) = \oint_{L_1(h)} x^{2i} y \, dx, \quad i = 0, 1, 2, 3.$$

Since $y^2 = x^2(1 - \frac{1}{2}x^2), 0 \le x \le \sqrt{2}$, along $L_1 = L_0|_{x \ge 0}$, we have

$$I_i(0) = \oint_{L_1} x^{2i} y \, dx = 2 \int_0^{\sqrt{2}} x^{2i+1} \sqrt{1 - \frac{1}{2}x^2} \, dx, \quad i = 0, 1, 2, 3,$$

$$I_i'(0) = \oint_{L_1} x^{2i} \, dt = 2 \int_0^{\sqrt{2}} \frac{x^{2i-1}}{\sqrt{1 - \frac{1}{2}x^2}} \, dx, \quad i = 1, 2, 3.$$

A direct calculation yields

$$I_0(0) = \frac{4}{3}, \qquad I_1(0) = \frac{16}{15}, \qquad I_2(0) = \frac{128}{105}, \qquad I_3(0) = \frac{512}{315},$$

$$I_1'(0) = 4, \qquad I_2'(0) = \frac{16}{3}, \qquad I_3'(0) = \frac{128}{15}.$$

Hence, it follows from (8.22) that

$$c_{01} = -\left(\frac{4}{3}a_0 + \frac{16}{15}a_1 + \frac{128}{105}a_2 + \frac{512}{315}a_3\right),$$

$$c_1 = a_0,$$

$$\bar{c}_2 = -\left(4a_1 + \frac{16}{3}a_2 + \frac{128}{15}a_3\right),$$

$$\bar{c}_3 = -\frac{1}{2}a_1.$$

We then have

$$
\frac{\partial(c_{01}, c_1, \bar{c}_2)}{\partial(a_0, a_1, a_2, a_3)} =
\begin{bmatrix}
\frac{4}{3} & \frac{16}{15} & \frac{128}{105} & \frac{512}{315} \\
1 & 0 & 0 & 0 \\
0 & 4 & \frac{16}{3} & \frac{128}{15}
\end{bmatrix},
$$

which obviously results in

$$
\operatorname{rank} \frac{\partial(c_{01}, c_1, \bar{c}_2)}{\partial(a_0, a_1, a_2, a_3)} = 3.
$$

Note that $c_{01} = c_1 = \bar{c}_2 = 0$ implies

$$
a_0 = 0, \qquad a_1 = \frac{32}{15}a_3, \qquad a_2 = -\frac{16}{5}a_3. \tag{9.10}
$$

Let

$$
M(h) = -\oint_{L_h} y\big[a_0 + a_1 x^2 + a_2 x^4 + a_3 x^6\big]\, dx.
$$

We claim that under the condition (9.10) there exists $h_0 > 0$ such that for $a_3 \neq 0$ the function $M(h)$ has $h_0 > 0$ as its zero with an odd multiplicity.

In fact, by symmetry and using (9.10) we have

$$
M(h) = -4 \int_0^{x_1(h)} \sqrt{2\big(h - G(x)\big)}\left[\frac{32}{15}x^2 - \frac{16}{5}x^4 + x^6\right] a_3\, dx
$$

$$
= -4 \int_0^{x_1(h)} \frac{a_3 F(x)}{\sqrt{2(h - G(x))}}\, dG(x), \qquad h > 0,
$$

where

$$
G(x) = \frac{1}{4}x^4 - \frac{1}{2}x^2, \qquad F(x) = \int_0^x \left(\frac{32}{15}x^2 - \frac{16}{5}x^4 + x^6\right) dx,
$$

and $x_1(h) > \sqrt{2}$ satisfies $G(x_1(h)) = h$ for $h > 0$.

Let $x_0 > \sqrt{2}$ be such that $F(x) > 1$ for $x > x_0$. Then we can write

$$
M(h) = a_3\big[m_1(h) + m_2(h)\big]
$$

where

$$
m_1(h) = -4 \int_0^{x_0} \frac{F(x)}{\sqrt{2(h - G(x))}}\, dG(x),
$$

$$
m_2(h) = -4 \int_{G(x_0)}^{h} \frac{F(x_1(z))}{\sqrt{2(h - z)}}\, dz.
$$

It is obvious that $m_1(h) \to 0$ as $h \to \infty$ and

$$m_2(h) < -4 \int_{G(x_0)}^{h} \frac{1}{\sqrt{2(h-z)}} dz = \frac{-8}{\sqrt{2}} \sqrt{h - G(x_0)} \to -\infty$$

as $h \to \infty$. It follows that $a_3 M(h) \to -\infty$ as $h \to \infty$ for $a_3 \neq 0$. On the other hand, by Theorem 8.8 together with condition (9.10) we have

$$a_3 M(h) = -\frac{32}{15} a_3^2 h^2 \ln |h| + O(h^2) > 0 \quad \text{for } 0 < h \ll 1.$$

Then, the claim follows. Hence, by Theorem 8.9 and its proof for $a_3 \neq 0$, system (9.9) can have 8 limit cycles for some $(\varepsilon, a_0, a_1, a_2, a_3)$ near $(0, 0, \frac{32}{15} a_3, -\frac{16}{5} a_3, a_3)$, of which 7 limit cycles are near the loop L_0.

The proof is complete.   □

9.3.2 System 2: 5 Limit Cycles

Next, consider a system of the form

$$\dot{x} = y + by^2 + \varepsilon f(x, y), \qquad \dot{y} = x - x^2 + kbx^2(x-1) + \varepsilon g(x, y), \qquad (9.11)$$

where k and b are small constants, and

$$f(x, y) = \sum_{i+j=1}^{3} a_{ij} x^i y^j, \qquad g(x, y) = \sum_{i+j=1}^{3} b_{ij} x^i y^j.$$

When $\varepsilon = 0$, system (9.11) has the Hamiltonian function

$$H(x, y, b) = \frac{1}{2}(y^2 - x^2) + \frac{1+kb}{3} x^3 + \frac{b}{3} y^3 - \frac{kb}{4} x^4. \qquad (9.12)$$

For b small, we have a family of periodic orbits defined by

$$L_h : H(x, y, b) = h, \quad x > 0, \qquad h^* = \frac{kb}{12} - \frac{1}{6} < h < 0.$$

Let

$$M(h, b) = \oint_{L_h} g \, dx - f \, dy.$$

Then, by Green's formula, we have

$$M(h, b) = \iint_{H \le h} (f_x + g_y) \, dx \, dy$$

$$= \iint_{H \le h} \left(\sigma_1 + \sigma_2 x + \sigma_3 y + \sigma_4 x^2 + \sigma_5 xy + \sigma_6 y^2 \right) dx \, dy$$

$$= \oint_{L_h} \bar{g}(x, y) \, dx,$$

where

$$\bar{g}(x, y) = \sigma_1 y + \sigma_2 xy + \frac{1}{2}\sigma_3 y^2 + \sigma_4 x^2 y + \frac{1}{2}\sigma_5 xy^2 + \frac{1}{3}\sigma_6 y^3, \tag{9.13}$$

$$\sigma_1 = a_{10} + b_{01}, \qquad \sigma_2 = 2a_{20} + b_{11}, \qquad \sigma_3 = a_{11} + 2b_{02},$$

$$\sigma_4 = 3a_{30} + b_{21}, \qquad \sigma_5 = 2(a_{21} + b_{12}), \qquad \sigma_6 = a_{12} + 3b_{03}.$$

Thus, by Theorem 8.4 and (9.12), we obtain

$$\bar{c}_0 = \sigma_1 I_{01}(b) + \sigma_2 I_{11}(b) + \frac{1}{2}\sigma_3 I_{02}(b)$$

$$+ \sigma_4 I_{21}(b) + \frac{1}{2}\sigma_5 I_{12}(b) + \frac{1}{3}\sigma_6 I_{03}(b),$$

$$\bar{c}_1 = -\sigma_1, \tag{9.14}$$

$$\bar{c}_2 = \sigma_2 \tilde{I}_{10}(b) + \sigma_3 \tilde{I}_{01}(b) + \sigma_4 \tilde{I}_{20}(b) + \sigma_5 \tilde{I}_{11}(b) + \sigma_6 \tilde{I}_{02}(b),$$

$$\bar{c}_3 = \frac{1}{2}\left((1 + kb)\sigma_2 + b\sigma_3 + \sigma_4 - \sigma_6\right),$$

where

$$I_{ij}(b) = \oint_{L_0} x^i y^j \, dx, \qquad \tilde{I}_{ij}(b) = \oint_{L_0} x^i y^j \, dt.$$

Obviously,

$$I_{02}(b) = \oint_{L_0} y^2 \, dx = \oint_{L_0} \left(x^2 - \frac{2}{3}kbx^3 - \frac{2}{3}x^3 - \frac{2}{3}by^3 + \frac{kb}{2}x^4 \right) dx = -\frac{2}{3}bI_{03}(b),$$

$$\tilde{I}_{01}(b) = \oint_{L_0} y \, dt = \oint_{L_0} \frac{1}{1 + by} \, dx = \oint_{L_0} \left(1 - \frac{by}{1 + by} \right) dx = -b\tilde{I}_{02}(b).$$

Let $\sigma_6 - b\sigma_3 = \tilde{\sigma}_3$ so that (9.14) becomes

$$\bar{c}_0 = \sigma_1 I_{01}(b) + \sigma_2 I_{11}(b) + \frac{1}{3}\tilde{\sigma}_3 I_{03}(b) + \sigma_4 I_{21}(b) + \frac{1}{2}\sigma_5 I_{12}(b),$$

$$\bar{c}_1 = -\sigma_1,$$

$$\bar{c}_2 = \sigma_2 \tilde{I}_{10}(b) + \tilde{\sigma}_3 \tilde{I}_{02}(b) + \sigma_4 \tilde{I}_{20}(b) + \sigma_5 \tilde{I}_{11}(b),$$

$$\bar{c}_3 = \frac{1}{2}\big((1+kb)\sigma_2 - \tilde{\sigma}_3 + \sigma_4\big).$$

(9.15)

We also have

$$I_{12}(b) = \oint_{L_0} x\left(x^2 - \frac{2}{3}kbx^3 - \frac{2}{3}x^3 - \frac{2}{3}by^3 + \frac{kb}{2}x^4\right)dx = -\frac{2}{3}bI_{13}(b),$$

$$\tilde{I}_{11}(b) = \oint_{L_0} xy\,dt = \oint_{L_0} \frac{x}{1+by}\,dx = \oint_{L_0}\left(x - \frac{bxy}{1+by}\right)dx = -b\tilde{I}_{12}(b),$$

and

$$I_{11}(b) = \oint_{L_0} xy\,dx = \oint_{L_0} y\,d\left(\frac{1}{2}x^2\right)$$

$$= \oint_{L_0} y\,d\left(\frac{1}{2}y^2 + \frac{b}{3}y^3 + \frac{1+kb}{3}x^3 - \frac{kb}{4}x^4\right)$$

$$= \oint_{L_0} y\big((1+kb)x^2 - kbx^3\big)dx = (1+kb)I_{21}(b) - kbI_{31}(b),$$

(9.16)

$$\tilde{I}_{10}(b) = \oint_{L_0} \frac{x}{y+by^2}\,dx$$

$$= \oint_{L_0} \frac{1}{y+by^2}\,d\left(\frac{1}{2}y^2 + \frac{b}{3}y^3 + \frac{1+kb}{3}x^3 - \frac{kb}{4}x^4\right)$$

$$= \oint_{L_0} \frac{(1+kb)x^2 - kbx^3}{y+by^2}\,dx = (1+kb)\tilde{I}_{20}(b) - kb\tilde{I}_{30}(b).$$

For j even, $I_{ij}(0) = 0$; and for j odd, the following holds:

$$I_{ij}(0) = 2\int_0^{\frac{3}{2}} x^{i+j}\left(\sqrt{1 - \frac{2}{3}x}\right)^j dx.$$

Then, we obtain

$$I_{01}(0) = \frac{6}{5}, \qquad I_{11}(0) = \frac{36}{35}, \qquad I_{21}(0) = \frac{36}{35},$$

$$I_{03}(0) = \frac{108}{385}, \qquad I_{13}(0) = \frac{1296}{5005}, \qquad I_{31}(0) = \frac{432}{385}.$$

So for j odd, $\tilde{I}_{ij}(0) = 0$; and for j even, we have

$$\tilde{I}_{10}(0) = \oint_{L_0} x \, dt = \oint_{L_0} \frac{x}{y} \, dx = 2 \int_0^{\frac{3}{2}} \frac{1}{\sqrt{1 - \frac{2}{3}x}} \, dx = 6,$$

$$\tilde{I}_{02}(0) = \oint_{L_0} y^2 \, dt = \oint_{L_0} y \, dx = I_{01}(0) = \frac{6}{5},$$

$$\tilde{I}_{20}(0) = \oint_{L_0} x^2 \, dt = \oint_{L_0} \frac{x^2}{y} \, dx = 2 \int_0^{\frac{3}{2}} \frac{x}{\sqrt{1 - \frac{2}{3}x}} \, dx = 6,$$

$$\tilde{I}_{30}(0) = \oint_{L_0} x^3 \, dt = \oint_{L_0} \frac{x^3}{y} \, dx = 2 \int_0^{\frac{3}{2}} \frac{x^2}{\sqrt{1 - \frac{2}{3}x}} \, dx = \frac{36}{5},$$

$$\tilde{I}_{12}(0) = \oint_{L_0} x y^2 \, dt = \oint_{L_0} x y \, dx = I_{11}(0) = \frac{36}{35}.$$

Thus, combining the above results yields

$$I_{01}(b) = \frac{6}{5} + O(b), \qquad I_{11}(b) = \frac{36}{35} + O(b), \qquad I_{03}(b) = \frac{108}{385} + O(b),$$

$$I_{02}(b) = -\frac{72}{385}b + O(b^2), \qquad I_{12}(b) = -\frac{864}{5005}b + O(b^2),$$

$$I_{21}(b) = \frac{36}{35} + O(b), \qquad I_{13}(b) = \frac{1296}{5005} + O(b), \qquad I_{31}(b) = \frac{432}{385} + O(b),$$

$$\tilde{I}_{10}(b) = 6 + O(b), \qquad \tilde{I}_{20}(b) = 6 + O(b), \qquad \tilde{I}_{02}(b) = \frac{6}{5} + O(b),$$

$$\tilde{I}_{30}(b) = \frac{36}{5} + O(b), \qquad \tilde{I}_{12}(b) = \frac{36}{35} + O(b),$$

$$\tilde{I}_{01}(b) = -\frac{6}{5}b + O(b^2), \qquad \tilde{I}_{11}(b) = -\frac{36}{35}b + O(b^2).$$

$$(9.17)$$

Let $\bar{c}_0 = \bar{c}_1 = \bar{c}_2 = 0$, which is equivalent to

$$\sigma_1 = 0,$$

$$\sigma_2 I_{11}(b) + \frac{1}{3}\tilde{\sigma}_3 I_{03}(b) + \sigma_4 I_{21}(b) + \frac{1}{2}\sigma_5 I_{12}(b) = 0, \qquad (9.18)$$

$$\sigma_2 \tilde{I}_{10}(b) + \tilde{\sigma}_3 \tilde{I}_{02}(b) + \sigma_4 \tilde{I}_{20}(b) + \sigma_5 \tilde{I}_{11}(b) = 0.$$

Solve for $\tilde{\sigma}_3$ and σ_4 from the second and third equations of (9.18) to obtain

$$\tilde{\sigma}_3 = \frac{3}{\tilde{I}_{20}I_{03} - 3\tilde{I}_{02}I_{21}}\left[(\tilde{I}_{10}I_{21} - \tilde{I}_{20}I_{11})\sigma_2 + \left(\tilde{I}_{11}I_{21} - \frac{1}{2}\tilde{I}_{20}I_{12}\right)\sigma_5\right],$$

$$\sigma_4 = \frac{1}{\tilde{I}_{20}I_{03} - 3\tilde{I}_{02}I_{21}}\left[(3\tilde{I}_{02}I_{11} - \tilde{I}_{10}I_{03})\sigma_2 + \left(\frac{3}{2}\tilde{I}_{02}I_{12} - \tilde{I}_{11}I_{03}\right)\sigma_5\right].$$

(9.19)

Further using (9.16) and (9.19) we have

$$\tilde{\sigma}_3 = 3kbA_1(b)\sigma_2 + 3A_2(b)\sigma_5,$$

$$\sigma_4 = \left[-(1+kb) + kbA_3(b)\right]\sigma_2 + A_4(b)\sigma_5,$$

(9.20)

where

$$A_1(b) = \frac{A(b)}{C(b)}, \qquad A_2(b) = \frac{\tilde{I}_{11}I_{21} - \frac{1}{2}\tilde{I}_{20}I_{12}}{C(b)},$$

$$A_3(b) = \frac{B(b)}{C(b)}, \qquad A_4(b) = \frac{\frac{3}{2}\tilde{I}_{02}I_{12} - \tilde{I}_{11}I_{03}}{C(b)},$$

and

$$A(b) = \tilde{I}_{20}I_{31} - \tilde{I}_{30}I_{21}, \qquad B(b) = \tilde{I}_{30}I_{03} - 3\tilde{I}_{02}I_{31}, \qquad C(b) = \tilde{I}_{20}I_{03} - 3\tilde{I}_{02}I_{21}.$$

By (9.17), it is easy to show that

$$A(0) = -\frac{1296}{1925} = A_0, \qquad B(0) = -\frac{3888}{1925} = B_0, \qquad C(0) = -\frac{3888}{1925} = C_0.$$

Taking $k = 0$, then from (9.12) the homoclinic loop L_0 satisfies

$$y^2 = \varphi(x) - \frac{2}{3}by^3,$$

where $\varphi(x) = x^2 - \frac{2}{3}x^3$. It is obvious that $\varphi(x) \geq 0$ when $b = 0$. Note that the above equation is invariant under the change $y \to -y$, $b \to -b$. It has two solutions $y = y_1(x, b) \geq 0$ and $y = y_2(x, b) \leq 0$ in y with the property $y_2(x, -b) = -y_1(x, b)$. Furthermore, it is easy to obtain

$$\frac{\partial y_i}{\partial b}\bigg|_{b=0} = -\frac{y_i^2}{3}\bigg|_{b=0} = -\frac{1}{3}\varphi(x), \qquad \frac{\partial^2 y_i}{\partial b^2}\bigg|_{b=0} = \frac{5\varphi^2(x)}{9y_i}\bigg|_{b=0}.$$

Hence, we have

$$\frac{\partial y_1}{\partial b}\bigg|_{b=0} = \frac{\partial y_2}{\partial b}\bigg|_{b=0} = -\frac{1}{3}\varphi(x),$$

$$\frac{\partial^2 y_1}{\partial b^2}\bigg|_{b=0} = \frac{5}{9}\varphi^{\frac{3}{2}}, \qquad \frac{\partial^2 y_2}{\partial b^2}\bigg|_{b=0} = -\frac{5}{9}\varphi^{\frac{3}{2}}.$$

(9.21)

By introducing

$$\psi_{ij}(x, y) = x^i y^j, \qquad \tilde{\psi}_{ij}(x, y, b) = \frac{x^i y^j}{y + by^2},$$

we obtain

$$I_{ij}(b) = \int_0^{\frac{3}{2}} \left(\psi_{ij}(x, y_1) - \psi_{ij}(x, y_2) \right) dx,$$

$$\tilde{I}_{ij}(b) = \int_0^{\frac{3}{2}} \left(\tilde{\psi}_{ij}(x, y_1, b) - \tilde{\psi}_{ij}(x, y_2, b) \right) dx.$$

Using $y_2(x, -b) = -y_1(x, b)$, it is easy to prove that the functions $\tilde{I}_{30}(b)$, $\tilde{I}_{02}(b)$, $\tilde{I}_{20}(b)$, $I_{21}(b)$, $I_{03}(b)$ and $I_{31}(b)$ are all even, while $\tilde{I}_{11}(b)$ and $I_{12}(b)$ are odd. Therefore, the functions $A(b)$, $B(b)$, $C(b)$, $A_1(b)$ and $A_3(b)$ are even in b and $A_2(b)$ and $A_4(b)$ are odd in b in the case of $k = 0$. In particular, we have

$$\tilde{I}'_{20}(0) = \tilde{I}'_{30}(0) = \tilde{I}'_{02}(0) = I'_{21}(0) = I'_{03}(0) = I'_{31}(0) = 0,$$

and

$$A'(0) = B'(0) = C'(0) = 0.$$

Note that

$$\left(\tilde{\psi}_{20}(x, y_i, b) \right)'_b = -\frac{\frac{\partial y_i}{\partial b} + y_i^2 + 2by_i \frac{\partial y_i}{\partial b}}{(y_i + by_i^2)^2} x^2, \qquad i = 1, 2,$$

$$\left(\tilde{\psi}_{20}(x, y_i, b) \right)''_b \Big|_{b=0} = -\frac{y_i \left(\frac{\partial^2 y_i}{\partial b^2} + 4y_i \frac{\partial y_i}{\partial b} \right) - 2\left(\frac{\partial y_i}{\partial b} + y_i^2 \right)^2}{y_i^3} x^2, \qquad i = 1, 2.$$

It follows from (9.21) that

$$\tilde{I}''_{20}(0) = \left(\int_0^{\frac{3}{2}} \left(\tilde{\psi}_{20}(x, y_1, b) - \tilde{\psi}_{20}(x, y_2, b) \right) dx \right)'' \Bigg|_{b=0}$$

$$= \int_0^{\frac{3}{2}} \left(\tilde{\psi}_{20}(x, y_1, b) - \tilde{\psi}_{20}(x, y_2, b) \right)''_b \Big|_{b=0} dx$$

$$= \frac{10}{3} \int_0^{\frac{3}{2}} x^2 \varphi^{\frac{1}{2}} dx = \frac{10}{3} \int_0^{\frac{3}{2}} x^3 \sqrt{1 - \frac{2}{3}x} \, dx = \frac{12}{7}.$$

Similarly, using

$$\left(\tilde{\psi}_{02}(x, y_i, b) \right)''_b \Big|_{b=0} = \frac{\partial^2 y_i}{\partial b^2} - 4y_i \frac{\partial y_i}{\partial b} + 2y_i^3,$$

$$\left(\widetilde{\psi}_{30}(x, y_i, b)\right)''_b\Big|_{b=0} = -\frac{y_i\left(\frac{\partial^2 y_i}{\partial b^2} + 4y_i\frac{\partial y_i}{\partial b}\right) - 2\left(\frac{\partial y_i}{\partial b} + y_i^2\right)^2}{y_i^3}x^3,$$

$$\left(\psi_{03}(x, y_i, b)\right)''_b\Big|_{b=0} = 6y_i\left(\frac{\partial y_i}{\partial b}\right)^2 + 3y_i^2\left(\frac{\partial^2 y_i}{\partial b^2}\right),$$

$$\left(\psi_{21}(x, y_i, b)\right)''_b\Big|_{b=0} = x^2\frac{\partial^2 y_i}{\partial b^2},$$

$$\left(\psi_{31}(x, y_1, b)\right)''_b\Big|_{b=0} = x^3\frac{\partial^2 y_i}{\partial b^2},$$

and (9.21) we obtain

$$\widetilde{I}''_{02}(0) = \frac{70}{9}\int_0^{\frac{3}{2}}\varphi^{\frac{3}{2}}\,dx = \frac{70}{9}\int_0^{\frac{3}{2}}x^3\left(1 - \frac{2}{3}x\right)^{\frac{3}{2}}\,dx = \frac{12}{11},$$

$$\widetilde{I}''_{30}(0) = \frac{10}{3}\int_0^{\frac{3}{2}}x^3\varphi^{\frac{1}{2}}\,dx = \frac{10}{3}\int_0^{\frac{3}{2}}x^4\sqrt{1 - \frac{2}{3}x}\,dx = \frac{144}{77},$$

$$I''_{21}(0) = \frac{10}{9}\int_0^{\frac{3}{2}}x^2\varphi^{\frac{3}{2}}\,dx = \frac{10}{9}\int_0^{\frac{3}{2}}x^5\left(1 - \frac{2}{3}x\right)^{\frac{3}{2}}\,dx = \frac{144}{1001},$$

$$I''_{31}(0) = \frac{10}{9}\int_0^{\frac{3}{2}}x^3\varphi^{\frac{3}{2}}\,dx = \frac{10}{9}\int_0^{\frac{3}{2}}x^6\left(1 - \frac{2}{3}x\right)^{\frac{3}{2}}\,dx = \frac{2592}{17017},$$

$$I''_{03}(0) = \frac{14}{3}\int_0^{\frac{3}{2}}\varphi^{\frac{5}{2}}\,dx = \frac{14}{3}\int_0^{\frac{3}{2}}x^5\left(1 - \frac{2}{3}x\right)^{\frac{5}{2}}\,dx = \frac{432}{2431}.$$

Thus, we have

$$A''(0)|_{k=0} = -\frac{10368}{85085} = \widetilde{A}_1, \qquad B''(0)|_{k=0} = -\frac{15831936}{6551545} = B_1,$$

$$C''(0)|_{k=0} = -\frac{1391904}{595595} = C_1.$$

Noting $A(0) = A_0$, $B(0) = B_0$, $C(0) = C_0$, for any k, it follows that

$$A(b) = A_0 + \frac{1}{2}\widetilde{A}_1 b^2 + O\left(b^4 + |bk|\right),$$

$$B(b) = B_0 + \frac{1}{2}B_1 b^2 + O\left(b^4 + |bk|\right),$$

$$C(b) = C_0 + \frac{1}{2}C_1 b^2 + O\left(b^4 + |bk|\right),$$

for $|b| + |k|$ small. Similarly,

$$\tilde{I}_{11}I_{21} - \frac{1}{2}\tilde{I}_{20}I_{12} = -\frac{94608}{175175}b + O\left(b^2(|b| + |k|)\right),$$

$$\frac{3}{2}\tilde{I}_{02}I_{12} - \tilde{I}_{11}I_{03} = -\frac{3888}{175175}b + O\left(b^2(|b| + |k|)\right).$$

Therefore, for $|b| + |k|$ small, we obtain

$$A_1 = \frac{A(b)}{C(b)} = \frac{1}{C_0}\left[A_0 + \frac{1}{2}\tilde{A}_1 b^2 + O\left(b^4 + |bk|\right)\right]$$

$$\times\left[1 - \frac{C_1}{2C_0}b^2 + O\left(b^4 + |bk|\right)\right]$$

$$= \frac{1}{C_0}\left[A_0 + \left(\frac{\tilde{A}_1}{2} - \frac{A_0 C_1}{2C_0}\right)b^2 + O\left(b^4 + |bk|\right)\right]$$

$$= \frac{1}{3} - \frac{755}{4641}b^2 + O\left(b^4 + |bk|\right),$$

$$A_2 = \frac{\tilde{I}_{11}I_{21} - \frac{1}{2}\tilde{I}_{20}I_{12}}{C(b)} = -\frac{94608}{175175C_0}b + O\left(b^2(|b| + |k|)\right)$$

$$= \frac{73}{273}b + O\left(b^2(|b| + |k|)\right),$$

$$A_3 = \frac{B(b)}{C(b)} = \frac{1}{C_0}\left[B_0 + \frac{1}{2}B_1 b^2 + O\left(b^4 + |bk|\right)\right]$$

$$\times\left[1 - \frac{C_1}{2C_0}b^2 + O\left(b^4 + |bk|\right)\right]$$

$$= \frac{1}{C_0}\left[B_0 + \left(\frac{B_1}{2} - \frac{B_0 C_1}{2C_0}\right)b^2 + O\left(b^4 + |bk|\right)\right]$$

$$= 1 + \frac{335}{17017}b^2 + O\left(b^4 + |bk|\right),$$

$$A_4 = \frac{\frac{3}{2}\tilde{I}_{02}I_{12} - \tilde{I}_{11}I_{03}}{C(b)} = -\frac{3888}{175175C_0}b + O\left(b^2(|b| + |k|)\right)$$

$$= \frac{1}{91}b + O\left(b^2(|b| + |k|)\right).$$

(9.22)

Next, we introduce a change of variables of the form

$$u = x - 1, \qquad v = \frac{1}{\sqrt{1 - kb}}y,$$

and use a time rescaling, $t \to \sqrt{1-kb}\,t$, so that for $|b|$ small, (9.11) becomes

$$
\dot{u} = v + b\sqrt{1-kb}\,v^2 + \frac{\varepsilon}{\sqrt{1-kb}} f(u+1, \sqrt{1-kb}\,v),
$$

$$
\dot{v} = -u + \frac{2kb-1}{1-kb}u^2 + \frac{kb}{1-kb}u^3 + \frac{\varepsilon}{1-kb}g(u+1, \sqrt{1-kb}\,v). \tag{9.23}
$$

Thus, for $\varepsilon = 0$ we have

$$
H(u, v, b) = \frac{1}{2}(u^2 + v^2) + \frac{b}{3}\sqrt{1-kb}\,v^3 + \frac{1-2kb}{3-3kb}u^3 - \frac{kb}{4-4kb}u^4.
$$

Let

$$
p(u, v) = \frac{1}{\sqrt{1-kb}} f(u+1, \sqrt{1-kb}\,v),
$$

$$
q(u, v) = \frac{1}{1-kb} g(u+1, \sqrt{1-kb}\,v). \tag{9.24}
$$

Using (9.13) and (9.24) a direct calculation leads to

$$
p_u + q_v = \sum_{i+j=0}^{2} c_{ij} u^i v^j,
$$

where

$$
c_{00} = \frac{1}{\sqrt{1-kb}}(\sigma_1 + \sigma_2 + \sigma_4), \qquad c_{10} = \frac{1}{\sqrt{1-kb}}(\sigma_2 + 2\sigma_4),
$$

$$
c_{01} = \sigma_3 + \sigma_5, \qquad c_{11} = \sigma_5, \qquad c_{02} = \sqrt{1-kb}\,\sigma_6, \qquad c_{20} = \frac{1}{\sqrt{1-kb}}\sigma_4.
$$

Now, applying (7.44) to (9.23) and noting $\sigma_6 - b\sigma_3 = \tilde{\sigma}_3$, we have

$$
M(h, b) = b_0(h - h^*) + b_1(h - h^*)^2 + O(|h - h^*|^3),
$$

where

$$
b_0 = \frac{2\pi}{\sqrt{1-kb}}(\sigma_1 + \sigma_2 + \sigma_4),
$$

$$
b_1 = \frac{\pi}{\sqrt{1-kb}}\left[\frac{2kb-1}{1-kb}\sigma_2 + (1-kb)\tilde{\sigma}_3 + \frac{3kb-1}{1-kb}\sigma_4 - b(1-kb)\sigma_5\right]. \tag{9.25}
$$

Further, under (9.18), let $b_0 = 0$, or equivalently $\sigma_4 = -\sigma_2$, which is then substituted into (9.20), together with (9.17) and (9.22), to yield

$$
\sigma_4 = -\sigma_2 = \frac{A_4(b)}{kb(A_3(b) - 1)}\sigma_5,
$$

$$
\tilde{\sigma}_3 = \frac{3A_2(b)(A_3(b) - 1) - 3A_1(b)A_4(b)}{A_3(b) - 1}\sigma_5. \tag{9.26}
$$

Then, by (9.15), (9.22), (9.25) and (9.26) we have

$$
\bar{c}_3 = \frac{1}{2}(kb\sigma_2 - \tilde{\sigma}_3) = \frac{A_4(b)(3A_1(b) - 1) - 3A_2(b)(A_3(b) - 1)}{2(A_3(b) - 1)}\sigma_5
$$

$$
= -\frac{180b}{335(b + O(|b|^3 + |k|))}\sigma_5,
$$

$$
b_1 = \frac{\pi}{\sqrt{1 - kb}}\left[\frac{kb}{1 - kb}\sigma_4 + (1 - kb)(\tilde{\sigma}_3 - b\sigma_5)\right]
$$

$$
= \frac{\pi\sigma_5}{(1 - kb)^{\frac{3}{2}}(A_3(b) - 1)}
$$

$$
\times \left\{A_4(b) + (1 - kb)^2\left[\left(3A_2(b) - b\right)\left(A_3(b) - 1\right) - 3A_1(b)A_4(b)\right]\right\}
$$

$$
= \frac{\pi b\sigma_5}{(1 - kb)^{\frac{3}{2}}(A_3(b) - 1)}A^*(b, k),
$$

where

$$
A^*(b, 0) = \frac{1}{b}\left[A_4(b) + (1 - kb)^2\left(\left(3A_2(b) - b\right)\left(A_3(b) - 1\right) - 3A_1(b)A_4(b)\right)\right]\big|_{k=0}
$$

$$
= \frac{1}{b}\left[A_4(b)\left(1 - 3A_1(b)\right) + \left(3A_2(b) - b\right)\left(A_3(b) - 1\right)\right]
$$

$$
= \left(\frac{1}{91} + O(b^2)\right)\left(\frac{755}{1547}b^2 + O(b^4)\right)
$$

$$
+ \left(-\frac{18}{91} + O(b^2)\right)\left(\frac{335}{17017}b^2 + O(b^4)\right)
$$

$$
= \left(\frac{25}{17017} + O(b^2)\right)b^2.
$$

Now we take

$$
\sigma_5 \neq 0, \qquad k = |b|^\alpha\left(1 + O(b)\right), \qquad \alpha \geq 3, \sigma_0 < b < 0, \tag{9.27}
$$

where σ_0 is a negative constant. Then using (9.22) and noting that $A^*(b, k) = A^*(b, 0) + O(k)$ we have

$$
\bar{c}_3 = -\frac{180}{335}\sigma_5\left(1 + O(b^2)\right), \qquad b_1 = \frac{5}{67}\pi b\sigma_5\left(1 + O(b)\right),
$$

which indicates $b_1\bar{c}_3 > 0$. Finally, by (9.15), (9.25) and (9.16) we obtain

$$\det \frac{\partial(b_0, \bar{c}_0, \bar{c}_1, \bar{c}_2)}{\partial(\sigma_1, \sigma_2, \tilde{\sigma}_3, \sigma_4)}$$

$$= \frac{2\pi}{\sqrt{1-kb}} \begin{vmatrix} 1 & 1 & 0 & 1 \\ I_{01}(b) & I_{11}(b) & \frac{1}{3}I_{03}(b) & I_{21}(b) \\ -1 & 0 & 0 & 0 \\ 0 & \tilde{I}_{10}(b) & \tilde{I}_{02}(b) & \tilde{I}_{20}(b) \end{vmatrix}$$

$$= \frac{2\pi}{\sqrt{1-kb}} \left[\tilde{I}_{02}(b)\big(I_{21}(b) - I_{11}(b)\big) - \frac{1}{3}I_{03}(b)\big(\tilde{I}_{20}(b) - \tilde{I}_{10}(b)\big) \right]$$

$$= \frac{2\pi kb}{\sqrt{1-kb}} \left[\tilde{I}_{02}(b)\big(I_{31}(b) - I_{21}(b)\big) - \frac{1}{3}I_{03}(b)\big(\tilde{I}_{30}(b) - \tilde{I}_{20}(b)\big) \right]$$

$$= \frac{2\pi kb}{\sqrt{1-kb}} B^*(b, k).$$

Since $\tilde{I}_{20}(b)$, $\tilde{I}_{30}(b)$, $\tilde{I}_{02}(b)$, $I_{21}(b)$, $I_{31}(b)$, $I_{03}(b)$ are all even, then the function $B^*(b, 0)$ is even in b. Therefore, we have

$$B^*(b, 0) = \left(\frac{86832}{6551545} + O(b^2) \right) b^2.$$

Then, by Theorem 9.1 we have proved the following theorem.

Theorem 9.9 Let (9.17) be satisfied. Then system (9.11) can have 5 limit cycles.

Chapter 10
Limit Cycle Bifurcations in Equivariant Systems

In this chapter, we consider plane systems with some symmetry and, in particular, investigate the bifurcation of limit cycles in equivariant systems, including Z_q-equivariant vector fields and Z_q-reversible-equivariant vector fields. An
S-symmetrical quadratic system and a Z_3-equivariant system are given particular
attention.

10.1 Symmetry of a Vector Field

Suppose we have a two-dimensional system

$$\frac{du}{dt} = V(u), \tag{10.1}$$

where $u = (x, y)$ and $V(u) = (P(x, y), Q(x, y))$.

Definition 10.1 Let $S : R^2 \to R^2$ be an invertible differentiable map. We say that
S is a symmetry of (10.1) if

$$DS(u) \cdot V(u) = V\big(S(u)\big), \quad u \in R^2. \tag{10.2}$$

In this case, we call (10.1) S-symmetrical.

We claim that (10.1) is S-symmetrical if and only if the system is invariant under
the transformation S.
In fact, by making the transformation $v = S(u)$, it then follows from (10.1) that

$$\frac{dv}{dt} = DS\big(S^{-1}(v)\big)V\big(S^{-1}(v)\big) \equiv \tilde{V}(v),$$

where S^{-1} denotes the inverse of S. Then (10.2) holds if and only if $V = \tilde{V}$. Hence,
the claim follows.

M. Han, P. Yu, *Normal Forms, Melnikov Functions and Bifurcations of Limit Cycles*,   353
Applied Mathematical Sciences 181,
DOI 10.1007/978-1-4471-2918-9_10, © Springer-Verlag London Limited 2012

Thus, for the S-symmetrical system (10.1) we have

$$S \circ \varphi^t = \varphi^t \circ S \tag{10.3}$$

where φ^t denotes the flow of (10.1). A consequence of (10.3) is that the symmetric S maps an orbit to another orbit preserving time. To be precise, let L_0 be an orbit of (10.1) and let

$$L_j = S(L_{j-1}) = S^j(L_0), \quad j = 1, \ldots, q-1,$$

where $q \geq 2$ is an integer. Then $L_0, L_1, \ldots, L_{q-1}$, are q different orbits if they do not have common points.

If S is linear, then (10.2) becomes

$$SV(u) = V(S(u)), \quad u \in R^2. \tag{10.4}$$

For example, if S is the reflection

$$S = \begin{bmatrix} 1 & 0 \\ 0 & -1 \end{bmatrix},$$

then (10.4) holds if and only if system (10.1) satisfies

$$P(x, y) = P(x, -y), \qquad Q(x, y) = -Q(x, -y). \tag{10.5}$$

In this case, the flow of (10.1) is symmetric with respect to the x-axis.

10.2 S-Symmetrical Quadratic Systems

For a quadratic polynomial system we have the following result.

Lemma 10.2 *Let P and Q in (10.1) be quadratic polynomials in (x, y) with the symmetry $S = \mathrm{diag}(1, -1)$. Then system (10.1) has no limit cycles.*

Proof Under the conditions, P and Q are quadratic polynomials satisfying (10.5). Hence, (10.1) has the form

$$\dot{x} = a_{00} + a_{10}x + a_{20}x^2 + a_{02}y^2, \qquad \dot{y} = y(b_{01} + b_{11}x).$$

For this system, the line $y = 0$ is invariant. Thus, any periodic orbit, if it exists, must lie above or below it. It suffices to prove that no limit cycles exist in the region $y > 0$. The conclusion is obviously true if $b_{11} = 0$ or $a_{02} = 0$. Let $b_{11}a_{02} \neq 0$. By introducing a change of variables

$$\tilde{x} = -b_{01} - b_{11}x, \qquad \tilde{y} = \ln y,$$

we obtain a new system of the form

$$\dot{\tilde{x}} = \tilde{a}_{00} + \tilde{a}_{10}\tilde{x} + \tilde{a}_{20}\tilde{x}^2 + \tilde{a}_{02}e^{2\tilde{y}}, \qquad \dot{\tilde{y}} = -\tilde{x}, \qquad (10.6)$$

which has a unique singular point if and only if $\tilde{a}_{00}\tilde{a}_{02} < 0$. The singular point is a focus or center if $\tilde{a}_{02} > 0$. Thus, without of loss of generality, we may suppose $\tilde{a}_{00} < 0, \tilde{a}_{02} > 0$. Then the singular point is a center surrounded by a family of periodic orbits, since when $\tilde{a}_{10} = 0$ the vector field is symmetric with respect to the $\tilde{y}$-axis. More precisely, any orbit of the system

$$\dot{\tilde{x}} = \tilde{a}_{00} + \tilde{a}_{20}\tilde{x}^2 + \tilde{a}_{02}e^{2\tilde{y}}, \qquad \dot{\tilde{y}} = -\tilde{x}, \qquad (10.7)$$

is periodic as long as it intersects the $\tilde{y}$-axis twice.

Further, we prove that (10.6) has no periodic orbit for $\tilde{a}_{10} \neq 0$. By contradiction, suppose (10.6) has a periodic orbit L for $\tilde{a}_{10} \neq 0$. Define

$$V^{\perp}(\tilde{x}, \tilde{y}) = \left(\tilde{x}, \tilde{a}_{00} + \tilde{a}_{10}\tilde{x} + \tilde{a}_{20}\tilde{x}^2 + \tilde{a}_{02}e^{2\tilde{y}}\right),$$

$$V_0(\tilde{x}, \tilde{y}) = \left(\tilde{a}_{00} + \tilde{a}_{20}\tilde{x}^2 + \tilde{a}_{02}e^{2\tilde{y}}, -\tilde{x}\right).$$

Then we have the inner product

$$V^{\perp} \cdot V_0 = -\tilde{a}_{10}\tilde{x}^2.$$

For definiteness, let $\tilde{a}_{10} > 0$. Then $V^{\perp} \cdot V_0 < 0$ for $\tilde{x} \neq 0$. This implies that the angle between the vectors V_0 and $V^{\perp}$ is always larger than π at any point on L with $\tilde{x} \neq 0$. Note that $V^{\perp}$ is normal to L outside. It follows that the positive orbit of (10.7) starting at any point on L is inside L and not closed. This is a contradiction.

The proof is complete. $\qquad\qquad\qquad\qquad\qquad\qquad\qquad\qquad\qquad\qquad\qquad$ $\square$

10.3 Z_q-Equivariant Vector Fields

Now let S be a rotation by angle $(2\pi/q)$ for an integer $q \geq 2$. Then

$$S = \begin{bmatrix} \cos\frac{2\pi}{q} & -\sin\frac{2\pi}{q} \\ \sin\frac{2\pi}{q} & \cos\frac{2\pi}{q} \end{bmatrix} \equiv A_q, \qquad (10.8)$$

which generates a group Z_q, i.e.,

$$Z_q = \left\{I, A_q, A_q^2, \ldots, A_q^{q-1}\right\}.$$

Definition 10.3 We call (10.1) Z_q-equivariant if it is A_q-symmetrical.

We use complex variables to give an equivalent form to (10.4).

Let $z = x + iy$, $\bar{z} = x - iy$, where $i = \sqrt{-1}$. Then (10.1) becomes

$$\frac{dz}{dt} = F(z, \bar{z}), \qquad \frac{d\bar{z}}{dt} = \bar{F}(z, \bar{z}), \tag{10.9}$$

where

$$F(z, \bar{z}) = P\left(\frac{z + \bar{z}}{2}, \frac{z - \bar{z}}{2i}\right) + iQ\left(\frac{z + \bar{z}}{2}, \frac{z - \bar{z}}{2i}\right). \tag{10.10}$$

We have the following result [140].

Theorem 10.4 *Let A_q be a rotation given by (10.8). Then (10.1) is Z_q-equivariant if and only if the function F in (10.9) has the form*

$$F(z, \bar{z}) = \sum_{l \geq 1} p_l(|z|^2)\bar{z}^{lq-1} + \sum_{l \geq 0} h_l(|z|^2)z^{lq+1}, \tag{10.11}$$

where p_l and h_l are complex functions.

Proof We have

$$\begin{pmatrix} z \\ \bar{z} \end{pmatrix} = \begin{bmatrix} 1 & i \\ 1 & -i \end{bmatrix}\begin{pmatrix} x \\ y \end{pmatrix},$$

$$P = \frac{1}{2}(F + \bar{F}), \qquad Q = \frac{1}{2i}(F - \bar{F}).$$

Then by noting $V = (P, Q)$ and introducing the complex vector $\tilde{V} = (F, \bar{F})$, we obtain the relation

$$V = T\tilde{V}T^{-1},$$

where

$$T = \frac{1}{2}\begin{bmatrix} 1 & 1 \\ -i & i \end{bmatrix}, \qquad T^{-1} = \begin{bmatrix} 1 & i \\ 1 & -i \end{bmatrix}.$$

Thus, (10.4) holds if and only if

$$\tilde{S}_q \circ \tilde{V} = \tilde{V} \circ \tilde{S}_q, \tag{10.12}$$

where $\tilde{S}_q = T^{-1}A_qT$. By (10.8), it is easy to show that

$$\tilde{S}_q = \begin{bmatrix} e^{\frac{2\pi}{q}i} & 0 \\ 0 & e^{-\frac{2\pi}{q}i} \end{bmatrix}.$$

Hence, (10.12) is equivalent to

$$\left(e^{\frac{2\pi i}{q}}F(z, \bar{z}), e^{-\frac{2\pi i}{q}}\bar{F}(z, \bar{z})\right) = \left(F(e^{\frac{2\pi i}{q}}z, e^{-\frac{2\pi i}{q}}\bar{z}), \bar{F}(e^{\frac{2\pi i}{q}}z, e^{-\frac{2\pi i}{q}}\bar{z})\right)$$

or

$$e^{\frac{2\pi i}{q}} F(z, \bar{z}) = F\left(e^{\frac{2\pi i}{q}} z, e^{-\frac{2\pi i}{q}} \bar{z}\right). \tag{10.13}$$

Using the formal expansion of F of the form

$$F(z, \bar{z}) = \sum_{j,k \geq 0} a_{jk} z^j \bar{z}^k, \tag{10.14}$$

equation (10.13) becomes

$$\sum_{j,k \geq 0} e^{\frac{2\pi i}{q}} a_{jk} z^j \bar{z}^k = \sum_{j,k \geq 0} a_{jk} e^{\frac{2\pi i}{q}(j-k)} z^j \bar{z}^k,$$

yielding

$$e^{\frac{2\pi i}{q}} = e^{\frac{2\pi i}{q}(j-k)} \quad \text{for } a_{jk} \neq 0, \ j, k \geq 0. \tag{10.15}$$

The equality in (10.15) holds if and only if there is an integer l such that

$$j - k = 1 + lq \quad \text{for } a_{jk} \neq 0, j, k \geq 0.$$

Thus, for $a_{jk} \neq 0$ we have equivalently

$$\begin{aligned} k &= j + lq - 1, \quad l \geq 1 \quad \text{if } k \geq j, \\ j &= k + lq + 1, \quad l \geq 0 \quad \text{if } k < j. \end{aligned} \tag{10.16}$$

Now it is clear that (7.11) holds if and only if (10.14) holds together with (10.16). This completes the proof. $\qquad\square$

Corollary 10.5 *Let* (10.1) *be a polynomial system of degree n. If it is Hamiltonian and invariant under rotation by angle $(2\pi/q)$ with $q \geq n + 2$, then* (10.9) *satisfies*

$$F(z, \bar{z}) = i \sum_{k=0}^{m} a_k |z|^{2k} z, \quad a_k \in R, 2m + 1 \leq n.$$

In this case, (10.1) *is called* trivial equivariant *and has the form*

$$\dot{x} = -y \sum_{k=0}^{m} a_k \left(x^2 + y^2\right)^k, \qquad \dot{y} = x \sum_{k=0}^{m} a_k \left(x^2 + y^2\right)^k.$$

Proof Since P and Q are of degree n in (x, y), the function $F(z, \bar{z})$ has degree n in $(z, \bar{z})$. Thus, since $q \geq n + 2$, it follows from Theorem 10.4 that

$$F(z, \bar{z}) = h_0(|z|^2) z = \sum_{l=0}^{m} A_l (|z|^{2l}) z, \quad 2m + 1 \leq n.$$

Further, note that $P_x + Q_y = 0$ if and only if $F_z + \bar{F}_{\bar{z}} = 0$. Thus, if (10.1) is Hamiltonian, then the complex numbers A_l must be purely imaginary, i.e., $A_l = a_l i$ with a_l real. Then the conclusion follows and the proof is completed. $\qquad \square$

Corollary 10.6 *Polynomial system* (10.1), *of degree* 5, *is* Z_2-*equivariant if and only if the function* V *in* (10.1) *satisfies* $V(-u) = -V(u)$; *it is* Z_2-*equivariant* $(3 \leq q \leq 6)$ *if and only if the function* F *in* (10.9) *has the form*

(1) $q = 3$, $F(z, \bar{z}) = (A_0 + A_1|z|^2 + A_2|z|^4)z + (A_3 + A_4|z|^2)\bar{z}^2 + A_5 z^4 + A_6 \bar{z}^5$,

(2) $q = 4$, $F(z, \bar{z}) = (A_0 + A_1|z|^2 + A_2|z|^4)z + (A_3 + A_4|z|^2)\bar{z}^3 + A_5 z^5$,

(3) $q = 5$, $F(z, \bar{z}) = (A_0 + A_1|z|^2 + A_2|z|^4)z + A_3\bar{z}^4$,

(4) $q = 6$, $F(z, \bar{z}) = (A_0 + A_1|z|^2 + A_2|z|^4)z + A_3\bar{z}^5$.

10.4 Cubic Z_q-Equivariant Systems

For the Hopf bifurcation of limit cycles in cubic Z_q-equivariant systems at the origin we have the following theorem.

Theorem 10.7 *For* $q \geq 3$, *any cubic* Z_q-*equivariant system has at most* 1 *limit cycle surrounding a unique elementary singular point in the origin.*

Proof By Theorem 10.4 and Corollary 10.6, for cubic Z_q-equivariant systems we have

$$q = 3, \quad F(z, \bar{z}) = (A_0 + A_1|z|^2)z + A_2\bar{z}^2,$$

$$q = 4, \quad F(z, \bar{z}) = (A_0 + A_1|z|^2)z + A_2\bar{z}^3,$$

$$q \geq 5, \quad F(z, \bar{z}) = (A_0 + A_1|z|^2)z.$$

Owing to the similarity we only consider the case $q = 3$. In this case, we introduce the variable rescaling $z \to \mu e^{i\theta} z$ with suitable real numbers μ and θ to obtain a system of the form

$$\dot{z} = (A_0 + A_1|z|^2)z + ic\bar{z}^2, \quad c \in R.$$

The corresponding real form of the above system is

$$\dot{x} = a_0 x - b_0 y + 2cxy + a_1 x^3 - b_1 x^2 y + a_1 xy^2 - b_1 y^3,$$
$$\dot{y} = b_0 x + a_0 y + cx^2 - cy^2 + b_1 x^3 + a_1 x^2 y + b_1 xy^2 + a_1 y^3. \qquad (10.17)$$

In order that the origin is elementary, we need $b_0 \neq 0$. Without loss of generality, we may assume $b_0 = 1$. The divergence of (10.17) is

$$\mathrm{div}_{(10.17)} = 2a_0 + 4a_1(x^2 + y^2),$$

which implies that the origin is a center for $a_0 = a_1 = 0$. Thus, one can show that the Poincaré map of (10.17) has the following form near the origin:

$$P(r, \mu) - r = 2\pi r \left[v_1 P_1(r) + r^2 v_3 P_2(r) \right],$$

where

$$v_1 = \frac{1}{2\pi} \left(e^{2\pi a_0} - 1 \right), \qquad v_3 = 5a_1/8, \quad P_j(r) = 1 + O(r), \quad j = 1, 2.$$

Then, we can prove that the function $P(r, \mu) - r$ has at most one positive zero near $r = 0$.

The proof is completed. $\square$

It was shown in Chap. 4 that there exists a cubic Z_q-equivariant system which has 4 limit cycles for $q = 3$ or 4 and has 12 limit cycles for $q = 2$. More results on the Hopf bifurcation of limit cycles of equivariant systems can be found in Chap. 4.

10.5 Reversing Symmetry and Z_q-Reversible-Equivariant Vector Fields

Below we turn to considering another kind of symmetry.

Definition 10.8 Let $S : R^2 \to R^2$ be an invertible differentiable map. We say that S is a reversing symmetry or time-reversal symmetry of (10.1) if

$$DS(u) \cdot V(u) = -V\big(S(u)\big), \quad u \in R^2. \tag{10.18}$$

In this case, we call (10.1) an S-reversibly-symmetrical dynamical system.

Similarly to the discussion in the equivariant case, it is easy to prove that system (10.1) is S-reversibly symmetrical if and only if it is invariant under the transformation S together with the time reversing $t \to -t$. Thus, similarly to (10.3) for S-reversibly-symmetrical system (10.1), we have

$$S \circ \varphi^t = \varphi^{-t} \circ S = \left(\varphi^t \right)^{-1} \circ S.$$

Therefore, the reversing symmetric S maps an orbit of (10.1) to another orbit of it, but now the time direction is reversed.

Taking

$$S = \begin{bmatrix} 1 & 0 \\ 0 & -1 \end{bmatrix},$$

then from (10.18) we obtain

$$P(x, y) = -P(x, -y), \qquad Q(x, y) = Q(x, -y). \tag{10.19}$$

In this case, the flow of (10.1) is symmetrical with respect to the x-axis with the orientation reversed on both sides of the axis.

Similarly to Lemma 10.2 we can prove that any quadratic system satisfying (10.19) does not have a limit cycle.

We call (10.1) Z_q-reversible-equivariant if it is A_q-reversibly symmetrical. We have a theorem similar to Theorem 10.4.

Theorem 10.9 *Let A_q be a rotation given by (10.8). Then (10.1) is Z_q-reversible-equivariant if and only if the function F in (10.9) has the form*

$$F(z,\bar{z}) = \sum_{l\geq 1} p_l(|z|^2)\bar{z}^{(2l-1)r-1} + \sum_{l\geq 0} h_l(|z|^2)z^{(2l+1)r+1}, \tag{10.20}$$

where $r = \frac{q}{2}$ is an integer and p_l and h_l are complex functions.

In fact, when (10.1) is Z_q-reversible-equivariant, instead of (10.15) we have

$$e^{\frac{2\pi i}{q}} = -e^{\frac{2\pi i}{q}(j-k)} = e^{\frac{2\pi i}{q}(j-k)-\pi i} \quad \text{for } a_{jk} \neq 0,$$

which implies

$$\frac{2}{q} = \frac{2(j-k)}{q} - 1 - 2l,$$

or

$$2(j-k-1) = (2l+1)q, \tag{10.21}$$

for $l \in \{0, \pm 1, \pm 2, \dots\}$. Equation (10.21) can be rewritten as

$$j = k + 1 + (2l+1)r, \quad l \geq 0 \quad \text{if } j \geq k+1,$$
$$k = j - 1 + (2l-1)r, \quad l \geq 1 \quad \text{if } j < k+1.$$

Then (10.20) follows from the above equations and (10.14).

Corollary 10.10 *Polynomial system (10.1), of degree 6 with A_q given by (10.8), is Z_2-reversible-equivariant if and only if the function V in (10.1) satisfies $V(-u) = V(u)$; it is Z_q-reversible-equivariant for $q = 4, 6, 8$ if and only if F has the form*

$$q = 4, \quad F(z,\bar{z}) = (A_0 + A_1|z|^2 + A_2|z|^4)\bar{z} + A_3\bar{z}^5 + (A_4 + A_5|z|^2)z^3,$$

$$q = 6, \quad F(z,\bar{z}) = (A_0 + A_1|z|^2 + A_2|z|^4)\bar{z}^2 + (A_3 + A_4|z|^2)z^4,$$

$$q = 8, \quad F(z,\bar{z}) = (A_0 + A_1|z|^2)\bar{z}^3 + A_2z^5.$$

From the results in [256] we know that any quadratic Z_2-reversible-equivariant system has at most 2 limit cycles.

10.6 Cubic Z_3-Equivariant Hamiltonian Systems

In this section, we study the number of limit cycles in a $(2\pi/3)$-equivariant system.

From the proof of Theorem 10.7, the Z_3-equivariant system (10.17) is Hamiltonian if and only if $a_0 = a_1 = 0$. Let $b_1 \neq 0$. Then by suitable time-rescaling we can assume $b_1 = -1$. Thus, a general cubic Z_3-equivariant Hamiltonian system can be written in the form

$$
\begin{aligned}
\dot{x} &= -y[b - (x^2 + y^2) - 2cx], \\
\dot{y} &= x[b - (x^2 + y^2)] + c(x^2 - y^2).
\end{aligned}
\tag{10.22}
$$

Further, we may suppose $c \geq 0$. When $c = 0$, the phase portraits of (10.22) are trivial. Let $c > 0$. Then there are 5 cases to consider [140]:

Case (1) $b > 0$;
Case (2) $b = 0$;
Case (3) $-c^2/4 < b < 0$;
Case (4) $b = -c^2/4$;
Case (5) $b < -c^2/4$.

In the last case, the origin is a global center. For the first four cases, the phase portraits of (10.22) are given in Fig. 10.1. Here, we consider Case (3) with $b = -3$, $c = 4$. We take the perturbed system in the form

$$
\begin{aligned}
\dot{x}_1 &= 3y_1 + 8x_1 y_1 + x_1^2 y_1 + y_1^3 + \varepsilon p_1(x_1, y_1), \\
\dot{y}_1 &= -3x_1 + 4x_1^2 - 4y_1^2 - x_1^3 - x_1 y_1^2 + \varepsilon q_1(x_1, y_1),
\end{aligned}
\tag{10.23}
$$

where $p_1(x_1, y_1)$ and $q_1(x_1, y_1)$ are polynomials of degree 5 such that (10.23) is $(2\pi/3)$-equivariant. By Corollary 10.6, we know that p_1 and q_1 have the forms

$$
\begin{aligned}
p_1 ={}& a_0 x_1 + (x_1 y_1^2 + x_1^3)a_1 + (x_1^2 - y_1^2)a_2 + (-y_1^4 + x_1^4)a_3 \\
& + (-6x_1^2 y_1^2 + y_1^4 + x_1^4)a_4 + (-10x_1^3 y_1^2 + x_1^5 + 5x_1 y_1^4)a_5 \\
& + (x_1^5 + x_1 y_1^4 + 2x_1^3 y_1^2)a_6 - b_0 y_1 + (-x_1^2 y_1 - y_1^3)b_1 + 2b_2 x_1 y_1 \\
& + (2x_1^3 y_1 + 2x_1 y_1^3)b_3 + (4x_1 y_1^3 - 4x_1^3 y_1)b_4 \\
& + (5x_1^4 y_1 - 10x_1^2 y_1^3 + y_1^5)b_5 + (-x_1^4 y_1 - 2x_1^2 y_1^3 - y_1^5)b_6, \\
q_1 ={}& a_0 y_1 + (y_1^3 + x_1^2 y_1)a_1 - 2a_2 x_1 y_1 + (-2x_1^3 y_1 - 2x_1 y_1^3)a_3 \\
& + (-4x_1 y_1^3 + 4x_1^3 y_1)a_4 + (-y_1^5 - 5x_1^4 y_1 + 10x_1^2 y_1^3)a_5 \\
& + (x_1^4 y_1 + y_1^5 + 2x_1^2 y_1^3)a_6 + b_0 x_1 + (x_1 y_1^2 + x_1^3)b_1 + (x_1^2 - y_1^2)b_2 \\
& + (-y_1^4 + x_1^4)b_3 + (-6x_1^2 y_1^2 + y_1^4 + x_1^4)b_4 + (x_1^5 - 10x_1^3 y_1^2 + 5x_1 y_1^4)b_5 \\
& + (x_1^5 + x_1 y_1^4 + 2x_1^3 y_1^2)b_6.
\end{aligned}
\tag{10.24}
$$

Fig. 10.1 The phase portraits
of (10.22)

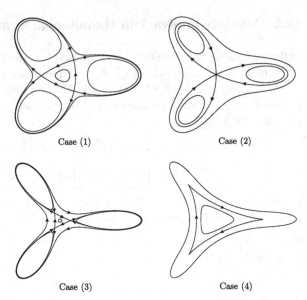

Case (1)   Case (2)

Case (3)   Case (4)

For $\varepsilon = 0$, (10.23) becomes

$$\dot{x}_1 = 3y_1 + 8x_1y_1 + x_1^2y_1 + y_1^3,$$
$$\dot{y}_1 = -3x_1 + 4x_1^2 - 4y_1^2 - x_1^3 - x_1y_1^2,$$

which has the Hamiltonian function

$$H_1(x_1, y_1) = \frac{3}{2}(x_1^2 + y_1^2) - \frac{4}{3}x_1^3 + 4x_1y_1^2 + \frac{1}{4}(x_1^4 + y_1^4) + \frac{1}{2}x_1^2y_1^2 - \frac{5}{12},$$

with the phase portraits as shown in Fig. 10.1 for Case (3). The system has three sad-
dles, $S_1(1, 0)$, $S_2(\frac{1}{2}, \frac{\sqrt{3}}{2})$, $S_3(-\frac{1}{2}, -\frac{\sqrt{3}}{2})$, and four centers, $C_1(3, 0)$, $C_2(-\frac{3}{2}, \frac{3\sqrt{3}}{3})$,
$C_3(-\frac{3}{2}, -\frac{3\sqrt{3}}{2})$ and $O(0, 0)$. It is easy to obtain

$$\alpha = H_1(C_i) = -\frac{8}{3}, \qquad \beta = H_1(S_i) = 0, \qquad H_1(O) = -\frac{5}{12}.$$

Also, the Hamiltonian system has a $(2\pi/3)$-equivariant 3-polycycle, $\Gamma^3 = L_{12} \cup$
$L_{23} \cup L_{31} \cup \{S_1, S_2, S_3\}$, where L_{12}, L_{23}, L_{31} are heteroclinic orbits satisfying

$$\alpha(L_{ij}) = S_i, \qquad \omega(L_{ij}) = S_j.$$

The system also has 3 homoclinic orbits, L_1, L_2, L_3, outside Γ^3 such that

$$\alpha(L_i) = \omega(L_i) = S_i, \qquad i = 1, 2, 3.$$

Then the 3-polycycle, Γ^3, together with the 3 homoclinic orbits, forms an equivariant compound cycle, denoted by $C_1^{(3)}$:

$$C_1^{(3)} = \Gamma^3 \cup \left(\bigcup_{i=1}^{3} L_i \right).$$

System (10.23) has the following 5 Melnikov functions:

$$M_i(h) = \oint_{L_{h,i}} (q_1\, dx_1 - p_1\, dy_1), \quad 0 < -h < \frac{8}{3}, \quad i = 1, 2, 3,$$

$$M_4(h) = \oint_{L_{h,4}} (q_1\, dx_1 - p_1\, dy_1), \quad 0 < -h < \frac{5}{12}, \tag{10.25}$$

$$M(h) = \oint_{\Gamma_h} (q_1\, dx_1 - p_1\, dy_1), \quad 0 < h < \infty,$$

where $L_{h,i}$, $L_{h,4}$ and Γ_h denote the periodic orbits surrounding the unique center C_i, the unique center at the origin and all singular points, respectively, defined by $H_1(x_1, y_1) = h$. By equivariance, we have $M_1(h) = M_2(h) = M_3(h)$.

Next, introducing the transformation

$$x = 6^{-\frac{1}{4}}(x_1 - 1), \qquad y = 6^{\frac{1}{4}} y_1, \tag{10.26}$$

into system (10.23) yields

$$\dot{x} = 2 \cdot 6^{\frac{1}{2}} y + \frac{5}{3} 6^{\frac{3}{4}} xy + x^2 y + \frac{1}{6} y^3 + \varepsilon p(x, y),$$

$$\dot{y} = 2 \cdot 6^{\frac{1}{2}} x + 6^{\frac{3}{4}} x^2 - \frac{5}{6} 6^{\frac{3}{4}} y^2 - 6x^3 - xy^2 + \varepsilon q(x, y), \tag{10.27}$$

with

$$p(x, y) = 6^{-\frac{1}{4}} p_1 \left(6^{\frac{1}{4}} x + 1, y \right),$$

$$q(x, y) = 6^{\frac{1}{4}} q_1 \left(6^{\frac{1}{4}} x + 1, y \right). \tag{10.28}$$

For $\varepsilon = 0$, (10.27) is Hamiltonian with

$$H(x, y) = H_1 \left(6^{\frac{1}{4}} x + 1, 6^{-\frac{1}{4}} y \right)$$

$$= \sqrt{6}\left(y^2 - x^2\right) - \frac{1}{3} 6^{\frac{3}{4}} x^3 + \frac{5}{6} 6^{\frac{3}{4}} xy^2 + \frac{3}{2} x^4 + \frac{1}{2} x^2 y^2 + \frac{1}{24} y^4.$$

The equation $H = h$ defines a periodic orbit L_h on the region $x > 0$ for $h \in (-\frac{8}{3}, 0)$. The limit L_0 of L_h as $h \to 0$ is a homoclinic loop. Under (10.26), $S_1(1, 0)$ and $C_1(3, 0)$ are moved to $S_1'(0, 0)$ and $C_1'(2 \cdot 6^{-\frac{1}{4}}, 0)$, respectively. Note that

$H(S_1') = 0$, $H(C_1') = -\frac{8}{3}$. The intersection of L_0 and the positive x-axis is $(x_0, 0)$ with $x_0 = \frac{1}{9}6^{\frac{3}{4}} + \frac{2}{9}15^{\frac{1}{2}}6^{\frac{1}{4}}$.

By (10.24) and (10.28) we have

$$p_x + q_y = 2a_0 + f_1 a_1 + f_2 a_3 + 4 f_2 a_4 + f_3 b_3 - 4 f_3 b_4 + f_4 a_6,$$

where

$$f_1 = 4\sqrt{6}x^2 + \frac{2}{3}\sqrt{6}y^2 + 8 \cdot 6^{\frac{1}{4}}x + 4,$$

$$f_2 = 2 \cdot 6^{\frac{3}{4}}x^3 - 6^{\frac{3}{4}}xy^2 + 6\sqrt{6}x^2 - \sqrt{6}y^2 + 6 \cdot 6^{\frac{1}{4}}x + 2,$$

$$f_3 = 6 \cdot 6^{\frac{1}{4}}x^2 y - \frac{1}{3}6^{\frac{1}{4}}y^3 + 12xy + 6^{\frac{3}{4}}y,$$

$$f_4 = 36x^4 + 12x^2 y^2 + y^4 + 24 \cdot 6^{\frac{3}{4}}x^3 + 4 \cdot 6^{\frac{3}{4}}xy^2 + 36\sqrt{6}x^2$$
$$+ 2\sqrt{6}y^2 + 24 \cdot 6^{\frac{1}{4}}x + 6.$$

Let

$$g = g_1 a_0 + g_2 a_2 + g_3 a_3 + 4 g_3 a_4 + g_4 b_3 - 4 g_4 b_4 + g_5 a_6,$$

where

$$g_1 = 2y,$$

$$g_2 = 4\sqrt{6}x^2 y + \frac{2}{9}\sqrt{6}y^3 + 8 \cdot 6^{\frac{1}{4}}xy + 4y,$$

$$g_3 = 2 \cdot 6^{\frac{3}{4}}x^3 y - \frac{1}{3}6^{\frac{3}{4}}xy^3 + 6\sqrt{6}x^2 y - \frac{1}{3}\sqrt{6}y^3 + 6 \cdot 6^{\frac{1}{4}}xy + 2y,$$

$$g_4 = 3 \cdot 6^{\frac{1}{4}}x^2 y^2 - \frac{1}{12}6^{\frac{1}{4}}y^4 + 6xy^2 + \frac{1}{2}6^{\frac{3}{4}}y^2,$$

$$g_5 = 36x^4 y + 4x^2 y^3 + \frac{1}{5}y^5 + 24 \cdot 6^{\frac{3}{4}}x^3 y + \frac{4}{3}6^{\frac{3}{4}}xy^3 + 36\sqrt{6}x^2 y$$
$$+ \frac{2}{3}\sqrt{6}y^3 + 24 \cdot 6^{\frac{1}{4}}xy + 6y.$$

Then $g_y = p_x + q_y$. Thus, by (10.25), for $h \in (-\frac{8}{3}, 0)$ we have

$$M_1(h) = \oint_{L_h} (q\,dx - p\,dy) = \oint_{L_h} g\,dx.$$

Introduce

$$I_1 = \oint_{L_0} g_1\,dx, \qquad I_2 = \oint_{L_0} g_2\,dx, \qquad I_3 = \oint_{L_0} g_3\,dx, \qquad I_4 = \oint_{L_0} g_4\,dx,$$

$$I_5 = \oint_{L_0} g_5\,dx, \qquad J_1 = \oint_{L_0}(f_1 - 4)\,dt, \qquad J_2 = \oint_{L_0}(f_2 - 2)\,dt,$$

$$J_3 = \oint_{L_0} f_3\,dt, \qquad J_4 = \oint_{L_0}(f_4 - 6)\,dt.$$

By Theorem 8.4, we obtain

$$M_1(h) = c_0(\delta) + c_1(\delta)h\ln|h| + c_2(\delta)h + c_3(\delta)h^2\ln h + O\left(h^2\right),$$

where

$$
\begin{aligned}
c_0(\delta) &= M_1(0)\\
&= I_1 a_0 + I_2 a_1 + I_3 a_3 + 4I_3 a_4 + I_4 b_3 - 4I_4 b_4 + I_5 a_6,\\
c_1(\delta) &= -\frac{a_{10}+b_{01}}{|\lambda|} = -\frac{2a_0 + 4a_1 + 2a_3 + 8a_4 + 6a_6}{2\sqrt{6}},\\
\bar{c}_2(\delta) &= \oint_{L_0}(p_x + q_y - a_{10} - b_{01})\,dt\\
&= J_1 a_1 + J_2 a_3 + 4J_2 a_4 + J_3 b_3 - 4J_3 b_4 + J_4 a_6,\\
c_3(\delta) &= -\frac{\sqrt{6}}{12}a_1 + \frac{\sqrt{6}}{32}a_3 + \frac{\sqrt{6}}{8}a_4 + \frac{\sqrt{6}}{4}a_6.
\end{aligned}
\tag{10.29}
$$

Consider the equations

$$c_0(\delta) = 0, \qquad c_1(\delta) = 0, \qquad \bar{c}_2(\delta) = 0,$$

which give

$$
\begin{aligned}
a_0 &= \frac{3I_3 J_1 - 2I_3 J_4 - 3I_2 J_2 + I_2 J_4 - I_5 J_1 + 2I_5 J_2}{I_2 J_2 - 2I_1 J_2 + I_1 J_1 - I_3 J_1}a_6\\
&\quad - \frac{2I_3 J_3 + I_4 J_1 - 2I_4 J_2 - I_2 J_3}{I_2 J_2 - 2I_1 J_2 + I_1 J_1 - I_3 J_1}b_3 + 4\frac{2I_3 J_3 + I_4 J_1 - 2I_4 J_2 - I_2 J_3}{I_2 J_2 - 2I_1 J_2 + I_1 J_1 - I_3 J_1}b_4,\\
a_1 &= \frac{3I_1 J_2 - I_1 J_4 + I_3 J_4 - I_5 J_2}{I_2 J_2 - 2I_1 J_2 + I_1 J_1 - I_3 J_1}a_6 - \frac{I_4 J_2 + I_1 J_3 - I_3 J_3}{I_2 J_2 - 2I_1 J_2 + I_1 J_1 - I_3 J_1}b_3\\
&\quad + 4\frac{I_4 J_2 + I_1 J_3 - I_3 J_3}{I_2 J_2 - 2I_1 J_2 + I_1 J_1 - I_3 J_1}b_4,\\
a_3 &= -4a_4 + \frac{-3I_1 J_1 + 2I_1 J_4 - I_2 J_4 + I_5 J_1}{I_2 J_2 - 2I_1 J_2 + I_1 J_1 - I_3 J_1}a_6\\
&\quad + \frac{I_4 J_1 + 2I_1 J_3 - I_2 J_3}{I_2 J_2 - 2I_1 J_2 + I_1 J_1 - I_3 J_1}b_3 - 4\frac{I_4 J_1 + 2I_1 J_3 - I_2 J_3}{I_2 J_2 - 2I_1 J_2 + I_1 J_1 - I_3 J_1}b_4.
\end{aligned}
\tag{10.30}
$$

Next, we discuss the properties of $M_1(h)$ near $h = -\frac{8}{3}$. First, we introduce a transformation

$$u = 6^{\frac{1}{4}}x - 2, \qquad v = 6^{\frac{1}{4}}y. \tag{10.31}$$

Thus, under (10.31), (10.27) becomes

$$\frac{du}{dt} = 6\sqrt{6}v + \frac{7}{3}\sqrt{6}uv + \frac{1}{6}\sqrt{6}u^2v + \frac{1}{36}\sqrt{6}v^3 + \varepsilon\tilde{p}(u,v),$$

$$\frac{dv}{dt} = -6\sqrt{6}u - 5\sqrt{6}u^2 - \frac{7}{6}\sqrt{6}v^2 - \sqrt{6}u^3 - \frac{1}{6}\sqrt{6}uv^2 + \varepsilon\tilde{q}(u,v), \tag{10.32}$$

where

$$\tilde{p}(u,v) = 6^{\frac{1}{4}}p\left(6^{-\frac{1}{4}}(u+2), 6^{-\frac{1}{4}}v\right),$$

$$\tilde{q}(u,v) = 6^{\frac{1}{4}}q\left(6^{-\frac{1}{4}}(u+2), 6^{-\frac{1}{4}}v\right). \tag{10.33}$$

For $\varepsilon = 0$, the Hamiltonian function of (7.32) is

$$\tilde{H}(u,v) = \sqrt{6}(u^2 + v^2) + \frac{5}{3}\sqrt{6}u^3 + \frac{7}{6}\sqrt{6}uv^2 + \frac{\sqrt{6}}{4}u^4 + \frac{\sqrt{6}}{12}u^2v^2$$

$$+ \frac{\sqrt{6}}{144}v^4 - \frac{8}{3}\sqrt{6}$$

$$= \sqrt{6}H\left(6^{-\frac{1}{4}}(u+2), 6^{-\frac{1}{4}}v\right),$$

and

$$L_h : H(x,y) = h \quad \text{for } h \in \left(-\frac{8}{3}, 0\right)$$

becomes

$$\tilde{L}_h : \tilde{H}(u,v) = h \quad \text{for } h \in \left(-\frac{8}{3}\sqrt{6}, 0\right).$$

Further, using the time rescaling, $\tau \to 6\sqrt{6}t$, (10.32) becomes

$$\frac{du}{d\tau} = v + \frac{7}{18}uv + \frac{1}{36}u^2v + \frac{1}{216}v^3 + \varepsilon p_2(u,v),$$

$$\frac{dy}{d\tau} = -u - \frac{5}{6}u^2 - \frac{7}{36}v^2 - \frac{1}{6}u^3 - \frac{1}{36}uv^2 + \varepsilon q_2(u,v), \tag{10.34}$$

where

$$p_2(u,v) = \frac{1}{6\sqrt{6}}\tilde{p}(u,v), \qquad q_2(u,v) = \frac{1}{6\sqrt{6}}\tilde{q}(u,v). \tag{10.35}$$

For $\varepsilon = 0$, the Hamiltonian function of (10.34) is

$$H_2(u, v) = \frac{1}{2}(u^2 + v^2) + \frac{5}{18}u^3 + \frac{7}{36}uv^2 + \frac{1}{24}u^4 + \frac{1}{72}u^2v^2 + \frac{1}{864}v^4 - \frac{4}{9}$$

$$= \frac{1}{6\sqrt{6}}\tilde{H}(u, v),$$

and

$$\tilde{L}_h : \tilde{H}(u, v) = h \quad \text{for } h \in \left(-\frac{8}{3}\sqrt{6}, 0\right)$$

becomes

$$L_h^* : H_2(u, v) = h \quad \text{for } h \in \left(-\frac{4}{9}, 0\right).$$

Thus, it is easy to obtain the Melnikov function of (10.34), $M_1^*(h)$, satisfying

$$M_1^*(h) = \oint_{H_2(u,v)=h} q_2 \, du - p_2 \, dv = \frac{1}{6\sqrt{6}}\tilde{M}_1(6\sqrt{6}h),$$

where $\tilde{M}_1(h) = \oint_{\tilde{H}(u,v)=h} \tilde{q} \, du - \tilde{p} \, dv$ is the Melnikov function of (10.32). Furthermore,

$$\tilde{M}_1(h) = \oint_{\tilde{H}(u,v)=h} \tilde{q} \, du - \tilde{p} \, dv$$

$$= 6^{\frac{1}{4}} \oint_{H(6^{-\frac{1}{4}}(u+2), 6^{-\frac{1}{4}}v)=\frac{h}{\sqrt{6}}} q\left(6^{-\frac{1}{4}}(u+2), 6^{-\frac{1}{4}}v\right) du$$

$$- p\left(6^{-\frac{1}{4}}(u+2), 6^{-\frac{1}{4}}v\right) dv$$

$$= \sqrt{6} \oint_{H(x,y)=\frac{h}{\sqrt{6}}} q(x, y) \, dx - p(x, y) \, dy$$

$$= \sqrt{6} M_1\left(\frac{h}{\sqrt{6}}\right).$$

Therefore,

$$M_1(h) = \frac{1}{\sqrt{6}}\tilde{M}_1(\sqrt{6}h) = 6M_1^*\left(\frac{1}{6}h\right).$$

By using (10.35), (10.33), (10.28) and (10.24), a direct calculation shows

$$p_{2u} + q_{2v} = \frac{\sqrt{6}}{18}(a_0 + 18a_1 + 27a_3 + 108a_4 + 243a_6)$$

$$+ \frac{\sqrt{6}}{6}(4a_1 + 9a_3 + 36a_4 + 108a_6)u$$

$$+ \left(\frac{3}{2}b_3 - 6b_4 \right)v + \frac{\sqrt{6}}{18}(2a_1 + 9a_3 + 36a_4 + 162a_6)u^2$$

$$+ (b_3 - 4b_4)uv + \frac{\sqrt{6}}{108}(2a_1 - 9a_3 - 36a_4 + 54a_6)v^2$$

$$+ \frac{\sqrt{6}}{18}(a_3 + 4a_4 + 9a_6)u^3 + \frac{1}{6}(b_3 - 4b_4)u^2 v$$

$$+ \frac{\sqrt{6}}{36}(-a_3 - 4a_4 + 12a_6)uv^2 + \frac{1}{108}(-b_3 + 3b_4)v^3$$

$$+ \frac{\sqrt{6}}{6}a_6 u^4 + \frac{\sqrt{6}}{18}a_6 u^2 v^2 + \frac{\sqrt{6}}{216}a_6 v^4.$$

Utilizing (7.44), we have

$$M_1^*(h) = \tilde{b}_0 \left(h + \frac{4}{9} \right) + \tilde{b}_1 \left(h + \frac{4}{9} \right)^2 + \tilde{b}_2 \left(h + \frac{4}{9} \right)^3 + O\left(\left| h + \frac{4}{9} \right|^4 \right),$$

$$\text{for } 0 < h + \frac{4}{9} \ll 1,$$

where

$$\tilde{b}_0 = \frac{\sqrt{6}}{9}\pi(a_0 + 18a_1 + 27a_3 + 108a_4 + 243a_6),$$

$$\tilde{b}_1 = -\sqrt{6}\pi \left(\frac{5}{9}a_1 + \frac{9}{8}a_3 + \frac{9}{2}a_4 + 9a_6 \right), \qquad (10.36)$$

$$\tilde{b}_2 = -\sqrt{6}\pi \left(\frac{6235}{11664}a_1 + \frac{173521}{176256}a_3 + \frac{173521}{44064}a_4 + \frac{80885}{7344}a_6 \right).$$

Then, under (10.30), (10.36) becomes

$$\tilde{b}_0 = \frac{2\sqrt{6}}{9}\pi \left(\frac{13I_5 J_1 + 8I_3 J_4 + 120I_2 J_2 - 120I_3 J_1 - 8I_5 J_2 - 13I_2 J_4 + 81I_1 J_1 + 18I_1 J_4 - 216I_1 J_2}{I_2 J_2 - 2I_1 J_2 + I_1 J_1 - I_3 J_1} \right)a_6$$

$$+ \frac{2\sqrt{6}}{9}\pi \left(\frac{8I_3 J_3 - 8I_4 J_2 + 13I_4 J_1 - 13I_2 J_3 + 18I_1 J_3}{I_2 J_2 - 2I_1 J_2 + I_1 J_1 - I_3 J_1} \right)(b_3 - 4b_4),$$

$$\tilde{b}_1 = \frac{\sqrt{6}}{72}\pi \left(\frac{648I_2 J_2 + 405I_1 J_1 - 1176I_1 J_2 - 648I_3 J_1 + 81I_5 J_1 - 81I_2 J_4 + 122I_1 J_4 - 40I_5 J_2 + 40I_3 J_4}{I_2 J_2 - 2I_1 J_2 + I_1 J_1 - I_3 J_1} \right)a_6$$

$$- \frac{\sqrt{6}}{72}\pi \left(\frac{122I_1 J_3 - 81I_2 J_3 + 81I_4 J_1 - 40I_4 J_2 + 40I_3 J_3}{I_2 J_2 - 2I_1 J_2 + I_1 J_1 - I_3 J_1} \right)(b_3 - 4b_4),$$

$$\tilde{b}_2 = \frac{\sqrt{6}}{1586304}\pi \left(\frac{\gamma}{I_2 J_2 - 2I_1 J_2 + I_1 J_1 - I_3 J_1} \right)a_6$$

$$- \frac{\sqrt{6}}{1586304}\pi \left(\frac{1561689I_4 J_1 + 2275418I_1 J_3 - 1561689I_2 J_3 - 847960I_4 J_2 + 847960I_3 J_3}{I_2 J_2 - 2I_1 J_2 + I_1 J_1 - I_3 J_1} \right)(b_3 - 4b_4).$$

where

$$\gamma = 32398440 I_1 J_2 + 17471160 I_3 J_1 - 127896093 I_1 J_1$$
$$- 17471160 I_2 J_2 - 1561689 I_5 J_1 - 2275418 I_1 J_4$$
$$+ 1561689 I_2 J_4 + 847960 I_5 J_2 - 847960 I_3 J_4.$$

Clearly, $\tilde{b}_0 = 0$ if and only if

$$a_6 = Q(b_3 - 4b_4),\tag{10.37}$$

where

$$Q = \frac{-8 I_4 J_2 + 13 I_4 J_1 - 13 I_2 J_3 + 18 I_1 J_3 + 8 I_3 J_3}{120 I_3 J_1 + 8 I_5 J_2 + 13 I_2 J_4 - 8 I_3 J_4 - 120 I_2 J_2 - 13 I_5 J_1 - 8 I_1 J_1 - 18 I_1 J_4 + 216 I_1 J_2}.$$

With (10.30) and (10.37), it follows from (7.29) and (10.36) that

$$\tilde{b}_1 = N_0(b_3 - 4b_4),$$
$$\tilde{b}_2 = N_1 \tilde{b}_1,$$
$$c_3(\delta) = N_2(b_3 - 4b_4),$$

where

$$N_0 = \frac{2\sqrt{6}}{9}\pi \left(\frac{81 I_4 J_1 - 24 I_3 J_3 + 8 I_5 J_3 + 24 I_4 J_2 + 162 I_1 J_3 - 81 I_2 J_3 - 8 I_4 J_4}{-13 I_5 J_1 - 8 I_3 J_4 - 120 I_2 J_2 + 120 I_3 J_1 + 8 I_5 J_2 + 13 I_2 J_4 - 81 I_1 J_1 - 18 I_1 J_4 + 216 I_1 J_2} \right),$$

$$N_1 = -\frac{9}{198288}$$
$$\times \left(\frac{2865051 I_1 J_3 - 2375880 I_4 J_2 + 2375880 I_3 J_3 + 91877 I_4 J_4 - 91877 I_5 J_3 + 2482650 I_4 J_1 - 2482650 I_2 J_3}{81 I_4 J_1 - 24 I_3 J_3 + 8 I_5 J_3 + 24 I_4 J_2 + 162 I_1 J_3 - 81 I_2 J_3 - 8 I_4 J_4} \right),$$

$$N_2 = \frac{\sqrt{6}}{96} \left(\frac{72 I_1 J_2 - 8 I_5 J_2 + 8 I_3 J_4 - 14 I_1 J_4 - 15 I_1 J_1 - 3 I_5 J_1 + 3 I_2 J_4 - 24 I_2 J_2 + 24 I_3 J_1}{120 I_3 J_1 + 8 I_5 J_2 + 13 I_2 J_4 - 8 I_3 J_4 - 120 I_2 J_2 - 13 I_5 J_1 - 8 I_1 J_1 - 18 I_1 J_4 + 216 I_1 J_2} \right)$$
$$\times \left(\frac{-8 I_4 J_2 + 8 I_3 J_3 + 13 I_4 J_1 - 13 I_2 J_3 + 18 I_1 J_3}{2 I_1 J_2 - I_2 J_2 - I_1 J_1 + I_3 J_1} \right) - \frac{\sqrt{6}}{96} \left(\frac{8 I_4 J_2 + 14 I_1 J_3 - 8 I_3 J_3 - 3 I_2 J_3 + 3 I_4 J_1}{2 I_1 J_2 - I_2 J_2 - I_1 J_1 + I_3 J_1} \right).$$

It is easy to find

$$\det \frac{\partial(\tilde{b}_0, c_0, c_1, \tilde{c}_2)}{\partial(a_0, a_1, a_3, a_6)} = \begin{vmatrix} \frac{\sqrt{6}}{9} & 2\sqrt{6} & 3\sqrt{6} & 27\sqrt{6} \\ I_1 & I_2 & I_3 & I_5 \\ -\frac{1}{\sqrt{6}} & -\frac{2}{\sqrt{6}} & -\frac{1}{\sqrt{6}} & -\frac{3}{\sqrt{6}} \\ 0 & J_1 & J_2 & J_4 \end{vmatrix} = -9.9241853.$$

Next, we consider the expansion of the Melnikov function, $M(h)$, of (10.23) with $0 < h \ll 1$.

Introduce

$$\bar{I}_1 = \int_{-\frac{1}{2}}^{1} (y_1 - x_1 s)\, dx_1,$$

$$\bar{I}_2 = \int_{-\frac{1}{2}}^{1} \left(y_1^3 + x_1^2 y - (x_1 y_1^2 + x_1^3)s\right) dx_1,$$

$$\bar{I}_3 = \int_{-\frac{1}{2}}^{1} \left(-2x_1^3 y_1 - 2x_1 y_1^3 - (x_1^4 + y_1^4)s\right) dx_1,$$

$$\bar{I}_4 = \int_{-\frac{1}{2}}^{1} \left((4x_1^3 y_1 - 4x_1 y_1^3) - (x_1^4 - 6x_1^2 y_1^2 + y_1^4)s\right) dx_1,$$

$$\bar{I}_5 = \int_{-\frac{1}{2}}^{1} \left((x_1^4 y_1 + y_1^5 + 2x_1^2 y_1^3) - (x_1^5 + x_1 y_1^4 + 2x_1^3 y_1^2)s\right) dx_1,$$

$$\bar{J}_1 = \int_{L_{21}} \left(4(x_1^2 + y_1^2 - 1)\right) dt,$$

$$\bar{J}_2 = \int_{L_{21}} \left(2x_1^3 - 6x_1 y_1^2 - 2\right) dt,$$

$$\bar{J}_3 = \int_{L_{21}} \left(-24x_1 y_1^2 + 8x_1^3 - 8\right) dt,$$

$$\bar{J}_4 = \int_{L_{21}} \left(6y_1^4 + 12x_1^2 y_1^2 + 6x_1^4 - 6\right) dt,$$

where

$$s = -\frac{(-24x_1 - 12x_1^2 - 9 + 4\sqrt{3}\sqrt{(x_1 + 2)(1 + 2x_1)^2} + x_1\sqrt{3}\sqrt{(x_1 + 2)(1 + 2x_1)^2})\sqrt{3}}{\sqrt{-9x_1^2 - 72x_1 - 27 + 12\sqrt{3}\sqrt{(x_1 + 2)(1 + 2x_1)^2}}\sqrt{(x_1 + 2)(1 + 2x_1)^2}},$$

$$y_1 = \frac{1}{3}\sqrt{-27 + 12\sqrt{6 + 36x_1^2 + 27x_1 + 12x_1^3} - 9x_1^2 - 72x_1}.$$

Then, by Theorem 8.13, we obtain

$$M(h) = 3(c_{01} + c_{02}) - \frac{6d_1}{|\lambda_1|} h \ln|h| + \left[3(c_{21} + c_{22}) + bd_1\right]h$$
$$+ 6c_3(S_1, \delta)h^2 \ln|h| + \cdots,$$

where

$$c_{01}(\delta) = c_0(\delta),$$
$$d_1(\delta) = -3c_1(\delta),$$
$$c_{21}(\delta) = c_2(\delta),$$

$$c_{02}(\delta) = a_0 \bar{I}_1 + a_1 \bar{I}_2 + a_3 \bar{I}_3 + a_4 \bar{I}_4 + a_6 \bar{I}_5,$$
$$c_{22}(\delta) = a_1 \bar{J}_1 + a_3 \bar{J}_2 + a_4 \bar{J}_3 + a_6 \bar{J}_4.$$

Under (10.30) and (10.37), we have

$$c_{01}(\delta) = d_1(\delta) = c_{21}(\delta) = 0,$$
$$c_{02}(\delta) = N_3(b_3 - 4b_4),$$
$$c_{22}(\delta) = N_4(b_3 - 4b_4),$$

where

$$N_3 = \frac{n_3}{120 I_3 J_1 + 8 I_5 J_2 + 13 I_2 \bar{J}_4 - 8 I_3 J_4 - 120 I_2 J_2 - 13 I_5 J_1 - 81 I_1 J_1 - 18 I_1 J_4 + 216 I_1 J_2},$$

$$\begin{aligned}
n_3 = {}& 81 I_1 J_3 \bar{I}_2 + 18 I_1 J_3 \bar{I}_5 - 216 I_1 J_3 \bar{I}_3 + 81 I_4 J_1 \bar{I}_1 + 13 I_4 J_1 \bar{I}_5 - 120 I_4 J_1 \bar{I}_3 \\
& + 120 I_2 J_3 \bar{I}_3 + 8 I_4 J_4 \bar{I}_3 - 8 I_5 J_3 \bar{I}_3 - 216 I_4 J_2 \bar{I}_1 + 18 I_4 J_4 \bar{I}_1 - 81 I_2 J_3 \bar{I}_1 \\
& - 18 I_5 J_3 \bar{I}_1 + 120 I_4 J_2 \bar{I}_2 - 13 I_4 J_4 \bar{I}_2 - 13 I_2 J_3 \bar{I}_5 + 8 I_3 J_3 \bar{I}_5 - 8 I_4 J_2 \bar{I}_5 \\
& + 216 I_3 J_3 \bar{I}_1 - 120 I_3 J_3 \bar{I}_2 + 13 I_5 J_3 \bar{I}_2,
\end{aligned}$$

$$N_4 = \frac{n_4}{120 I_3 J_1 + 8 I_5 J_2 + 13 I_2 \bar{J}_4 - 8 I_3 J_4 - 120 I_2 J_2 - 13 I_5 J_1 - 81 I_1 J_1 - 18 I_1 J_4 + 216 I_1 J_2},$$

$$\begin{aligned}
n_4 = {}& 81 I_1 J_3 \bar{J}_1 - 216 I_1 J_3 \bar{J}_2 + 18 I_1 J_3 \bar{J}_4 + 13 I_4 J_1 \bar{J}_4 - 120 I_4 J_1 \bar{J}_2 \\
& - 120 I_3 J_3 \bar{J}_1 + 120 I_2 J_3 \bar{J}_2 - 13 I_4 J_4 \bar{J}_1 - 13 I_2 J_3 \bar{J}_4 + 8 I_3 J_3 \bar{J}_4 \\
& - 8 I_4 J_2 \bar{J}_4 - 8 I_5 J_3 \bar{J}_2 + 120 I_4 J_2 \bar{J}_1 + 8 I_4 J_4 \bar{J}_2 + 13 I_5 J_3 \bar{J}_1.
\end{aligned}$$

Hence, for $0 < h \ll 1$,

$$\begin{aligned}
(b_3 - {}& 4b_4)M(h) \\
& = \left(3N_3 + 3N_4 h + 6N_2 h^2 \ln|h|\right)(b_3 - 4b_4)^2 + O\left(h^2\right) \\
& = \left(-42.80665767 + 3.63797256h - 11.59214480h^2 \ln h\right)(b_3 - 4b_4)^2 + O\left(h^2\right) \\
& < 0.
\end{aligned}$$

By (10.24), we have

$$p_{1x_1} + q_{1y_1} = 2a_0 + \tilde{f}_1 a_1 + \tilde{f}_2 a_3 + 4\tilde{f}_2 a_4 + \tilde{f}_3 b_3 - 4\tilde{f}_3 b_4 + \tilde{f}_4 a_6,$$

where

$$\tilde{f}_1 = 4x_1^2 + 4y_1^2, \qquad \tilde{f}_2 = 2x_1^3 - 6x_1 y_1^2,$$
$$\tilde{f}_3 = 6x_1^2 y_1 - 2y_1^3, \qquad \tilde{f}_4 = 6x_1^4 + 12x_1^2 y_1^2 + 6y_1^4.$$

By introducing

$$\tilde{g}_0 = 2y_1, \qquad \tilde{g}_1 = 4x_1^2 y_1 + \frac{4}{3}y_1^3, \qquad \tilde{g}_2 = 2x_1^3 y_1 - 2x_1 y_1^3,$$

$$\tilde{g}_3 = 3x_1^2 y_1^2 - \frac{1}{2}y_1^4, \qquad \tilde{g}_4 = 6x_1^4 y_1 + 4x_1^2 y_1^3 + \frac{6}{5}y_1^5,$$

we obtain

$$\tilde{g} = \tilde{g}_0 a_0 + \tilde{g}_1 a_1 + \tilde{g}_2 a_3 + 4\tilde{g}_2 a_4 + \tilde{g}_3 b_3 - 4\tilde{g}_3 b_4 + \tilde{g}_4 a_6,$$

and

$$\tilde{g}_y = p_1 x_1 + q_1 y_1,$$

which yields $M(h) = \oint_{H_1 = h} \tilde{g}\, dx_1$ by (10.25).

Next, taking $h = \frac{65531}{12}$, we have

$$M\left(\frac{65531}{12}\right) = \oint_{H_1 = \frac{65531}{12}} \tilde{g}\, dx_1$$
$$= \hat{J}_0 a_0 + \hat{J}_1 a_1 + \hat{J}_2 a_3 + 4\hat{J}_2 a_4 + \hat{J}_3 b_3 - 4\hat{J}_3 b_4 + \hat{J}_4 a_6,$$

where

$$\hat{J}_i = \oint_{H_1 = \frac{65531}{12}} \tilde{g}_i\, dx_1, \quad i = 0, \ldots, 4.$$

Then it follows from (10.30) and (10.37) that

$$(b_3 - 4b_4)M\left(\frac{65531}{12}\right) = -8.370791584 \times 10^6 (b_3 - 4b_4)^2 < 0.$$

Similarly, taking $h = \frac{240005}{12}$, we get

$$(b_3 - 4b_4)M\left(\frac{240005}{12}\right) = -2.407368277 \times 10^8 (b_3 - 4b_4)^2 < 0,$$

which implies that there probably exist no zeros of M for $h > 0$ under (10.30) and (10.37).

Taking $a_4 = 1$ and $(b_3, b_4) = (1, \frac{1}{2})$, it is easy to obtain from (10.30) and (10.37) that

$$(a_0, a_1, a_3, a_6) = (22.14620575, -8.329054789, -14.59838107, 1.703428246).$$

Thus, for this case, we have $c_3(\delta)\bar{b}_1 = N_0 N_2 (b_3 - 4b_4)^2 = 4.350438661(b_3 - 4b_4)^2 > 0$. Then, by Theorem 9.1, we conclude that there exist 5 limit cycles inside L_0. Since system (10.23) is $(2\pi/3)$-equivariant, we have the following theorem.

Fig. 10.2 The distribution of
15 limit cycles of (10.23)

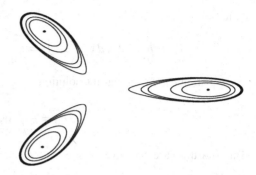

Theorem 10.11 ([159]) *System (10.23) can have* 15 *limit cycles, with the distribution shown in Fig.* 10.2.

Next we consider some special cases of (7.1). First, consider the following cubic $(2\pi/3)$-equivariant near-Hamiltonian system:

$$\dot{x}_1 = 3y_1 + 8x_1y_1 + x_1^2y_1 + y_1^3 + \varepsilon p_3(x_1, y_1),$$
$$\dot{y}_1 = -3x_1 + 4x_1^2 - 4y_1^2 - x_1^3 - x_1y_1^2 + \varepsilon q_3(x_1, y_1),$$
$$(10.38)$$

where p_3 and q_3 satisfy

$$p_3 = a_0x_1 - b_0y_1 + a_2x_1^2 + 2b_2x_1y_1 - a_2y_1^2 + a_1x_1^3$$
$$\quad - b_1x_1^2y_1 + a_1x_1y_1^2 - b_1y_1^3,$$
$$q_3 = b_0x_1 + a_0y_1 + b_2x_1^2 - 2a_2x_1y_1 - b_2y_1^2 + b_1x_1^3$$
$$\quad + a_1x_1^2y_1 + b_1x_1y_1^2 + a_1y_1^3.$$
$$(10.39)$$

For $\varepsilon = 0$, the Hamiltonian function of (10.38) is

$$H_1(x_1, y_1) = \frac{3}{2}(x_1^2 + y_1^2) - \frac{4}{3}x_1^3 + 4x_1y_1^2 + \frac{1}{4}(x_1^4 + y_1^4) + \frac{1}{2}x_1^2y_1^2 - \frac{5}{12}.$$

By (10.39), we have

$$p_{3x_1} + q_{3y_1} = 2a_0 + 4a_1(x_1^2 + y_1^2), \qquad \bar{g} = 2a_0y_1 + \left(4x_1^2y_1 + \frac{4}{3}y_1^3\right)a_1.$$

As before, for $0 < -h \ll \frac{8}{3}$,

$$M_1(h) = c_{01} - \frac{d_1}{|\lambda_1|}h\ln|h| + O(h),$$

where

$$c_{01}(\delta) = I_1a_0 + I_2a_1, \qquad -\frac{d_1}{|\lambda_1|} = -\frac{2a_0 + 4a_1}{3},$$

with

$$I_1 = \oint_{H_1=0} 2y_1\, dx_1, \qquad I_2 = \oint_{H_1=0} \left(4x_1^2 y_1 + \frac{4}{3}y_1^3\right) dx_1.$$

The equation $c_{01}(\delta) = 0$ has the solution

$$a_0 = -\frac{I_2}{I_1}a_1. \tag{10.40}$$

Thus, for this case, we have

$$-\frac{d_1}{|\lambda_1|} = -\frac{2a_0 + 4a_1}{3} = \frac{2I_2 - 4I_1}{3I_1}a_1.$$

Hence, for $a_1 \neq 0$ and a_0 near $-\frac{I_2}{I_1}a_1$ the system can have 1 limit cycle near L_1.

Next, we discuss the properties of $M(h)$ with $0 < h \ll 1$. By Theorem 8.13,

$$M(h) = 3(c_{01} + c_{02}) - \frac{6d_1}{|\lambda_1|}h \ln|h| + O(h),$$

where

$$c_{02}(\delta) = \int_{L_{21}} q_3\, dx_1 - p_3\, dy_1,$$

$$d_1(\delta) = (p_{3x_1} + q_{3y_1})(S_1, 0, \delta) = 2a_0 + 4a_1.$$

Since $a_0 = -\frac{I_2}{I_1}a_1$, we obtain

$$c_{01}(\delta) = 0,$$

$$c_{02}(\delta) = \left(\bar{I}_2 - \frac{\bar{I}_1}{I_1}I_2\right)a_1 = \hat{N}_0 a_1,$$

$$d_1(\delta) = \left(4 - 2\frac{I_2}{I_1}\right)a_1 = \hat{N}_1 a_1.$$

Therefore, with the condition (10.40) and for $0 < h \ll 1$ we have

$$M(h) = (3\hat{N}_0 + 2\hat{N}_1 h \ln h)a_1 + O(h)$$

$$= (-30.81936396 + 39.34098556 h \ln|h|)a_1 + O(h).$$

It then follows that $a_1 M(h) < 0$ for $0 < h \ll 1$.

Further, taking $h = \frac{65531}{12}$, we have

$$M\left(\frac{65531}{12}\right) = \oint_{H_1 = \frac{65531}{12}} \bar{g}\, dx_1.$$

Fig. 10.3 The distribution of
4 limit cycles

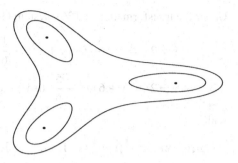

By introducing

$$\hat{J}_0 = \oint_{H_1 = \frac{65531}{12}} 2y_1 \, dx_1, \qquad \hat{J}_1 = \oint_{H_1 = \frac{65531}{12}} \left(4x_1^2 y_1 + \frac{4}{3} y_1^3 \right) dx_1,$$

we obtain

$$a_1 M\left(\frac{65531}{12} \right) = (\hat{J}_0 a_0 + \hat{J}_1 a_1) a_1 = \left(\hat{J}_1 - \hat{J}_0 \frac{I_2}{I_1} \right) a_1^2$$

$$= (1.285388534 \times 10^5) a_1^2 > 0.$$

Therefore, there exists $\hat{h} \in (0, \frac{65531}{12})$ such that $M(\hat{h}) = 0$. Hence, we have proved the following theorem.

Theorem 10.12 ([159]) *The cubic system* (10.38) *can have 4 limit cycles, with the distribution as shown in Fig.* 10.3.

Now we consider the following $(2\pi/3)$-equivariant polynomial near-Hamiltonian system of degree 4:

$$\dot{x}_1 = 3y_1 + 8x_1 y_1 + x_1^2 y_1 + y_1^3 + \varepsilon p_4(x_1, y_1),$$
$$\dot{y}_1 = -3x_1 + 4x_1^2 - 4y_1^2 - x_1^3 - x_1 y_1^2 + \varepsilon q_4(x_1, y_1), \tag{10.41}$$

where

$$p_4 = a_0 x_1 - b_0 y_1 + a_2 x_1^2 + 2b_2 x_1 y_1 - a_2 y_1^2 + a_1 x_1^3 - b_1 x_1^2 y_1$$
$$+ a_1 x_1 y_1^2 - b_1 y_1^3 + (a_3 + a_4) x_1^4 + (2b_3 - 4b_4) x_1^3 y_1 - 6a_4 x_1^2 y_1^2$$
$$+ (2b_3 + 4b_4) x_1 y_1^3 - (a_3 - a_4) y_1^4,$$

$$q_4 = b_0 x_1 + a_0 y_1 + b_2 x_1^2 - 2a_2 x_1 y_1 - b_2 y_1^2 + b_1 x_1^3 + a_1 x_1^2 y_1 + b_1 x_1 y_1^2$$
$$+ a_1 y_1^3 + (b_3 + b_4) x_1^4 - (2a_3 - 4a_4) x_1^3 y_1 - 6b_4 x_1^2 y_1^2$$
$$- (2a_3 + 4a_4) x_1 y_1^3 - (b_3 - b_4) y_1^4. \tag{10.42}$$

Under the transformation (10.26), system (10.41) becomes

$$\dot{x} = 2 \cdot 6^{\frac{1}{2}} y + \frac{5}{3} \cdot 6^{\frac{3}{4}} xy + x^2 y + \frac{1}{6} y^3 + \varepsilon \tilde{p}_4(x, y),$$

$$\dot{y} = 2 \cdot 6^{\frac{1}{2}} x + 6^{\frac{3}{4}} x^2 - \frac{5}{6} \cdot 6^{\frac{3}{4}} y^2 - 6x^3 - xy^2 + \varepsilon \tilde{q}_4(x, y),$$
(10.43)

with

$$\tilde{p}_4(x, y) = 6^{-\frac{1}{4}} p_4\left(6^{\frac{1}{4}} x + 1, y\right), \qquad \tilde{q}_4(x, y) = 6^{\frac{1}{4}} q_4\left(6^{\frac{1}{4}} x + 1, y\right).$$
(10.44)

By using (10.44) and (10.42), we have

$$\tilde{p}_{4x} + \tilde{q}_{4y} = 2a_0 + f_1 a_1 + f_2 a_3 + 4 f_2 a_4 + f_3 b_3 - 4 f_3 b_4,$$

where

$$f_1 = 4\sqrt{6} x^2 + \frac{2}{3}\sqrt{6} y^2 + 8 \cdot 6^{\frac{1}{4}} x + 4,$$

$$f_2 = 2 \cdot 6^{\frac{3}{4}} x^3 - 6^{\frac{3}{4}} xy^2 + 6\sqrt{6} x^2 - \sqrt{6} y^2 + 6 \cdot 6^{\frac{1}{4}} x + 2,$$

$$f_3 = 6 \cdot 6^{\frac{1}{4}} x^2 y - \frac{1}{3} 6^{\frac{1}{4}} y^3 + 12xy + 6^{\frac{3}{4}} y.$$

Let

$$g = g_1 a_0 + g_2 a_2 + g_3 a_3 + g_4 b_3 - 4 g_4 b_4,$$

where

$$g_1 = 2y,$$

$$g_2 = 4\sqrt{6} x^2 y + \frac{2}{9}\sqrt{6} y^3 + 8 \cdot 6^{\frac{1}{4}} xy + 4y,$$

$$g_3 = 2 \cdot 6^{\frac{3}{4}} x^3 y - \frac{1}{3} 6^{\frac{3}{4}} xy^3 + 6\sqrt{6} x^2 y - \frac{1}{3}\sqrt{6} y^3 + 6 \cdot 6^{\frac{1}{4}} xy + 2y,$$

$$g_4 = 3 \cdot 6^{\frac{1}{4}} x^2 y^2 - \frac{1}{12} 6^{\frac{1}{4}} y^4 + 6xy^2 + \frac{1}{2} 6^{\frac{3}{4}} y^2.$$

Then $g_y = \tilde{p}_{4x} + \tilde{q}_{4y}$. Using the notation

$$I_1 = \oint_{L_0} g_1 \, dx, \qquad I_2 = \oint_{L_0} g_2 \, dx, \qquad I_3 = \oint_{L_0} g_3 \, dx, \qquad I_4 = \oint_{L_0} g_4 \, dx,$$

$$J_1 = \oint_{L_0} (f_1 - 4) \, dt, \qquad J_2 = \oint_{L_0} (f_2 - 2) \, dt, \qquad J_3 = \oint_{L_0} f_3 \, dt,$$

then as before, for $0 < -h \ll \frac{8}{3}$, we obtain

$$M_1(h) = c_0(\delta) + c_1(\delta) h \ln |h| + c_2(\delta) h + c_3(\delta) h^2 \ln h + O(h^2),$$

where

$$c_0(\delta) = M(0)$$
$$= I_1 a_0 + I_2 a_1 + I_3 a_3 + 4 I_3 a_4 + I_4 b_3 - 4 I_4 b_4,$$
$$c_1(\delta) = -\frac{a_{10} + b_{01}}{|\lambda|} = -\frac{2a_0 + 4a_1 + 2a_3 + 8a_4}{2\sqrt{6}},$$
$$\bar{c}_2(\delta) = \oint_{L_0} (p_x + q_y - a_{10} - b_{01})\, dt$$
$$= J_1 a_1 + J_2 a_3 + 4 J_2 a_4 + J_3 b_3 - 4 J_3 b_4.$$

The equations

$$c_0(\delta) = 0, \qquad c_1(\delta) = 0, \qquad \bar{c}_2(\delta) = 0$$

have the solutions

$$a_0 = -\left(\frac{2I_3 J_3 + I_4 J_1 - 2I_4 J_2 - I_2 J_3}{I_2 J_2 - 2I_1 J_2 + I_1 J_1 - I_3 J_1}\right) b_3$$
$$+ 4\left(\frac{2I_3 J_3 + I_4 J_1 - 2I_4 J_2 - I_2 J_3}{I_2 J_2 - 2I_1 J_2 + I_1 J_1 - I_3 J_1}\right) b_4,$$

$$a_1 = -\left(\frac{I_4 J_2 + I_1 J_3 - I_3 J_3}{I_2 J_2 - 2I_1 J_2 + I_1 J_1 - I_3 J_1}\right) b_3$$
$$+ 4\left(\frac{I_4 J_2 + I_1 J_3 - I_3 J_3}{I_2 J_2 - 2I_1 J_2 + I_1 J_1 - I_3 J_1}\right) b_4, \qquad (10.45)$$

$$a_3 = -4a_4 + \left(\frac{I_4 J_1 + 2I_1 J_3 - I_2 J_3}{I_2 J_2 - 2I_1 J_2 + I_1 J_1 - I_3 J_1}\right) b_3$$
$$- 4\left(\frac{I_4 J_1 + 2I_1 J_3 - I_2 J_3}{I_2 J_2 - 2I_1 J_2 + I_1 J_1 - I_3 J_1}\right) b_4,$$

which implies that

$$c_3(\delta) = -\frac{\sqrt{6}}{12} a_1 + \frac{\sqrt{6}}{32} a_3 + \frac{\sqrt{6}}{8} a_4 = \tilde{N}_2(b_3 - 4b_4), \qquad (10.46)$$

where

$$\tilde{N}_2 = \frac{\sqrt{6}}{96}\left(\frac{-8I_3 J_3 + 8I_4 J_2 + 14I_1 J_3 - 3I_2 J_3 + 3I_4 J_1}{I_2 J_2 - 2I_1 J_2 + I_1 J_1 - I_3 J_1}\right).$$

One can easily show

$$\det \frac{\partial(c_0, c_1, \bar{c}_2)}{\partial(a_0, a_1, a_3)} = \begin{vmatrix} I_1 & I_2 & I_3 \\ -\frac{1}{\sqrt{6}} & -\frac{2}{\sqrt{6}} & -\frac{1}{\sqrt{6}} \\ 0 & J_1 & J_2 \end{vmatrix} = -22.73946152.$$

Hence, by Theorem 8.7, there can exist 3 limit cycles near L_0.

To see if more limit cycles exist we need to consider the expansion of $M(h)$ near the center C_1'. As before, we use the transformation (10.30) and the time rescaling $\tau \to 6\sqrt{6}t$, so that (10.41) becomes

$$\frac{du}{d\tau} = v + \frac{7}{18}uv + \frac{1}{36}u^2v + \frac{1}{216}v^3 + \varepsilon\bar{p}_4,$$
$$\frac{dv}{d\tau} = -u - \frac{5}{6}u^2 - \frac{7}{36}v^2 - \frac{1}{6}u^3 - \frac{1}{36}uv^2 + \varepsilon\bar{q}_4,$$

(10.47)

where

$$\bar{p}_4(u, v) = \frac{1}{6\sqrt{6}}p_4\big(6^{-\frac{1}{4}}(u+2), 6^{-\frac{1}{4}}v\big),$$
$$\bar{q}_4(u, v) = \frac{1}{6\sqrt{6}}q_4\big(6^{-\frac{1}{4}}(u+2), 6^{-\frac{1}{4}}v\big).$$

(10.48)

Then we can write

$$M_1(h) = 6M_1^*\left(\frac{1}{6}h\right),$$

where $M_1^*(h)$ is the Melnikov function of (10.47).

It then follows from (10.48), (10.43) and (10.42) that

$$\begin{aligned}
\bar{p}_{4u} + \bar{q}_{4v} &= \frac{\sqrt{6}}{18}(a_0 + 18a_1 + 27a_3 + 108a_4) + \frac{\sqrt{6}}{6}(4a_1 + 9a_3 + 18a_4)u \\
&+ \left(\frac{3}{2}b_3 - 6b_4\right)v + \frac{\sqrt{6}}{18}(2a_1 + 9a_3 + 18a_4)u^2 + (b_3 - 4b_4)uv \\
&+ \sqrt{6}\left(\frac{1}{54}a_1 - \frac{1}{12}a_3 - \frac{1}{3}a_4\right)v^2 + \frac{\sqrt{6}}{18}(a_3 + 4a_4)u^3 \\
&+ \frac{1}{6}(b_3 - 4b_4)u^2v + \frac{\sqrt{6}}{36}(-a_3 - 4a_4)uv^2 + \frac{1}{108}(-b_3 + 4b_4)v^3.
\end{aligned}$$

Then using (7.42), we obtain

$$\tilde{b}_0 = \frac{\sqrt{6}}{9}\pi(a_0 + 18a_1 + 27a_3 + 108a_4).$$

Thus, under (10.45) the following holds:

$$\tilde{b}_0 = \tilde{N}_3(b_3 - 4b_4),$$

where

$$\tilde{N}_3 = \frac{\sqrt{6}}{9}\pi\left(\frac{8I_3J_3 - 8I_4J_2 + 13I_4J_1 - 13I_2J_3 + 18I_1J_3}{I_2J_2 - 2I_1J_2 + I_1J_1 - I_3J_1}\right).$$

It is obvious that

$$c_3(\delta)\tilde{b}_0 = \tilde{N}_2\tilde{N}_3(b_3 - 4b_4)^2 = -11.10696874(b_3 - 4b_4)^2 < 0.$$

This implies that $M(h)$ has the same sign near the endpoints of the interval $(-\frac{8}{3}, 0)$. Hence, we cannot find $\tilde{h} \in (-\frac{8}{3}, 0)$ such that $M(\tilde{h}) = 0$.

Next, we consider the expansion of the Melnikov function $M(h)$ of (10.41) with $0 < h \ll 1$. By introducing

$$\bar{I}_1 = \int_{-\frac{1}{2}}^{1} (y_1 - x_1 s) \, dx_1,$$

$$\bar{I}_2 = \int_{-\frac{1}{2}}^{1} (y_1^3 + x_1^2 y - (x_1 y_1^2 + x_1^3)s) \, dx_1,$$

$$\bar{I}_3 = \int_{-\frac{1}{2}}^{1} (-2x_1^3 y_1 - 2x_1 y_1^3 - (x_1^4 + y_1^4)s) \, dx_1,$$

$$\bar{I}_4 = \int_{-\frac{1}{2}}^{1} (4x_1^3 y_1 - 4x_1 y_1^3 - (x_1^4 - 6x_1^2 y_1^2 + y_1^4)s) \, dx_1,$$

$$\bar{J}_1 = \int_{L_{21}} (4(x_1^2 + y_1^2 - 1)) \, dt,$$

$$\bar{J}_2 = \int_{L_{21}} (2x_1^3 - 6x_1 y_1^2 - 2) \, dt,$$

$$\bar{J}_3 = \int_{L_{21}} (-24x_1 y_1^2 + 8x_1^3 - 8) \, dt,$$

we have

$$M(h) = 3(c_{01} + c_{02}) - \frac{6d_1}{|\lambda_1|} h \ln |h|$$
$$+ [3(c_{21} + c_{22}) + bd_1]h + 6c_3(S_1, \delta)h^2 \ln |h| + \cdots,$$

where

$$c_{01}(\delta) = c_0(\delta),$$

$$d_1(\delta) = -3c_1(\delta),$$

$$c_{21}(\delta) = c_2(\delta),$$

$$c_{02}(\delta) = a_0 \bar{I}_1 + a_1 \bar{I}_2 + a_3 \bar{I}_3 + a_4 \bar{I}_4,$$

$$c_{22}(\delta) = a_1 \bar{J}_1 + a_3 \bar{J}_2 + a_4 \bar{J}_3,$$

which in turn, by using (10.45), yields

$$c_{01}(\delta) = d_1(\delta) = c_{21}(\delta) = 0,$$

$$c_{02}(\delta) = \tilde{N}_4(b_3 - 4b_4),$$

$$c_{22}(\delta) = \tilde{N}_5(b_3 - 4b_4),$$

where

$$\tilde{N}_4 = \left(\frac{-2I_3 J_3 \bar{I}_1 + I_2 J_3 \bar{I}_1 + 2I_4 J_2 \bar{I}_1 - I_4 J_1 \bar{I}_1 - I_4 J_2 \bar{I}_2 - I_1 J_3 \bar{I}_2 + I_3 J_3 \bar{I}_2 + I_4 J_1 \bar{I}_3 - I_2 J_3 \bar{I}_3 + 2I_1 J_3 \bar{I}_3}{I_2 J_2 - 2I_1 J_2 + I_1 J_1 - I_3 J_1} \right),$$

$$\tilde{N}_5 = \left(\frac{-I_4 J_2 \bar{J}_1 - I_1 J_3 \bar{J}_1 + I_3 J_3 \bar{J}_1 + I_4 J_1 \bar{J}_2 - I_2 J_3 \bar{J}_2 + 2I_1 J_3 \bar{J}_2}{I_2 J_2 - 2I_1 J_2 + I_1 J_1 - I_3 J_1} \right).$$

Therefore,

$$M(h) = \left(3\tilde{N}_4 + 3\tilde{N}_5 h + 6\tilde{N}_2 h^2 \ln h \right)(b_3 - 4b_4) + O\left(h^2\right) \quad \text{for } 0 < h \ll 1.$$

Further, noting

$$\tilde{N}_2 = 0.7236056092, \qquad \tilde{N}_4 = 2.180188697, \qquad \tilde{N}_5 = 4.461867637,$$

we have

$$(b_3 - 4b_4)M(h)$$
$$= \left(6.540566091 + 13.38560291h + 4.341633655h^2 \ln h \right)(b_3 - 4b_4)^2 > 0.$$

Then, by (10.42), we obtain

$$p_{4x_1} + q_{4y_1} = 2a_0 + \tilde{f}_1 a_1 + \tilde{f}_2 a_3 + 4\tilde{f}_2 a_4 + \tilde{f}_3 b_3 - 4\tilde{f}_3 b_4,$$

where

$$\tilde{f}_1 = 4x_1^2 + 4y_1^2, \qquad \tilde{f}_2 = 2x_1^3 - 6x_1 y_1^2, \qquad \tilde{f}_3 = 6x_1^2 y_1 - 2y_1^3.$$

Again by introducing

$$\tilde{g}_0 = 2y_1, \qquad \tilde{g}_1 = 4x_1^2 y_1 + \frac{4}{3} y_1^3, \qquad \tilde{g}_2 = 2x_1^3 y_1 - 2x_1 y_1^3,$$

$$\tilde{g}_3 = 3x_1^2 y_1^2 - \frac{1}{2} y_1^4,$$

and

$$\tilde{g} = \tilde{g}_0 a_0 + \tilde{g}_1 a_1 + \tilde{g}_2 a_3 + 4\tilde{g}_2 a_4 + \tilde{g}_3 b_3 - 4\tilde{g}_3 b_4,$$

we have $\tilde{g}_y = p_{4x_1} + q_{4y_1}$, which yields $M(h) = \oint_{H_1 = h} \tilde{g} \, dx_1$.

Now, taking $h = \frac{65531}{12}$, we have

$$M\left(\frac{65531}{12} \right) = \oint_{H_1 = \frac{65531}{12}} \tilde{g} \, dx_1 = \hat{J}_0 a_0 + \hat{J}_1 a_1 + \hat{J}_2 a_3 + 4\hat{J}_2 a_4 + \hat{J}_3 b_3 - 4\hat{J}_3 b_4,$$

Fig. 10.4 The distribution of
10 limit cycles

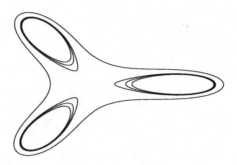

where

$$\hat{J}_i = \oint_{H_1 = \frac{65531}{12}} \tilde{g}_i \, dx_1, \quad i = 0, \ldots, 3.$$

Then, it follows from (10.45) that

$$(b_3 - 4b_4) M\left(\frac{65531}{12}\right) = \tilde{N}_6 (b_3 - 4b_4)^2,$$

where

$$\tilde{N}_6 = \frac{n_6}{I_2 J_2 - 2 I_1 J_2 + I_1 J_1 - I_3 J_1}, \quad \text{and}$$

$$n_6 = -2\hat{J}_0 I_3 J_3 - \hat{J}_0 I_4 J_1 + 2\hat{J}_0 I_2 J_3 - \hat{J}_2 I_2 J_3 + \hat{J}_2 I_4 J_1 + 2\hat{J}_2 I_1 J_3$$

$$+ \hat{J}_3 I_2 J_2 - 2\hat{J}_3 I_1 J_2 + \hat{J}_3 I_1 J_1 - \hat{J}_3 I_3 J_1 - \hat{J}_1 I_4 J_2 - \hat{J}_1 I_1 J_3 + \hat{J}_1 I_3 J_3.$$

With the aid of Maple, we finally obtain

$$(b_3 - 4b_4) M\left(\frac{65531}{12}\right) = -1.695726443 \times 10^6 (b_3 - 4b_4)^2 < 0,$$

which implies $M(\tilde{h}) = 0$ for some $\tilde{h} \in (0, \frac{65531}{12})$. Summarizing the above results gives the following theorem.

Theorem 10.13 ([159]) *System* (10.41) *can have* 10 *limit cycles, with the distribution shown in Fig.* 10.4.

For more results on equivariant systems, see [89, 95–97, 128, 131, 134–136, 138, 139, 204, 205, 209, 210, 243, 257, 259].

References

1. Abed, E.H., Fu, J.H.: Local feedback stabilization and bifurcation control, II. Stationary bifurcation. Syst. Control Lett. **8**, 467–473 (1987)
2. Algaba, A., Freire, E., Gamero, E.: Hypernormal form for the Hopf-zero bifurcation. Int. J. Bifurc. Chaos **8**, 1857–1887 (1998)
3. Algaba, A., Freire, E., Gamero, E., Garcia, C.: Dynamical analysis, quasi-homogeneous normal forms. J. Comput. Appl. Math. **150**, 193–216 (2003)
4. Doedel, E.J.: AUTO-07P: Continuation and Bifurcation Software for Ordinary Differential Equations. Concordia University, Montreal, Canada (August 2007)
5. Arnold, V.I.: Lectures on bifurcations in versal families. Russ. Math. Surv. **27**, 54–123 (1972)
6. Arnold, V.I.: Loss of stability of self-oscillations close to resonance and versal deformations of equivariant vector fields. Funct. Anal. Appl. **11**, 85–92 (1977)
7. Arnold, V.I.: Geometric Methods in the Theory of Ordinary Differential Equations. Springer, New York (1983)
8. Ashkenazi, M., Chow, S.N.: Normal forms near critical points for differential equations and maps. IEEE Trans. Circuits Syst. **35**, 850–862 (1988)
9. Atherton, D.P.: Stability of Nonlinear Systems. Research Studies Press, New York (1981)
10. Baider, A.: Unique normal forms for vector fields and Hamiltonians. J. Differ. Equ. **78**, 33–52 (1989)
11. Baider, A., Churchill, R.: Unique normal forms for planar vector fields. Math. Z. **199**, 303–310 (1988)
12. Baider, A., Sanders, J.A.: Unique normal forms: the nilpotent Hamiltonian case. J. Differ. Equ. **92**, 282–304 (1991)
13. Baider, A., Sanders, J.A.: Further reduction of the Takens–Bogdanov normal forms. J. Differ. Equ. **99**, 205–244 (1992)
14. Bautin, N.N.: On the number of limit cycles which appear with the variation of coefficients from an equilibrium position of focus or center type. Mat. Sb. (N. S.) **30**(72), 181–196 (1952)
15. Belitskii, G.: Invariant normal forms of formal series. Funct. Anal. Appl. **13**, 46–47 (1979)
16. Belitskii, G.: Normal forms relative to a filtering action of a group. Trans. Mosc. Math. Soc. **2**, 1–39 (1981)
17. Berns, D., Moiola, J.L., Chen, G.R.: Controlling oscillation amplitudes via feedback. Int. J. Bifurc. Chaos **10**, 2815–2822 (2000)
18. Beyn, W.J., Champneys, A., Doedel, E., Govaerts, W., Kuznetsov, Yu.A., Sandstede, B.: Numerical continuation, and computation of normal forms. In: Fiedler, B. (ed.) Handbook of Dynamical Systems, vol. 2, pp. 149–219. North-Holland, Amsterdam (2002)
19. Bhatia, N.P., Szegö, G.P.: Stability Theory of Dynamical Systems. Springer, Berlin (1970)
20. Bi, Q., Yu, P.: Computation of normal forms of differential equations associated with non-semisimple zero eigenvalues. Int. J. Bifurc. Chaos **8**(12), 2279–2319 (1998)

M. Han, P. Yu, *Normal Forms, Melnikov Functions and Bifurcations of Limit Cycles*,
Applied Mathematical Sciences 181,
DOI 10.1007/978-1-4471-2918-9, © Springer-Verlag London Limited 2012

21. Bi, Q., Yu, P.: Symbolic computation of normal forms for semi-simple cases. J. Comput. Appl. Math. **102**, 195–220 (1999)
22. Bogdanov, R.I.: Versal deformations of a singular point on the plane in the case of zero eigenvalues. Funct. Anal. Appl. **9**, 144–145 (1975)
23. Broer, H.W., Chow, S.-N., Kim, Y.I., Vegter, G.: The Hamiltonian double-zero eigenvalues. Fields Inst. Commun. **4**, 1–20 (1995)
24. Binyamini, B., Novikov, D., Yakovenko, S.: On the number of zeros of Abelian integrals: a constructive solution of the infinitesimal Hilbert 16th problem. arXiv:0808.2952v3, 1–59 (2009)
25. Birkhoff, G.D.: Dynamical Systems. AMS, Providence (1927)
26. Campbell, S.A., Stone, E.: Analysis of the chatter instability in a nonlinear model for drilling. ASME J. Comput. Nonlinear Dyn. **1**, 294–306 (2006)
27. Carr, J.: Applications of Center Manifold Theory. Springer, New York (1981)
28. Caubergh, M., Dumortier, F.: Hopf–Takens bifurcations and centers. J. Differ. Equ. **202**(1), 1–31 (2004)
29. Caubergh, M., Dumortier, F.: Hilbert's 16th problem for classical Liénard equations of even degree. J. Differ. Equ. **244**(6), 1359–1394 (2008)
30. Caubergh, M., Dumortier, F., Luca, S.: Cyclicity of unbounded semihyperbolic 2-saddle cycles in polynomial Liénard systems. Discrete Contin. Dyn. Syst. **27**(3), 963–980 (2010)
31. Caubergh, M., Françoise, J.P.: Generalized Liénard equations, cyclicity and Hopf–Takens bifurcations. Qual. Theory Dyn. Syst. **5**(2), 195–222 (2004)
32. Caubergh, M., Roussarie, R.: Relations between Abelian integrals and limit cycles. In: Normal Forms, Bifurcations and Finiteness Problems in Differential Equations, pp. 1–32. Kluwer Academic, Dordrecht (2004)
33. Chen, G.R.: Controlling Chaos and Bifurcation in Engineering and Systems. CRC Press, Boca Raton (2000)
34. Chen, G.R., Dong, X.: From Chaos to Order. World Scientific, Singapore (1998)
35. Chen, H.B., Liu, Y.R.: Linear recursion formulas of quantities of singular point and applications. Appl. Math. Comput. **148**, 163–171 (2004)
36. Chen, H.B., Liu, Y., Yu, P.: Center and isochronous center at infinity in a class of planar systems. Dyn. Contin. Discrete Impuls. Syst., Ser. B, Appl. Algorithms **15**(1), 57–74 (2008)
37. Chen, G.R., Moiola, J.L., Wang, H.O.: Bifurcation control: theories, methods, and applications. Int. J. Bifurc. Chaos **10**, 511–548 (2000)
38. Chen, L.S., Wang, M.S.: The relative position, and the number, of limit cycles of a quadratic differential system. Acta Math. Sin. **22**, 751–758 (1979)
39. Chen, D., Wang, H.O., Chen, G.: Anti-control of Hopf bifurcation. IEEE Trans. Circuits Syst. I **48**, 661–672 (2001)
40. Chen, G., Wu, Y., Yang, X.: The number of limit cycles for a class of quintic Hamiltonian systems under quintic perturbations. J. Aust. Math. Soc. **73**, 37–53 (2002)
41. Chen, Z., Yu, P.: Hopf bifurcation control for an Internet congestion model. Int. J. Bifurc. Chaos **15**(8), 2643–2651 (2005)
42. Chen, Z., Yu, P.: Controlling and anti-controlling Hopf bifurcations in discrete maps using polynomial functions. Chaos Solitons Fractals **26**(4), 1231–1248 (2005)
43. Chiang, H.D., Conneen, T.P., Flueck, A.J.: Bifurcation and chaos in electric power systems. J. Franklin Inst. B **331**, 1001–1036 (1994)
44. Chicone, C., Jacobs, M.: Bifurcation of critical periods for plane vector fields. Trans. Am. Math. Soc. **312**, 433–486 (1989)
45. Chow, S.N., Drachman, B., Wang, D.: Computation of normal forms. J. Comput. Appl. Math. **29**, 1290–1430 (1990)
46. Chow, S.N., Hale, J.K.: Methods of Bifurcation Theory. Springer, New York (1982)
47. Chow, S.-N., Li, C.-C., Wang, D.: Normal Forms and Bifurcation of Planar Vector Fields. Cambridge University Press, Cambridge (1994)
48. Chow, S.-N., Li, C., Yi, Y.: The cyclicity of period annulus of degenerate quadratic Hamiltonian system with elliptic segment. Ergod. Theory Dyn. Syst. **22**(4), 1233–1261 (2002)

49. Christopher, C.J., Li, C.: Limit cycles of differential equations. In: Advanced Courses in Mathematical CRM Barcelona. Birkhäuser, Basel (2007)

50. Christopher, C.J., Lloyd, N.G.: Polynomial systems: a lower bound for the Hilbert numbers. Proc. R. Soc. Lond. A **450**, 219–224 (1995)

51. Christopher, C.J., Lloyd, N.G.: Small-amplitude limit cycles in polynomial Liénard systems. Nonlinear Differ. Equ. Appl. **3**(2), 183–190 (1996)

52. Christopher, C.J., Lynch, S.: Small-amplitude limit cycle bifurcations for Liénard systems with quadratic or cubic damping or restoring forces. Nonlinearity **12**(4), 1099–1112 (1999)

53. Chua, L.O., Desoer, C.A., Kuh, E.S.: Linear and Nonlinear Circuits. McGraw-Hill, New York (1987)

54. Chua, L.O., Kokubu, H.: Normal forms for nonlinear vector fields—Part I: Theory and algorithm. IEEE Trans. Circuits Syst. **35**, 863–880 (1988)

55. Chua, L.O., Kokubu, H.: Normal forms for nonlinear vector fields—Part II: Applications. IEEE Trans. Circuits Syst. **36**, 51–70 (1988)

56. Dhooge, A., Govaerts, W., Kuznetsov, Yu.A.: MATCONT: A MATLAB package for numerical bifurcation analysis of ODEs. ACM Trans. Math. Softw. **29**, 141–164 (2003)

57. Dhooge, A., Govaerts, W., Kuznetsov, Yu.A., Meijer, H.G.E., Sautois, B.: New features of the software MatCont for bifurcation analysis of dynamical systems. Math. Comput. Model. Dyn. Syst. **14**, 145–175 (2008)

58. Dumortier, F.: Compactification and desingularization of spaces of polynomial Liénard equations. J. Differ. Equ. **224**(2), 296–313 (2006)

59. Dumortier, F., Li, C.: Quadratic Liénard equations with quadratic damping. J. Differ. Equ. **139**(1), 41–59 (1997)

60. Dumortier, F., Llibre, J., Artés, J.: Qualitative Theory of Planar Differential Systems Springer, Berlin (2006)

61. Dumortier, F., Panazzolo, D., Roussarie, R.: More limit cycles than expected in Liénard equations. Proc. Am. Math. Soc. **135**, 1895–1904 (2007)

62. Dumortier, F., Roussarie, R.: Abelian integrals and limit cycles. J. Differ. Equ. **227**, 116–165 (2006)

63. Elphick, C., Tirapegui, E., Brachet, M.E., Coullet, P., Iooss, G.: A simple global characterization for normal forms of singular vector fields. Physica D **29**, 95–127 (1987)

64. Farr, W.W., Li, C.Z., Labouriau, I.S., Langford, W.F.: Degenerate Hopf bifurcation formulas and Hilbert's 16th problem. SIAM J. Math. Anal. **20**, 13–30 (1989)

65. Françoise, J.P.: Successive derivatives of the first return map, application to the study of quadratic vector fields. Ergod. Theory Dyn. Syst. **16**(1), 87–96 (1996)

66. Gasull, A., Torregrosa, J.: Small-amplitude limit cycles in Liénard systems via multiplicity. J. Differ. Equ. **159**(1), 186–211 (1999)

67. Gasull, A., Torregrosa, J.: A new approach to the computation of the Lyapunov constants. The geometry of differential equations and dynamical systems. Comput. Appl. Math. **20**(1–2), 149–177 (2001)

68. Gazor, M., Yu, P.: Infinite order parametric normal form of Hopf singularity. Int. J. Bifurc. Chaos **18**(11), 3393–3408 (2008)

69. Golubitsky, M.S., Schaeffer, D.G.: Singularities and Groups in Bifurcation Theory. Springer, New York (1985)

70. Gonzalez-Vega, L., Rouillier, F., Roy, M.-F., Trujillo, G.: Symbolic Recipes for Real Solutions, in Some Tapas of Computer Algebra. Springer, Heidelberg (1999)

71. Gu, G., Sparks, A.G., Banda, S.S.: Bifurcation based nonlinear feedback control for rotating stall in axial flow compressors. Int. J. Control **6**, 1241–1257 (1997)

72. Guckenheimer, J., Holmes, P.: Nonlinear Oscillations, Dynamical Systems, and Bifurcations of Vector Fields, 4th edn. Springer, New York (1993)

73. Haase, A.T.: Population biology of HIV-1 infection: virus and CD4+ T cell demography and dynamics in lymphatic tissues. Annu. Rev. Immunol. **17**, 625–656 (1999)

74. Han, M.: On the number of limit cycles bifurcating from a homoclinic or heteroclinic loop. Sci. China Ser. A **36**(2), 113–132 (1993)

75. Han, M.: Bifurcations of invariant tori and subharmonic solutions for periodic perturbed systems. Sci. China Ser. A **37**(11), 1325–1336 (1994)
76. Han, M.: Cyclicity of planar homoclinic loops and quadratic integrable systems. Sci. China Ser. A **40**(12), 1247–1258 (1997)
77. Han, M.: Bifurcations of limit cycles from a heteroclinic cycle of Hamiltonian systems. Chin. Ann. Math., Ser. B **19**(2), 189–196 (1998)
78. Han, M.: Liapunov constants and Hopf cyclicity of Liénard systems. Ann. Differ. Equ. **15**(2), 113–126 (1999)
79. Han, M.: On Hopf cyclicity of planar systems. J. Math. Anal. Appl. **245**, 404–422 (2000)
80. Han, M., Chen, J.: The number of limit cycles bifurcating from a pair of homoclinic loops. Sci. China Ser. A **30**(5), 401–414 (2000)
81. Han, M.: Periodic Solution and Bifurcation Theory of Dynamical Systems. Science Publication, Beijing (2002) (in Chinese)
82. Han, M.: Bifurcation theory of limit cycles of planar systems. In: Canada, A., Drabek, P., Fonda, A. (eds.) Handbook of Differential Equations, Ordinary Differential Equations, vol. 3. Elsevier, Amsterdam (2006). Chapter 4
83. Han, M., Hu, S., Liu, X.: On the stability of double homoclinic and heteroclinic cycles. Nonlinear Anal. **53**, 701–713 (2003)
84. Han, M., Lin, Y., Yu, P.: A study on the existence of limit cycles of a planar system with 3rd-degree polynomials. Int. J. Bifurc. Chaos **14**(1), 41–60 (2004)
85. Han, M., Luo, D., Zhu, D.: Uniqueness of limit cycles bifurcating from a singular closed orbit. II. Acta Math. Sin. **35**(4), 541–548 (1992) (in Chinese)
86. Han, M., Shang, D., Wang, Z., Yu, P.: Bifurcation of limit cycles in a 4th-order near-Hamiltonian systems. Int. J. Bifurc. Chaos **17**(11) (2007)
87. Han, M., Wang, Z., Zang, H.: Limit cycles by Hopf and homoclinic bifurcations for near-Hamiltonian systems. Chin. J. Contemp. Math. **28**(4), 423–434 (2007)
88. Han, M., Wu, Y., Bi, P.: Bifurcation of limit cycles near polycycles with n vertices. Chaos Solitons Fractals **22**(2), 383–394 (2004)
89. Han, M., Wu, Y., Bi, P.: A new cubic system having eleven limit cycles. Discrete Contin. Dyn. Syst. **12**(4), 675–686 (2005)
90. Han, M., Yang, C.: On the cyclicity of a 2-polycycle for quadratic systems. Chaos Solitons Fractals **23**, 1787–1794 (2005)
91. Han, M., Yang, J., Tarta, A., Yang, G.: Limit cycles near homoclinic and heteroclinic loops. J. Dyn. Differ. Equ. **20**, 923–944 (2008)
92. Han, M., Yang, J., Yu, P.: Hopf bifurcations for near-Hamiltonian systems. Int. J. Bifurc. Chaos **19**(12), 4117–4130 (2009)
93. Han, M., Ye, Y.: On the coefficients appearing in the expansion of Melnikov function in homoclinic bifurcations. Ann. Differ. Equ. **14**(2), 156–162 (1998)
94. Han, M., Ye, Y., Zhu, D.: Cyclicity of homoclinic loops and degenerate cubic Hamiltonians. Sci. China Ser. A **42**(6), 605–617 (1999)
95. Han, M., Zhang, T., Zang, H.: On the number and distribution of limit cycles in a cubic system. Int. J. Bifurc. Chaos **14**(12), 4285–4292 (2004)
96. Han, M., Zhang, T., Zang, H.: Bifurcation of limit cycles near equivariant compound cycles. Sci. China Ser. A **50**(4), 503–514 (2007)
97. Han, M., Zhu, H.: The loop quantities and bifurcations of homoclinic loops. J. Differ. Equ. **234**(2), 339–359 (2007)
98. Hassard, B.D., Kazarinoff, N.D., Wan, Y.-H.: Theory and Applications of Hopf Bifurcation. Cambridge University Press, Cambridge (1981)
99. Hilbert, D.: Mathematical problems (M. Newton, transl.). Bull. Am. Math. Soc. **8**, 437–479 (1902); reprinted in Bull. Am. Math. Soc. (N. S.), **37**, 407–436 (2000)
100. Hopf, E.: Abzweigung einer periodischen Losung von stationaren Losung einers differentialsystems. Ber. Math. Phys. Kl. Sachs Acad. Wiss. Leipzig **94**, 1–22 (1942); and Ber. Math. Phys. Kl. Sachs Acad. Wiss. Leipzig Math.-Nat. Kl. **95**, 3–22 (1942)
101. Hou, Y., Han, M.: Melnikov functions for planar near-Hamiltonian systems and Hopf bifurcations. J. Shanghai Norm. Univ. (Nat. Sci.) **35**(1), 1–10 (2006)

102. Huseyin, K.: Multiple-Parameter Stability Theory and Its Applications. Oxford University Press, Oxford (1980)
103. Huseyin, K., Lin, R.: An intrinsic multiple-scale harmonic balance method for non-linear vibration and bifurcation problems. Int. J. Non-Linear Mech. **26**, 727–740 (1991)
104. Huseyin, K., Yu, P.: On bifurcations into non-resonant quasi-periodic motions. Appl. Math. Model. **12**, 189–201 (1988)
105. Iooss, G.: Bifurcation of Maps and Applications. North-Holland, New York (1979)
106. Ilyashenko, Yu.: Centennial history of Hilbert's 16th problem. Bull., New Ser., Am. Math. Soc. **39**(3), 301–354 (2002)
107. Ilyashenko, Yu., Panov, A.: Some upper estimations of the number of limit cycles of planar vector fields with application to the Liénard equation. Mosc. Math. J. **1**(4), 583–599 (2001)
108. Jacobson, N.: Lie Algebras. Wiley, New York (1966)
109. James, E.M., Lloyd, N.G.: A cubic system with eight small-amplitude limit cycles. IMA J. Appl. Math. **47**, 163–171 (1991)
110. Janeway, C., Travers, P.: Immunobiology: The Immune System in Health and Disease. Garland, New York (2005)
111. Jiang, Q., Han, M.: Melnikov functions and perturbation of a planar Hamiltonian system. Chin. Ann. Math., Ser. B **20**(2), 233–246 (1999)
112. Jiang, J., Han, M., Yu, P., Lynch, S.: Limit cycles in two types of symmetric Liénard systems. Int. J. Bifurc. Chaos **17**(6), 2169–2174 (2007)
113. Jiang, X., Yuan, Z., Yu, P., Zou, X.: Dynamics of an HIV-1 therapy model of fighting a virus with another virus. J. Biol. Dyn. **3**(4), 387–409 (2009)
114. Joyal, P., Rousseau, C.: Saddle quantities and applications. J. Differ. Equ. **78**, 374–399 (1989)
115. Kahn, P.B., Zarmi, Y.: Nonlinear dynamics: a tutorial on the method of normal forms. Am. J. Phys. **68**(10), 907–919 (2000)
116. Kang, W.: Bifurcation and normal form of nonlinear control systems, parts I and II. SIAM J. Control Optim. **36**, 193–232 (1998)
117. Kepler, T., Perelson, A.: Drug concentration heterogeneity facilitates the evolution of drug resistance. Proc. Natl. Acad. Sci. USA **95**, 11514–11519 (1998)
118. Kolutsky, G.: One upper estimate on the number of limit cycles in even degree Liénard equations in the focus case. arXiv:0911.3516v1 [math.DS], November 18 (2009)
119. Krause, P.C., Wasynczuk, O., Sudhoff, S.D.: Analysis of Electrical Machinery and Drive Systems. IEEE Press and Wiley Interscience, New York (2002)
120. Kuznetsov, Yu.A.: Numerical normalization techniques for all codimension 2 bifurcations of equilibria in ODEs. SIAM J. Numer. Anal. **36**, 1104–1124 (1999)
121. Kuznetsov, Yu.A.: Elements of Applied Bifurcation Theory, 3rd edn. Springer, New York (2004)
122. Kuznetsov, Yu.A.: Practical computation of normal forms on center manifolds at degenerate Bogdanov–Takens bifurcations. Int. J. Bifurc. Chaos **15**(11), 3535–3546 (2005)
123. Langford, W.F.: Periodic and steady-state mode interactions lead to tori. SIAM J. Appl. Math. **37**, 22–48 (1979)
124. Langford, W.F., Zhan, K.: Strongly resonant Hopf bifurcation and vortex induced vibrations. In: IUTAM Symposium, Nonlinearity and Chaos in Engineering Dynamics. Springer, New York (1994)
125. Leblanc, V.G., Langford, W.F.: Classification and unfoldings of 1:2 resonant Hopf bifurcation. Arch. Ration. Mech. Anal. **136**, 305–357 (1996)
126. Levy, J.: HIV and the Pathogenesis of AIDS. AMS, Washington (1998)
127. Li, J.: Chaos and Melnikov Method. Chongqing University Press, Chongqing (1989)
128. Li, J.: Hilbert's 16th problem and bifurcations of planar polynomial vector fields. Int. J. Bifurc. Chaos **13**, 47–106 (2003)
129. Li, J., Bai, J.X.: The cyclicity of multiple Hopf bifurcation in planar cubic differential systems: $M(3) \geq 7$. Preprint, Kunming Institute of Technology (1989)
130. Li, J., Chan, H.S.Y., Chung, K.W.: Bifurcations of limit cycles in a Z_3-equivariant planar vector field of degree 5. Int. J. Bifurc. Chaos **11**, 2287–2298 (2001)

131. Li, J., Chan, H.S.Y., Chung, K.W.: Investigations of bifurcations of limit cycles in Z_2-equivariant planar vector fields of degree 5. Int. J. Bifurc. Chaos **12**(10), 2137–2157 (2002)

132. Li, J., Huang, Q.: Bifurcations of limit cycles forming compound eyes in the cubic system. Chin. Ann. Math., Ser. B **8**, 391–403 (1987)

133. Li, J., Li, C.F.: Planar cubic Hamiltonian systems and distribution of limit cycles of (E_3). Acta Math. Sin. **28**, 509–521 (1985)

134. Li, J., Lin, Y.: Global bifurcations in a perturbed cubic system with Z_2-symmetry. Acta Math. Appl. Sin. **8**(2), 131–143 (1992)

135. Li, J., Liu, Z.: Bifurcation set and limit cycles forming compound eyes in a perturbed Hamiltonian system. Publ. Math. **35**, 487–506 (1991)

136. Li, J., Liu, Z.: Bifurcation set and compound eyes in a perturbed cubic Hamiltonian system, in ordinary and delay differential equations. In: Pitman Research Notes in Mathematics Ser., vol. 272, pp. 116–128. Longman, Harlow (1991)

137. Li, C., Liu, L., Yang, J.: A cubic system with thirteen limit cycles. J. Differ. Equ. **246**, 3609–3619 (2009)

138. Li, J., Zhang, M.Q.: Bifurcations of limit cycles in a Z_8-equivariant planar vector field of degree 7. J. Differ. Equ. Dyn. Syst. **16**(4), 1123–1139 (2004)

139. Li, J., Zhang, M., Li, S.: Bifurcations of limit cycles in a Z_2-equivariant planar polynomial vector field of degree 7. Int. J. Bifurc. Chaos **16**(4), 925–943 (2006)

140. Li, J., Zhao, X.: Rotation symmetry groups of planar Hamiltonian systems. Ann. Differ. Equ. **5**, 25–33 (1989)

141. Li, J., Zhou, H.: On the control of parameters of distributions of limit cycles for a Z_2-equivariant perturbed planar Hamiltonian polynomial vector field. Int. J. Bifurc. Chaos **15**, 137–155 (2005)

142. Liao, X.X., Wang, L., Yu, P.: Stability of Dynamical Systems. Elsevier, Amsterdam (2007)

143. Liao, X.X., Yu, P.: Absolute Stability of Nonlinear Control Systems. Springer, New York (2008)

144. Liénard, A.: Etude des oscillations entretenues. Rev. Gén. électr. **23**, 901–912 (1928)

145. Lins, A., De Melo, W., Pugh, C.C.: On Liénard equation. Lect. Notes Math. **597**, 335–357 (1977)

146. Liu, Y., Li, J.: Theory of values of singular point in complex autonomous differential system. Sci. China Ser. A **33**, 10–24 (1990)

147. Liu, Y., Li, J.: New results on the study of Z_q-equivariant planar polynomial vector fields. Qual. Theory Dyn. Syst. **9**, 167–219 (2010)

148. Liu, Y., Li, J., Huang, W.: Singular Point Values, Center Problem and Bifurcations of Limit Cycles of Two Dimensional Differential Autonomous Systems. Science Press, Beijing (2008)

149. Llibre, J.: Integrability of polynomial differential systems. In: Canada, A., Drabek, P., Fonda, A. (eds.) Handbook of Differential Equations, Ordinary Differential Equations, vol. 1. Elsevier, Amsterdam (2004). Chapter 5

150. Llibre, J., Rodríguez, G.: Configurations of limit cycles and planar polynomial vector fields. J. Differ. Equ. **198**(2), 374–380 (2004)

151. Lloyd, N.G.: Limit cycles of polynomial systems. In: Bedford, T., Swift, J. (eds.) New Directions in Dynamical Systems. London Mathematical Society Lecture Notes, vol. 40, pp. 192–234 (1988)

152. Lloyd, N.G., Blows, T.R., Kalenge, M.C.: Some cubic systems with several limit cycles. Nonlinearity **1**, 653–669 (1988)

153. Lloyd, N., Pearson, J.: Symmetry in planar dynamical systems. J. Symb. Comput. **33**, 357–366 (2002)

154. Lorenz, E.N.: Deterministic nonperiodic flow. J. Atmos. Sci. **20**, 130–141 (1963)

155. Luca, S., Dumortier, F., Caubergh, M., Roussarie, R.: Detecting alien limit cycles near a Hamiltonian 2-saddle cycle. Discrete Contin. Dyn. Syst. **25**(4), 1081–1108 (2009)

156. Luo, D., Wang, X., Zhu, D., Han, M.: Bifurcation Theory and Methods of Dynamical Systems. Advanced Series in Dynamical Systems, vol. 15. World Scientific, Singapore (1997)

157. Lynch, S.: Generalized cubic Liénard equations. Appl. Math. Lett. **12**, 1–6 (1999)
158. Lynch, S., Christopher, C.J.: Limit cycles in highly non-linear differential equations. J. Sound Vib. **224**(3), 505–517 (1999)
159. Ma, H., Han, M.: Limit cycles of a Z_3-equivariant near-Hamiltonian system. Nonlinear Anal. **71**(9), 3853–3871 (2009)
160. Mandadi, V., Huseyin, K.: Non-linear bifurcation analysis of non-gradient systems. Int. J. Non-Linear Mech. **15**, 159–172 (1980)
161. Mañosas, F., Villadelprat, J.: A note on the critical periods of potential systems. Int. J. Bifurc. Chaos **16**, 765–774 (2006)
162. Marsden, J.E., McCracken, M.: The Hopf Bifurcation and Its Applications. Springer, New York (1976)
163. Mebatsion, T., Finke, S., Weiland, F., Conzelmann, K.: A CXCR4/CD4 pseudotype rhabdovirus that selectively infects HIV-1 envelope protein-expressing cells. Cell **90**, 841–847 (1997)
164. Melnikov, V.K.: On the stability of the center for time periodic perturbations. Trans. Mosc. Math. Soc. **12**, 1–57 (1963)
165. Murdock, J.A.: Normal Forms and Unfoldings for Local Dynamical Systems. Springer, New York (2003)
166. Nayfeh, A.H.: Methods of Normal Forms. Wiley, New York (1993)
167. Nayfeh, A.H., Harb, A.M., Chin, C.M.: Bifurcations in a power system model. Int. J. Bifurc. Chaos **12**, 497–512 (1996)
168. Nayfeh, A.H., Mook, D.T.: Nonlinear Oscillations. Wiley, New York (1979)
169. Nicolis, G., Prigogine, I.: Self-Organizations in Non-Equilibrium Systems. Wiley-Interscience, New York (1977)
170. Nolan, G.P.: Harnessing viral devices as pharmaceuticals: fighting HIV-1s fire with fire. Cell **90**, 821–824 (1997)
171. Nelson, P., Mittler, J., Perelson, A.: Effect of drug efficacy and the eclipse phase of the viral life cycle on estimates of HIV-1 viral dynamic parameters. J. AIDS **26**, 405–412 (2001)
172. Nowak, M., May, R.: Virus Dynamics. Oxford University, Oxford (2000)
173. Ogata, K.: Modern Control Engineering. Prentice-Hall, Englewood Cliffs (1970)
174. Ono, E., Hosoe, S., Tuan, H.D., Doi, S.: Bifurcation in vehicle dynamics and robust front wheel steering control. IEEE Trans. Control Syst. Technol. **6**, 412–420 (1998)
175. Perelson, A., Kirschner, D., De Boer, R.: Dynamics of HIV infection of CD_4^+T cells. Math. Biosci. **114**, 81–125 (1993)
176. Perelson, A., Nelson, P.: Mathematical models of HIV dynamics in vivo. SIAM Rev. **41**, 3–44 (1999)
177. Perelson, A., Neumann, A., Markowitz, M., Leonard, J., Ho, D.: HIV-1 dynamics in vivo: virion clearance rate, infected cell life-span, and viral generation time. Science **271**, 1582–1586 (1996)
178. Poincaré, H.: Les Méthodes Nouvelles de la Mécanique Céleste (1892–1899)
179. Pontryagin, L.S.: On dynamical systems close to Hamiltonian systems. J. Exp. Theor. Phys. **4**, 234–238 (1934)
180. Rayleigh, J.: The Theory of Sound. Dover, New York (1945)
181. Revilla, T., Garcia-Ramos, G.: Fighting a virus with a virus: a dynamic model for HIV-1 therapy. Math. Biosci. **185**, 191–203 (2003)
182. Romanovski, V.G., Shafer, D.S.: The Center and Cyclicity Problems, A Polynomial Algebra Approach. Birkhäuser, Boston (2008)
183. Roussarie, R.: On the number of limit cycles which appear by perturbation of separatrix loop of planar vector fields. Bol. Soc. Bras. Mat. **17**(2), 57–101 (1986)
184. Roussarie, R.: Putting a boundary to the space of Liénard equations. Discrete Contin. Dyn. Syst. **17**(2), 441–448 (2007)
185. Rousseau, C., Toni, B.: Local bifurcation of critical periods in vector fields with homogeneous nonlinearities of the third degree. Can. Math. Bull. **36**, 473–484 (1993)
186. Sanders, J.A.: Normal form theory and spectral sequences. J. Differ. Equ. **192**, 536–552 (2003)

187. Schlomiuk, D.: Algebraic and geometric aspects of the theory of polynomial vector fields. In: Schlomiuk, D. (ed.) Bifurcations and Periodic Orbits of Vector Fields. NATO ASI Series C, vol. 408, pp. 429–467. Kluwer Academic, London (1993)

188. Schnell, M.J., Johnson, E., Buonocore, L., Rose, J.K.: Construction of a novel virus that targets HIV-1 infected cells and control HIV-1 infection. Cell **90**, 849–857 (1997)

189. Shi, S.: A concrete example of the existence of four limit cycles for plane quadratic systems. Sci. Sin. **11**, 1051–1056 (1979) (in Chinese); **23**, 153–158 (1980) (in English)

190. Smale, S.: Dynamics retrospective: great problem, attempts that failed. Physica D **51**, 261–273 (1991)

191. Smale, S.: Mathematical problems for the next century. Math. Intell. **20**(2), 7–15 (1998)

192. Stilianakis, N.I., Schenzle, D.: On the intra-host dynamics of HIV-1 infections. Math. Biosci. **199**, 1–25 (2006)

193. Takens, F.: Unfoldings of certain singularities of vector fields: generalized Hopf bifurcations. J. Differ. Equ. **14**, 476–493 (1973)

194. Takens, F.: Singularities of vector fields. Publ. Math. IHÉS **43**, 47–100 (1974)

195. Ushiki, S.: Normal forms for singularities of vector fields. Jpn. J. Appl. Math. **1**, 1–37 (1984)

196. Van der Pol, B.: On relaxation-oscillations. Philos. Mag. **2**(7), 978–992 (1926)

197. Varchenko, A., Khovansky, A.: Asymptotics of integrals over vanishing cycles and the Newton polyhedron. Dokl. Akad. Nauk USSR **283**(3), 521–525 (1985); Sov. Math. Dokl. (Eng. Transl. DAN), **32**, 122–127 (1985)

198. Wagner, E.K., Hewlett, M.J.: Basic Virology. Blackwell, New York (1999)

199. Wang, D.M.: A class of cubic differential systems with 6-tuple focus. J. Differ. Equ. **87**, 305–315 (1990)

200. Wang, S.: Hilbert's 16th problem and computation of limit cycles. PhD Thesis, The University of Western Ontario, Canada (2004)

201. Wang, H.O., Abed, E.G.: Bifurcation control of a chaotic system. Automatica **31**, 1213–1226 (1995)

202. Wang, D., Li, J., Huang, M., Jiang, Y.: Unique normal forms of Bogdanov–Takens singularity. J. Differ. Equ. **163**, 223–238 (2000)

203. Wang, S., Yu, P.: Bifurcation of limit cycles in a quintic Hamiltonian system under sixth-order perturbation. Chaos Solitons Fractals **26**(5), 1317–1335 (2005)

204. Wang, S., Yu, P.: Existence of 121 limit cycles in a perturbed planar polynomial Hamiltonian vector field of degree 11. Chaos Solitons Fractals **30**(3), 606–621 (2006)

205. Wang, S., Yu, P., Li, J.: Bifurcation of limit cycles in Z_{10}-equivariant vector fields of degree 9. Int. J. Bifurc. Chaos **16**(8), 2309–2324 (2006)

206. Wiggins, S.: Introduction to Applied Nonlinear Dynamical Systems and Chaos. Springer, New York (1990)

207. Wu, Z.: Nonlinear modes and direct computation method for normal forms of multi-degree-freedom nonlinear systems. PhD Thesis, Tianjin University, China (1996)

208. Wu, J., Faria, T., Huang, Y.S.: Synchronization and stable phase-locking in a network of neurons with memory. Math. Comput. Model. **30**, 117–138 (1999)

209. Wu, Y., Han, M.: On the study of limit cycles of the generalized Rayleigh–Liénard oscillator. Int. J. Bifurc. Chaos **14**(8), 2905–2914 (2004)

210. Wu, Y., Han, M., Liu, X.: On the study of limit cycles of a cubic polynomials system under Z_4-equivariant quintic perturbation. Chaos Solitons Fractals **24**(4), 999–1012 (2005)

211. Xia, H., Wolkowicz, G.S.K., Wang, L.: Transient oscillations induced by delayed growth response in the chemostat. J. Math. Biol. **50**, 489–530 (2005)

212. Xu, Z., Hangan, H., Yu, P.: Analytical solutions for invisicid Gaussian impinging jets. ASME J. Appl. Mech. **75**, 021019 (2008)

213. Xu, F., Yu, P., Liao, X.: Global analysis on n-scroll-chaotic attractors of modified Chua's circuit. Int. J. Bifurc. Chaos **19**(1), 135–157 (2009)

214. Yang, J., Han, M.: Limit cycles near a double homoclinic loop. Ann. Differ. Equ. **23**(4), 536–545 (2007)

215. Yang, J., Han, M., Li, J., Yu, P.: Existence conditions of thirteen limit cycles in a cubic system. Int. J. Bifurc. Chaos **20**(8), 2569–2577 (2010)

216. Yao, W., Yu, P.: Bifurcation of small limit cycles in Z_5-equivariant planar vector fields of order 5. J. Math. Anal. Appl. **328**(1), 400–413 (2007)
217. Ye, Y.Q.: Theory of Limit Cycles. Transl. Math. Monographs, vol. 66. AMS, Providence (1986)
218. Yu, P.: Computation of normal forms via a perturbation technique. J. Sound Vib. **211**(1), 19–38 (1998)
219. Yu, P.: Simplest normal forms of Hopf and generalized Hopf bifurcations. Int. J. Bifurc. Chaos **9**(10), 1917–1939 (1999)
220. Yu, P.: A method for computing center manifold and normal forms. In: Int. Conf. on Diff. Eqns. (Berlin 1999), vol. 2, pp. 832–837. World Scientific, Singapore (2000)
221. Yu, P.: Symbolic computation of normal forms for resonant double Hopf bifurcations using multiple time scales. J. Sound Vib. **247**(4), 615–632 (2001)
222. Yu, P., Han, M.: Limit cycles in 3rd-order planar system. In: International Congress of Mathematicians, Beijing, China, August 20–28, 2002
223. Yu, P.: Analysis on double Hopf bifurcation using computer algebra with the aid of multiple scales. Ann. Differ. Equ. **27**, 19–53 (2002)
224. Yu, P.: Computation of the simplest normal forms with perturbation parameters based on Lie transform and rescaling. J. Comput. Appl. Math. **144**(1–2), 359–373 (2002)
225. Yu, P.: Bifurcation dynamics in control systems. In: Chen, R., Hill, D.J., Yu, X. (eds.) Chaos and Bifurcation Control: Theory and Applications, vol. 293, pp. 99–126. Springer, New York (2003)
226. Yu, P.: A simple and efficient method for computing center manifold and normal forms associated with semi-simple cases. Dyn. Contin. Discrete Impuls. Syst., Ser. B, Appl. Algorithms **10**(1–3), 273–286 (2003)
227. Yu, P.: Closed-form conditions of bifurcation points for general differential equations. Int. J. Bifurc. Chaos **15**(4), 1467–1483 (2005)
228. Yu, P.: Bifurcation, limit cycle and chaos of nonlinear dynamical systems. In: Sun, J., Luo, A. (eds.) Bifurcation and Chaos in Complex Systems, vol. 1, pp. 1–125. Elsevier, New York (2006)
229. Yu, P.: Computation of limit cycles—the second part of Hilbert's 16th problem. Fields Inst. Commun. **49**, 151–177 (2006)
230. Yu, P.: Local and global bifurcations to limit cycles in a class of Liénard equation. Int. J. Bifurc. Chaos **17**(1), 183–198 (2007)
231. Yu, P., Bi, Q.: Analysis of non-linear dynamics and bifurcations of a double pendulum. J. Sound Vib. **217**(4), 691–736 (1998)
232. Yu, P., Chen, G.: Hopf bifurcation control using nonlinear feedback with polynomial functions. Int. J. Bifurc. Chaos **14**(5), 1683–1704 (2004)
233. Yu, P., Chen, R.: The simplest parametrized normal forms of Hopf and generalized Hopf bifurcations. Ann. Differ. Equ. **50**(1–2), 297–313 (2007)
234. Yu, P., Chen, R.: Computation of focus values with applications. Ann. Differ. Equ. **51**(3), 409–427 (2008)
235. Yu, P., Chen, R.: Bifurcation of limit cycles in a 5th-order Z_6-equivariant planar vector field. Preprint
236. Yu, P., Corless, R.M.: Symbolic computation of limit cycles associated with Hilbert's 16th problem. Commun. Nonlinear Sci. Numer. Simul. **14**(12), 4041–4056 (2009)
237. Yu, P., Han, M.: Twelve limit cycles in a 3rd-order planar system with Z_2 symmetry. Commun. Pure Appl. Anal. **3**(3), 515–526 (2004)
238. Yu, P., Han, M.: Twelve limit cycles in a cubic case of the 16th Hilbert problem. Int. J. Bifurc. Chaos **15**(7), 2191–2205 (2005)
239. Yu, P., Han, M.: Small limit cycles from fine focus points in cubic order Z_2-equivariant vector fields. Chaos Solitons Fractals **24**(1), 329–348 (2005)
240. Yu, P., Han, M.: Limit cycles in generalized Liénard systems. Chaos Solitons Fractals **30**(5), 1048–1068 (2006)
241. Yu, P., Han, M.: On limit cycles of the Liénard equations with Z_2 symmetry. Chaos Solitons Fractals **31**(3), 617–630 (2007)

242. Yu, P., Han, M.: Critical periods of planar reversible vector field with 3rd-degree polynomial functions. Int. J. Bifurc. Chaos **19**(1), 419–433 (2009)

243. Yu, P., Han, M., Yuan, Y.: Analysis on limit cycles of Z_q-equivariant polynomial vector fields with degree 3 or 4. J. Math. Anal. Appl. **322**(1), 51–65 (2006)

244. Yu, P., Huseyin, K.: Static and dynamic bifurcations associated with a double zero eigenvalues. Dyn. Stab. Syst. **1**, 73–86 (1986)

245. Yu, P., Huseyin, K.: A perturbation analysis of interactive static and dynamic bifurcations. IEEE Trans. Autom. Control **33**, 28–41 (1988)

246. Yu, P., Huseyin, K.: Parametrically excited nonlinear systems: a comparison of certain methods. Int. J. Non-Linear Mech. **33**(6), 967–978 (1998)

247. Yu, P., Leung, A.Y.T.: The simplest normal form of Hopf bifurcation. Nonlinearity **16**(1), 277–300 (2003)

248. Yu, P., Leung, A.Y.T.: A perturbation method for computing the simplest normal forms of dynamical systems. J. Sound Vib. **261**(1), 123–151 (2003)

249. Yu, P., Leung, A.: Normal forms of vector fields with perturbation parameters. Chaos Solitons Fractals **34**(2), 564–579 (2007)

250. Yu, P., Leung, A.: The simplest normal form and its application to bifurcation control. Chaos Solitons Fractals **33**(3), 845–863 (2007)

251. Yu, P., Yuan, Y., Xu, J.: Study of double Hopf bifurcation and chaos for an oscillator with time delayed feedback. Commun. Nonlinear Sci. Numer. Simul. **7**(1–2), 69–91 (2002)

252. Yu, P., Yuan, Y.: A matching pursuit technique for computing the simplest normal forms of vector fields. J. Symb. Comput. **35**(5), 591–615 (2003)

253. Yu, P., Yuan, Y.: An efficient method for computing the simplest normal forms of vector fields. Int. J. Bifurc. Chaos **13**(1), 19–46 (2003)

254. Yu, P., Zhu, S.: Computation of the normal forms for general M-DOF systems using multiple time scales. Part I: Autonomous systems. Commun. Nonlinear Sci. Numer. Simul. **10**(8), 869–905 (2005)

255. Yuan, Y., Yu, P.: Computation of simplest normal forms of differential equations associated with a double-zero eigenvalues. Int. J. Bifurc. Chaos **11**(5), 1307–1330 (2001)

256. Zhang, Z., Ding, T., Huang, W., Dong, Z.: Qualitative Theory of Differential Equations. Transl. Math. Monogr., vol. 101. AMS, Providence (1992)

257. Zhang, T.H., Han, M., Zang, H., Meng, X.Z.: Bifurcations of limit cycles for a cubic Hamiltonian system under quartic perturbations. Chaos Solitons Fractals **22**, 1127–1138 (2004)

258. Zhang, W.N., Hou, X.R., Zeng, Z.B.: Weak centers and bifurcation of critical periods in reversible cubic systems. Comput. Math. Appl. **40**, 771–782 (2000)

259. Zhang, T., Zang, H., Han, M.: Bifurcations of limit cycles in a cubic system. Chaos Solitons Fractals **20**, 629–638 (2004)

260. Zhu, S.: Computation of normal forms of differential equations using perturbation methods. Ph.D. Thesis, University of Western Ontario, Canada (2001)

261. Zoladek, H.: Eleven small limit cycles in a cubic vector field. Nonlinearity **8**, 843–860 (1995)

Index